PLUTONIUM FUTURES— THE SCIENCE

Previous Proceedings in the Series of Conferences on Plutonium Futures – The Science

Year		Publisher	ISBN
Second	2000	AIP Conference Proceedings Vol. 532	1-56396-948-3
First	1997	Los Alamos National Laboratory	LA-13338-C

Other Related Titles from AIP Conference Proceedings

654 Space Technology and Applications International Forum - STAIF 2003: Conference on Thermophysics in Microgravity; Conference on Commercial/Civil Next Generation Space Transportation; 20th Symposium on Space Nuclear Power and Propulsion; Conference on Human Space Exploration; 1st Symposium on Space Colonization
Edited by Mohamed S. El-Genk, February 2003, 0-7354-0114-4
CD-ROM: 0-7354-0115-2

644 Exotic Clustering: 4th Catania Relativistic Ion Studies; CRIS 2002
Edited by Salvatore Costa, Antonio Insolia, and Cristina Tuvè, November 2002, 0-7354-0099-7

610 Nuclear Physics in the 21st Century: International Nuclear Physics conference, INPC 2001
Edited by Eric Norman, Lee Schroeder, and Gordon Wozninak, April 2002, 0-7354-0056-3

529 Capture Gamma-Ray Spectroscopy and Related Topics: 10th International Symposium
Edited by Stephen Wender, July 2000, 1-56396-952-1

513 Nuclear and Condensed Matter Physics: VI Regional CRRNSM Conference
Edited by Antonino Messina, April 2000, 1-56396-929-7

495 Experimental Nuclear Physics in Europe: ENPE 99, Facing the Next Millennium
Edited by Berta Rubio, Manuel Lozano, and William Gelletly, November 1999, 1-56396-907-6

455 ENAM 98: Exotic Nuclei and Atomic Masses
Edited by B. M. Sherrill, D. J. Morrissey, and C. N. Davids, December 1998, 1-56396-804-5

447 Nuclear Fission and Fission-Product Spectroscopy: Second International Workshop
Edited by G. Fioni, H. Faust, S. Oberstedt, F.-J. Hambsch, October 1998, 1-56396-823-1

To learn more about these titles, or the AIP Conference Proceedings Series, please visit the webpage **http://proceedings.aip.org/proceedings**

PLUTONIUM FUTURES—THE SCIENCE

Third Topical Conference on Plutonium and Actinides

Albuquerque, New Mexico *6-10 July 2003*

EDITOR
Gordon D. Jarvinen
Los Alamos National Laboratory
Los Alamos, New Mexico

SPONSORING ORGANIZATIONS
Los Alamos National Laboratory
G. T. Seaborg Institute for Transactinium Science
Associate Director for Weapons Engineering and Manufacturing
Nonproliferation and International Security (NIS) Division
Associate Director for Strategic Research
American Nuclear Society

Melville, New York, 2003
AIP CONFERENCE PROCEEDINGS ■ VOLUME 673

Editor:

Gordon D. Jarvinen
Mail Stop E505
Glenn T. Seaborg Institute for Transactinium Science
Los Alamos National Laboratory
P. O. Box 1663
Los Alamos, NM 87545
USA

E-mail: gjarvinen@lanl.gov

L.C. Catalog Card No. 2003106917
ISBN 0-7354-0140-3
ISSN 0094-243X

Printed in the United States of America

CONTENTS

SESSION ONE
MATERIAL SCIENCE/CONDENSED MATTER PHYSICS

SESSION TWO
ACTINIDE COMPOUNDS AND COMPLEXES

SESSIONS THREE AND FOUR
THE NUCLEAR FUEL CYCLE

SESSION FIVE
MATERIAL SCIENCE AND PLUTONIUM PROPERTIES

SESSION SIX
ACTINIDES IN THE ENVIRONMENT AND LIFE SCIENCES

SESSION SEVEN
DETECTION AND ANALYSIS

POSTER SESSION
MATERIAL SCIENCE/CONDENSED MATTER PHYSICS

ACTINIDE COMPOUNDS AND COMPLEXES

NUCLEAR FUEL CYCLE

ACTINIDES IN THE ENVIRONMENT AND LIFE SCIENCES

DETECTION AND ANALYSIS

Preface

In the more than 60 years since Glenn Seaborg and his coworkers first generated plutonium in the laboratory, plutonium has proven to be one of the most complex elements in the periodic table. The metal exhibits six solid allotropes at ambient pressure and its phases are notoriously unstable with temperature, pressure, chemical additions and time. Plutonium sits near the middle of the actinide series, which marks the emergence of 5f electrons in the valence shell. Elements to the left of plutonium have delocalized (bonding) electrons, while elements to the right of plutonium are localized (non-bonding). Plutonium is trapped in the middle, and for the delta-phase metal, the electrons seem to be in a unique state of being neither fully bonding or localized, which leads to novel electronic interactions and unusual physical and chemical behavior. The concept of localized or delocalized 5f electrons also pervades the bonding descriptions of many of the plutonium molecules and compounds. Indeed, in the summer of 2003, an understanding of the electronic structure of the pure element and its compounds continues to challenge both theorists and experimentalists in all areas of plutonium science.

Over 1,000 tons of plutonium exist throughout the world in the form of nuclear fuel, nuclear weapons, inventories of various types, and legacy materials. Regardless of one's views of how this condition came to be, or what actions are taken in the future, it is a certainty that large quantities of plutonium must be managed for many decades. Thus it is clear that the plutonium challenge is not only scientific, but also political and socioeconomic as well.

In view of the global nature of plutonium research efforts, the "Plutonium Futures – The Science" conference was established to increase awareness of the importance of the scientific underpinnings of plutonium research, and facilitate communication among its international practitioners. The 2003 conference is the third in this series, and has attracted more than 180 contributed presentations covering the latest results in plutonium condensed matter physics, materials science, compounds and complexes, environmental behavior, detection and analysis, separations and purification, nuclear fuel cycles, and waste isolation and disposal.

In 2003, many exciting new developments in plutonium science and technology will be presented. For example, the latest results on plutonium-based superconductivity; new advances in actinide separations and nuclear fuel fabrication; the local and long-range structure of key alloys, compounds and molecular systems; the multiphase behavior of pure 238-plutonium metal; new insights on the problem of delta-phase metastability of Pu-Ga and Pu-Al systems; the latest insights on the role of natural and intrinsic colloids in the environmental transport of actinides; and the pervasive effects of alpha-particle self-irradiation in solids and solutions will be described.

In these conference transactions we aim to enhance the dialog among scientists on the fundamental properties of plutonium and their technological consequences. Moreover, we hope that this conference will stimulate the next generation of scientists and students to study the fundamental properties of plutonium. We are encouraged by the great response we received to the third conference, with contributors from the scientific community around the world.

The Los Alamos National Laboratory, currently celebrating its 60th anniversary, and the Los Alamos branch of the Glenn T. Seaborg Institute are proud to sponsor this conference in cooperation with the American Nuclear Society. It is our hope to continue the renaissance of scientific interest in plutonium and other actinides that has emerged over the past decade, which in turn will allow us to continue to make great strides in solving the Cold War legacy problems, and allow us to take full advantage of the enormous energy potential of plutonium for the benefit of all mankind.

Gordon D. Jarvinen
David L. Clark

Program Co-chairs

SESSION ONE:
MATERIAL SCIENCE/
CONDENSED
MATTER PHYSICS

Recent Highlights in Actinide Research at ITU

R. Schenkel and G. H. Lander

European Commission, JRC, Institute of Transuranium Elements,
Postbox 2340, 76125 Karlsruhe, Germany

ITU in Karlsruhe is a Laboratory operated by the European Commission, with wide expertise and equipment in the actinide field. The "core" programs are devoted to nuclear fuel safety, studies connected with partitioning and transmutation, safeguards, and basic actinide research. Many of these core programs involve European industry and other national consortia. The basic actinide research is centered around the Actinide UserLab, a facility with funding to bring European researchers (especially students) to ITU to allow them to gain experience with the science of actinides. In addition, and of increasing importance, there are a number of newer studies that use our competence and expertise. Examples of these are analytical techniques for examining particles containing radioactive isotopes, the use of sensitive methods to detect the presence of uranium enrichment or plutonium processing, and the use of alpha-emitting isotopes in the treatment of cancer.

Some of the subjects covered briefly in this talk will be the following:

(1) The continuing investigation of Np, Pu, and Am, so-called 1:1:5 materials, which have been found to be superconducting at relatively high temperatures. The form of superconductivity still remains an open question (in collaboration with Los Alamos National Laboratory).

(2) A survey of high-pressure phases for the actinide metals emphasizes the crucial importance of the α-uranium structure in distinguishing materials with itinerant or localized *5f* electrons (in collaboration with Oak Ridge National Laboratory and the ESRF, Grenoble).

(3) Laser-induced photofission has been observed using a table-top laser generating 10^{19} W/cm^2; opening possible new avenues for transmutation (in collaboration with the University of Jena).

(4) The first successful separation of an actinide (Am) and a lanthanide (Nd) by pyroprocessing in a test of a new partition processing using Al cathodes and a molten chloride salt (supported by the Japanese CRIEPI project).

(5) Radiolysis effects in UO_2 pellets doped with ^{238}Pu and ^{233}U to enhance the α activity and simulate long-term fuel storage have been studied. Enhanced U dissolution was observed in all cases, but the effect is *not* proportional to the fuel radioactivity. A tendency to saturation is observed.

(6) As part of our activities in preparing for the enlargement of the European Union, ITU is currently executing a series of projects with the candidate countries; these projects involve methods of combating illicit trafficking of nuclear materials. This aspect of safeguards is reinforced by a strong in-house effort to develop analytical methods for such materials. For example, neutron coincidence counting for Pu assay relies on ^{240}Pu, but the results are affected by any ^{244}Cm. A new method to detect ^{244}Cm at the 10 ppb level has been developed. Further examples will be given of particle analysis and forensic science, both of which have a growing role in environmental studies and/or monitoring safeguards.

(7) A new cancer treatment method based on the use of α-emitting isotopes was developed and is currently undergoing preclinical and clinical testing in different hospitals in both Europe and the U.S. (including a collaboration with Sloan Kettering Hospital in New York).

web site http://itu.jrc.cec.eu.int

CP673, *Plutonium Futures — The Science,* edited by G. D. Jarvinen

Synchrotron-Radiation-Based Photoelectron and X-Ray Absorption Spectroscopy of Cerium and Plutonium

B. W. Chung,[1] K. T. Moore,[1] S. A. Morton,[1] D.K. Shuh,[2] and J. G. Tobin[1]

[1]*Lawrence Livermore National Laboratory*
[2]*Lawrence Berkeley National Laboratory*

INTRODUCTION

The electronic structure of the actinides is of considerable interest as a result of the possibly correlated nature of the 5*f* electrons. At standard pressure, there is a profound difference between the early and late actinide metals, in the sense that the 5*f* electrons are delocalized (itinerant) for the elements up to and including α Pu, where they are localized and atomic-like for the elements beyond Pu. A similar trend has been reported in the lanthanides where the localization-delocalization transition takes place around the first element with an appreciable occupation of the 4*f* shell, Cerium.[1] Cerium exhibits an alpha-gamma phase transition characterized by a volume expansion of 20%, which is analogous to the 26% volume expansion upon going from alpha to delta in Pu.[2] Both cerium and plutonium exhibit similar physical properties, including multiple allotropic phases, and undergo large volume changes during transformation. This behavior has been hypothesized to be related to changes in the degree of localization of *f*-electrons in 4*f* electrons of Ce and 5*f* electrons of Pu. These properties make both Ce and Pu excellent candidate materials to study the variations of *f*-electronic behavior associated with phase changes. Additionally, Cerium is an ideal surrogate material for plutonium, considering that the study of Pu to this date has been compromised by its radioactivity properties and acute toxicity. Although the electronic structures of Ce and Pu have been investigated extensively,[3–9] a complete understanding of their *f*-electronic structure is still lacking.

This summary abstract describes our initial results and program being developed to perform Photoelectron Spectroscopy (PES) and X-Ray Absorption Spectroscopy (XAS) upon *f*-electron materials, particularly plutonium and cerium, at the Advanced Light Source in Berkeley, California, USA. Our purpose is to study the electronic properties as a function of temperature and phase at temperatures in the range of 0 K to slightly above room temperature (300 K).

RESULTS AND DISCUSSION

The first XAS and PES experiments on Ce were performed using Beamline 7 at the Advanced Light Source in Berkeley, California. Bulk cerium samples and Ce films were investigated using PES and XAS, through total electron yield (TEY). The structures of the bulk cerium metal samples were characterized ex situ with x-ray diffraction, thus establishing the phase transition temperature before the spectroscopic experiments. The surface of bulk γ Ce metal was scraped in situ with a diamond file and never exposed to pressure greater than 10^{-9} Torr. The PES and XAS spectra of Ce were analyzed and compared to Pu spectra obtained in a previous experiment.[3,6]

Three examples of our XAS results are shown in Figure 1. TEY measurements were carried out on an evaporated Ce film, a bulk Ce sample at T = 170 K and a bulk Ce sample at T = 300 K. These results are consistent with Wieliczka et al.[9]

CP673, *Plutonium Futures — The Science,* edited by G. D. Jarvinen

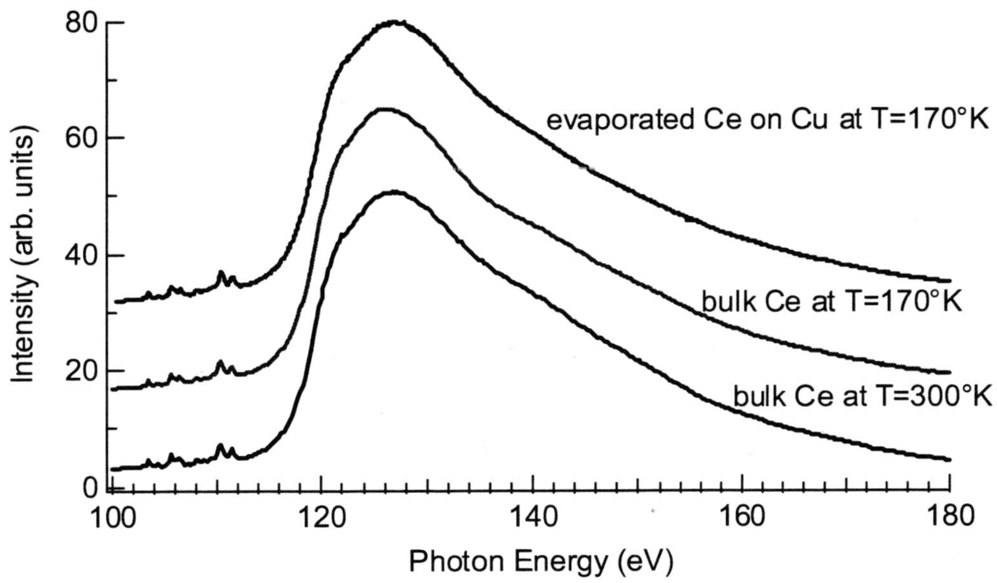

FIGURE 1. XAS spectra of Ce samples collected in the total electron yield mode.

The valence band PES spectrum of evaporated film of Ce taken at the region of the giant 4*d*->4*f* resonance at a temperature of 170 K, well above the gamma-alpha phase transition temperature, is shown in Figure 2. The PES spectrum shows a doublet A1 and A2 close to E_F and B at a binding energy of near 2 eV similar to previous works.[7–9] The doublet near the Fermi energy represents mainly $4f^1$ character and is probably a result of spin-orbit splitting. This doublet is often described as the Kondo peak,[10] although this description is still controversial. In previous work, Weschke et al.[7] have claimed that when going from γ-Ce to α-Ce metal, the doublet gains in relative intensity with a larger increase of the spectral weight at A1 and that α Ce has a γ-like surface layer. However, neither Weschke et al.[7] nor Wieliczka et al.[9] provided any structural measurements such as electron or x-ray diffraction and used temperature as a measure of phase state.

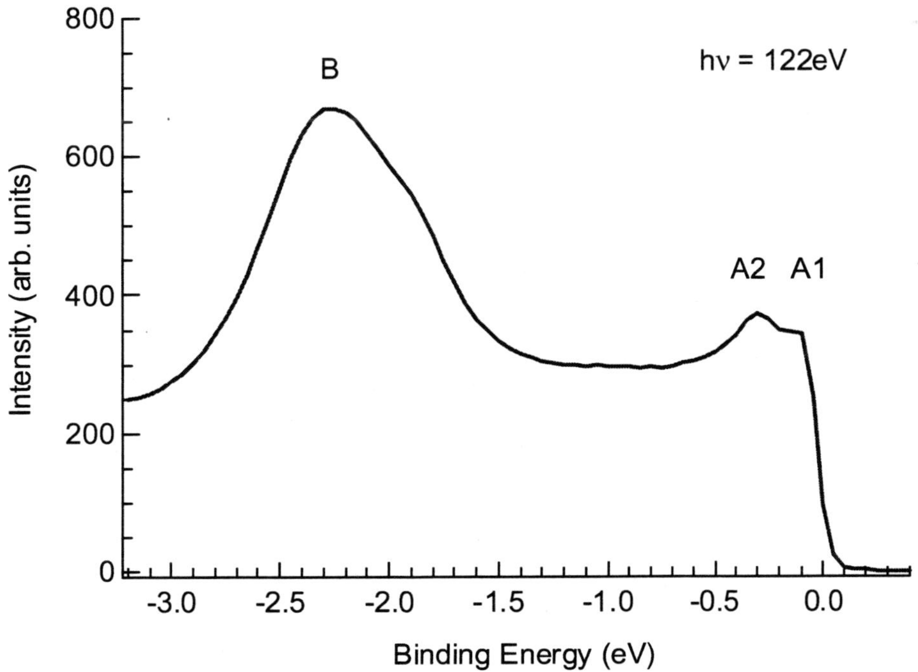

FIGURE 2. Valence band PES spectrum of Ce film at 170 K.

The valence electronic structure of Pu is dependent upon its phase.[3,4] Additionally, there is substantial evidence for δ-like reconstruction of the surface for α Pu.[4] These characteristics are similar to the described spectroscopic results of cerium metal, suggesting that the nature of the *f*-electrons in both materials is similar. The delocalization-localization transition of both cerium and plutonium metal is manifested in the spectral weight change near the Fermi energy.

In the future, we intend to perform spectroscopic investigations of different phases of both materials and probe in detail the *f*-electron contribution to the observed spectra by measuring the spin dependency of each state. The central thrust of our effort will be to use spin-resolving measurements with the MiniMott detector. For Pu samples, due to its radioactivity, we are presently constructing a new dedicated Actinide Spectrometer with spin resolving capability and sophisticated safety features. One particular reason for investigating the complex structure-electronic property relationship of cerium is to understand the nature of 4*f* electrons, an issue still in debate. The similar bulk properties and phase transformations of cerium and plutonium are a strong indication that the nature of *f* electrons is also similar. This makes cerium an excellent plutonium surrogate. In particular, clarifying the role of *f*-electrons, 4*f* for Ce and 5*f* for Pu, may be the key to understanding the observed electronic properties and complex phase transformations in lanthanide and actinide series of the periodic table.

ACKNOWLEDGMENTS

This work was performed under the auspices of the U. S. Department of Energy by the University of California, Lawrence Livermore National Laboratory under Contract No. W-7405-ENG-48.

REFERENCES

1. Johansson, B., Phys. Rev. B 11, 2740 (1975).
2. Meot-Reymond, S., and Fournier, J. M., J Alloys Comp. 232, 119 (1996).
3. Terry, J., Schulze, R. K., Farr, J. D., Zocco, T., Heinzelman, K., Rotenberg, E., Shuh, D. K., Van der Laan, G., Arena, D. A., and Tobin, J. G., Surface Science 499, L141 (2002).
4. Gouder, T., Havela, L., Wastin, F., and Rebizant, J., Europhys. Lett. 55, 705 (2001).
5. Arko, A. J., Joyce, J. J., Morales, L., Wills, J., and Lashley, J., Phys. Rev. B. 62, 1773 (2000).
6. Tobin, J. G., Chung, B. W., Waddill, G. D., Schulze, R. K., Terry, J., Farr, J. D., Zocco, T., Shuh, D. K., Heinzelman, K., Rotenberg, E., and Van der Laan, G., unpublished.
7. Weschke, E., Laubschat, C., Simmons, T., Domke, M., Strebel, O., and Kaindl, G., *Phys. Rev. B* 44, 8304 (1991).
8. Weschke, E., Hohr, A., Kaindl, G., Molodtsov, S. L., Danzenbacher, S., Richter, M., and Laubschat, C., *Phys. Rev. B* 58, 3682 (1998).
9. Wieliczka, D., Weaver, J. H., Lynch, D. W., and Olson, C. G., *Phys. Rev. B* 26, 7056 (1982).
10. Patthey, F., Delley, B., Schneider, W. D., Beck, H., Haer, Y., and Delley, B., *Phys. Rev. B* 42, 8864 (1990).

Local Structure in Plutonium Alloys Stabilized in δ-Phase

B. Ravat, L. Jolly, C. Valot, and N. Baclet

CEA, Centre de Valduc, 21120 Is sur Tille, France

Plutonium metal can take six different phases between ambient temperature and its melting point (640°C). The complexity of plutonium is mainly due to its 5f electrons that are delocalized in the α-phase (monoclinic) but they are partially localized in the δ-phase (face-centered cubic). The δ-phase, stable between 319°C and 451°C, can be maintained at room temperature by alloying plutonium with so-called deltagen elements (Ga, Al, Am, Ce...).[1] However, the stabilizing mechanisms induced by such additions are still not well understood. As shown in PuCe alloys, the determination of the local structure of such alloys is very useful to understand the stabilization mechanisms.[2,3] In this work, we report a structural study on PuGa and PuAm alloys in the δ-phase by the use of two complementary techniques: X-ray diffraction (XRD) and X-ray absorption spectroscopy (EXAFS). This kind of experimental approach leads to important information regarding the local structure around the gallium, americium, and plutonium atoms.

The electronic effects in PuAm alloys have been followed with different techniques such as XRD and resistivity. Therefore, electronic effects have been suggested by a positive deviation of the cell parameter versus Am content from the Vegard's law, as shown by XRD.[4] A complete localization of the 5f electrons of Pu atoms has been correlated by electrical and magnetic measurements for an Am content of about 25 at. %.[5] However, the techniques used until now give only "overall" information. EXAFS then appears to be a necessary complementary technique to precisely characterize the alloys and to provide information that helps to improve and to provide "overall" results that are more precise.

Therefore, EXAFS measurements have been performed at the LIII-edge of Pu (18057 eV) and at the LIII-edge of Am (18504 eV) in the transmission mode on four alloys: $Pu_{0.95}Am_{0.05}$, $Pu_{0.85}Am_{0.15}$, $Pu_{0.75}Am_{0.25}$ and $Pu_{0.57}Am_{0.43}$ (samples prepared at the Karlsruhe ITU, Germany). Data have been recorded at the beam line ROBL BM 20 (ESRF Grenoble, France) which is a specific line for radioactive materials.

The results highlight an increase in both Am-Pu distances and Pu-Pu distances with the Am content. An increase in cell parameter with Am content has been already observed by XRD, which is a sensitive technique for long-range order. This can first be attributed to a steric effect, the Am atomic volume being more important than the Pu one. Moreover, the Am-Pu distances increase more than the Pu-Pu ones. This suggests that the localization of the 5f electrons appears preferentially on Pu atoms that are close to the Am ones.

For the PuGa alloy, several analyses have been already performed to probe the environment surrounding the gallium atoms to compare the Ga-Pu distances to Pu-Pu ones. Conradson[6] highlighted the fact that the Ga-Pu distance is approximately 0.13 Å shorter than the Pu-Pu for the first shell and 0.05 for the second one. Experiments at ambient temperature performed by Faure et al.[7] showed a small difference with Conradson results for the first shell. (difference between Pu-Pu and Ga-Pu distances: 0.1Å). Unfortunately, the second shell could not be extracted. Therefore, some recent experiments have been carried out on a PuGa alloy at low temperature (~ 30K) in order to reduce the damping of the EXAFS oscillations resulting from thermal disorder. EXAFS measurements have been performed at the LIII edge of Pu in the transmission mode and at the K edge of the Ga (10,367 eV) in fluorescence mode on the D44 beam line at the LURE (Orsay, France) radiation synchrotron facility using the X-ray beam emitted by the DCI storage ring. Our investigations showed a Ga-Pu distance 0.1Å shorter than the Pu-Pu for the first shell and a distance for the second one close to the Pu-Pu distance, highlighting the fact that the lattice collapse around the gallium may occur only for the Pu atoms' nearest neighbors.

CP673, *Plutonium Futures — The Science,* edited by G. D. Jarvinen

ACKNOWLEDGEMENTS

We wish to thank F. Wastin, E. Colineau, J. Rebizant, and G. Lander (ITU), V. Briois, S. Belin, and E. Elkaïm (LURE), and T. Reich and C. Hennig (ESRF).

REFERENCES

1. Lallement, R., and Pascard, R., "Plutonium: Métallurgie et Propriétés Physiques du Métal, Nouveau Traité de Chimie Minérale," (P. Pascal) XV, 5, Masson et Cie Ed. (1970), pp. 217–292.
2. Dormeval, M., Baclet, N., Valot, C., Rofidal, P., and Fournier, J. M., "Crystalline and Electronic Structure of Pu-Ce and Pu-Ce-Ga Alloys Stabilized in the δ-Phase," Journal of Alloys and Compounds (2002), in press.
3. Dormeval, M., Valot, C., and Baclet, N., "X-Ray Diffraction Investigation of Pu-Ce and Pu-Ce-Ga Alloys Stabilized in the δ-Phase," Journal de Physique IV France, 10 (2000).
4. Dormeval, M., Baclet, N., and Fournier, J.-M., "Plutonium Futures—the Science," a 4-day topical Conference on Plutonium and Actinides, July 10–13, 2000, Santa Fe, New Mexico.
5. Dormeval, M., "Structure Electronique d'Alliages Pu-Ce(-Ga) et Pu-Am(-Ga) Stabilisés en Phase δ," Thèse de l'Université de Bourgogne (2001).
6. Conradson, S., "Where is the Gallium? Searching the Plutonium Lattice with XAFS," Los Alamos Science 26, 356–363 (2000).
7. Faure, Ph., Deslandes, B., Bazin, D., Tailland, C., Doukhan, R., Fournier, J. M., and Falanga, A., "Lattice Collapse Around Gallium in PuGa Alloys as Revealed by X-Ray Absorption Spectroscopy," Journal of Alloys and Compounds 244, 131–139 (1996).

Investigating the δ/α' Martensitic Phase Transformation in Pu-Ga Alloys

Kerri Blobaum,[1] Jeff Haslam,[1] April Brough,[2] Mark Wall,[1] and Adam Schwartz[1]

[1]*Materials Science and Technology Division, Lawrence Livermore National Laboratory, Livermore, CA*
[2]*Brigham Young University, Provo, UT*

INTRODUCTION

In pure plutonium, the brittle monoclinic α phase is stable at ambient temperatures. By adding a small amount of an alloying element such as Ga or Al, however, the high-temperature, face-centered cubic δ phase can be stabilized at room temperature. When Pu-Ga alloys are subjected to low-temperature excursions, some of the δ phase transforms martensitically to α', a monoclinic α phase with Ga trapped in the α' lattice. This transformation is unusual because it does not go to completion (the maximum amount of α' formation in a 1.9 at. % alloy is only ~30% by volume) and the densities of the α' and δ phases differ significantly (20%). Previous work has established time-temperature-transformation (TTT) diagrams for various alloys, and many Pu-Ga alloys exhibit an atypical metallurgical phenomenon of two "noses" in the TTT curves (see Figure 1).[1] The reversion of α' to δ on heating is also notable because there is a large transformation hysteresis; α' forms significantly below room temperature on cooling, but does not revert to δ until above ambient temperatures.

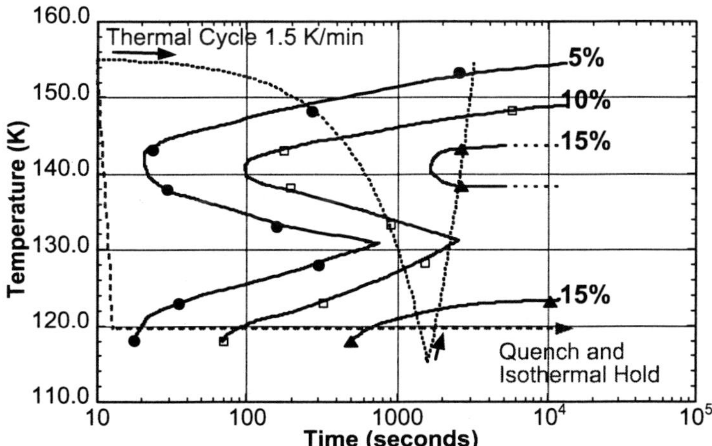

FIGURE 1. TTT curves for a 1.9 at. % Pu-Ga alloy.[1] The contour lines indicate the amount of transformation (5%, 10%, and 15%) to the α' phase. The dashed curves overlaid on the TTT diagram indicate the thermal paths of cycling and annealing experiments.

The martensitic δ → α' phase transformation is not well understood. Here, we report the development of experimental techniques for investigating the stability and transformation of Pu-Ga alloys. Because the resistivities of the α' and δ phases differ significantly, four-point probe resistometry is an effective and sensitive technique for monitoring the transformation. These measurements, however, are challenging because the resistivities of both

CP673, *Plutonium Futures — The Science,* edited by G. D. Jarvinen

phases drop off sharply at low temperatures. We will also use differential scanning calorimetry (DSC) as a tool to study the thermodynamics and kinetics of the Pu-Ga alloy phase transformations. This technique will allow us to quantify transformation temperatures as well as the amount of heat released or absorbed in the transformation. Additionally, metallography and transmission electron microscopy will be used to investigate the structure and morphology of the α' and δ phases; possible morphological differences between α' formed in the upper and lower "noses" of the TTT curves are of particular interest.

RESULTS

The resistivity of a 1.9 at. % Pu-Ga alloy was measured while it was thermally cycled between 70 K and 625 K in vacuum. The beginning of α' formation on cooling (martensite start, M_s) is exhibited as an abrupt increase in resistivity at ~154 K. The reversion to the δ phase on heating (austenite start, A_s) is apparent from a significant change in the slope of the resistivity curve at ~310 K. Numerous cycles show that the hysteresis, as well as the M_s and A_s temperatures, are reproducible. Several cycles at 1.5 K/min are shown in Figure 2. Additional cycles at 0.3 K/min and 5 K/min indicate that M_s is not a strong function of cooling rate.

Isothermal annealing experiments at 175 K show a possible incubation period before the formation of the α' phase. Based on the TTT diagrams, an incubation period might be expected; the duration depends on composition and temperature. Additional work is anticipated in this area.

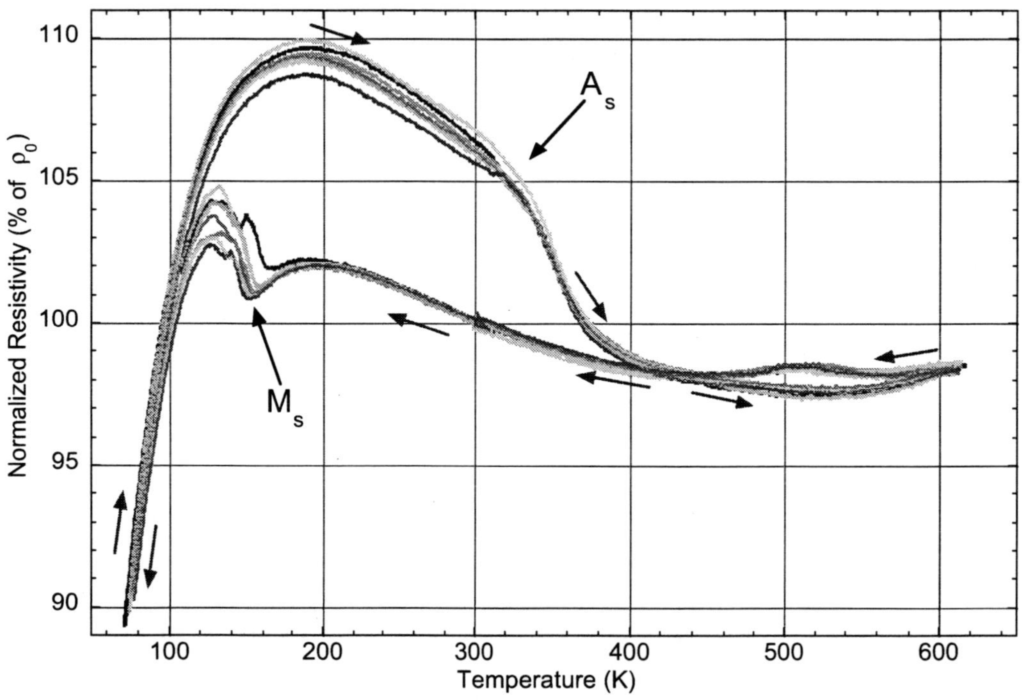

FIGURE 2. Thermal cycles of a 1.9 at. % Pu-Ga alloy at 1.5 K/min. Multiple cycles demonstrate a reproducible hysteresis behavior. Martensite start (M_s) and austenite start (A_s) temperatures are indicated.

DISCUSSION

Based on the preliminary results presented here, we are beginning to gain a better understanding of the δ/α' martensitic transformation in Pu-Ga alloys. Additional resistivity experiments with different alloys will allow us to further characterize the phase transformation and investigate the TTT diagrams. Observation of a possible incubation period in the annealing experiments raises many questions about the formation of α': Is the growth nucleation-controlled? Is it possible that α' is not detected by the resistivity technique until the particles reach a certain size? Future experiments with faster quenching rates may provide answers to these questions.

Elucidating the mechanisms, thermodynamics, and kinetics of the δ/α' phase transformation are important for understanding how Pu-Ga alloys can change over time and under a variety of thermal conditions. Basic studies such as this provide the scientific framework for future work in the area of stockpile stewardship.

ACKNOWLEDGMENTS

This work was performed under the auspices of the U.S. Department of Energy by the University of California, Lawrence Livermore National Laboratory under Contract No. W–7405–ENG-48.

REFERENCE

1. Orme, J. T., Faiers, M. E., and Ward, B. J., *Plutonium 1975 and Other Actinides*, North-Holland Publishing Company (1975), pp. 761–773.

Plutonium-Based Superconductivity: The Audacity of the 5f Electrons?

J. L. Sarrao,* L. A. Morales,* J. D. Thompson,* B. L. Scott,* G. R. Stewart,*† F. Wastin,† J. Rebizant,‡ P. Boulet,‡ E. Colineau,‡ and G. H. Lander*‡

*Los Alamos National Laboratory, Los Alamos, NM 87545, USA
†Department of Physics, University of Florida, Gainesville, FL 32611, USA
‡European Commission, JRC, Institute for Transuranium Elements, Postfach 2340, 76125 Karlsruhe, Germany

Plutonium and its alloys and compounds comprise a set of materials that is of fundamental scientific interest. The electronic structure of these materials "sets" the physical properties. For plutonium alloys in particular, the interplay between electronic structure and metallurgical properties is poorly understood.[1] Plutonium's 5f electrons are thought to be poised on a boundary between localized and itinerant behavior; their theoretical treatment pushes the limits of current electronic structure calculations.[2] In part, this situation is enabled by the lack of sufficiently definitive experiments against which theory can be tested and iteratively refined. Our recent research efforts have been focused on the synthesis of single-crystal plutonium compounds in which the electronic behavior could be tuned from localized to itinerant hybridizations. As a result, the research landscape was changed by the recent discovery of superconductivity, with a transition temperature, T_c, exceeding 18.5 K, in single crystals of $PuCoGa_5$. Broken symmetries, such as superconductivity, provide critical constraints on how we understand the electronic structure of the higher temperature, uniform state, δ-Pu. Structurally, $PuCoGa_5$ is a layered derivative of δ-Pu, and its superconductivity, the first ever in a Pu-based material, is a consequence of the 5f electrons interacting with their environment and provides a previously unavailable key to revealing the 5f configuration in Pu.

Low-temperature magnetic susceptibility and specific measurements on $PuCoGa_5$ are shown in Figure 1. Temperature-dependent magnetic susceptibility and specific heat measurements clearly show a sharp diamagnetic transition about 18.5 K. At low temperature, the magnitude of the susceptibility corresponds to almost 100% of perfect diamagnetism. The inferred Sommerfeld coefficient, γ, of 77 mJ $mol^{-1}K^{-2}$ for $PuCoGa_5$ is indicative of a modest quasiparticle enhancement. Taken together, these data provide ample evidence for bulk superconductivity.

A T_c of 18 K is unusual for an intermetallic compound; only a small number of intermetallics have a T_c above 18 K (MgB_2 has a T_c of 39 K).[3] The upper critical field in $PuCoGa_5$ is correspondingly large, as derived from data shown in Figure 2. The initial slope dH_{c2}/dT of -59 $kOeK^{-1}$ gives an orbital critical field of 740 kOe. The lower critical field $H_{c1} \sim 350$ Oe is derived from data shown in the lower inset of Figure 2.

Further measurements yielded a critical current, J_c, greater than 10^4 A cm^{-2} for $T > 0.9T_c$. Such a result for J_c is competitive with the best available applied superconductors. Measurements on the same sample as a function of time revealed that T_c decreased at a rate of approximately 0.2 K per month, however over that same time period Jc increased by a factor of nearly two. Self-irradiation damage in $PuCoGa_5$ seems to enhance the critical current by creating effective flux pinning sites.

Figure 3 shows the magnetic susceptibility and electrical resistivity data for $PuCoGa_5$. The temperature dependence of the electrical resistivity is similar to that of $UMGa_5$ and is indicative of spin-disorder scattering. The susceptibility of $PuCoGa_5$ shows local-moment behavior expected for Pu^{3+}. Local-moment susceptibility is also found in $CeCoIn_5$, however isostructural $UCoGa_5$ displays paramagnetism indicating itinerant f-electron behavior.[4,5] These data suggest that the hybridization of the 5f-electrons in $PuCoGa_5$ lies between that of its Ce and U based analogues. Furthermore, since the 5f-electrons of the actinides are intermediate between the more localized 4f electrons of the rare earths and the itinerant d electrons of the transition metals, the observed unconventional superconductivity of $PuCoGa_5$ bridges the two classes of spin-fluctuation-mediated superconductors: the heavy-fermion superconductors and the high-T_c copper oxides.

CP673, *Plutonium Futures — The Science,* edited by G. D. Jarvinen

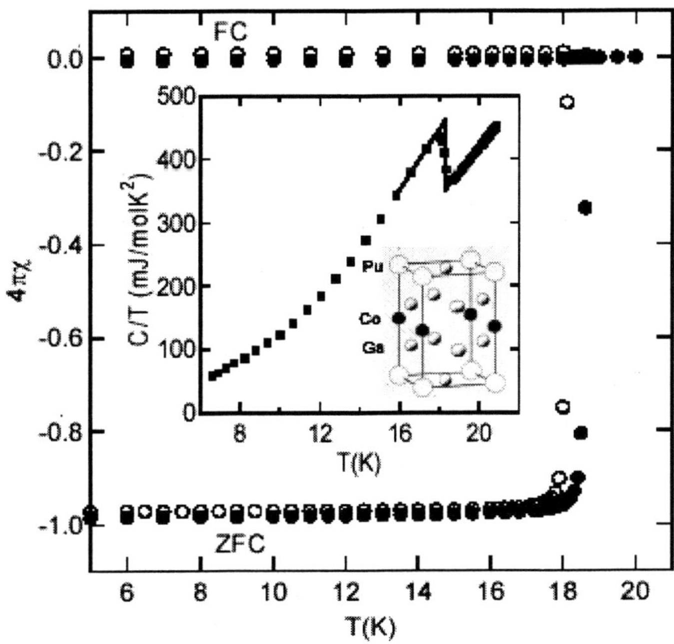

FIGURE 1. Crystal structure and evidence for superconductivity in $PuCoGa_5$. $PuCoGa_5$ crystallizes in the P4/mmm space group—Pu: 1a, (0,0,0); Co: 1b, (0,0,0.5); Ga1: 1c, (0.5,0.5,0); Ga2: 4i, (0,0.5,0.312). Zero-field cooled (ZFC) and field-cooled (FC) magnetic susceptibility of $PuCoGa_5$ were measured in 10 Oe for a fresh (filled symbols) and an aged (empty symbols) single crystal. Magnetization measurements were performed in a Quantum Design SQUID magnetometer with the sample sealed in an alumina holder designed to minimize background signal and prevent spread of radioactive contamination. The inset shows heat capacity, plotted as heat capacity divided by temperature, vs temperature for $PuCoGa_5$. Heat capacity measurements were made in a Quantum Design Physical Property Measurement System on a 27 mg single crystal. Self-heating limited the lowest measurement temperature to 6.6 K, although the calorimeter reached a base temperature of 1.5 K.

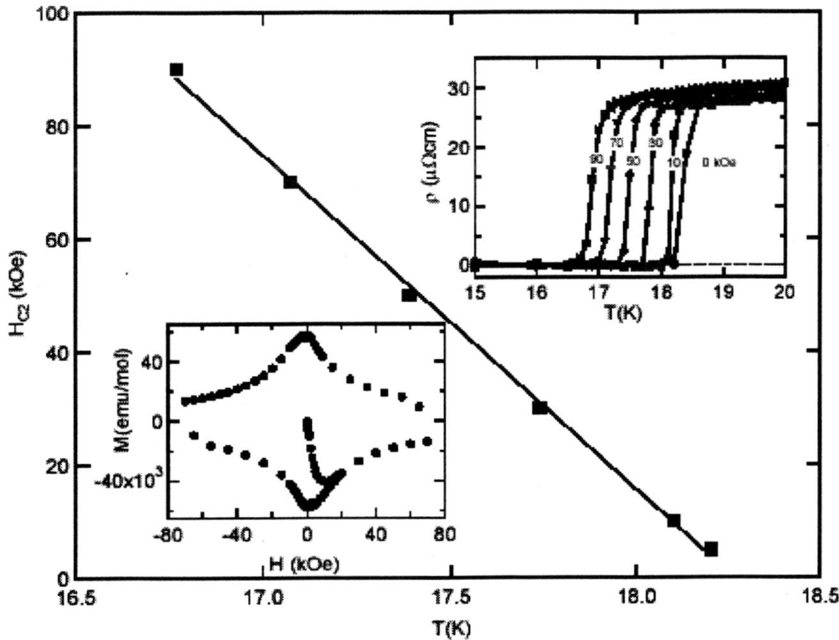

FIGURE 2. Upper critical field of $PuCoGa_5$ as a function of temperature. The upper inset shows the field-dependent resistivity data from which the field temperature phase diagram was deduced. The lower inset shows a representative magnetization loop for $PuCoGa_5$, measured at 5 K.

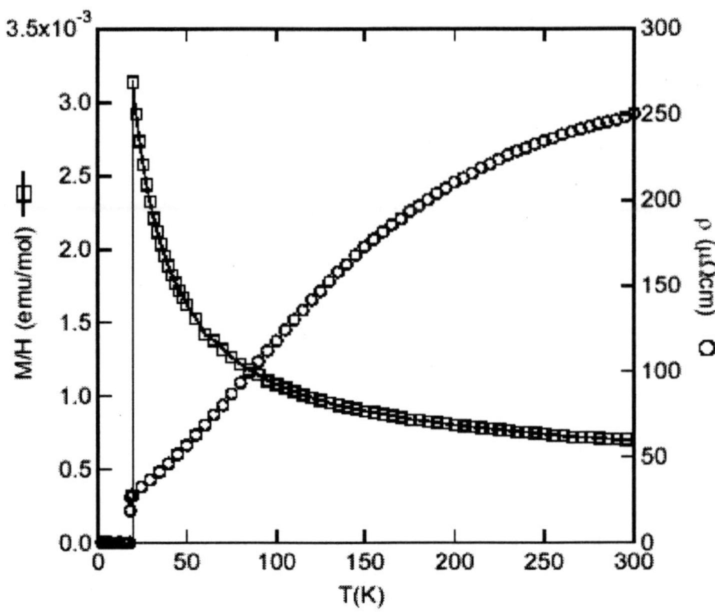

FIGURE 3. Normal-state properties of $PuCoGa_5$. The electrical resistivity is indicated by circles. It increases approximately as $T^{1.35}$ from just above T_c to 50 K. The magnetic susceptibility (squares) as a function of temperature follows $\chi = \chi_0 + C/(T-\theta)$ with an effective moment $\mu = (8C)^{1/2} = 0.68$ μB and an interaction temperature, $\theta \sim -2$ K.

REFERENCES

1. Hecker, S. S., "The Complex World of Plutonium Science," MRS Bulletin 26, 672–678 (2001).
2. Savrasov, S. Y., Kotliar, G., and Abrahams, E., "Correlated Electrons in δ-Plutonium within a Dynamical Mean-Field Picture," Nature 410, 793–795 (2001).
3. Nagamatsu, J., Nakagawa, N., Muranaka, T., Zenitani, Y., and Akimitsu, J., "Superconductivity at 39 K in Magnesium Diboride," Nature 410, 63–64 (2001).
4. Grin, Yu. N., Rogl, P., and Hiebl, K., "Structural Chemistry and Magnetic Behavior of Ternary Uranium Gallides U(Fe,Co,Ni,Ru,Rh,Pd,Os,Ir,Pt)-Ga_5," J. Less Common Met. 121, 497–505 (1986).
5. Petrovic, C., Pagliuso, P. G., Hundley, M. F., Movshovich, R., Sarrao, J. L., Thompson, J. D., Fisk, Z., and Monthoux, P., "Heavy-Fermion Superconductivity in $CeCoIn_5$ at 2.3 K," J. Phys-Condens Mat. 13, L337–L342 (2001).

Phonon Dispersion in Actinides Measured with Inelastic X-Ray Scattering: New Opportunities to Solve Some Old Problems

M. E. Manley,[1] G. H. Lander,[1,2] H. Sinn,[3] A. Alatas,[3] W. L. Hults,[1]
R. J. McQueeney,[1] J. C. Lashley,[1] J. L. Smith,[1] and J. Willit[3]

[1]*Los Alamos National Laboratory, Los Alamos, New Mexico 87545*
[2]*European Commission, JRC, Institute for Transuranium Elements, Postfach 2340, D-76125 Karlsruhe, Germany*
[3]*Argonne National Laboratory, Argonne, Illinois 60439*

Measuring the dispersion (energy wave-vector relation) of phonons in crystals gives the most detailed description of the atomic vibrational motion. Because neutrons simultaneously possess both energies (E ~ 1 meV) and wavelengths (λ ~ 1 Å) similar to phonons, they have been the method of choice for these measurements. However, despite an interest in the unusual display of complex behavior seen in *f*-electron elements, neutron results have remained elusive for many actinides. This has been because many isotopes have large neutron absorption cross sections, and even in cases where a suitable isotope exists, single crystals large enough for neutron scattering (>0.1 cm^3) were rarely realized. Recent developments in high-resolution inelastic x-ray scattering (IXS) at the Advanced Photon Source promise to reinvigorate this field. They are now capable of obtaining a 21.6 keV (λ = 0.57 Å) beam with a 2-meV energy width ($\Delta E/E$ ~ 0.0000001). The technique is isotope independent and requires only very small samples. We recently demonstrated the power of the technique by measuring the dispersion of phonons in a uranium crystal from a scattering volume of only ~2 x 10^{-3} mm^3.

Before the development of inelastic neutron scattering techniques in the 1950s and 1960s, there were few data on the dispersion of phonons in crystals, and therefore, little was know about the detailed motion of atoms in solids. Since then, large advances have been made in our understanding of phonons. Information from phonon dispersion measurements is now routinely used to explain a wide range of phenomena including superconductivity, soft-mode phase transition mechanisms, and phase stability. However, despite the general interest in *f*-electron elements,[1-3] details about the phonon dispersion relationships exist only for uranium.[4] This is because many isotopes have large neutron absorption cross sections, and even in cases where a suitable isotope exists, crystals large enough for neutron scattering (at least 0.1 cm^3) are unavailable. Work on uranium, for example, is limited by the fact that only one large crystal has ever been produced.[3]

We have begun an effort to change this by taking advantage of the development of high-resolution inelastic x-ray scattering (IXS) at the Advanced Photon Source at Argonne National Laboratory, Argonne, Illinois. In our first demonstration, we measured the phonon dispersion curves in uranium. At first, we did not think this would work. The incident energy chosen to give the maximum flux with a resolution of ~2 meV[5] has the unfortunate consequence that it is just 700 eV *above* the L_2 edge of uranium at 20.95 keV. The calculated low-penetration depth of ~6 μm and accompanying large fluorescence initially suggested the experiment would not succeed, or at least would be very difficult, and indeed the calculation shown in Figure 5 of Reference 5 is equally pessimistic. It shows that the phonon signal for actinide ($Z > 90$) elements will be between one and two orders of magnitude smaller than for light ($Z < 20$) elements. To our surprise, however, our test experiment was a considerable success. Phonon count rates of ~1 ct/s were obtained from an equivalent scattering mass of ~40 μg of a heavy element. Figure 1, taken from Reference 6, shows the raw data. These data are also compared with the curves fit to neutron data in Figure 2.[6] The small differences have been attributed to the low quality of the large crystal used in all of the neutron work.[6] The success of this work opens up many new possibilities. For example, we have a single crystal of plutonium[7] that we are now preparing to measure.

CP673, *Plutonium Futures — The Science,* edited by G. D. Jarvinen

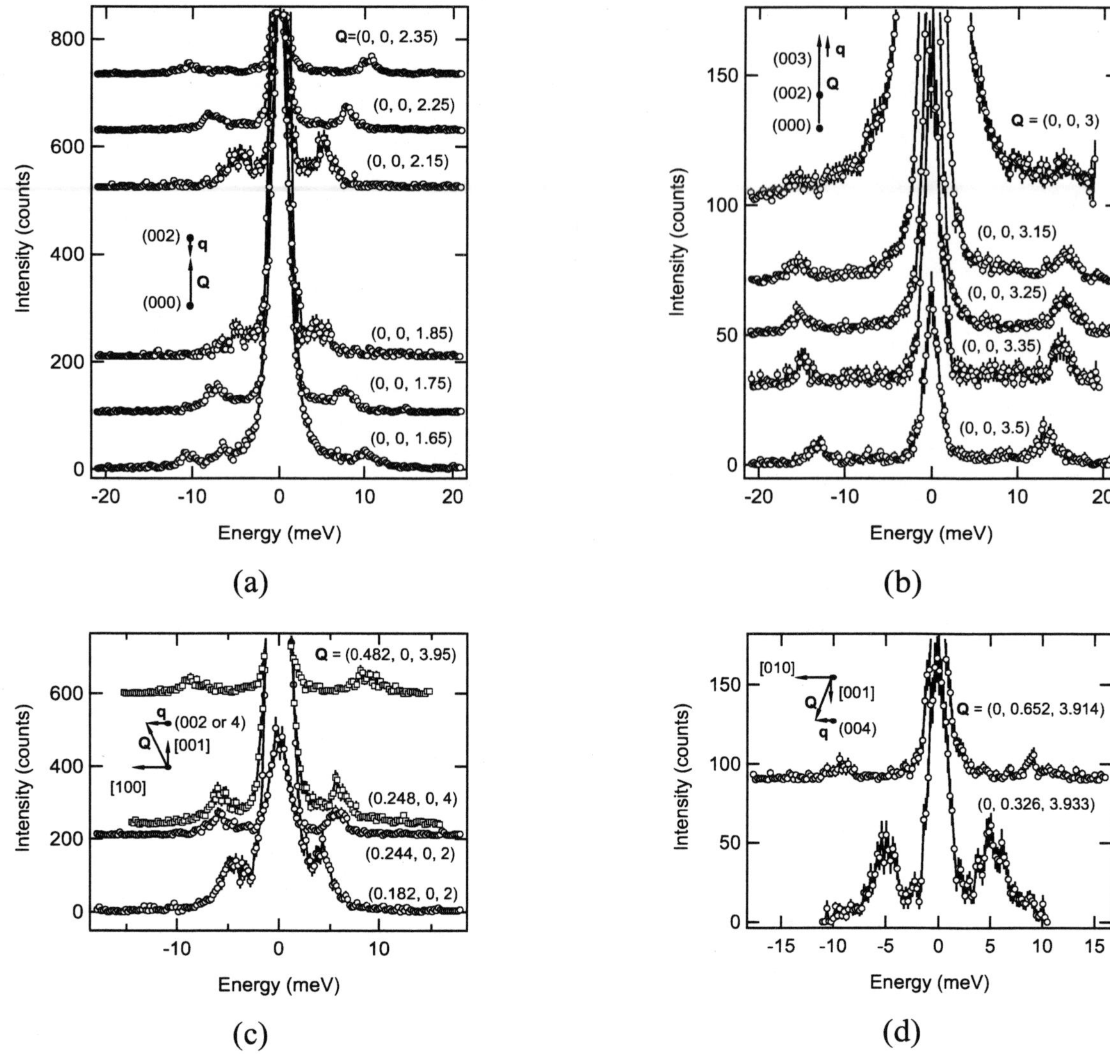

FIGURE 1. Raw data counted 60 seconds per point, with the data offset for clarity. Frames (a) and (b) show longitudinal modes and frames (c) and (d) show transverse modes. The geometry in the reciprocal space is shown as an insert in each frame. The scattering vector, Q, is on the c-axis for the longitudinal modes and is tilted slightly off axis (in either [100] or [010]) near a reciprocal lattice point to obtain transverse modes. The phonon wave vector, q, conserves momentum according to $Q = G + q$, where G is a reciprocal lattice vector pointing to the nearest reciprocal lattice point.

It is difficult to speculate what will be discovered in making these measurements on plutonium. Plutonium is more complicated than any element ever measured. For example, its phase diagram contains six different solid-state phases at ambient pressure, including one phase with a negative coefficient of thermal expansion.[1] Moreover, the elastic constants measured on single crystals of δ-stabilized ^{239}Pu show most unusual properties, with an elastic anisotropy higher than for any other fcc material.[8] Perhaps reviewing the lessons learned from studying a closely related and perhaps the second most complicated element, uranium, would be a good place to start.

Phonon-dispersion curve measurements performed on the uranium crystal after the first room-temperature experiments[4] were largely motivated by the discovery of several charge-density wave transitions at low temperatures.[3] The charge-density wave transition mechanism was observed explicitly in the phonon dispersion curves. At about [0.500] in Figure 2 there is a famous dip in one of the curves that was fit to neutron data. As the temperature is decreased towards the first charge density wave transition (~42 K), this dip decreases in energy as it approaches zero. This can be understood as a wave with a fixed wavelength and direction gradually slowing its motion until eventually it freezes in and becomes a static distortion. The distortion results in a new structure, and therefore a new phase is formed.

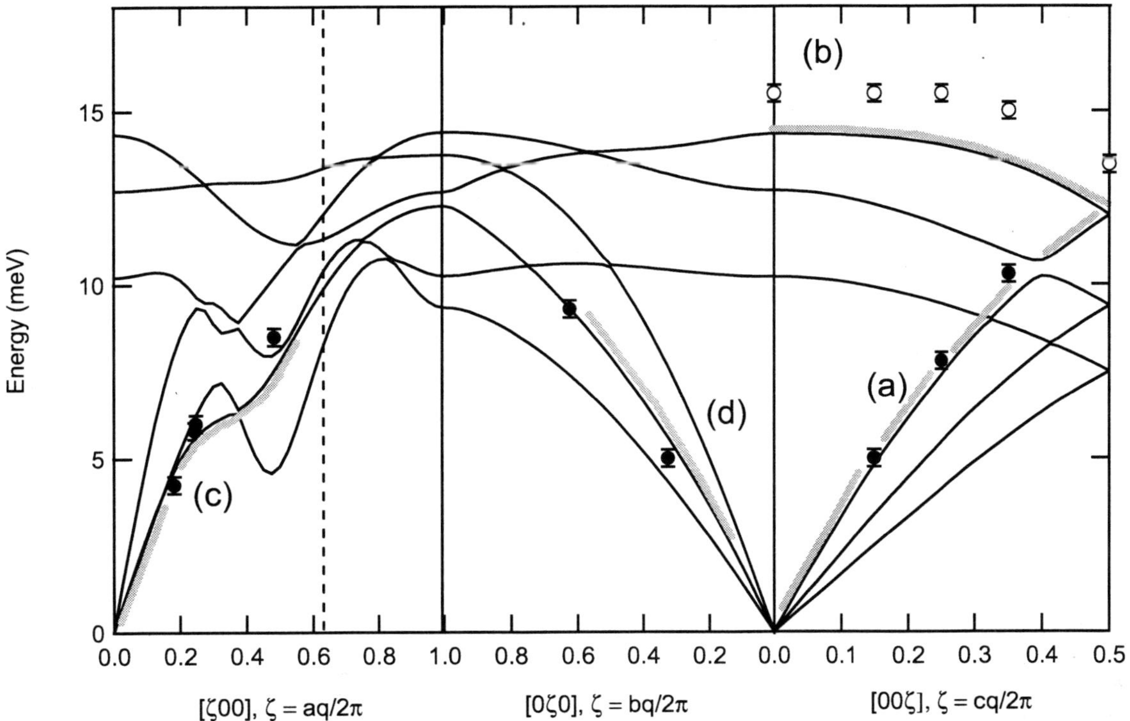

FIGURE 2. The solid lines are phonon dispersion curves as determined from a force constant model fit of neutron data in Reference 7. The data points are from the present IXS experiment (labels correspond to Frames (a)–(d) in Figure 1). The grey lines indicate the phonon branches selected by the scattering geometries in our experiment. Unlike our data, the longitudinal optic [00ζ] modes from the neutron data[7] fall on the line of the model.

More recent measurements of the phonon density of states on polycrystalline samples of uranium revealed a large harmonic softening over the entire α-phase temperature range.[9] The lack of anharmonicity implies that the interatomic potential landscape of uranium is continuously altered by electronic excitations as the temperature is raised.[9] In addition, evidence extracted from neutron-diffraction data[10] suggests that this unusual phonon softening may occur in many of the light actinides, including plutonium. These results present a fundamental challenge to the way we understand solids. First, present state-of-the-art electronic structure calculations are based on the assumption that electronic excitations have no effect on phonon frequencies. Second, there is a continual exchange between electronic energy and phonon entropy, which implies that our usual notions of electronic and phonon contributions to, for example, specific heat, are no longer valid. In addition to these points, there is also the practical problem that a large temperature effect means that more measurements are required within each phase to fully understand the behavior.

For the reasons discussed above, there are no direct measurements of the phonon dispersion curves in any actinide other than uranium. Information on vibrations have been limited to elastic properties[8] (long wavelength limit of phonons) and mean square vibrational amplitudes[10] (a moment of the phonon density of states). This has made it extremely difficult to develop accurate models for the behavior of these materials. The equation of state for plutonium is a particularly poignant case. Understanding the plutonium equation of state is crucial to the stockpile stewardship program and therefore to the mission of the laboratory. For this reason, considerable effort has been put into this problem at LANL as evidenced by the 2000 issue of Los Alamos Science.[1] Despite all this effort, however, there is still much we are unsure of when it comes to plutonium.[1] In the first article of Reference 1 (*An update: Plutonium and Quantum Criticality*) it is suggested that the strange behavior of plutonium may be related to quantum criticality. As pointed out by Laughlin et al.,[11] it is virtually impossible to calculate from first principles the properties of systems whose behavior is dominated by a quantum-phase transition. If this is true, then to understand plutonium we will have to rely heavily on experimental data.[1]

Measurement of the details of how plutonium atoms move about their equilibrium positions is required if our understanding of its phase transitions is to advance. The largest contribution to the entropy of a solid at high temperatures comes from atomic vibrations. Because phase transitions that occur with increasing temperature are driven by entropy, without this information we cannot expect to understand why there are six solid-state phases in plutonium at ambient pressure. Vibrational entropy changes across phase transitions can be calculated from the phonon dispersion curves. The development of IXS presents a new and unique opportunity to solve this problem for plutonium.

REFERENCES

1. Los Alamos Science, Challenges in Plutonium Science, edited by N. G. Cooper, Number 26, Vol. I (2000).
2. Holland-Moritz, E., and Lander, G. H., "Neutron Inelastic Scattering from Actinides and Anomalous Lanthanides," in Handbook on the Physics and Chemistry of Rare Earths, edited by K. A. Gschneidner, Jr., L. Eyring, G. H. Lander, and G. R. Choppin, Elsevier Science (1994), Vol. 19, Chapter 130.
3. Lander, G. H., Fisher, E. S., and Bader, S. D., Adv. Phys. 43, 1–111 (1994).
4. Crummett, W. P., Smith, H. G., Nicklow, R. M., and Wakabayashi, N., Phys. Rev. B 19, 6028 (1979).
5. Sinn, H., J. Phys. Cond. Matter 13, 7525 (2001).
6. Manley, M. E., Lander, G. H., Sinn, H., Alatas, A., Hults, W. L., McQueeney, R. J., and Smith, J. L., Phys. Rev. B. 67, 0923XX (2003).
7. Lashley, J. C., Stout, M. G., Peyera, R. A., and Embury, J. D., Scripta Materialia 44, 2815 (2001).
8. Ledbetter, H. M., and Moment, R. L., Acta Metallurgica 24, 891 (1976).
9. Manley, M. E., Fultz, B., McQueeney, R. J., Brown, C. M., Hults, W. L., Smith, J. L., Thoma, D. J., Osborn, R., and Robertson, J. L., Phys. Rev. Lett. 86, 3076 (2001).
10. Lawson, A. C., Martinez, B., Roberts, J. A., Bennett, B. I., and Richardson, J. W. Jr., Philos. Mag. B 80, 53 (2000).
11. Laughlin, R. B., Lonzarich, G. G., Monthoux, P., and Pines, D., Adv. In Physics 50, 361 (2001).

A New Paradigm for the Determination of the 5f Electronic Structure of Pu and the Actinides

James G. Tobin,*[1] B. W. Chung,[1] R. K. Schulze,[2] and D. K. Shuh[1]

[1]*Lawrence Livermore National Laboratory, Livermore, CA USA 94550*
[2]*Los Alamos National Laboratory, Los Alamos, NM USA 87545*

The valence electronic structures of the actinide metals and alloys in general and plutonium (Pu) in particular remain mired in controversy.[1] Interestingly, the various phases of Pu metal provide a mirocosm of the metallic actinides as a whole. Thus, unravelling the nuances of the interplay of electronic and geometric structures in Pu will illuminate the properties of all transuranic metals. In a sense, the behavior of the Pu 5f electrons is completely counter-intuitive. The dense phase, α, has some semblance of delocalization in the 5f valence bands and can be treated theoretically within single electron models such as the Local Density Approximation (LDA). The α phase is monoclinic, which is a low symmetry ordering. The less-dense δ-phase is fcc and exhibits evidence of localized and/or correlated electronic behavior. The fcc is a high-symmetry phase that is normally associated with superior wavefunction overlap in d-state metals. But herein is the key: the linear combinations of the 5fs do not produce the nicely lobed wavefunctions with symmetry about the x, y, and z and diagonal axes, as occurs for d states. Instead, the linear combinations of f states have oddly lobed and badly directed wavefunctions that match very poorly with the high symmetry of the fcc structure. In the rare earths and lanthanides, this odd lobing is of no consequence: generally, the 4f valence states are inside the outer 5d, 6p, and 6s electrons and can be treated as isolated, atomic-like orbitals that do not participate in chemical bonding. In the actinides, the 5fs are less well shielded. Shielding is undoubtedly a key issue and relates to the volume changes. In fact, the 20% volume increase of Pu between the α and δ phases is a reflection of the discontinuous 30% volume jump between Pu and Am, as one moves along the row of actinide elements. Finally, we have not even begun to deal with the issue of dilute alloy formation and its impact upon the electronic structure of metallic actinides. The upshot of this is that the valence electronic structure of Pu in particular and the actinides in general are only poorly understood.

Within this context and despite recent intensive experimental effort,[1–3] the electronic structure of Pu, particularly δ-Pu, remains ill defined. An evaluation of our previous synchrotron-radiation-based investigation of α-Pu and δ-Pu[1] has led to a new paradigm for the interpretation of photoemission spectra of U, Np, α-Pu, and δ-Pu . This approach is founded upon a model in which spin and spin-orbit splittings are included in the picture of the 5f states[4] and upon the observation of chiral/spin-dependent effects in nonmagnetic systems.[5,6] By extending a quantitative model developed for the interpretation of core level spectroscopy in magnetic systems,[7] it is possible to predict the contributions of the individual component states within the 5-f manifold. This has led to a remarkable agreement between the results of the model and the previously collected spectra of U, Np, and Pu , particularly δ-Pu,[1–3,8] and to a prediction of what we might expect to see in future spin-resolving experiments.[9]

ACKNOWLEDGMENTS

This work was performed under the auspices of the U.S. Department of Energy by the University of California, Lawrence Livermore National Laboratory under Contract No. W-7405-ENG-48.

CP673, *Plutonium Futures — The Science*, edited by G. D. Jarvinen

REFERENCES

1. Terry, J., Schulze, R. K., Farr, J. D., Zocco, T., Heinzelman, K., Rotenberg, E., Shuh, D. K., van der Laan, G., Arena, D. A., and Tobin, J. G., Surface Science Letters 499, L141 (2002).
2. Gouder, T., Havela, L., Wastin, F., and Rebizant, J., Europhys. Lett. 55, 705 (2001); MRS Bulletin 26, 684 (2001); Phys. Rev. Lett. 84, 3378 (2000).
3. Arko, A. J., Joyce, J. J., Morales, L., Wills, J., Lashley, J., Wastin, F., and Rebizant, J., Phys. Rev. B 62, 1773 (2000).
4. Savrosov, S. Y., and Kotliar, G., Phys. Rev. Lett. 84, 3670 (2000).
5. Roth, Ch. et al., Phys. Rev. Lett. 73, 1963 (1994).
6. Starke, K. et al., Phys. Rev. B 53, 10544 (1996).
7. Tobin, J. G., and Schumann, F. O., Surface Science 478, 211 (2001).
8. Naegele, J. R., "Photoem. of Solids," Landolt-Bornstein III/B,183 (1994).
9. Tobin, J., Arena, D. A., Chung, B., Roussel, P., Terry, J., Schulze, R. K., Farr, J. D., Zocco, T., Heinzelman, K., Rotenberg, E., and Shuh, D. K., "Photoelectron Spectroscopy of Plutonium at the Advanced Light Source," UCRL-JC-145703, J. Nucl. Sci. Tech./ Proc. of Actinides 2001 (accepted, 2002).

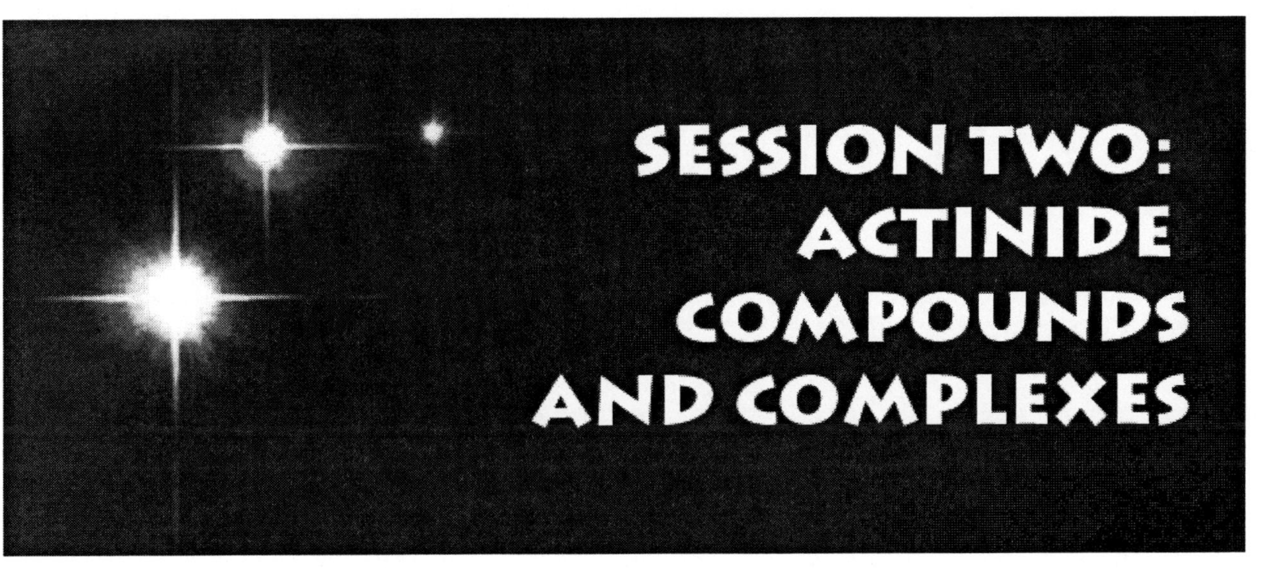
SESSION TWO:
ACTINIDE
COMPOUNDS
AND COMPLEXES

Insight into the Coordination Chemistry of Plutonium Compounds: Synthesis and Structural Characteristics of Pu(III) Oxalate and Pu(VI) Hydrous Oxides

Wolfgang Runde,[1] Amanda Bean,[2] and Brian L. Scott[1]

Los Alamos National Laboratory, [1]Chemical and [2]Nuclear Material Divisions, Los Alamos, NM 87544

INTRODUCTION

Modern advances in plutonium (Pu) process and separation chemistry, prediction of weapons aging, and long-term prediction of nuclear waste storage and disposition rely on the understanding of Pu materials on a molecular and atomistic scale. In recent years, synthesis and structural characterization of transuranium compounds with both extended structures and molecular coordination has intensified, however, structural studies of inorganic Pu compounds remain rare. In fact, most references in the structural databases, CSD and ICSD, describe diffraction powder studies and Pu alloys, but contain only a handful of data that are based on the characterization of inorganic Pu phases using single crystal x-ray diffraction (XRD). The single crystal structures reported in the literature are limited to sulfate, nitrate, and carbonate complexes of Pu(IV) and the aquo ion of Pu(III).[1] Single crystal structures of Pu(V) and Pu(VI) compounds remain rare. Compounds that have been used for separation for over fifty years, such as oxalates and hydroxides,[2] remain structurally a mystery, despite their recognized importance. The absence of detailed structural data limits the ability to enhance separation yields and hinders prediction of the environmental behavior of Pu chemistry and its impact on nuclear waste performance. We are overcoming the obstacle of synthesizing quality Pu single crystals by using mild hydrothermal conditions. In this presentation, we will review the preparation of historically important Pu oxalate compounds and present the structural details we have recently explored. We also provide some insight into the rich structural chemistry of hydrous Pu(VI) oxide phases, compare the structural and spectroscopic characteristics with its chemical analogue U(VI), and discuss the importance of secondary hydroxide and silicate phases for Pu environmental behavior.

RESULTS

Several Pu(IV) and (VI) oxalate solution complexes have been characterized, and powder diffraction data are available for the Pu(III) oxalate, $Pu_2(C_2O_4)_3 \cdot 10H_2O$, the Pu(IV) compounds of general formula $Pu(C_2O_4)_2 \cdot nH_2O$ (n = 2 or 6), and the Pu(VI) compound $PuO_2C_2O_4 \cdot 3H_2O$.[2] We synthesized two new Pu(III) oxalates, $Pu(C_2O_4)_{1.5}(H_2O)_3$ and $KPu(C_2O_4)_2H_2O \cdot 2H_2O$. In the blue binary Pu(III) compound, plutonium is coordinated bidentately to three oxalate ligands and three terminal waters with Pu–O bond distances between 2.47 and 2.56 Å. In contrast, the $[PuO_9]$ polyhedra in $KPu(C_2O_4)_2H_2O \cdot 2H_2O$ are built up by eight oxygen atoms from four oxalates and one water molecule bound to the plutonium atom. The unique alternating bridging pattern of the oxalate groups creates an open three-dimensional framework with channels along the c axis of about 4.9 Å x 11.8 Å diameter.

The hydrolysis of U(VI) is the probably most studied complexation reaction of the light actinide elements due to its prominent alteration phases in spent nuclear fuel subjected to oxidative dissolution. The chemistry of solid hydrated U(VI) oxides is extremely complex, and about 20 different phases have been structurally characterized.[3] The characteristic feature of bridging μ_3- and μ_2-OH (or O) are found in the extended solid phases as well as in the structures of molecular polynuclear U(VI) hydroxo complexes. Although the solubility of hydrous Pu(VI) oxide has been studied, the data interpretation lack accurate determination of the solid equilibrium phase due to its amorphous

CP673, *Plutonium Futures — The Science*, edited by G. D. Jarvinen

character.[4] Under ambient conditions a red-brown gelatinous solid phase precipitates upon addition of hydroxide to an acidic or neutral Pu(VI) solution. Applying hydrothermal conditions enabled us to form a crystalline Pu(VI) precipitate in basic solutions (pH 10–14). In the presence of large amounts of KIO_4 (used to maintain plutonium in its hexavalent oxidation state at increased temperature), a mixed hydroxo-iodato phase, $(PuO_2)_2(IO_3)(OH)_3$, is obtained. Ribbons of edge-sharing $[PuO_7]$ polyhedra are connected via iodate and bridging μ_2-OH groups with the Pu–OH distances ranging between 2.31 Å and 2.39 Å. Two iodate ligands are coordinated in the equatorial plane around the plutonium with Pu–O(IO_2) of 2.48 and 2.58 Å. The axial Pu=O bond length of 1.75(1) Å is distinctive for hexavalent plutonium. At lower iodate concentrations, iodate is not incorporated in the structure, and we obtained the first Pu(VI) structure containing $(PuO_2)(O)(OH)_2$ sheets with K^+ and I^- ions between the layers. The sheets are built of edge-sharing $[PuO_7]$ polyhedra with equatorial Pu-O distances of 2.20 Å (μ_3-O), 2.42 Å (μ_3-OH), and 2.51 Å (μ_3-OH). This network of edge-sharing pentagonal bipyramids is very similar to many uranium(VI) complex oxyhydroxide minerals.[3] However, the water molecules do not interleave the actinide layers but are attracted closely to the layer with only appr. 1 Å above the equatorial plane.

DISCUSSION

Both oxalate and hydrous oxide phases of plutonium have significant historic and future importance in process and separation chemistry, as well as in the long-term prediction of nuclear weapons aging and nuclear waste storage and disposition. Hydrous oxide phases form as alteration products from UO_2 and have been suggested as host matrix for the light actinide elements Np and Pu. Long-term stability calculations predict, however, the formation of U(VI) silicate phases, which are claimed to be unable to incorporate Pu, concurring with the lack of unknown solid Pu(VI) silicate phases. We will present the first structurally characterized Pu(VI) silicate phase, which is isostructural with the U(VI) mineral boltwoodite,[5] $(K_{0.56}Na_{0.42})[(UO_2)(SiO_3OH)](H_2O)_{1.5}$, and which formed as a second phase besides the Pu(VI) hydrous oxide. This mineral is likely to be formed in substantial quantities when spent nuclear fuel is corroded in silica-bearing waters at the proposed repository at Yucca Mountain.

REFERENCES

1. (a) Clark D. L., Conradson S. D., Keogh D. W., Palmer P. D., Scott B. L., and Tait C. D., Inorg. Chem. 37, 2893–2899 (1998); (b) Matonic, J. H., Scott, B. L., and Neu, M.P., Inorg. Chem. 40, 2638–2639 (2001).
2. Cleveland, J. M., The Chemistry of Plutonium, American Nuclear Society, LaGrange Park, Illinois (1979).
3. Volume 38, URANIUM: Mineralogy, Geochemistry and the Environment, edited by P. C. Burns and R. Finch, Mineralogical Society of America, Washington, DC (1999), pp. 23–180.
4. Chemical Thermodynamics, Vol. 4. Chemical Thermodynamics of Neptunium and Plutonium, NEA/OECD, Elsevier, North-Holland (2001).
5. Burns, P. C. Can Mineral. 36, 1069–1075 (1998).

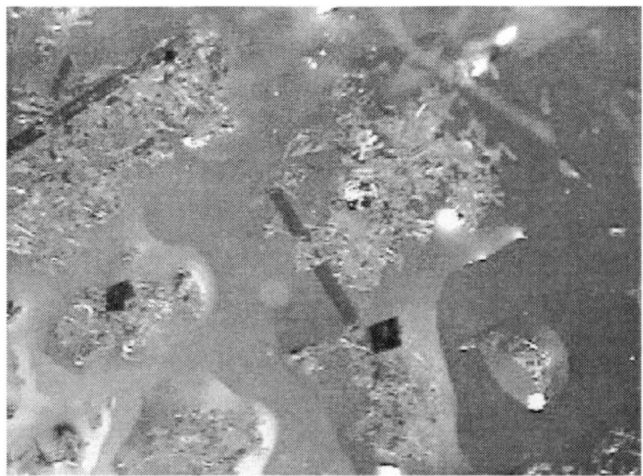

FIGURE 1. Microphotograph of blue $Pu(C_2O_4)_{1.5}(H_2O)_3$ crystals.

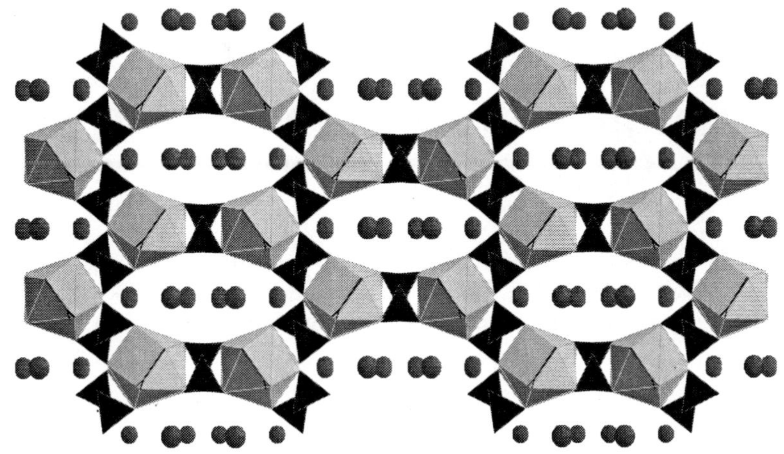

FIGURE 2. View along the *c* axis of $KPu(C_2O_4)_2H_2O{\bullet}2H_2O$.

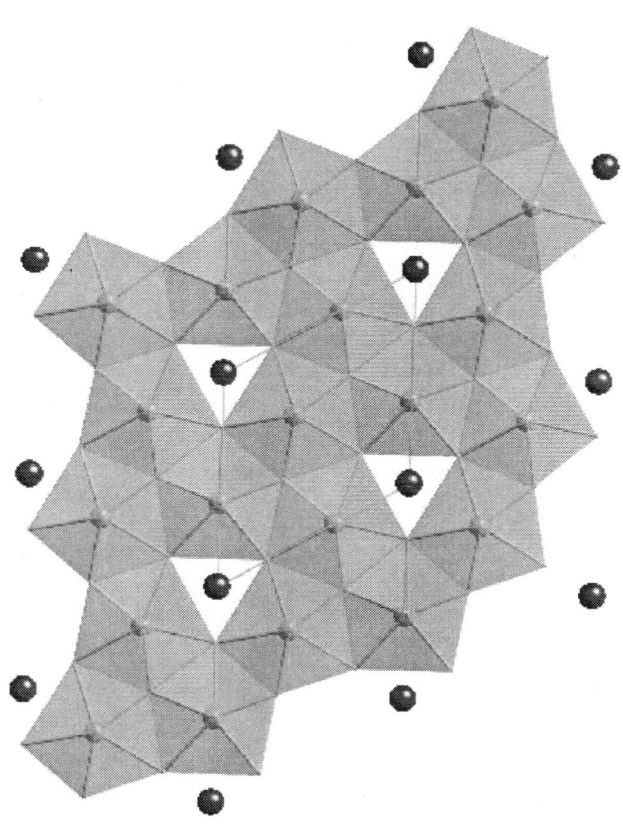

FIGURE 3. Sheet of $[PuO_7]$ polyhedra that occurs in the structure of the synthesized Pu(VI) hydrous oxide.

A Study of Colloid Generation and Disproportionation of Pu(IV) in Aquatic Solution by LIBD and LPAS

C. Bitea, C. Walther, J. I. Yun, Ch. Marquardt, A. Seibert, V. Neck, Th. Fanghänel, and J. I. Kim

Institut für Nukleare Entsorgung, Forschungszentrum Karlsruhe, 76021 Karlsruhe, Germany

INTRODUCTION

The tetravalent plutonium ion in aquatic solution is known to be unstable even at low pH in dilute concentrations because of its tendency towards colloid formation or disproportionation. Such particular properties of Pu(IV) have turned into great difficulties for the appraisal of its thermodynamic solubility in aquatic systems and thus resulted in a large number of controversial results.[1] Undergoing chemical reactions of Pu(IV) in dilute concentrations in the low pH region can only be recognized by sensitive speciation approaches, which, however, have not been hitherto available. The latest development of laser spectroscopic speciation approaches provides the possibility now of assessing the chemical reactions of interest down to the Pu(IV) concentration of a few micromol/l or even lower.

The present work describes a notable example how a combination of two different laser spectroscopic speciation approaches can be applied for the study of aquatic chemical reactions of Pu(IV) in dilute concentrations beyond the expediency of a conventional spectroscopic approach. The two methods concerned are laser-induced breakdown detection (LIBD)[2] for colloid quantification and laser-induced photoacoustic spectroscopy (LPAS)[3] for measuring optical properties of an ionic state. The chemical behavior of the Pu^{4+} ion is investigated in the downward concentration of 10^{-4} mol/l at low pH (0.3–2.0) for the colloid formation or disproportionation. The results lead to evaluate the solubility of Pu(IV) colloids, which corresponds very closely to the solubility of Pu(IV) hydroxide or oxyhydrate, as is the case observed on Np(IV).[3]

RESULTS AND DISCUSSION

The ^{242}Pu solution is prepared in 0.5 M HCl by electrochemical reduction of a mixture of oxidation states to Pu(III) and subsequent oxidation to Pu(IV) under UV spectroscopic control. The Pu concentration is assayed by a liquid scintillation spectrometer. The solution pH is varied from 0.3 to 2.0 by appropriate dilution with 0.5 M NaCl to a desired value. The use of NaOH is avoided because it contains latent colloidal impurities, which are difficult to be separated to attain a desired purity suitable for the present work. The experiment is undertaken under ambient atmosphere at room temperature.

The formation of Pu(IV) colloid kernels is observed by LIBD at a certain pH when gradually increasing low pH to higher pH. The breakdown events detected by acoustical monitoring of plasma generation on colloids[2] remains at a background level as in pure water at low pH. Once pH is reached to the point where Pu(IV) colloid kernels are beginning to form, the breakdown events become distinctively noticeable (Figure 1) at $\geq$1% breakdown probability.

When the Pu(IV) concentration is gradually diluted, the pH of colloid kernel formation is increased accordingly. The crossing points of Pu(IV) colloid formation as a function of pH are illustrated in Figure 2, which are very narrowly scattered around a slope of –2. This particular slope infers the reaction: $Pu^{4+} + 2H_2O = Pu(OH)_2^{2+} + 2H^+$, which further undergoes colloid formation and thus $Pu(OH)_2^{2+}$ is in equilibrium with colloids. Knowing the hydrolysis constant of $Pu(OH)_2^{2+}$,[1] the solubility product of Pu colloids (presumably oxyhydrate) can be derived from Figure 2, which appears to be: log Ksp = –59.0 $\pm$ 0.3 at zero ionic strength.

CP673, *Plutonium Futures — The Science,* edited by G. D. Jarvinen

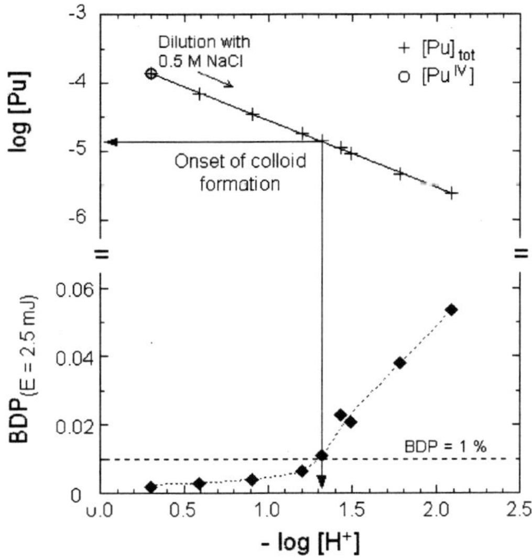

FIGURE 1. By dilution with 0.5 M NaCl, the Pu concentration is decreased and the pH is increased (top) until the solubility is exceeded and Pu(IV) colloid kernels are formed (crossing point). The colloids are detected directly by LIBD (bottom).

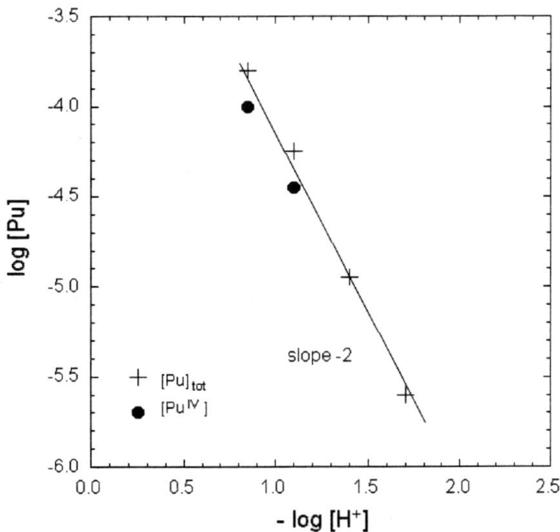

FIGURE 2: For decreasing Pu concentration, the crossing point of colloid formation is shifted to a higher pH. From the slope = –2, we infer the Pu(IV) to be present as dihydroxo-complex in equilibrium with colloids (solubility product log Ksp = –59.0 ± 0.3).

Because the amount of Pu(IV) colloids is relatively small (<5% of the total Pu concentration) at a given pH of colloid kernel formation, the rest of the Pu(IV) ionic species undergoes disproportionation with time, i.e., 3Pu(IV) + $2H_2O$ = 2Pu(III) + Pu(VI) + $4H^+$. This reaction is observed then by LPAS for Pu(IV) and Pu(III) as shown in Figure 3 at a Pu concentration of 3.15 x 10^{-5} mol/l. Although a relatively small amount of Pu(IV) colloids present in this solution at pH 1.2 cannot be appraised by LPAS, such a minute concentration of colloids (<10^{-6} mol/l Pu(IV)) can, on the other hand, be detected by LIBD as shown in Figure 2. Therefore, a combination of LIBD and LPAS facilitates a sound assessment of the chemical reactions of Pu(IV) involved at a rim of its solubility-constrained pH for the total Pu(IV) concentration down to 10^{-6} mol/l and for its colloids about 100 times less (down to a particle concentration of 10^{-8} mol/l). The present experiment shows how complicated the thermodynamic assessment of the Pu(IV) solubility is. For such reasons, there is a wide scattering of the Pu(IV) solubility product published in the literature either for its oxide or for hydroxide, the difference being a few orders of magnitude.[1] As a consequence, the solubility assessment of Pu(IV) has until now remained with considerable uncertainties. The present experiment demonstrates how to constrict such uncertainties in assistance of the novel spectroscopic approaches.

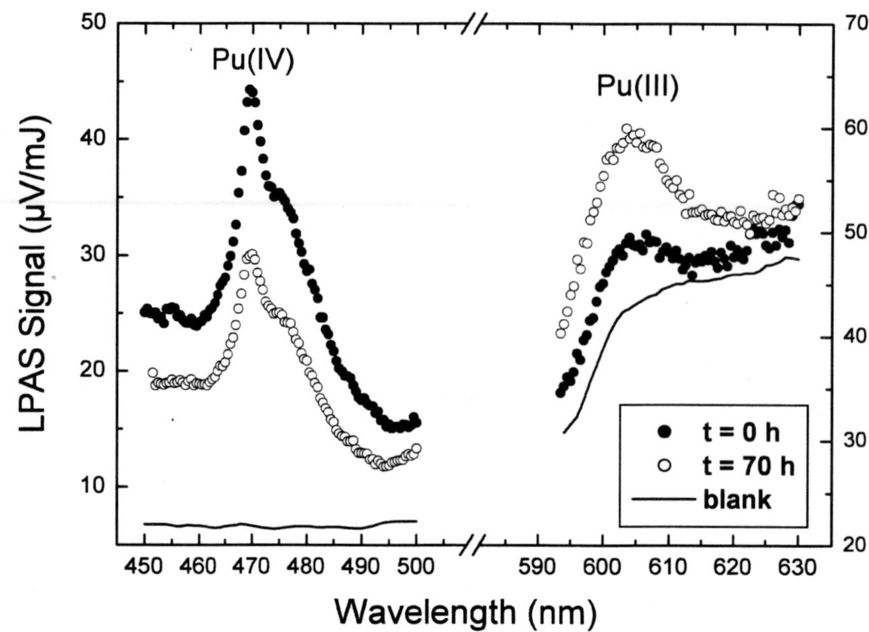

FIGURE 3. Disproportionation of Pu(IV) with time at a concentration of 3.15 x 10^{-5} mol/l is monitored directly by LPAS. The signal at 470 nm indicating Pu(IV) decreases while the Pu(III) signal at 605 nm increases by the same amount.

REFERENCES

1. Neck, V., and Kim, J. I., Radiochimica Acta 89, 1 (2001).
2. Walther, C., Bitea, C., Hauser, W., Kim, J. I., and Scherbaum, F. J., Nucl. Instrum. Meth. Phys. Res. B195, 374 (2002).
3. Neck, V., Kim, J. I., Seidel, B. S., Marquardt, C., Dardenne, K., Jensen, M. P., and Hauser, W., Radiochimica Acta 89, 439 (2001).

http://www.lanl.gov/orgs/nmtdo/PuConf2003/Pu3call.html

Investigation of Pu(IV) Complexation by TcO_4^- Anion in Perchloric Media

Ph. Moisy,[1] L. Abiad,[1] C. Madic,[1] and P. Turq[2]

[1] *CEA Valrhô DEN/DRCP/SCPS/LCA; BP 17171, 30207 Bagnols sur Ceze - France*
[2] *Université P & M Curie (Paris VI); 7, pl. Jusssieu 75005 Paris - France*

INTRODUCTION

Technetium is an artificial radio element arising primarily from the nuclear industry with the accumulation of ^{99}Tc in spent fuel. Numerous liquid/liquid extraction studies have shown that Tc(+VII) is coextracted with the actinides U(VI) and Pu(IV) in the organic phase (e.g., tributyl phosphate: TBP). The formation of TcO_4^- complexes with metallic cations such as Pu(+IV), U(+VI), Th(+IV) or Zr(+IV) accounts for the presence of technetium coextracted in the organic phase. The following species have been identified in the organic phase during studies of nitric acid solutions: $UO_2(TcO_4)(NO_3)\cdot 2TBP$ and $M(TcO_4)(NO_3)_3\cdot n TBP$, where M = Th and Zr, and n = 1 or 2.

The authors investigated the complexation of plutonium at the oxidation state (+IV) by the pertechnetate anion in an aqueous solution (perchloric acid) using visible/NIR absorption spectrophotometry and liquid/liquid extraction by TBP.

RESULTS AND DISCUSSION

Absorption spectrophotometry observations of a 2-mM solution of Pu(IV) in 2 M $HClO_4$ at TcO_4^- concentrations ranging from 0.2 mM to 1.8 M confirmed the complexation of Pu(IV) by the TcO_4^- anion. Chemometric techniques were used for quantitative processing of the spectrum evolution.[1] Although the mathematical treatment allowed us to rule out 1:1 and 1:2 stoichiometry for these complexes, it was not sufficient to discriminate between two ($Pu(TcO_4)_3^+$) and ($Pu(TcO_4)_4$) complexes with 1:3 and 1:4 stoichiometry.

In order to identify the stoichiometry of these complexes (1:3 or 1:4), we examined the partitioning of trace amounts of Pu(IV) between an organic phase (30 vol % TBP) and an aqueous phase (2 M $HClO_4$) at TcO_4^- concentrations ranging from 15.9 mM to 0.78 M. Using TBP, a neutral extractant, we were able to demonstrate 1:3 stoichiometry for the complex over the TcO_4^- concentration range. The calculated β_3 equilibrium constant at 25°C was approximately 2.3.

Applying the van't Hoff law to the spectrophotometric measurements revealed that the entropy term ($-T\Delta S$) controls the complexation reaction. The characteristic data for the formation of the 1:3 complex are the following: $\Delta H = 3.5$ kcal/mol and $-T\Delta S = -4.3$ kcal/mol.

The study of Pu(IV) partitioning in largely excess TcO_4^- over a range of TBP concentrations suggests that the 1:4 limit complex (which is extractable by TBP) is probably solvated in the organic phase as both $Pu(TcO_4)_4\cdot TBP$ and $Pu(TcO_4)_4\cdot 2TBP$.

REFERENCES

1. Pochon, P., Moisy, Ph., Donnet, L., de Brauer, C., and Blanc, P., "Investigation of Neptunium(VI) Complexation by $SiW_{11}O_{39}^{8-}$ by Visible/Near Infrared Spectrophotometry and Factor Analysis," Phys. Chem. Chem. Phys. 2, 3813–3818 (2000).

CP673, *Plutonium Futures — The Science,* edited by G. D. Jarvinen

Thermochemistry of Transuranium Actinide Oxide Molecules Investigated by FTICR-MS

John K. Gibson,[a] Richard G. Haire,[a] Marta Santos,[b] Joaquim Marçalo,[b] and António Pires de Matos[b]

[a]*Chemical Sciences Division, Oak Ridge National Laboratory, Oak Ridge, Tennessee, USA*
[b]*Departamento de Química, Instituto Tecnológico e Nuclear, 2686-953 Sacavém, Portugal*

INTRODUCTION

Chemical reactions involving gas-phase transuranium actinide (An) ions have been studied by Fourier Transform Ion Cyclotron Resonance Mass Spectrometry (FTICR-MS). Our initial studies focused on oxidation and charge-exchange reactions for Np and Pu ions,[1] but Am ion chemistry has been recently studied. Oxidation reactions are represented by equations (1) and (2), where RO represents an oxygen-donor such as CO_2 (R = CO), and the charge-exchange reactions by equation (3), where X represents an electron donor such as ferrocene.

(1) $An^+ + RO \rightarrow AnO^+ + R$ [BDE(An^+-O) $\geq$ BDE(R-O)]

(2) $AnO^+ + RO \rightarrow AnO_2^+ + R$ [BDE(OAn^+-O) $\geq$ BDE(R-O)]

(3) $AnO_2^+ + X \rightarrow AnO_2 + X^+$ [IE(AnO_2) > IE(X)]

The occurrence and rates of reactions (1) and (2) for a variety of oxidants have provided oxide-ion bond dissociation energies, BDE(An^+-O) and BDE(OAn^+-O). By determining whether reaction (3) proceeds with reagents, X, having different ionization energies permits the bracketing of IE(AnO_2). Because the IE(AnO) are too low to determine by charge exchange, we have employed a reactivity technique developed by Schwarz and coworkers to derive monoxide ionization energies.[2]

The measured values for oxide ions provide neutral oxide bond energies via equations (4) and (5).

(4) BDE(An-O) = BDE(An^+-O) + IE(AnO) – IE(An)

(5) BDE(OAn-O) = BDE(OAn^+-O) + IE(AnO_2) – IE(AnO)

RESULTS

Reactions (1) and (2) were studied with Np and Pu using several oxidants having BDE(R-O) in the range of 167–752 kJ/mol. The measured oxidation efficiencies relative to the theoretical collisional rate (k/k_{ADO}) were between ~1 to below 0.001, the latter being the reaction detection limit. The charge exchange reaction (3) was studied using reagents with IE(X) in the range of 6.7–8.8 eV. Reactions of AnO^+ with 1,3-butadiene and isoprene were used to determine IE(AnO).[2] Key results are summarized in Table 1. Our value for IE(NpO) of 5.7 eV is in accord with that from Reference 3, which is shown in Table 1. Because IE(NpO_2) is too low to be measured by charge exchange, the value from Hildenbrand et al.[3] is given in Table 1, with the caveat that the actual energy is likely to be >5.0 eV.

CP673, *Plutonium Futures — The Science,* edited by G. D. Jarvinen

DISCUSSION

Before the present study, the thermochemistries of neptunium and plutonium oxide molecules were based on electron impact (EI) experiments. The values obtained through FTICR-MS results significantly adjust key thermodynamic quantities, particularly in the case of plutonium dioxide.

TABLE 1. Measured and Derived Thermodynamic Quantities.*

	An = Np	An = Pu
$BDE(An^{+}-O)$	≥ 752	≥ 632
$IE(AnO)$	5.7 ± 0.1[3]	5.8 ± 0.2
$BDE(An-O)$	$\geq 697 \pm 10$	$\geq 603 \pm 20$
$BDE(OAn^{+}-O)$	580 ± 70	520 ± 20
$IE(AnO_2)$	5.0 ± 0.5[3]	7.03 ± 0.12
$BDE(OAn-O)$	512 ± 86	639 ± 30

*Bond dissociation energies are in kJ/mol. Ionization energies are in eV.

Based on the oxidation of Np^+ by formaldehyde, we establish lower limits for $BDE(Np^+-O)$ and BDE(Np-O) that are consistent with the literature values of 773 and 718 kJ/mol, respectively.[4] Experimental information is not available for comparison with our value for $BDE(ONp^+-O)$, but the derived BDE(ONp-O) is significantly smaller than anticipated from systematic variations across the actinide series. We propose that this anomaly results from the reported $IE(NpO_2)$ of 5.0 eV, which is unexpectedly smaller than IE(NpO) = 5.7 eV. Using an estimate for the actual $IE(NpO_2)$ of ~6 eV, a reasonable value for BDE(ONp-O) of ~610 kJ/mol is obtained.

Results of early studies of plutonium oxide molecules by EI have been summarized by Hildenbrand et al.[3] Capone et al.[5] have reported recently on another EI study of PuO and PuO_2. Our lower limit for $BDE(Pu^+-O)$ is in better agreement with the earlier value of 683 kJ/mol[3] than the recent value of 618 kJ/mol[5]. Based on the reactions of PuO^+ with dienes, the value for IE(PuO) of 6.6 eV[5] appears high and we recommend a value of 5.8 ± 0.2 eV, consistent with earlier EI results.[3]

The discrepancies between the EI and FTICR-MS results are particularly significant for plutonium dioxide. Direct comparison is made between our results and the recent EI values,[5] which are similar to the values given by Hildenbrand et al.[3] Capone et al.[5] obtained $BDE(OPu^+-O) = 261 \pm 22$ kJ/mol, but our observation of the oxidation of PuO^+ by O_2 establishes definitively a lower limit of 498 kJ/mol for this BDE. Our assignment of 520 ± 20 kJ/mol in Table 1 is based on oxidation efficiencies (rates). Capone et al.[5] reported a remarkably large $IE(PuO_2)$ of 10.1 ± 0.1 eV. By charge exchange we have established that $IE(PuO_2) = 7.03 \pm 0.12$ eV. The large errors in the EI values for $BDE(OPu^+-O)$ and $IE(PuO_2)$ cancel each other to give BDE(OPu-O) = 598 ± 24 kJ/mol[5], which is in reasonable agreement with our value in Table 1.

The FTICR-MS studies were extended to Am and have provided the first quantitative measurements of gas-phase americium oxidation kinetics and thermodynamics. Our FTICR-MS experiments with Np, Pu and Am ions have demonstrated this technique as being a valuable new approach for elucidating gas-phase actinide chemistry. The fundamental information obtained is important for developing new technological applications and understanding the science that involve phenomena such as the oxidative vaporization of plutonium.

ACKNOWLEDGEMENTS

This work was supported by the Division of Chemical Sciences, Geosciences, and Biosciences, U.S. Department of Energy (Contract DE-AC05-00OR22725 with Oak Ridge National Laboratory, managed and operated by UT-Batelle, LLC), by Fundação para a Ciência e a Tecnologia (Contract POCTI/35364/QUI/2000) and by Fundação Luso-Americana (Grant L-V-343/2001).

REFERENCES

1. Santos, M., Marçalo, J., Pires de Matos, A., Gibson, J. K., and Haire, R. G., J. Phys. Chem. A 106, 7190 (2002).
2. Cornehl, H. H., Wesendrup, R., Harvey, J. N., and Schwarz, H., J. Chem. Soc. Perkin Trans. 2, 2283 (1997).
3. Hildenbrand, D. L., Gurvich, L. V., and Yungman, V. S., The Chemical Thermodynamics of Actinide Elements and Compounds. Part 13—The Gaseous Actinide Ions, IAEA, Vienna (1985).
4. Lias, S. G., Bartmess, J. E., Liebman, J. F., Holmes, J. L., Levin, R. D., and Mallard, W. G., Gas-Phase Ion and Neutral Thermochemistry, American Chemical Society, Washington, DC (1988).
5. Capone, F., Colle, Y., Hiernaut, J. P., and Ronchi, C., J. Phys. Chem. A 103, 10899 (1999).

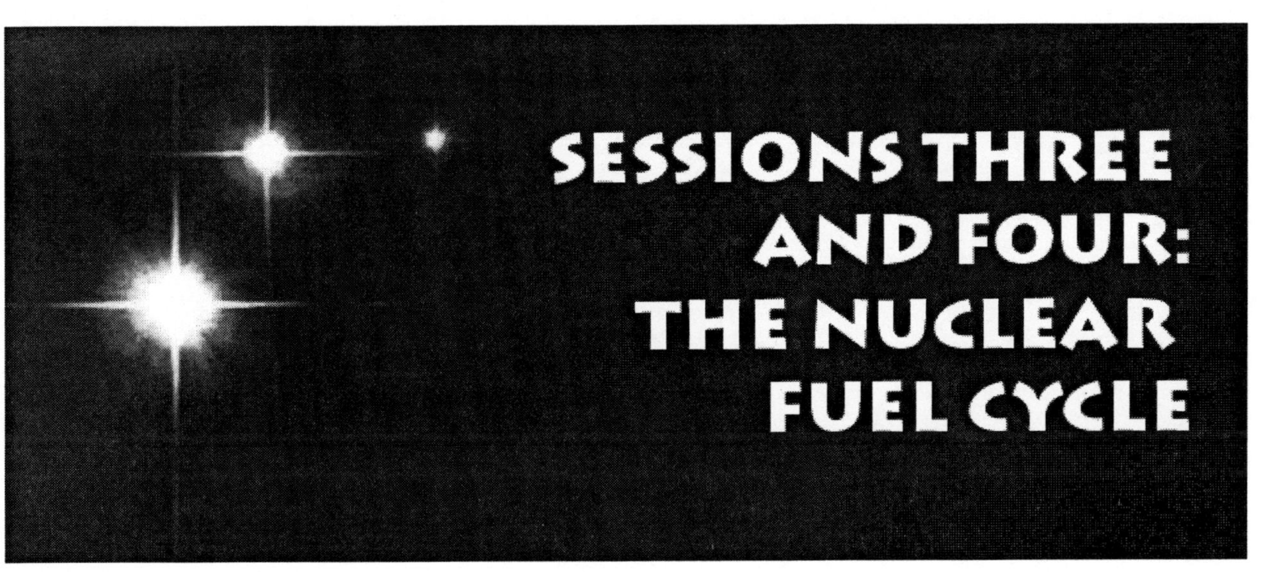
SESSIONS THREE
AND FOUR:
THE NUCLEAR
FUEL CYCLE

Plutonium, Politics and Policy: A New "Atoms for Peace"

Vic Reis

Science Applications International Corp.

On December 8, 1953, before the first nuclear power reactor generated any electricity (and not too long after the Soviet Union exploded it first thermonuclear bomb,) President Eisenhower gave his famous "Atoms for Peace" speech to the United Nations, expressing his vision for a future world in which nuclear technology could play a critical role in the peace and prosperity of the world. In this paper we will review the situation at the time of Eisenhower's speech, why he made it, and look at what ensued in the subsequent 50 years. Then we will suggest a vision for a new peaceful and prosperous global future and analyze what role nuclear technology, and in particular plutonium might play in such a world, consistent with current U.S. policy.

CP673, *Plutonium Futures — The Science,* edited by G. D. Jarvinen

Bicyclic and Acyclic Diamides: Comparison of Their Aqueous Phase Binding Constants with Tetra- and Hexavalent Actinides

Sergei I. Sinkov,[1] Brian M. Rapko,[1] Gregg J. Lumetta,[1]
James E. Hutchison,[2] and Bevin W. Parks[2]

[1]*Pacific Northwest National Laboratory, Richland, WA 99352*
[2]*Department of Chemistry, University of Oregon, Eugene, OR 97403*

INTRODUCTION

Diamides have been studied extensively as agents for selective extraction of trivalent f-block metal ions from aqueous solutions.[1] They form the basis for separation processes designed to partition the minor actinides from spent nuclear fuel for subsequent transmutation (with concomitant closure of the nuclear fuel cycle) or immobilization (for disposal). One possible drawback associated with the current generation of diamide extractants concerns the relatively modest actinide distribution coefficients obtained even at high (>1 M) extractant concentrations. In a recently published report, we described the extraction characteristics for a new type of diamide molecule.[2] This new diamide (structure **1b**, Figure 1) exhibits a 10^6 to 10^7 enhancement over previously examined diamides (e.g., tetraalkylmalonamides) in the extraction of trivalent lanthanides and actinides from aqueous solution.

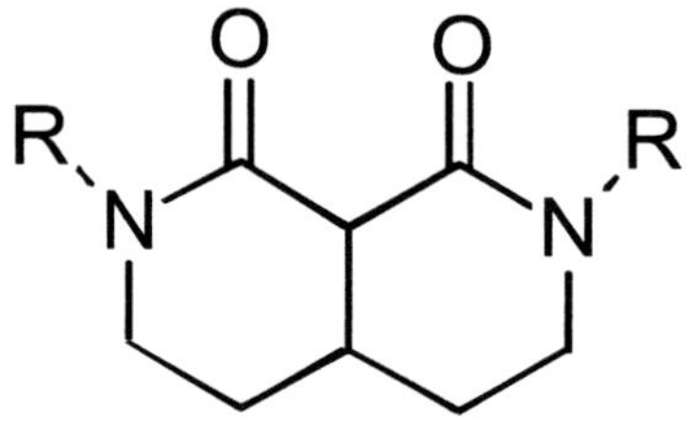

FIGURE 1. New bicyclic diamide ligand architecture optimized for bidentate lanthanide/actinide binding. R = methyl (**1a**), R = octyl (**1b**).

It was assumed that this extraction enhancement reflects superior ligand-metal binding of the bicyclic diamide as compared to acyclic analogs such as tetraalkylmalonamides. A fuller understanding of such structure-function aspects of actinide extraction by amides is needed for the intelligent design of extractants with superior characteristics to those currently available.

One of the possible approaches to quantify the magnitude of ligand-to-metal binding is the application of UV-Vis optical absorbance spectroscopy to monitor spectral changes induced by **1a** in the absorption spectra of actinide ions in aqueous solution. The spectral information obtained by varying the ligand-to-metal ratio not only allows for the estimation of the number of light-absorbing species in solution,[3] but can be used for measurement of the binding (or formation) constants of a ligand with metal ions.[4] This report presents a comparison of the formation constants for the dimethyl bicyclic diamide (**1a**) and for N,N,N',N'-tetramethylmalonamide (TMMA) with Pu(IV), Pu(VI) and U(VI) as measured in the aqueous phase at 1.0 M ionic strength (HNO_3).

CP673, *Plutonium Futures — The Science,* edited by G. D. Jarvinen

RESULTS AND DISCUSSION

Pu(IV)

The spectrophotometric titration was carried out by exposing a single portion of a $^{239,240}Pu(IV)$ stock solution (better than 99.5% valence purity) to successive introduction and dissolution of either solid crystals (**1a**) or tiny aliquots of a concentrated stock solution (TMMA). Neither precipitation nor perturbations in the oxidation state of Pu(IV) were observed with either ligand throughout the titration procedure at a metal concentration of 8 mM. With **1a**, the most significant spectral changes observed with increasing ligand concentration were found in the region of the main Pu(IV) peak (a 476 → 497 nm shift) and in the red part of the Pu(IV) spectrum (a 660 → 684 nm shift and a 2.4 times intensification of the absorption band). The spectral changes observed with TMMA were similar, but less pronounced and not identical to those observed with **1a** and occurred only at much higher ligand-to-metal ratios. An analysis of the number of species present, based on a Singular Value Decomposition procedure,[3] advocated the presence of not less than 4 complexed species for the Pu(IV)-**1a** system and the presence of 3 major complexed species, with a possible (but minor) contribution from a fourth species, for the TMMA-Pu(IV) system.

Pu(VI)

Not only was the most intense and sharp absorption peak of Pu(VI) at 833.6 nm monitored for expected complexation effects but a wider range (330–950 nm) was examined to evaluate possible changes in the oxidation state of the Pu (initially better than 99.5% hexavalent Pu). As with Pu(IV), the **1a** ligand produced pronounced spectral changes in the Pu(VI) spectrum, with two, new, well-resolved peaks at 840.6 nm and 846.3 nm emerging with even a moderate excess of ligand. However, after reaching a ligand-to-metal molar ratio of 80, further changes in the spectrum of hexavalent Pu were observed, with all three peaks of Pu(VI) gradually disappearing and a number of new spectral features emerging. In the visible range of the spectrum, the new major peak maxima were now at 497 nm and 684 nm. This new spectral signature clearly indicates the reduction of Pu(VI) to Pu(IV) in the presence of **1a**. A careful redox speciation analysis of the Pu(VI&IV) spectra has shown that this Pu(VI) reduction also is accompanied by the formation of Pu(V) (a weak spectral feature of variable intensity at 568 nm). After accounting for the presence of Pu(IV) and Pu(V), the portion of the spectral set in which Pu(VI) is still predominant in terms of concentration was processed by the nonlinear spectra processing routine SQUAD[4] and the respective formation constants refined from this analysis are shown in Table 1. In contrast to **1a**, the complexation of Pu(VI) with TMMA did not induce any redox reaction (even after an overnight contact time), and showed significantly weaker complexation effects, with the 1:2 complex seen not as a separate peak but only as a weak shoulder in the 842–846 nm range.

U(VI)

Both ligands induced significant spectral changes in the UO_2^{2+} spectrum, accompanied by a bathochromic shift and intensification of the absorption bands. An analysis as described above indicates that **1a** again acts as a much stronger binding agent compared to TMMA. In both cases no spectral evidence was found for the formation of higher complexes (1:3 and 1:4 stoichiometry).

Table 1 summarizes all the formation constants refined in the course of this project.

In conclusion, UV-visible spectroscopy has allowed quantification of binding constants for both bicyclic and acyclic diamide ligands with Pu(IV), Pu(VI) and U(VI) in acidic aqueous solution. Significant findings from this study include (1) that the alternation of the diamide structure from acyclic to bicyclic can increase the overall formation constants with Pu(IV) by as much as 7 orders of magnitude, (2) that the comparison of An(VI)-ligand binding affinities reveals enhanced binding to U(VI) versus Pu(VI), and (3) that in the presence of Pu(VI) [but not U(VI)] the bicyclic diamide triggers the reduction of Pu(VI) to Pu(V) and Pu(IV). Further work will focus on evaluating the formation constants of TMMA and **1a** with actinides in the +3 and +5 oxidation states.

TABLE 1. Conditional Formation Constants of Pu(IV), Pu(VI) and U(VI) Complexes with 1a and Tetramethylmalonamide in 1.0 M HNO_3 at 23 ± 0.5°C. Values in brackets represent one sigma standard deviations.

Complexant	Species M_iL_j	$\log\beta_{ij}$		
		Pu(IV)	**Pu(VI)**	**U(VI)**
1a	1:1	4.42 (0.08)	2.36 (0.04)	2.85 (0.02)
	1:2	7.95 (0.14)	3.44 (0.07)	4.61 (0.05)
	1:3	10.36 (0.13)	-	-
	1:4	11.48 (0.14)	-	-
TMMA	1:1	2.56 (0.08)	0.44 (0.01)	1.00 (0.01)
	1:2	4.08 (0.11)	-0.03 (0.03)	0.97 (0.03)
	1:3	5.01 (0.14)	-	-
	1:4	4.50 (0.38)	-	-

REFERENCES

1. Mathur, J. N., Murali, M. S., and Nash K. L., "Actinide Partitioning—A Review," Solvent Extr. Ion Exch. 19(3), 357 (2001).
2. Lumetta, J. L, Rapko, B. M., Garza P. A., Hay, B. P., Gilbertson, R. D., Weakley T. J. R., and Hutchison, J. E., "Deliberate Design of Ligand Architecture Yields Dramatic Enhancement of Metal Ion Affinity," J. Am. Chem. Soc. 124, 5644 (2002).
3. Bozhenko, E. I., KARP-95, Second International Symposium on Knowledge Acquisition, Representation and Processing, Auburn University, Alabama, USA, September 27 1995, (1995), p 10.
4. Leggett, D. J., in Computational Methods for the Determination of Formation Constants, Chapter 6, edited by D. J. Leggett, Plenum Press, New York (1985).

The Interactions of Iron and Plutonium Ions in Nitric Acid/Tri-Butyl Phosphate Systems and Process Flowsheets

Robin J. Taylor, David A. Woodhead, Caroline Biourge, Chris Mason, O. Danny Fox, Bill Carr, and Steve D. Cope

Research & Technology, BNFL Sellafield, Seascale, CA20 1PG, U.K.
220, AEA Technology, Harwell, Didcot, OX11 0QJ, U.K.
Sellafield M&O Services, BNFL Sellafield, Seascale, CA20 1PG, U.K.

INTRODUCTION

The separation of Pu in spent fuel processing by the reductive stripping of Pu(IV) from a TBP solvent phase using ferrous sulphamate (FeSM) is a very well established technology. Much fundamental data are available, and plants using this process have been successfully operated for many years, although most current plants use U(IV)/ hydrazine to reduce Pu(IV). With FeSM flowsheets, a feature of plant operations is the "box effect"—the process requires a substantial excess of FeSM in order to stabilize the Pu(III) ions. It is believed that this is due to nitrous acid catalysed reoxidation of Pu(III) in the solvent phase and the setting up of complex reaction cycles.[1] However, it has been very difficult to quantitatively reproduce this super-stoichiometric consumption of Fe(II) in process models. Progress has recently been made in process modelling, which coupled with solvent extraction experiments, plant data, and rig trials with miniature mixer-settler cascades has greatly improved our understanding of the box effect. These results have enabled the design and small scale testing of potential flowsheets in which FeSM is partially replaced by hydroxylamine nitrate (HAN) to improve the efficiency of the process. HAN is a salt-free reagent and does not interact significantly with Tc, unlike for hydrazine.[1]

RESULTS AND DISCUSSION

Process flowsheet simulation initially failed to replicate the box effect, although it was confirmed that the cycling occurred in the mixers rather than the settlers. Therefore, a single-stage model was used to test the sensitivities of the Fe(II)/Fe(III) ratio against a number of parameters. Modelling indicated that iron consumption is not particularly sensitive to the HNO_2 distribution coefficient; the kinetics of the aqueous phase HNO_2/ Pu(III), Pu(IV)/ Fe(II) and HNO_2/sulphamate reactions nor to the addition of HNO_2 to the solvent feed. The key factors were shown to be the stage efficiency of the mixer-settler (related to mass transfer); the form of the Pu(III) – HNO_2 reaction kinetics (in the TBP phase) used and the relative rates of backwashing (faster) versus extraction (slower). The extraction of Fe(II) in to the solvent phase was also shown to be a possible contributor to the box effect, but subsequent experimental work indicated that the distribution coefficients for Fe(II) were too low (<5E-5). Therefore, this possibility could be discounted. By adjusting these model parameters, the model could reproduce the box effect.

The flowsheet model was then validated against real plant data (ferrous/ferric ratio) obtained from the Magnox Reprocessing plant. A good replication of plant data was obtained (e.g., see Figure 1). The single-stage model was also validated against simple experiments in which a 20% TBP phase containing Pu(IV) was contacted with an aqueous HNO_3 phase containing FeSM. Pu mass transfer was shown to be a key factor. These experiments provided additional data on the interactions of Pu and Fe in these systems. Additional plant data on Fe entrainment in the solvent product was also obtained. Indications were that levels of entrainment were normal, and any relationship with the box effect is thought unlikely.

CP673, *Plutonium Futures — The Science,* edited by G. D. Jarvinen

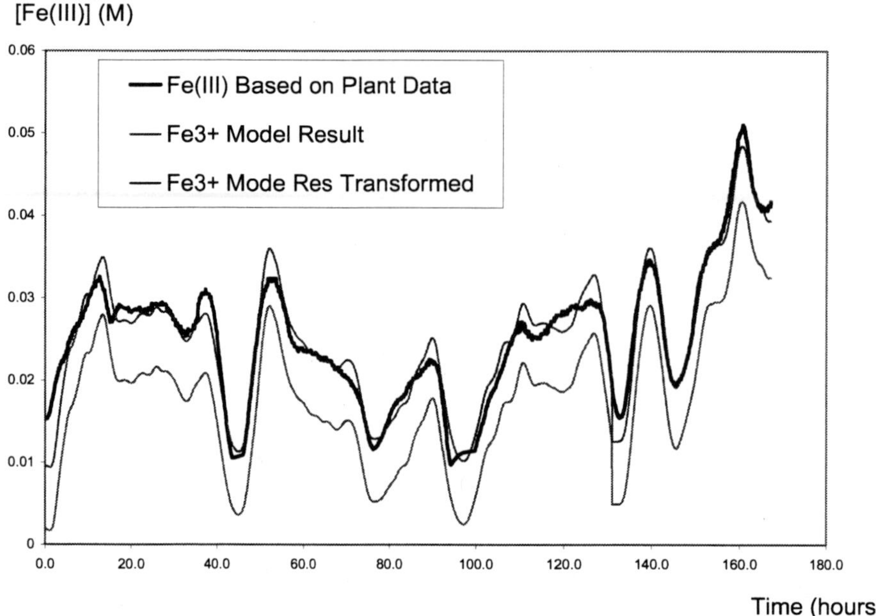

FIGURE 1. Model vs plant Fe(III) production.

Having developed and validated a model of the current flowsheet, extending our understanding of the process, hydroxylamine nitrate (HAN) data were added to the model. This enabled the simulation of flowsheets in which the FeSM levels were reduced and replaced by HAN. The key observation from the modelling was that the Pu was still controlled by Fe(II) reduction, with little or no contribution from HAN, but that HAN reduced Fe(III) in the final stages of the U strip mixer-settlers, thus regenerating Fe(II).

Flowsheets were then tested using a miniature mixer-settler cascade. The flowsheets tested were (a) an FeSM-only flowsheet designed to replicate plant performance as closely as possible [Fe:Pu = 27:1], (b) low Fe–HAN flowheets [with Fe:Pu ratios 4.6:1 and 1.5:1]. Additional changes were made to the temperatures, acidity profiles, and reagents feeds in the low Fe–HAN flowsheets, in order to optimize conditions. Product and interstage samples (for analysis of Tc, Np, U, Pu, HNO_3, HNO_2, Fe(II)/(III) ratios) were taken during each trial, and profiles across the whole U/Pu separation contactor were taken at the end of each trial (for Tc, Np, U, Pu, HNO_3). An example profile is included as Figure 2.

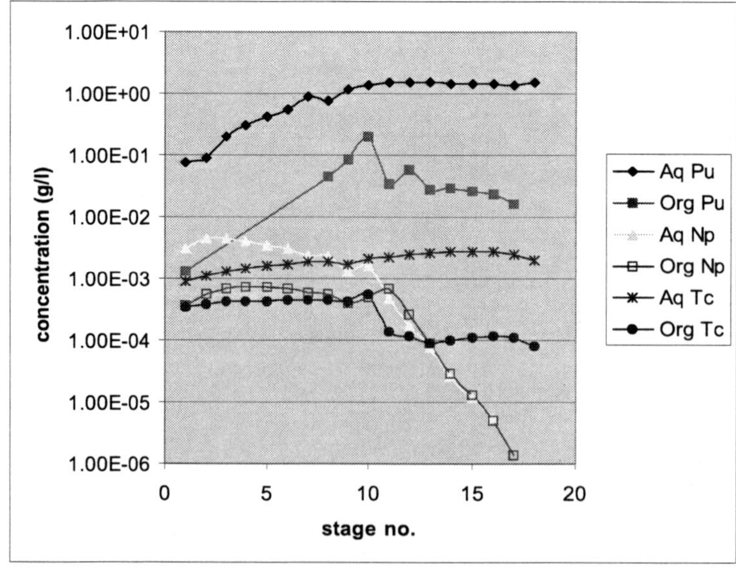

FIGURE 2. Example of Tc, U, Np, and Pu profiles (flowsheet trial Fe:Pu = 4.6:1).

All the trials worked well, even the very low Fe flowsheet, with similar decontamination factors (DFs) observed in all cases, most of which compared favourably with plant DFs. Some replication of the box effect was observed, although the residence times in these miniature mixer-settlers were very low compared with plant DFs. The Pu in U DF was low in all the flowsheets tested, with no obvious cause. Some small amounts of Np followed the Pu stream in the low Fe–HAN flowsheets. No untoward effects of Tc were observed. These flowsheets were then modelled.

In conclusion, a broad-based R&D programme, incorporating modelling; experiment; miniature scale flowsheet trials and plant data, has significantly extended our scientific understanding of a well-established process. It has also provided a technical basis for potential process improvements in which FeSM is partially replaced by HAN to improve process efficiencies.

REFERENCES

1. Miles, J. H., "Separation of Plutonium and Uranium," in Science and Technology of Tributyl Phosphate, Volume III. Applications of Tributyl Phosphate in Nuclear Fuel Reprocessing, edited by W. W. Schulz, L. L. Burger, J. D. Navratil, and K. P. Bender, CRC Press, Inc., Florida, 1990.

Development of Reprocessing Process by Plutonium Cocrystallization

Toshiaki Kikuchi,[a] Tomozo Koyama,[b] and Shunji Homma[c]

[a]*Mitsubishi Materials Corporation*
[b]*Japan Nuclear Cycle Development Institute*
[c]*Saitama University*

INTRODUCTION

This paper introduces a new fuel reprocessing process using plutonium cocrystallization. The process that consists of two-stage crystallization is much simpler than conventional reprocessing using solvent extraction, and thus it is expected to reduce the waste, to save the cost of development and operation, and to improve operation safety. The development is underway to realize an effective reprocessing for LWR fuel.

PAST ATTEMPT FOR APPLYING CRYSTALLIZATION TECHNIQUE TO THE SPENT–FUEL REPROCESSING

The application of crystallization techniques to spent nuclear fuel reprocessing was first reported in the 1980s.[1] Originally this technique was intended to be applied to uranium purification after solvent extraction. Removal of impurities, such as fission products or other actinide elements, is possible by cooling saturated uranyl nitrate hydrate (UNH: $UO_2(NO_3)_2 6H_2O$) solution and recovering UNH as crystals. Impurities remain in the solution because a concentration of UNH is much higher than those of impurities.

Research has further expanded to apply this technology to reprocessing for FBR spent fuel.[2] Uranium is recovered by crystallization from a solution dissolving FBR fuel. The U/Pu ratio in the residual solution after crystallization is adjusted to the U/Pu ratio specified for the product by controlling the recovery of uranium. UNH crystals can be used for FBR blanket fuel. The residual solution after crystallization is processed by the solvent extraction to remove impurities for direct preparation of MOX fuel.

PLUTONIUM COCRYSTALLIZATION ALONG WITH URANIUM CRYSTALLIZATION

In the course of the development of the crystallization application for the FBR reprocessing above mentioned, the unique phenomenon has been observed. The experiments using the solution dissolving real spent fuel revealed that the plutonium nitrate hydrate (PuNH: $PuO_2(NO_3)_2 6H_2O$) was cocrystallized with the UNH crystals under the condition that the PuNH concentration is far lower than its solubility. Meanwhile, in other experiments using Pu(IV) mixed with UNH-nitric-acid solution, no plutonium crystallization has occurred. This phenomenon leads to the notion that the main reprocessing stage can be established using only the crystallization technique and not using solvent extraction.

CP673, *Plutonium Futures — The Science,* edited by G. D. Jarvinen

A separation of FP from a dissolved solution by crystallization depends on a relationship between its quantity and solubility. Even though after single crystallization separation, a concentration of each FP is normally enough smaller than its solubility limit, especially in the case of LWR spent fuel (a quantity of FP in the LWR spent fuel is normally several times less than that of the FBR spent fuel). Thus, the recycling of the residual solution to the dissolution process can be attained, enabling the increase of the uranium and plutonium recovery at the co-crystallization stage. In single crystallization because the practical recovery of uranium and plutonium is limited up to approximately 90%, a recovery of residual uranium and plutonium, such as by introducing some additional recovery process, is an important consideration from an economical viewpoint. This residual solution recycling contributes not only to reduce the burden of the additional recovery process, but also to reduce a nitric acid quantity consumed at the dissolution process.

The application for the LWR fuel reprocessing only by crystallization can be established as the following flow-sheet (Figure 1).

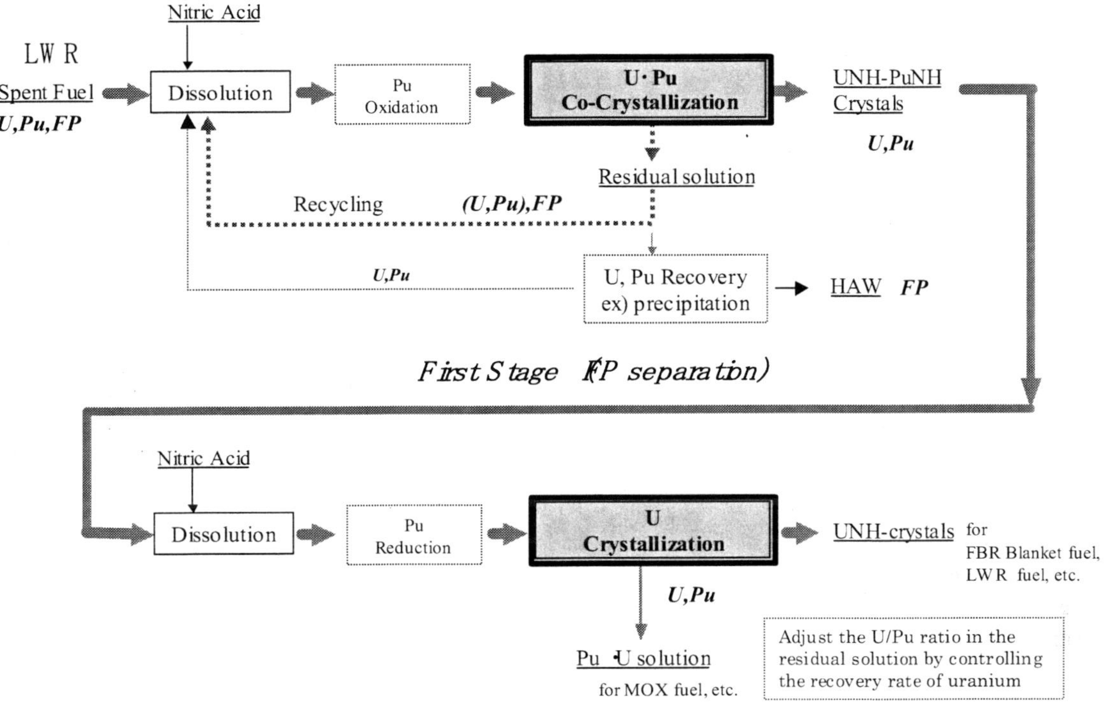

FIGURE 1. 1A block flow of crystallization process.

First Stage

- Oxidize the plutonium in a solution of dissolved nuclear fuel to hexavalent (VI) in a suitable way.
- Recover PuNH along with UNH by the cocrystallization (almost 90% of uranium and plutonium will be separated as crystals).
- Recycle the residual solution containing FP to the spent-fuel dissolution process. Assuming that the residual solution recycling can be attained two times, we can estimate that approximately 97% of uranium and plutonium can be separated in this first stage (a residual small fraction of uranium and plutonium should be recovered by a conventional separation technique such as precipitation). A number of the residual solution recycling depends on the quantity of each impurity (FP, etc.) in relevant spent fuel.

- Remove crystals of UNH and PuNH by filtration. Dissolve the crystals for the U/Pu partition at the second stage.

Second Stage

- Reduce plutonium in a dissolved solution to the tetravalent (IV) in a suitable way.
- Recover only UNH by the crystallization process. Adjust the U/Pu ratio in the residual solution after crystallization to the U/Pu ratio specified for the product by controlling the recovery rate of uranium.
- Remove UNH crystals by filtration.
- Recover uranium and plutonium from the residual solution.

Because the crystallization technique requires no organic solvents, liquid waste is expected to be significantly lower than that from conventional reprocessing using solvent extraction processes. A large amount of organic waste is generated from solvent extraction as a result of the degradation of organic solvent by radiation.

This application introduced in this paper contributes not only to reduction of process wastes and safety of operation, but also to

- cost reduction from more simplified process design than other types of applications, and
- an increase of nuclear proliferation resistance.

In addition, the recovery of uranium at the second crystallization step can be controlled by changing the cooling condition, thereby the U/Pu ratio in the residual solution can be adjusted to meet the specification of MOX fuel; therefore, this process enables the reprocessing of a wide variety of spent fuels at the same plant.

The feasibility study to elucidate the reaction mechanism of PuNH cocrystallization and to develop a process system, including equipment design, is now being conducted in Japan as industry-university joint research.

REFERENCES

1. Henrich, E., Schmieder, H. et al., "Combination of TBP Extraction and Nitrate Crystallization for Spent Nuclear Fuel Reprocessing," I Chem. Symposium Series 103 (1987)
2. Nomura, K. et al., "Study on U Crystallization for Advanced Reprocessing in JNC," Proceedings NUCEF2001 (2001).

Investigation of Plutonium in Uranium Products During Spent Fuel Treatment

B. R. Westphal, D. Vaden, L. W. Scott, S. R. Sherman, T. Q. Hua, and J. R. Krsul

Argonne National Laboratory
Idaho Falls, ID 83403

INTRODUCTION

The electrometallurgical treatment of Experimental Breeder Reactor-II (EBR-II) spent nuclear fuel at Argonne National Laboratory comprises a set of operations designed to separate uranium from radioactive fission products. Two primary operations are used for the recovery of uranium from the spent fuel; electrorefining and cathode processing.[1] During electrorefining, spent fuel is dissolved in an LiCl-KCl eutectic salt so that the transport of uranium to a cathode is possible. In addition to the LiCl-KCl, other chloride species are present in the electrorefiner salt as a result of the oxidation of fission products, bond sodium, and actinides present in the spent fuel. Once collected from the cathode, uranium is further processed in a cathode processor by vacuum distillation for the removal of adhering salts. The resultant uranium ingots are currently being stored pending a decision by DOE on their final disposition.

Uranium products from the treatment of EBR-II spent fuel either derive from driver or blanket-type spent fuel. Driver fuel was used in the core region of the reactor and contains a highly enriched uranium-zirconium alloy. Blanket fuel, consisting of depleted uranium, encircled the driver-core. Disposition of the blanket uranium products will most likely be dependent on the plutonium levels associated with the final ingots. At present, the blanket products would not meet LLW requirements with respect to their plutonium content. Thus, an investigation into the source of plutonium in the uranium products was initiated to understand its occurrence and potentially affect its contribution.

RESULTS

Because two pieces of equipment are being used for the treatment of spent fuel, the first step in the study was to distinguish where the plutonium was originating. Two different types of experiments were performed with products from the electrorefiner to determine if the plutonium was a result of the electrorefining operation. The results of these tests indicated that the electrorefiner was not the source but rather the cathode processor.

The next step then was to establish the species of plutonium as a result of cathode processing. Various experiments were performed in the cathode processor to determine if the plutonium contamination of the uranium products was as a chloride, an oxide, or a metal. Although other species are possible, their presence was not considered likely. The results of these tests point to metallic plutonium through the following reaction:

$$PuCl_3 \text{ (salt)} + U \text{ (ingot)} = UCl_3 \text{ (salt)} + Pu \text{ (ingot)} \quad (1)$$

The reaction is driven to the right by the distillation of uranium trichloride with respect to plutonium trichloride, leaving metallic plutonium in the uranium ingot. Plutonium trichloride is approximately a factor of ten less volatile than uranium trichloride at distillation temperatures in the cathode processor.[2]

CP673, *Plutonium Futures — The Science,* edited by G. D. Jarvinen

DISCUSSION

In order to control and/or reduce the plutonium in the blanket uranium products, the molar ratio of uranium trichloride to plutonium trichloride must be influenced during the cathode processor operation as shown by the following thermodynamic relationship:

$$e^{(\Delta G^o/-RT)} = \frac{A_{UCl_3} A_{Pu}}{A_{PuCl_3} A_U}, \tag{2}$$

where A_i are molar activities of species i, ΔG^o is the standard free energy at temperature T, and R is the gas constant. Because spent fuel is continuously being processed in the electrorefiner, reducing the amount of accumulated plutonium trichloride in the salt is not currently feasible. Alternatively, increasing the amount of uranium trichloride is possible and has shown promise.[3,4] The addition of uranium trichloride, either directly or indirectly through the oxidation of uranium metal, should decrease the plutonium content of the final uranium product. Testing in the cathode processor with the addition of cupric chloride to oxidize uranium has demonstrated that the plutonium levels of blanket uranium products can be reduced significantly.

REFERENCES

1. Westphal, B. R., and Mariani, R. D., "Uranium Processing during the Treatment of Sodium-Bonded Spent Nuclear Fuel," JOM 52(9), 21–25 (2000).
2. Roine, A., HSC Chemistry 4.0 Thermochemical Database, Finland: Outokumpu Research Oy (1999).
3. McKenzie, D. E., Elsdon, W. L., and Fletcher, J. W., "The Extraction of Plutonium from Neutron-Irradiated Uranium by Uranium Trichloride and by Magnesium Chloride," Can. J. Chem. 36, 1233–1240 (1958).
4. Chellew, N. R., and Steunenberg, R. K., "Extraction of Plutonium from Uranium-Plutonium Alloy with Uranium Trichloride," Nucl. Appl. 3, 142–146 (1967).

Application of Fibrous "Filled" Sorbents Polyorgs for Concentration of Plutonium and Other Radionuclides from Solutions

I. G. Tananaev, G. V. Myasoedova, and B. F. Myasoedov

Vernadsky Institute of Geochemistry and Analytical Chemistry, Russian Academy of Sciences

INTRODUCTION

Problems related with man-caused radioactivity and its impact upon the environment has become a global issue. As the result, a new radioecology discipline was formed. The reprocessing of radioactive waste has become one of the most important issues in this field as well as the cleansing of natural waters from long-lived radionuclides, including plutonium and technetium. In isolating radioactive elements from aqueous solutions, sorption methods are often used. Ion exchangers and complexing sorbents—synthetic or natural polymers produced by chemical bonding of ligands and impregnation—are employed in the practice. Complexing sorbents have higher selectivity towards ions that are determined by the nature of chemical active groups in polymer. The application of complexing sorbents in the form of fiber is the best perspective for their employment. Fibrous complexing sorbents have greater advantages compared to granular ones because along with high selectivity for metal ions, they have large surfaces that provide better kinetic properties. They are used in analytical chemistry and in technology, for example, on the stages of rare and noble metals concentrations[1–3] as well as the extraction of transuranium elements from solutions.[4,5]

Along with traditional fibrous sorbents obtained by the modification of prepared matrices, the fibrous "filled" materials consist of two polymers: the fine porous polyacrylonitrile filaments (diameter of 30–40 μm) and powder polymeric sorbent ("filler"), which is retained in the fiber matrix. The properties of these sorbents with respect to metals are conditioned by the nature of the filler functional groups.

Plutonium(IV) and technetium(VII) sorption from solutions by the "filled" fiber sorbents were studied.

EXPERIMENTAL

Techniques

A liquid-to-solid ratio (V:m) of 100 cm^3/ g was used in the batch experiments. The contact time (τ_c) was from 5 min to 24 h at 20°C. The distribution coefficients (K_d) of the Pu and Tc species have been determined by measured activity of the solution before and after sorption. Isotopes of ^{239}Pu and ^{99}Tc and others were used. For the identification of the elements' oxidation states, the spectrophotometer SHIMADZU UV-vis-160A was used. The fibrous "filled" sorbents type POLIORGS has been used. POLIORGS are the family of a chelating material produced at the Vernadsky Institute RAS during more that 30 years for stable and radioactive elements' isolation from aqueous solutions.[3,6,7]

CP673, *Plutonium Futures — The Science,* edited by G. D. Jarvinen

Plutonium

For a concentration of plutonium, the "filled" fibrous complexing sorbents POLYORGS 4-n were selected with 3(5)-methylpyrazole groups and POLYORGS 17-n with 1,3(5)- dimethylpyrazole groups, which possess the anion exchanger properties. The Pu sorption was also tested by an anion exchanger with quaternary ammonium groups in the form of "filled' fibrous material (AV-17-n). It was shown that only Pu among other actinide elements is sorbed completely by these "filled" fiber sorbents from acid nitric solutions at 1–5 M within 10 min at 20°C. The distribution coefficients of Pu(IV) (K_d^{Pu}) were determined as $1,5·10^3$, 10^3, and $9·10^2$ cm^3/g by POLYORGS 17-n, 4-n, and AV-17-n, respectively. The values of K_d^{Pu} were not changed in the presence of $NaNO_3$ concentration up to 1 M. For the stripping of Pu(IV), 0.5–1.0 M HNO_3 solutions were used. The Am(III), Eu(III), U(VI), Pa(V), Np(V), Tc(VII), and Cs(I) were very bad or not absorbed by these materials. The separation factors of Pu(IV) from other elements ($f = K_d^{Pu}/K_d^{M}$; M—and elements above) are listed in Table 1.

TABLE 1. The Separation Factors of Pu(IV) from Other Elements at V:m = 100 cm^3/g; τ_c = 2h; 1–5 M HNO_3.

Radionuclides	Separation Factors of Pu(IV) from Other Elements	
	POLIORGS-4-n	**POLIORGS-17-n**
Am(III)	$1·10^5$ - $1·10^6$	$1·10^5$ - $1·10^6$
Eu(III)	$1·10^5$ - $1·10^6$	10^2
U(VI)	$1·10^5$ - $1·10^6$	$0,75·10^2$
Pa(V)	$0,92·10^2$	10^2
Np(V)	$0,57·10^2$	$0,58·10^2$
Tc(VII)	$0,92·10^2$	$0,74·10^2$
Cs(I)	$1·10^5$ - $1·10^6$	$3·10^2$

Technetium

Tc(VII) sorption by fibrous "filled" sorbents from acid, neutral, and basic solutions was studied. It was shown that AV-17-n and POLYORGS-17-n quantitatively isolate Tc from low acidic as well as from neutral and alkaline media. The distribution coefficients (K_d`s) for pertechnetate ions in batches from acid, neutral, and basic solutions were determined. An investigation of the sorption degree as a function of time showed that the sorption process is almost finished in 10–20 min at room temperature. Technetium(VII) desorbed with 3M HNO_3 solutions. The calculated K_d^{Tc} are listed in Table 2.

TABLE 2. Distribution Coefficients on Tc(VII) Sorption from Solutions; V:m = 100 cm^3/g; τ_c = 2h.

Solution	Sorbents	
	AV-17-n	**POLYORGS-17-n**
	$[HNO_3]$,M	
10^{-3}	$3,7·10^4$	$1,0·10^3$
10^{-2}	$1,3·10^4$	$2,5·10^3$
0.1	$1,2·10^3$	$2,2·10^3$
0.2	671	309
0.5	382	189
3.0	<1	<1
	[NaOH],M	
0.1	$1.0·10^3$	277
1.0	900	-
	$[NaNO_3]$,M	
0.1	861	432

It was shown that a maximum Tc sorption by fibrous sorbents AV-17-n and POLYORGS 17-n takes place in the solution with a pH of about 3–5 and in 0.1 M NaOH. Distribution coefficients are decreasing with the increasing of acid concentration in the solution from 0.001 M to 0.5 M HNO_3 and from 0.1 M to 1 M NaOH.

DISCUSSION

The results obtained showed that "filled" fibrous sorbents POLYORGS are perspective for the isolation of plutonium from nitric acid solutions and for this separation from other radionuclides. The POLYORGSs are perspective also for concentrations of Technetium from low acid and alkaline solutions. It was noticed that "filled" fibrous sorbents possess good filter and fast kinetic properties provided by fibrous properties and by the small size of filler particles. "Filled" fibrous materials are prepared in nonfabric material form and are convenient in practice for sorption concentrating both in static and dynamic conditions. For example, these sorbents could be used in the form of discs placed on a perforated surface or into special cells.

REFERENCES

1. Kurashvili, S. E., Barash, A. N., and Yakovleva, N. Ya., Zh. Prikl. Khim.(in Russian) 65(5), 991 (1992).
2. Myasoedova, G. V., Nikashina, V. A., Molochnikova, N. P., and Lileeva, L. V., Zh. Anal. Khim. (in Russian) 55(6), 586 (2000).
3. Myasoedova, G. V., Fresenius J. Anal. Chem. 341, 586 (1991).
4. Molochnikova, N. P., Scherbinina, N. I., Myasoedova, G. V., and Myasoedov, B.F., Radiokhimiya (in Russian) 39(3), 279 (1997).
5. Myasoedova, G. V., Molochnikova, N. P., Lileeva, L. V., and Myasoedov, B.F., Radiokhimiya (in Russian) 41(5), 456 (1999).
6. Myasoedova, G. V., and Savvin, S. B., Chelating Sorbents (in Russian) Moscow (1984), 171 p.
7. Myasoedova, G. V., and Savvin, S. B., Crit. Rev. Anal. Chem. 17, 1–63 (1986).

The Separation of Americium and Plutonium Achieved by Facilitated Transport through Fixed Site Carrier Membranes Utilizing CMPO Ligands

Scott Sportsman, Elizabeth Bluhm, and Kent Abney

Los Alamos National Laboratory
Nuclear Materials Technology Division
Actinide Process Chemistry Group (NMT-2)

The processing of nuclear waste streams requires the separation of actinide radionuclides from lanthanides as well as from other transuranium (TRU) wastes. To meet these requirements, several separations strategies and flow sheets have evolved utilizing liquid-liquid extraction schemes using carbamoylphosphine oxide (CMPO) based ligands, particularly using n-octyl(phenyl)-N,N-diisobutylcarbamoylmethylphosphine oxide in the TRUEX process.[1] Unfortunately, the use of liquid-liquid extraction processes requires large amounts of solvents and extractants, which generate large amounts of waste products that are frequently contaminated with radioactive materials.

During the past few decades, membrane separations have become a more feasible method of performing difficult separations.[2] This increase in the number of membrane operations has been achieved through the improved manufacture of membranes, which has reduced the membrane thickness and produced defect-free membranes, as well as through an improved understanding of the interactions between the membrane material and the components involved in the desired separation.

The Nuclear Materials Technology (NMT) Division at Los Alamos National Laboratory (LANL) employs many unit operations involving the aqueous processing of radioactive materials. The recovery of these radioactive materials and minimization of waste streams has driven the research in numerous separation processes, including extraction chromatography[3] and various membrane processes.[4–8] In this ongoing effort to reduce the amount of extractant and generated solvent wastes while achieving an improved separation, a new membrane process utilizing a CMPO derivative is being developed for specific applications at LANL.

An electroless gold deposition technique was used to deposit a gold coating on track-etched polycarbonate membranes,[9] resulting in pores smaller than 5 nm in diameter. The extractant ligand, an octyl(4-t-butylphenyl) CMPO was synthesized with a thiol group attached to the octyl chain. The CMPO derivative was then chemisorbed through the thiol groups to the gold surface of the membrane and its pore spaces. The ligand chemisorption forms a self-assembled monolayer (SAM) coating the gold surface, which creates an avenue for selective, facilitated transport through the membrane.

Transport results have shown that this SAM increases the rate of transport of various metal ions through the membrane, despite the reduction of pore size during the membrane preparation. The SAM also introduces a significant improvement in the selectivity of various actinides allowing the separation of actinides from other metals, as well as the separation of various TRU species from other actinides.

REFERENCES

1. Nash, K. L., and Choppin, G. R., Separation Science and Technology 32, 255–274 (1997).
2. Drioli, E., and Romano, M., Industrial & Engineering Chemistry Research 40, 1277–1300 (2001).
3. Barr, M. E., Schulte, L. D., Jarvinen, G. D., Espinoza, J., Ricketts, T. E., Valdez, Y., Abney, K. D., and Bartsch, R. A., Journal of Radioanalytical and Nuclear Chemistry 248, 457–465 (2001).

CP673, *Plutonium Futures — The Science,* edited by G. D. Jarvinen

4. Bluhm, E. A., Bauer, E., Chamberlin, R. M., Abney, K. D., Young, J. S., and Jarvinen, G. D., Langmuir 15, 8668–8672 (1999).
5. Bluhm, E. A., Schroeder, N. C., Bauer, E., Fife, J. N., Chamberlin, R. M., Abney, K. D., Young, J. S., and Jarvinen, G. D., Langmuir 16, 7056–7060 (2000).
6. Barr, M. E., Jarvinen, G. D., Moody, E. W., Vaughn, R., Silks, L. A., and Bartsch, R. A., Separation Science and Technology 37, 1065–1078 (2002).
7. McCleskey, T. M., Ehler, D. S., Young, J. S., Pesiri, D. R., Jarvinen, G. D., and Sauer, N. N., Journal of Membrane Science 210, 273–278 (2002).
8. Ames, R. L., Bluhm, E. A., Abney, K. D., Way, J. D., Bunge, A. L., and Schreiber, S. B., Abstracts of Papers of the American Chemical Society 223, U673–U673 (2002).
9. Menon, V. P., and Martin, C. R., Analytical Chemistry 67, 1920–1928 (1995).

Advances in Code Validation for MOX Use in LWRs through Benchmark Experiments in the VENUS Critical Facility

P. D'hondt

SCK•CEN, Boeretang 200, B-2400 MOL, Belgium

A typical commercial thermal reactor produces around 250 kg of plutonium per GWye output. The majority of the spent fuel discharged from the world's 430 or so commercial nuclear reactors has been designated for interim storage and eventual direct disposal. Those utilities that have opted for the so-called "once-through" fuel cycle have chosen this option after considering the particular circumstances (political, economic, strategic, logistic, historic, etc.) that apply locally. With different local circumstances, other utilities have chosen to reprocess their spent fuel.

Current commercial reprocessing plants are all designed to separate the plutonium remaining at discharge for reuse. Historically, the rationale was to recover sufficient plutonium to enable a buildup of fast reactors, which were expected to be deployed as uranium reserves become scarce and prices rose. For a variety of reasons, but principally the low price of uranium ore, fast reactors have not yet been deployed commercially and projected timescales for doing so have been put back everywhere.

There are several technical options for plutonium management involving reuse in existing LWRs. Partial core loading of MOX assemblies in PWRs is already well established on a commercial basis, with 37 reactors in Europe (2 in Belgium, 22 in France, 10 in Germany, and 3 in Switzerland) currently operating with part MOX loading. There is less experience of MOX usage in BWRs, with just two BWRs in Germany currently using MOX fuel.

Partial MOX loading in a PWR involves the substitution of a fraction of the UO_2 assemblies with the same mechanical design. This avoids issues of thermal-hydraulic and mechanical-handling incompatibility. The plutonium concentration in the MOX assemblies is usually adjusted so that the reactivity lifetime of the MOX fuel coincides with that of the UO_2 fuel, in which case, the average discharge burnups of the two fuel types will be the same.

The presence of MOX fuel batches in an LWR affects the nuclear-design characteristics of the core in a complicated fashion. Consequently, there was a need in the late 80s to develop or to improve the in-core management codes and to benchmark and validate them against experimental data. An insufficient validation might induce a dramatic increase of the uncertainty factor with a possible reduction of the reactor power. Based on the experience accumulated during 25 years of collaboration SCK•CEN together with Belgonucléaire decided to implement a series of Benchmark experiments in the VENUS critical facility in MOL, Belgium, in order to give to organizations concerned with MOX fuel the possibility to calibrate and to improve their neutronic calculation tools.

The VENUS critical facility is a water-moderated zero-power reactor. It consists of an open (nonpressurized) stainless-steel cylindrical vessel including a set of grids that maintain fuel rods in a vertical position.

After a fuel configuration has been loaded, criticality is reached by raising the water level within the vessel. Because the neutron flux is very low, no water circulation is needed in order to keep the fuel rods at low (room) temperature. The reactor shutdown is induced by emptying the vessel.

A view of the VENUS critical facility is given in Figure 1. Because of the high flexibility, a series of experiments related to plutonium use in LWRs have been running in VENUS since 1990.

In early 1990 the Pu recycling in LWRs was investigated. The program, called VIP (VENUS International Programme), used fuel with high Pu and Gd content. The aim of the VIP program was the validation of reactor codes with respect to MOX fuel for both PWRs and BWRs. It focused on the criticality and fission-rate distribution calculation.

CP673, *Plutonium Futures — The Science,* edited by G. D. Jarvinen

FIGURE 1. View of the VENUS Critical Facility.

After VIP, an international program called VIPO (Void coefficient measurement In Plutonium mixed Oxide lattice) was devoted to determine the influence of a void bubble in an LWR reactor using high Pu enrichment (i.e., from 10% to 15%) and the validation of related computer codes.

After VIPO, another program called VIPEX-PWR was executed. This program, which was an extension of the former VIP-PWR program aimed at determining core physics parameters of MOX assemblies that are of interest for reactor operation such as the fraction of delayed neutrons ∃eff, the ^{241}Am-effect, the moderator density effect, the control rod worth and flux tilt within MOX rods.

After VIPIX-PWR, we have investigated recent BWR configurations in the BWR-NBN program. This program was very similar to the VIP-BWR program but for a 9 x 9 BWR configuration.

We also investigated the use of weapons-grade Plutonium in LWRs in VENUS. This was possible through the availability of plutonium fuel with plutonium vectors close to weapon-grade Pu. This program, called IMP (Investigation of Military Plutonium) was an SCK•CEN internal program.

Recently, a new international program was implemented in VENUS. This program, called REBUS, aims at establishing an experimental benchmark for validation of reactor physics codes for the calculation of the loss of reactivity caused by burnup for PWR fuel, both for UO_2 and MOX fuel bundles. The rational of this REBUS program lies in the fact that present criticality safety calculations of irradiated fuel often have to model the fuel as fresh fuel because no precise experimental confirmation exists of the decrease of reactivity as a result of accumulated burnup. On other occasions, only actinide depletion is allowed to be taken into account, and the influence of fission products has to be disregarded. The fact that this so-called "burnup credit" cannot (completely) be taken into account has serious economical implications for transport, storage, and reprocessing of irradiated fuel. For long-term geological storage, it is almost imperative to apply burnup credit.

All mentioned programs and their outcome will be highlighted, they have demonstrated that VENUS is a very flexible and easy-to-use tool for the investigation of neutronic data as well as for the study of licensing, safety, and operation aspects.

Such data permit the validation of the reactor physics codes for MOX use in LWRs. Through such validation, it is nowadays possible to safely exploit 37 PWRs and 2 BWRs with partial MOX core loadings.

At present, there is no experience of MOX uses in VVERs, though a program to use weapons-grade plutonium in VVERs is underway. The accumulated knowledge and experience is nowadays so sufficient that this can be regarded as a fully established option with low technical risk.

Overview of Nuclear Fuel Fabrication Efforts at Los Alamos for the Advanced Fuel Cycle Initiative

Robert W. Margevicius

NMT-11
Los Alamos National Laboratory
Los Alamos, NM 87545

The mission of the Advanced Fuel Cycle Initiative (AFCI) is to develop and implement spent-fuel treatment and transmutation technologies to enhance the performance of the proposed repository and reduce the cost of geologic disposal for the United States. The AFCI is closely coupled with the Generation IV Nuclear Energy Systems Program (Gen IV) that seeks to deploy a new generation of nuclear power plants by 2030 to provide a transmutation capability. Together, these two programs enable an expanded role for nuclear power as a sustainable resource that will address long-term U.S. energy security, environmental, and economical concerns. The AFCI also provides for an effective transition from the current once-through nuclear fuel cycle to the future sustainable cycle. Such a closed nuclear fuel cycle will provide a number of benefits, including reducing the cost of geologic disposal of commercial spent fuel, recovering energy value from spent fuel, reducing civilian inventories of plutonium, reducing the toxicity of high-level waste, and eliminating or significantly delaying the need for a second repository.

The AFCI fuels development effort will provide proliferation-resistant fuels for use in advanced fuel cycles for both the current LWRs and ALWRs (Series One) and for Generation IV nuclear power and transmutation systems (Series Two). At this time, the composition of Series One proliferation-resistant fuel is expected to include uranium, plutonium, neptunium, and possibly other minor actinides or burnable poisons. The goal of the Series One development effort is a lead test assembly (LTA) irradiation in approximately ten years. Test reactors will be used (e.g., Advanced Test Reactor, ATR) to perform qualification irradiation in the intervening years.

The Series-Two fuels development focuses on metal alloy fuels and nitride ceramic fuels; NMT-11 at Los Alamos is responsible for the latter fuel form. The goal of this effort is to identify promising fuel forms meeting the requirements of high burn-up, good performance, and reprocessability. Nonfertile and low-fertile nitrides are being fabricated for irradiation experiments in the ATR in the next few years. An international effort to perform irradiation tests in the fast-flux Phenix reactor in France is underway.

This presentation will give an overview of the AFCI program and discuss current efforts to fabricate fuels for Series-One and Series-Two irradiations. The details of the fabrication will be given, as will results of chemical and physical characterization.

CP673, *Plutonium Futures — The Science*, edited by G. D. Jarvinen

Plutonium and Minor Actinide Fuels—The Good, the Bad, and the Future?

Kenneth Chidester[1] and Wolfgang Stoll[2]

[1]*Los Alamos National Laboratory, Los Alamos, NM 87545*
[2]*Consultant, Hanau, Germany*

Whatever the future split in energy-producing systems may be in the U.S. and abroad, nuclear energy will continue to contribute substantially to the production of electric power. The current system of light water reactor (LWR) nuclear power plants (NPP) has already accumulated hundreds of tons of spent fuel and will continue to do so for at least a few more decades. Whatever new reactor systems are under consideration, whether they be advanced LWRs or next-generation reactors, spent-fuel forms will continue to be generated. Without action to reduce spent-fuel volume and radioactivity, the current planned repository will reach capacity in just a few years. The capacity of a final repository for spent fuel could be extended and the concern on the long-term seclusion of those radionuclides from the biocycle could be alleviated if the plutonium and minor actinides could be separated from the spent fuel and irradiated while harvesting their fission power.

Current and advanced LWRs could recycle plutonium with small amounts of minor actinides as MOX fuel to the extent of self-generation, but the minor actinides have limited fission and absorption cross sections and would accumulate. Only high fluences of neutrons with high energy (fast neutrons) can shorten and equilibrate this shortcoming. After recycling through the LWRs the residual plutonium and minor actinide material quickly reach an equilibrium level that can only be reduced further by exposure to higher energy and flux neutrons. Fast reactor (FR) systems can increase the neutron flux by about one order of magnitude; accelerator-driven systems (ADS) provide another one to two orders above that.

Although in the medium time range only fast reactor systems could mature into the industrial dimension, the ADSs are so much more promising from their actinide and fission product destruction power, that despite the high initial R&D costs, they may be the best follow-on generation machines to solve this problem.

The common denominator for fuel in both systems is a distinct separation of fuel and coolant in the classical way, by inserting fuel bundles consisting of fuel rods. The proven FR fuel is a column of U/Pu-oxide in pellet or particulate matter, contained in a steel tube of about 7-mm diameter. The limits of this system are swelling of the stainless-steel cladding and the long-term behavior of the oxide fuel. Although ferritic steel may be the cladding overcoming most of the swelling problems, the fuel could be improved by using actinide nitrides in both the FR and AD systems. In almost all nuclear research centers worldwide, there is interest in nitride fuels. The high melting point, monophase stability, and high irradiation tolerance of the nitrides make them ideal candidates for high-fluence, high-temperature fuels.

Previous programs at LANL worked on mixed Pu/U-nitride fuel fabrication, and in the 1980s a lot of experience was gained on U-nitride fuel for a space power reactor. In the current work under the Advanced Fuel Cycle (AFC) program at Los Alamos, nitrides of plutonium, americium, and neptunium have been synthesized and pellets fabricated for irradiation in the Advanced Test Reactor (ATR) at the Idaho Engineering and Environmental Laboratory near Idaho Falls, Idaho. Test details will be provided in the full paper. It is the aim of this program to not only provide fuel rods for irradiation testing, but also to tackle the still-open question of an optimum fuel fabrication route and a suitable reprocessing scheme.

Also, other effects like C-14 formation during irradiation, the question of secondary waste in multiple—and at least semi-remote—handling of those actinides, the overall safety improvement of the handling, transport and storage facilities and the nonproliferation aspects of those systems are under close consideration and will guide future work.

CP673, *Plutonium Futures — The Science,* edited by G. D. Jarvinen

The current AFC program is aimed at a complete fuel management system that consists of recovering the actinides, implementing multiple recycles in LWRs and fast neutron systems (FRs or ADSs) using fuels developed specifically to optimize the destruction of the actinides. Los Alamos is well suited to make significant contributions in the areas of actinide fuel materials development and recycle technologies. The complete infrastructure and authorization basis for working with the actinides is in place along with a cadre of scientists who have experience working with these materials.

Self-Radiation Effects in Plutonium-Bearing Glasses

W. J. Weber, J. P. Icenhower, and N. J. Hess

Pacific Northwest National Laboratory, P.O. Box 999, Richland, Washington 99352 USA

Three compositionally identical Pu-bearing reference glasses were prepared on July 27, 1982, each containing 1 wt % PuO_2;[1] however, the $^{238}Pu/^{239}Pu$ isotopic ratio was different in each glass. As a result, the α activities in the as-prepared glasses varied by nearly a factor of 200. The sample designations, nominal isotopic compositions, activities, and accumulated doses (as of September 1, 2002) are summarized in Table 1. The activities shown in Table 1 are the actual activities measured, which are within 15% of the intended values. In the 20 some years since their preparation, several studies have been performed on these glasses. The final results of the most recent studies are summarized in this paper.

TABLE 1. Isotopic Composition, Activity, and Dose in Pu-Bearing Reference Glasses.

Sample	Pu Isotopic Composition	Activity (Bq/g)	Dose (α decays/g)
DRG-P1	1.0 wt % $^{239}PuO_2$	1.9×10^7	1.2×10^{16}
DRG-P2	0.9 wt % $^{239}PuO_2$	4.7×10^8	2.8×10^{17}
	0.1 wt % $^{238}PuO_2$		
DRG-P3	0.1 wt % $^{239}PuO_2$	4.2×10^9	2.5×10^{18}
	0.9 wt % $^{238}PuO_2$		

The stored energy and macroscopic swelling measured for these glasses are shown in Figure 1 as a function of accumulated dose. The results show that the stored energy rapidly increases at very low doses (equivalent to 20 years storage for DOE HLW glasses) and increases to relatively large values at high doses (equivalent to 1 million years storage for DOE HLW glasses). The stored energy could provide a driving force for enhanced dissolution kinetics by decreasing the energy barrier for dissolution. The swelling does not increase as rapidly with dose, which indicates that the accumulation of defects giving rise to the stored energy precedes the rearrangement of the glass network that manifests as volume changes. These results also suggest that for this range of α activities there is no significant effect of the α activity (i.e., dose rate) on the swelling.

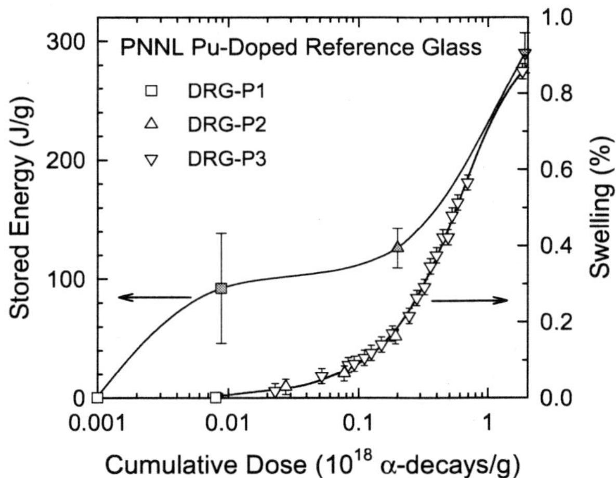

FIGURE 1. Stored energy and macroscopic swelling as a function of cumulative dose.

CP673, *Plutonium Futures — The Science,* edited by G. D. Jarvinen

Analysis of extended x-ray absorption fine structure (EXAFS) spectra for the glasses indicates that the local environment around the cations exhibits different degrees of disorder because of the accumulated α-decay dose.[2] In general, cations with short cation-oxygen bonds show little effect from self-radiation, whereas cations with long cation-oxygen bonds show a greater degree of disorder with accumulated α-decay dose.

A comprehensive set of dissolution kinetics experiments have been performed on these glasses using the single-pass flow-through (SPFT) test method to measure the forward reaction rate as a function of temperature and pH.[3] The forward rate is the most conservative estimate of the radionuclide release rate from the glasses. Because of the high flow-through rate (60 mL/day) used in these tests, the solution in contact with the glass remains very dilute in dissolved glass components, and the buildup of α-radiolysis products, which can enhance dissolution rates in these glasses,[1] is minimized. Using dilute pH buffers also controls solution pH. With this test method, it is possible to unambiguously compare the results between glasses with different radiation damage levels and quantitatively assess differences in their dissolution rate. Although the elemental concentrations of most elements were above the detection threshold, the Pu concentrations were just at the detection threshold and could not be quantified. The normalized elemental release rates as a function of ^{238}Pu concentration are shown in Figure 2 for experiments conducted at 83°C and pH(25°C) = 9. Rates based on the release of Al, Cs, and Na cluster between 1 to 2 g m^{-2} d^{-1} for the three glass specimens. Release rates based on B and U are slightly lower (0.5 to 1 g m^{-2} d^{-1}) compared with the other elements. The doses corresponding to these dissolution tests are given in Table 1. Within experimental error, there is no significant effect of accumulated dose on the forward dissolution rate over the range from 1.2 x 10^{16} to 2.5 x 10^{18} α decays/g. Thus, the effects of radiation damage on the dissolution rate appear to be negligible. The significance of these results will be discussed in detail.

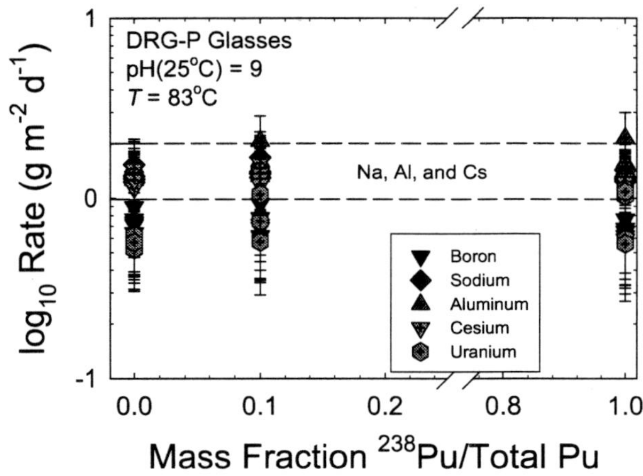

FIGURE 2. Normalized log dissolution rate as a function of the mass fraction of ^{238}Pu in the glass.

REFERENCES

1. Weber, W. J., Wald, J. W., and McVay, G. L., J. Amer. Ceram. Soc. 68(9), C253 (1985).
2. Hess, N. J., Weber, W. J., and Conradson, S. D., J. Nucl. Materials 254, 175 (1998).
3. McGrail, B. P., Ebert, W. L., Bakel, A. J., and Peeler, D. K., J. Nucl. Materials 249, 175 (1997).

Solubility Of Plutonium and Surrogates in Nuclear Glass Matrices

X. Deschanels, C. Lopez, C. Denauwer, and J. M. Bart

CEA / VALRHO / MARCOULE, Nuclear Energy Division, Confinement Research and Engineering Department, Waste Confinement and Vitrification
E-mail: xavier.deschanels@cea.fr

INTRODUCTION

The minor actinides streams from reprocessing plants are currently incorporated in borosilicate nuclear glass matrices. The resulting glass must be perfectly homogeneous, i.e., all the constituent elements must be uniformly distributed in the glass matrix. The work discussed here is part of a study of actinides solubility in borosilicate glass, undertaken to determine the extent of actinides solubility in the glass and to understand the mechanisms controlling actinides solubilization.

RESULTS

The study was performed with a borosilicate glass composition as follows (in wt %), SiO_2 58.84; B_2O_3 18.15; Na_2O 7.00; Al_2O_3 4.28; CaO 5.23; Li_2O 2.56; ZnO 3.24; and ZrO_2 0.70. Glass samples doped with actinides or actinides surrogates were produced under air or argon atmosphere from oxides, nitrates, and carbonate precursors, or from a glass frit. The chemical elements Ce, Hf, and Nd were added as CeO_2, HfO_2, and Nd_2O_3, and Pu was added as a nitric solution. The composition of the Pu-doped glass was modified; zirconium was removed to allow the EXAFS analysis, and some other compounds were added to the glass at a low level to simulate the fission products. The glass samples were processed at temperatures between 1,100°C and 1,400°C. The glass homogeneity was characterized by optical microscopy, scanning electron microscopy (SEM), and with EDX microanalysis. EXAFS spectroscopy was used in order to obtain local information on the glass network (see Table 1).

TABLE 1. EXAFS Spectroscopy Results in Relation to the Solubility Limit of the Actinide or Surrogate in the Glass. N is the number of oxygen atoms around the absorbing atom, $D_{C\text{-}O}$ is the length of the cation-oxygen bond.

	Structural parameters (EXAFS)		
Samples	**N**	**$D_{C\text{-}O}$ (Å)**	**Solubility (% mol.)**
Nd doped glass	8.2 ± 1.6	2.49 ± 0.01	2.52 ± 0.06
Ce doped glass	8.2 ± 1.6	2.44 ± 0.01	1.6 ± 0.06
Th doped glass	6.9 ± 1.4	2.37 ± 0.01	0.26 ± 0.02
Pu doped glass	6.8 ± 1.4	2.25 ± 0.01	>0.09 ± 0.02
Hf doped glass	7.5 ± 1.5	2.08 ± 0.01	0.11 ± 0.02

The observed Ce, Hf, Nd and Pu solubility limits are plotted in Figure 1. The most important solubility increase is observed in the case of Ce. Contrary to Hf and Nd, which have only one stable oxidation state, Ce can be present in the glass as Ce^{4+} or Ce^{3+}. Characterization by XANES spectroscopy and chemical titration showed that higher melting temperatures modify the equilibrium between Ce^{3+} and Ce^{4+}. The Ce^{3+} percentage in the glass increases from 53.8 ± 0.6% at 1,100°C to 88.1 ± 0.3% at 1,400°C. The redox effects on Ce solubility are being studied in order to obtain glasses elaborated at 1,200°C only containing the Ce^{3+} reduced form.

CP673, *Plutonium Futures — The Science,* edited by G. D. Jarvinen

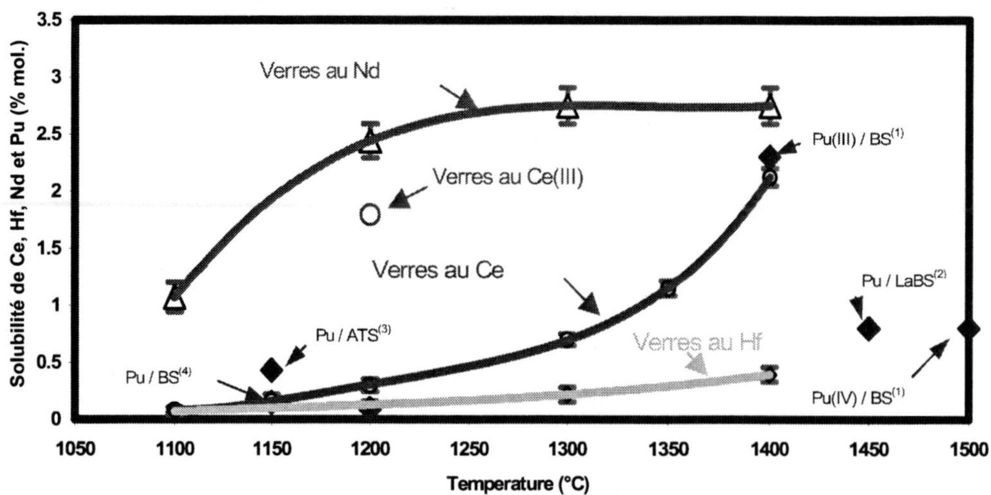

FIGURE 1. Ce, Hf, and Nd solubility vs temperature in a borosilicate glass. Experimental results obtained with Pu surrogates or Pu are compared with Pu solubility data cited in References 1–4.

The solubility was found to be less than 1.2 wt % Ce_2O_3 when the glass is melted at 1,200°C under oxidizing conditions. Chemical titration confirmed that the Ce^{3+} percentage in this glass is 67.7 ± 1.3%. For the same glass melted under an argon atmosphere with aluminum nitride as a reducing agent, the solubility of Ce was found to be at least 12 wt % Ce_2O_3. XANES spectroscopy confirmed that Ce has been completely reduced in this sample. These experimental results correlate the solubility variation with the Ce oxidation state.

The solubility limit of Pu was found to be higher than 1 wt % PuO_2 at 1,200°C. XANES spectroscopy showed that more than 90% of Pu is at the oxidation state +IV in this sample. A study is in progress to determine the solubility limit of plutonium into a glass melted at 1,400°C.

CONCLUSIONS AND DISCUSSION

An investigation of Ce, Hf, and Nd solubility in a borosilicate glass demonstrated that it increases with the melting temperature. The modification of the Ce oxidation state at a higher temperature has a favorable effect on Ce solubility because the reduced form is more soluble in the glass than the oxidized form. Studies are in progress to obtain a similar result for the Pu-doped glasses.

More generally, the elements at the lower oxidation state +III (Nd, Ce) are more soluble than those at the higher oxidation state +IV (Th, Pu, Hf). A correlation has been established between the length of the cation-oxygen bond and the solubility limit of the elements: the higher the solubility, the larger the length of the chemical bond (see Table 1).

REFERENCES

1. Feng, X., Li, H., Li, L. L. D. L., Darab, J. G., Schweiger, M. J., Vienna, J. D., Bunker, B. C., Allen, P. G., Bucher, J. J., Craig, I. M., Edelstein, N. M., Shuh, D. K., Ewing, R. C., Wang, L. M., and Vance, E. R., "Distribution and Solubility of Radionuclides in Waste Forms for Disposition of Plutonium and Spent Nuclear Fuels: Preliminary Results," Ceramic Trans. 93, 409–420 (1999).
2. Bibler, N. E., Ramsey, W. G., Meaker, T. F., and Pareizs, J. M., "Durabilities and Microstructure of Radioactive Glasses for Immobilization of Excess Actinides at the Savannah River Site," Material Research Society Symposium Proceedings 412, 65–72 (1996).
3. Bates, J. K., Ellison, A. G. J., Emery, J. W., and Hoh, J. C., "Glass as a Waste Form for the Immobilization of Plutonium," Materials Research Society Symposium Proceedings 412, 57–64 (1996).
4. Bates, J. K., Emery, J. W., Hoh, J. C., and Johnson, T. R., "Performance of High Plutonium-Containing Glasses for the Immobilization of Surplus Fissile Materials," Environmental Issues and Waste Management Technologies in the Ceramic and Nuclear Industries I—Ceramic Trans. 61, 447–454 (1995).

Plutonium Partitioning in Zirconolite and Pyrochlore Containing Multiphase Ceramics

S. V. Stefanovsky,[1] A. G. Ptashkin,[1] S. V. Yudintsev,[2]
Y. M. Kulyako,[3] and S. A. Perevalov[3]

[1]*SIA Radon*
[2]*Institute of Geology of Ore Deposits RAS*
[3]*Institute of Geochemistry and Analytical Chemistry RA*

INTRODUCTION

Zirconolite is one of the most promising host phases for plutonium.[1–3] Because zirconolite-based ceramics are usually not single phase, plutonium may be partitioned among zirconolite and co-existing phases such as perovskite, zirconia, and rutile.[4] In this work, we present some new data on plutonium partitioning in zirconolite-based ceramics.

Three ceramic samples were studied. Composition of Sample 1 was taken from reference data (in wt %): CaO – 13.4, Gd_2O_3 – 5.9, ZrO_2 – 29.4, PuO_2 – 8.8, TiO_2 – 40.8, and Al_2O_3 – 1.7.[2] Composition of Samples 2 and 3 corresponded to the nominal formula $PuZrTiAlO_7$. Oxide mixtures not containing Pu oxide were milled in an agate mortar, soaked with Pu nitrate solution, and dried. Then, mixtures for Samples 1 and 2 were compacted under a pressure of 200 MPa in pellets and placed in platinum crucibles, followed by melting at 1,550°C in air for 30 min. Carbon powder was admixed to the mixture for Sample 3, followed by melting at the same temperature in a glassy carbon crucible to obtain reducing conditions. The samples were examined with XRD and SEM/EDS. All interatomic-distance values are given in nm.

RESULTS

Sample 1 consists of zirconolite with a uniform composition over the bulk of the sample (Figures 1 and 2). However, rare grains of altered plutonium oxide were also found. Major zirconolite peaks are positioned at $d_{221,\bar{4}02}$ = 0.2936, d_{004} = 0.2797, $d_{\bar{2}23}$ = 0.2508, d_{223} = 0.2287, $d_{\bar{6}21}$ = 0.1801, and d_{225} = 0.1740.

Sample 2 has a complex composition. Peaks at d_{111} = 0.3095, d_{200} = 0.2683, d_{220} = 0.1893, d_{311} = 0.1613, d_{222} = 0.1545, d_{400} = 0.1336, d_{331} = 0.1226, d_{420} = 0.1195, and d_{422} = 0.1091 are due to cubic fluorite-structure oxide (space group *Fm3m*). Reflections d = 0.3003, 0.2560, 0.1842, 0.1589, and 0.1490 are probably due to the pyrochlore structure phase. Attribution of some peaks such as 0.3382, 0.2945, and 0.1811 is indefinitive. Combining XRD and SEM/EDS data, it may be concluded that this sample is composed of altered PuO_2-based grains forming a cubic solid solution $(Pu,Zr,Ti)O_2$ (lightest on the SEM image) phase with a pyrochlore or zirconolite structure, $Pu_2(Zr,Ti)_2O_7$ (darker on the SEM image), and an Al-Ti-rich phase, probably Al_2TiO_5 (the darkest on the SEM image and peak at 0.3382 nm on the XRD pattern). The latter phase content is low. Plutonium is partitioned between the first two phases.

Sample 3 is composed of a major Pu-Zr-Ti-O phase (light on the SEM image) and a minor Pu-Al-Ti-O phase with suggested perovskite structure ($PuAl_{1-x}Ti_xO_3$) (dark on the SEM image). All plutonium oxide has reacted, and an oxide phase in the sample has not been found. The first phase is responsible for XRD reflections d = 0.2983, 0.2894, 0.2546, 0.2265, 0.2225, 0.1831, 0.1794, and 0.1542. Peaks at 0.2718, 0.2614, and 0.1564 may be assigned to a minor perovskite phase.

CP673, *Plutonium Futures — The Science,* edited by G. D. Jarvinen

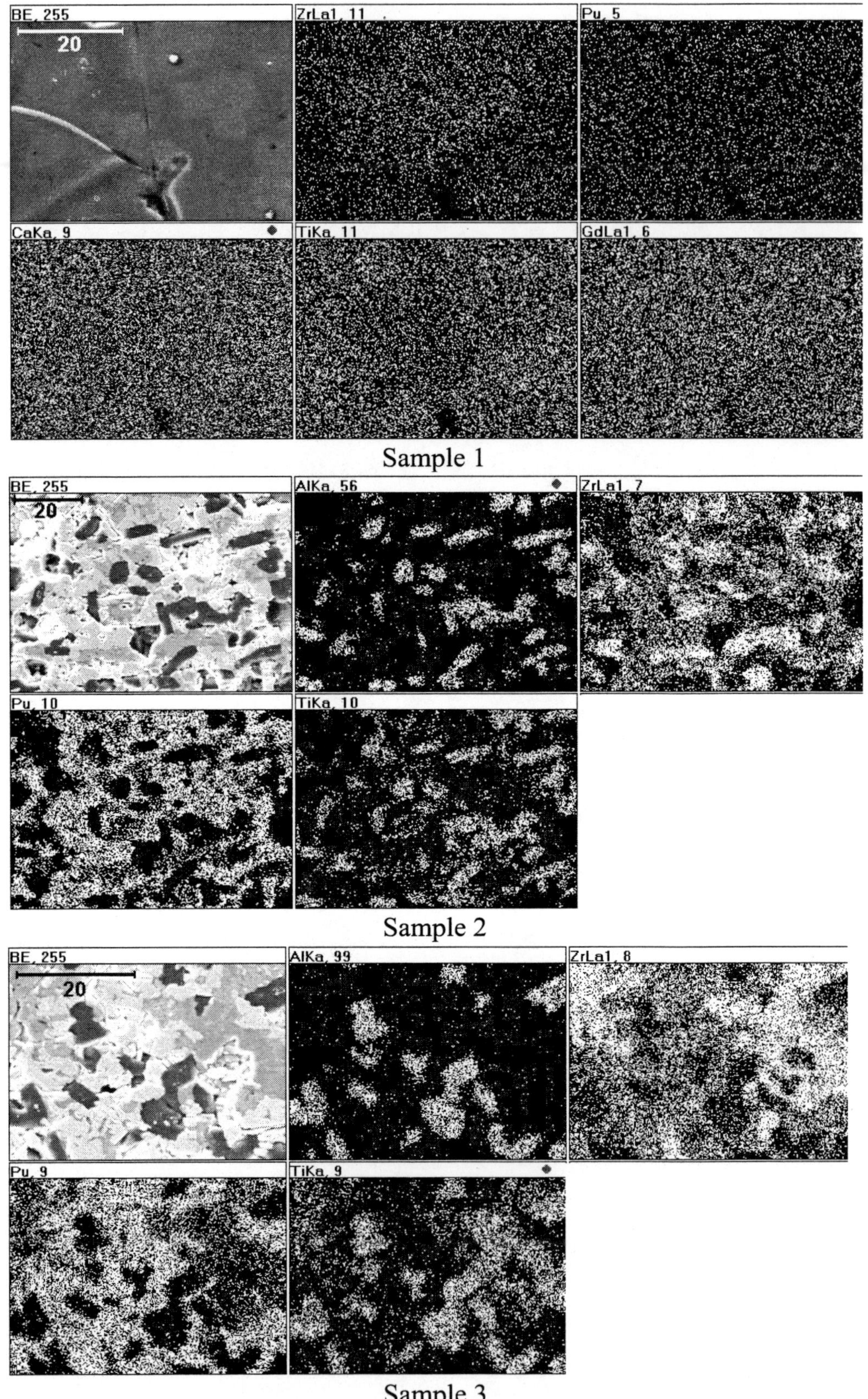

FIGURE 1. Backscattered SEM images of Samples 1, 2, and 3.

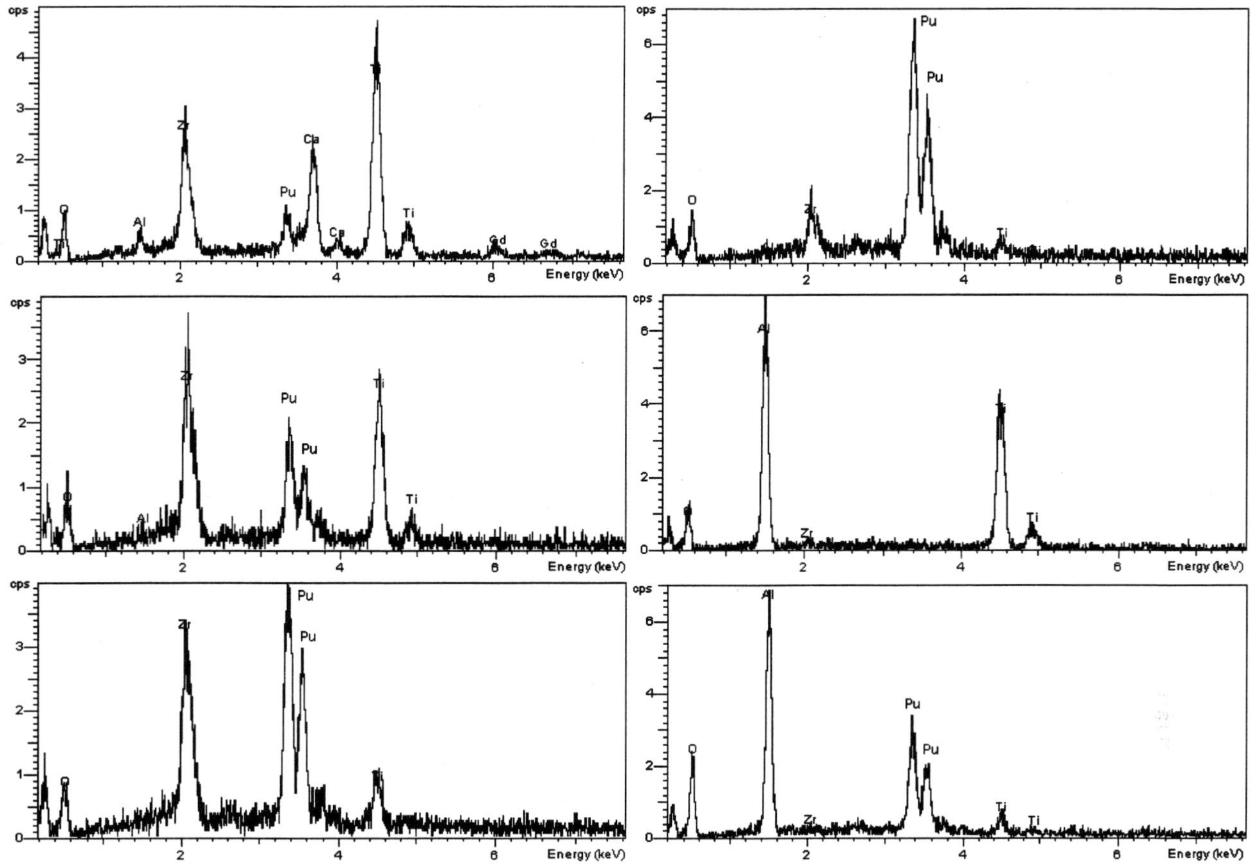

FIGURE 2. EDX spectra showing compositions of phases in the samples: 1 (upper left), 2 (upper right, middle left and right), and 3 (bottom left and right).

DISCUSSION

As follows from References 2 and 3 and our data, Pu-bearing ceramics are normally not single phase. The reason is the multivalent state of Pu, which ranges in ceramics between 3+ and 4+.[2] The composition of Sample 1 suggests that all Pu, after melting under oxidizing conditions, exists in a tetravalent form as Pu(IV) and substitutes for Zr^{4+} ions in their sites. Nevertheless, some Pu oxide grains and traces of a perovskite structure phase occurred. Because the plutonium oxide and the perovskite phase contents are low, their effect on Pu partitioning is negligible, and the major fraction of Pu enters zirconolite.

At high Pu content (Samples 2 and 3), the ceramics produced did not contain or contained only minor Pu-substituted zirconolite ($PuZrTiAlO_7$), whereas Sm, Eu, and Gd analogs exist.[3,5] Occurrence of a major fraction of Pu as Pu(IV) results in the formation in Sample 2 melted under oxidizing conditions of fluorite and pyrochlore structure phases with a very limited isomorphic capacity with respect to aluminum; and the latter forms aluminum titanate as an extra phase. Quantitative evaluation of the co-existing phase contents may be only approximate—fluorite structure solid solution: ~50%, pyrochlore ~30%, and aluminum titanate ~20%. Because the latter phase does not contain Pu (Figure 2), the major host phases for Pu are cubic oxide and "pyrochlore."

In Sample 3, melted under reducing conditions, along with pyrochlore-zirconolite, an extra perovskite-type phase occurred. Pu is partitioned among both phases, but the pyrochlore structure phase is enriched with Pu, and its content in the ceramic is higher (60%–70% of total bulk). Therefore, the major host for Pu in this ceramic is the pyrochlore phase.

The data obtained may be used as a study of inhomogeneous partitioning of plutonium and other fissile elements in multiphase ceramics, which can be important for plutonium leaching and criticality control.

REFERENCES

1. Vance, E. R., Jostsons, A., Day, R. A., Ball, C. J., Begg, B. D., and Angel, P. J., Mat. Res. Soc. Symp. Proc. 412, 41 (1996).
2. Buck, E. C., Ebbinghaus, B., Bakel, A. J., and Bates, J. K., Mat. Res. Soc. Symp. Proc. 465, 1259 (1997).
3. Jostsons, A., Vance, L., and Ebbinghaus, B., "Immobilization of Surplus Plutonium in Titanate Ceramics," in Proc. Int. Conf. on Future Nuclear Systems GLOBAL'99 "Nuclear Technology—Bridging the Millenia," Jackson Hole, Wyoming (1999), CD-ROM.
4. Chizhevskaya, S. V., and Stefanovsky, S. V., in Plutonium Futures—The Science, Conf. Trans. Santa Fe, New Mexico, July 10–13, AIP Conf. Proc. (2000), pp. 148–150.
5. Stefanovsky, S. V., Cherniavskaya, N. E., Ochkin, A. V., and Yudintsev, S. V., in 14th Radiochem. Conf. Marianske Lazne, Czech Rep. (2002).

SESSION FIVE:
MATERIAL SCIENCE
AND PLUTONIUM
PROPERTIES

An Overview of Plutonium Aging

Joseph C. Martz, Luis A. Morales, Kathleen B. Alexander

Los Alamos National Laboratory, MS G754, Los Alamos, NM 87545

INTRODUCTION

Planning for future refurbishment and manufacturing needs in the US nuclear weapons complex critically depends on credible estimates for the lifetimes of the nuclear components. A scientific understanding of the aging of plutonium primary components (called pits) is necessary for accurate lifetime predictions.

An evaluation of plutonium aging starts with an identification of the key properties required to ensure safe and reliable performance. Once these properties have been identified, diagnostic tools are developed to measure them with sufficient precision. Next, we identify the aging mechanisms that could potentially alter these properties with time.

The three most important potential aging effects in plutonium are the radiogenic decay of the various plutonium isotopes, the possible thermodynamic instability of the plutonium alloy itself, and the corrosion of the plutonium surface during both storage and function. In many cases, these aging effects accumulate slowly over decades, and not necessarily in a linear fashion. Only when key properties have sufficiently changed would we anticipate a measurable impact on safety or performance. Designers specify the limits of acceptable change for each of these properties. By combining these limits with the measured or predicted rates of change due to aging effects, we arrive at estimates for lifetimes.

A fundamental aspect in the accumulation of radiation damage in plutonium is the existence of a threshold beyond which further damage results in rapid swelling and density change of the material. Experience from materials similar to plutonium in reactor environments shows that the initial damage results in little change in density, but after an "incubation period", void swelling begins. The length of this incubation is unknown for plutonium.

RADIATION DAMAGE

The principal decay mechanism for most plutonium isotopes is alpha-particle decay. The parent atom spontaneously decays into a helium nucleus (i.e, alpha particle) and a uranium atom. Both of these particles are highly energetic. This initial decay event is rapid and results in considerable, local disruption of the crystalline lattice. This single decay moves roughly 20,000 other atoms off of their lattice sites, though about 90% of these displaced atoms return to a normal lattice position almost instantaneously. The remaining 10% of these atoms are more permanently displaced in the lattice as interstitials and vacancies. The ultimate disposition of these more permanent defects are the principal concern in our evaluation.

Details of the initial decay and atomic rearrangements are exceptionally difficult to experimentally measure. Hence, we depend entirely on modeling the damage process. The complexity of electronic structure and bonding in plutonium complicates efforts to develop an accurate lattice potential. Our best models still leave us with a substantial uncertainty in prediction of both the onset and rate of void swelling. Ultimately, experimental data of the damage that remains from the initial decay collision cascades will be necessary to establish confidence in these models and to reduce the uncertainty. The need for this data is the principal objective in our experiment to accelerate-age plutonium. An alloy of typical plutonium isotopic content mixed with 7.5% of the Pu-238 isotope will accumulate radiation damage at a rate 16 times faster than unspiked material alone. Critically, acceleration of the input of radiation damage must be matched by acceleration of the subsequent annealing and diffusion of that damage. We accomplish this subsequent acceleration by raising the temperature at which the samples are stored.

CP673, *Plutonium Futures — The Science,* edited by G. D. Jarvinen

Data from this program is key to refining our estimates and searching for the onset of non-linear aging mechanisms such as void swelling.

THERMODYNAMIC STABILITY OF PLUTONIUM ALLOYS

Another concern is the thermodynamic stability of the plutonium alloy itself. Unalloyed plutonium exists in the monoclinic, α-phase at room temperature – a brittle material that is difficult to fabricate. It is well-known that the face-centered cubic δ-phase, which can be retained to room temperature by small amounts of aluminum or gallium, is preferred. For a variety of reasons, gallium has become the alloying agent of choice.

Both the US and former Soviet Union have studied the long-term stability of these δ-stabilized alloys.[1] Whereas US researchers predicted a stable alloy at room temperature, Soviet researchers found that the thermodynamically stable condition was not δ-stabilized material, but a mixture of α-plutonium and the intermetallic compound Pu_3Ga. If the δ-stabilized material is not the thermodynamically-favored condition, then the rate and mechanism of this potential phase-change must be studied. The Soviet researchers found that this process was indeed extremely slow – conversion to the stable $\alpha + Pu_3Ga$ phases was predicted to take 11,000 years at room temperature. However, the influence of the radiation-damage processes on the question of phase stability is still unknown and represents another uncertainty in our evaluation of plutonium aging.

CORROSION

Finally, corrosion of plutonium is potentially the most catastrophic of all aging effects.[2] Fortunately, corrosion is both limited by the availability of corrosive agents and relatively easily studied. Whereas plutonium will readily oxidize given sufficient exposure to air or other oxidizing environments, it is hydrogen-catalyzed corrosion that is of greatest concern.[2] Most important from an aging perspective is the maintenance of well-sealed environments and the exclusion of foreign contaminants during production.

CONCLUSIONS

On the basis of careful evaluation of the effects described above (as well as a few other, less-prominent concerns), initial estimates of minimum lifetimes have been derived. The properties that have been measured to date, such as density and compressive strength have shown only small changes. Hence, we feel reasonably comfortable that the lifetimes of components will last at least several tens of years. The principal uncertainty in these estimates relates to the non-linear behavior due to the incubation periods inherent in radiation damage models. Additional uncertainty arises from the intrinsic scatter in much of the experimental data as well as uncertainties on the influence of certain changes on performance. Moreover, continuing research will strengthen the linkage between the plutonium microstructure and changes resulting from aging, key properties, and performance.

---

[1] S.S. Hecker and L.F. Timofeeva, Los Alamos Science, 26, vol 1. p. 244-251 (2000)

[2]J. M. Haschke and J.C. Martz, "Plutonium Storage", in Encyclopedia of Environmental Analysis and Remediation, John Wiley and Sons, New York (1998) pp. 3740-3755

Advanced Transmission Electron Microscopy of Pu Alloys

Adam J. Schwartz, Mark A. Wall, Wilhelm G. Wolfer, and Kevin T. Moore

Chemistry & Materials Science Directorate
Lawrence Livermore National Laboratory
L-355
7000 East Avenue
Livermore, CA 94550

INTRODUCTION

The characterization of microstructural changes in Pu–Ga alloys resulting from storage and aging phenomena is an important technical challenge to the nuclear Stockpile Stewardship program. We have identified at least two age-related phenomena that may occur in Pu alloys, dimensional changes due to the initial transient, helium accumulation, and void swelling, and phase instability. The initial transient is a well-known effect that results from the initial cascade damage. This form of dimensional change tends to saturate within approximately two years. A second contributor to dimensional change is the buildup of helium as a result of the alpha decay. Helium is generated at a rate of approximately 40 parts per million per year. Positron annihilation results by Howell[1] indicate that the helium atoms will quickly fill a nearby vacancy and diffuse through the lattice as a helium-filled vacancy. Void swelling is potentially the most severe mechanism of dimensional change in Pu alloys. It has been observed in all materials exposed to irradiation, but has yet to be seen in naturally aged Pu.

Phase instability is a potential concern because of the fact that the δ phase is thermodynamically metastable at room temperature. Timofeeva[2] has shown that the δ phase will decompose to δ phase and Pu_3Ga, given enough time at ambient temperature. At subambient temperatures, the δ phase undergoes a displacive or martensitic phase transformation to the monoclinic α' phase, which is approximately 20% more dense. Phase transformations such as these would result in density changes, dimensional changes, and changes in mechanical properties.

Traditional characterization techniques such as optical microscopy, x-ray diffraction, and scanning electron microscopy are insensitive to many of the age-related microstructural changes. In this investigation, we have applied advanced transmission electron microscopy (TEM) to investigate the microstructure and bonding of Pu alloys. A 300 keV Phillips CM300FEG with a field emission gun electron source and Gatan Imaging Filter are used for the investigations.

RESULTS

Advanced TEM has been used to characterize the helium bubble distribution in new and old Pu alloys. Figure 1a shows a TEM micrograph of He bubbles in a 42-year old material. These results have been coupled to rate equation modeling to predict the evolution of He bubble number density and average size with size as shown in Figure 1b.

The structure and composition of a naturally occurring Pu-Fe intermetallic phase has been determined by combining high-energy TEM with simulated and experimental electron-diffraction simulation and energy-dispersive spectroscopy. A phase belonging to the space group *I4/mcm* has been identified in a Pu-Ga alloy containing trace amounts of Fe and Ni using electron diffraction and energy-dispersive X-ray spectroscopy (EDXS) in a transmission electron microscope. The plane group symmetry of six experimental diffraction patterns shows that the structure of this phase was at least body-centered orthorhombic.

CP673, *Plutonium Futures — The Science,* edited by G. D. Jarvinen

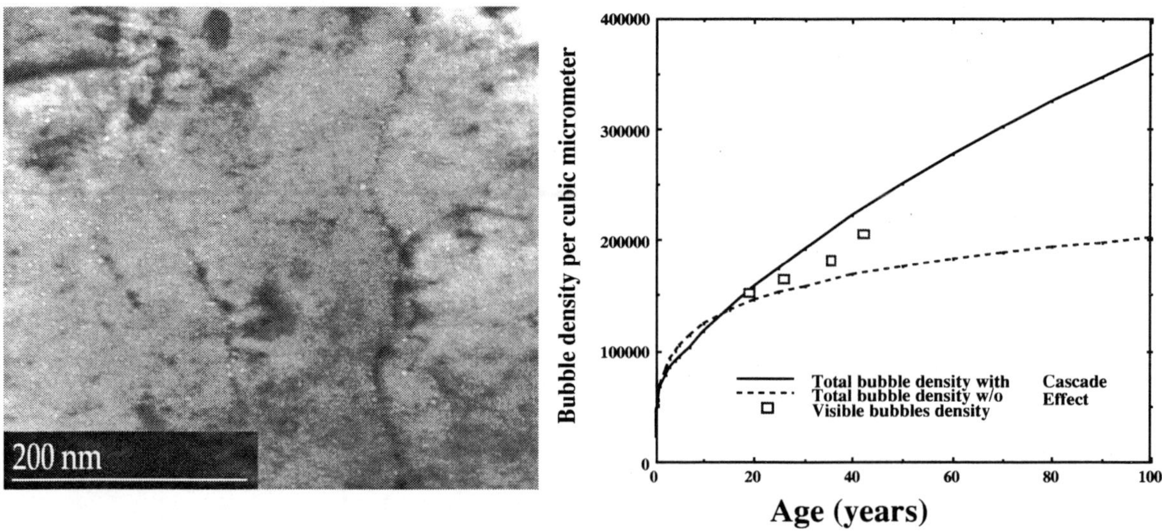

FIGURE 1. (a) Bright field TEM micrograph of the 42-year old material. The image is taken in the under-focus condition so that the bubbles appear as dark rings surrounding bright dots. (b) Model prediction of the bubble number density as a function of age.

Simulated diffraction patterns, generated from the body-centered tetragonal structure of ζ Pu_6Fe with the space group *I4/mcm*, match the experimental diffraction patterns closely. These results present the first crystallographic evidence for the existence of ζ Pu_6Fe in a Pu-Ga alloy. Figure 2a is a TEM micrograph of the Fe-containing phase in the δ-Pu lattice. Figure 2b is an energy dispersive spectrum showing high levels of Fe. The Pu/Fe ratio of the phase yielded by EDXS was 12.5% and the Pu/(Fe+Ni) ratio was 15.9%. These results suggest that Ni substitutes for Fe in the ζ Pu_6Fe lattice. By coupling electron diffraction, simulation, and EDS, we are able to determine that this is the body-centered tetragonal phase ζ-Pu_6Fe.

Using high-energy, electron-energy loss spectroscopy (HE-EELS), TEM, and synchrotron-radiation-based x-ray absorption spectroscopy (XAS), we are evaluating the bonding of the 5f states of Pu. The advantage of this approach is that the HE-EELS experiments are performed in a TEM and are coupled with imaging and diffraction data; therefore, the measurements are completely phase specific. Figure 3a is a TEM image of an α' particle imbedded in a δ matrix. The microstructure has been produced by cooling the Pu-Ga alloy to 150 K for 10 hours. Figure 3d is an electron energy loss spectrum taken from an α' particle embedded in a δ matrix. Implications of these results to our understanding of bonding will be discussed.

ACKNOWLEDGMENTS

This work was performed under the auspices of the U. S. Department of Energy by the University of California, Lawrence Livermore National Laboratory under Contract No. W-7405-ENG-48.

REFERENCES

1. Howell, R. H., Sterne, P. A., Hartley, J., and Cawan, T.E., Applied Surface Science 149 103–105 (1999).
2. Hecker, S. S., and Timofeeva, L. F., Los Alamos Science 26, 244 (2000).

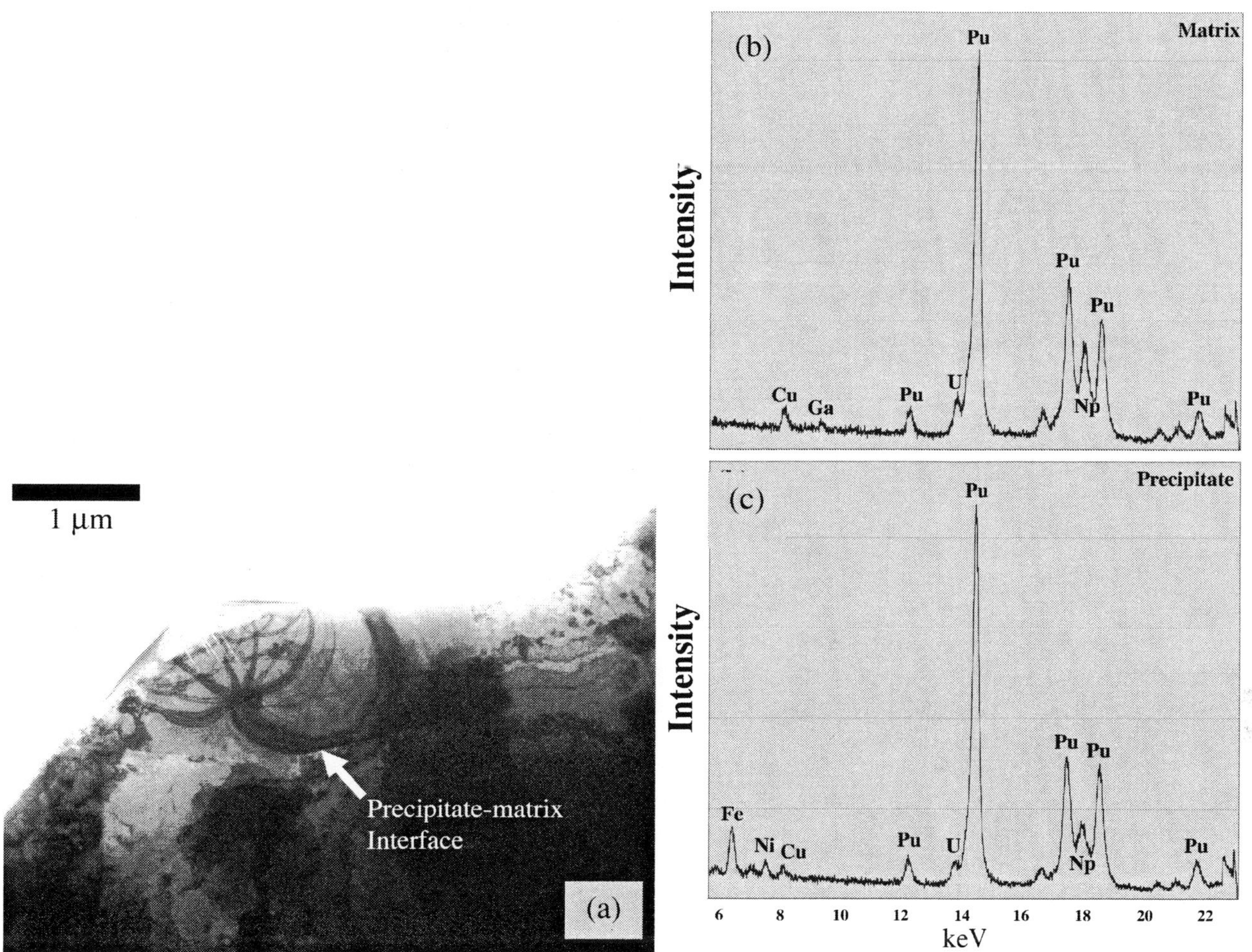

FIGURE 2. (a) A bright-field TEM image of one of the ζ Pu_6Fe precipitates contained in a fcc Pu matrix. The precipitate-matrix interface is marked with an arrow, (b) Two EDXS spectra taken from; (a) one of the ζ Pu_6Fe precipitates and (b) the fcc Pu matrix. Notice that an Fe peak is found in the spectra for ζ Pu_6Fe, but is absent in the spectra for the Pu matrix.

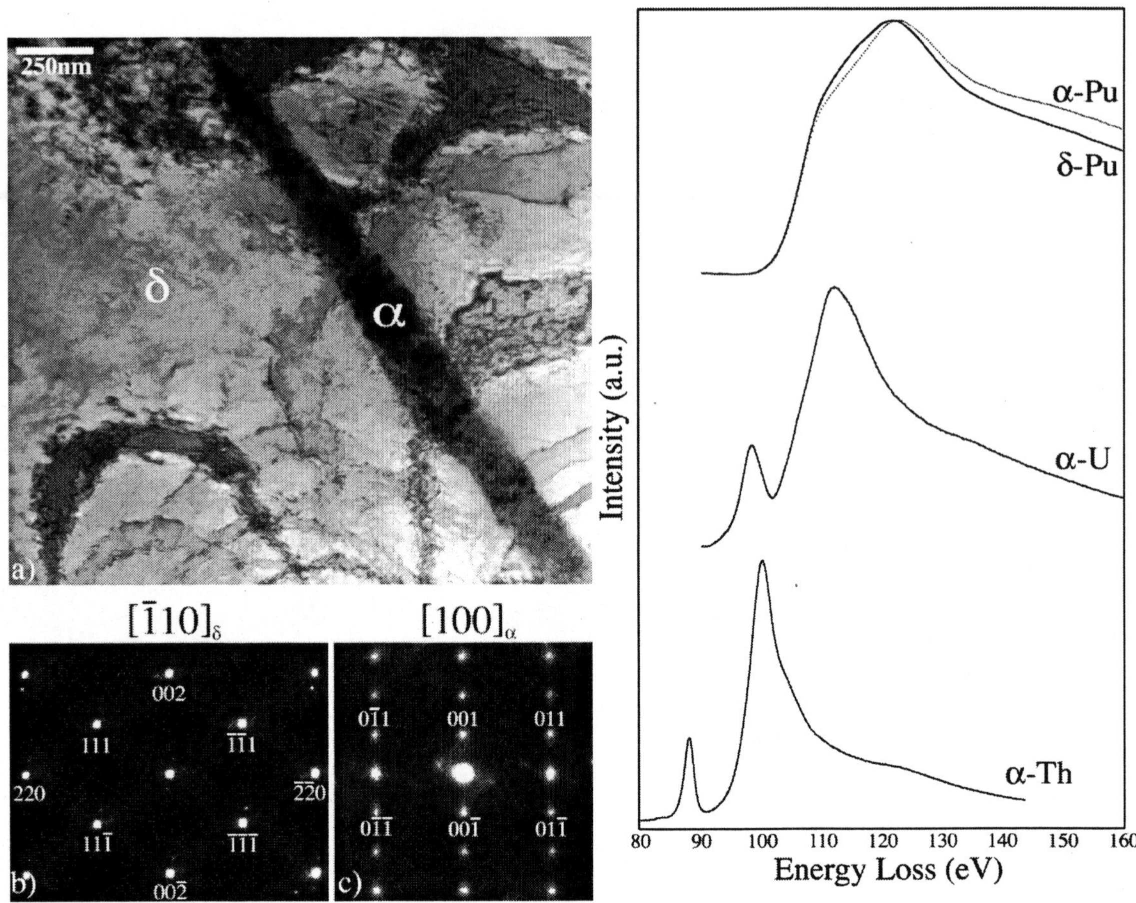

FIGURE 3. (a) A bright-field TEM image acquired near a $\left[\bar{1}10\right]_\delta \parallel \left[100\right]_\alpha$ zone axis showing an α plate in a δ matrix. (b) A $\left[\bar{1}10\right]$ diffraction pattern from δ and (c) a $\left[100\right]$ diffraction pattern from α, each with a number of reflections indexed, (d) the $O_{4,5}$ (5d→5f) absorption edges from α-Th, α-U, α-Pu and δ-Pu acquired by HE-EELS in a TEM. These spectra were collected at an accelerating voltage of 297 keV with an energy resolution of 0.8 eV.

Pu Has No Future: A Real-Time Measurement of Pu Aging

A. Migliori, D. A. Miller, J. C. Lashley, F. Freibert, J. B. Betts, and M. Ramos

Los Alamos National Laboratory, Los Alamos, New Mexico 87545

Abstract. Recent RUS measurements on polycrystalline samples of δ–Pu have established that RUS can detect changes in elastic moduli of 0.3 parts per million, so that for the first time ever we accurately measured aging effects in real time for Ga-stabilized fcc Pu. Our measurements were performed on alloys made from of nominally pure ^{239}Pu as well as ones made with a 7.1% ^{238}Pu fraction to accelerate aging.

INTRODUCTION

Sound propagation is governed by adiabatic elastic moduli that we are able to measure using Resonant Ultrasound Spectroscopy (RUS).[1] RUS is a well-established technique that is capable of the highest absolute accuracy for any routine elastic modulus measurement technique. It is based on analysis of the mechanical resonances of small samples. By using many resonances with different strain patterns, we are able to obtain the entire elastic tensor in one measurement. For the isotropic polycrystal samples used here, only two moduli are required. RUS also can have very high precision if the Q of the samples is high. For all the samples in this study, Q was over 5,000, making it possible for us to see change in moduli of order 1 part per million (PPM).

MEASUREMENTS

All the samples used for this study were prepared from purified starting materials. Of course, radioactive decay produces impurities with time, so all samples used were less than 4 years old. Samples were mounted in an RUS system of our design, with each sample contacting transducers on the corners of rectangular parallelepiped shape used for measurements. A Si diode thermometer was used to monitor temperature, but no temperature control was implemented. Instead, we used the known stability (<0.05 K fluctuations) of the glovebox to control stability. The sample containing 2.36 at. % Ga and pure ^{239}Pu was 0.3765 x 0.4733 x 0.4854 cm and weighed 1.37 gm, while the 3.3 at. % Ga sample containing 7.1% ^{238}Pu was 0.250 x 0.302 x 0.304 cm and weighed 0.36 gm.

RESULTS AND DISCUSSION

The small temperature fluctuations in the glovebox had to be accounted for. We made very careful measurements of the elastic moduli of homogenized and unhomogenized Pu samples to ensure that homogenization effects could not affect the results, shown in Figure 1. We also made the most accurate measurements ever of several Pu alloys to determine the temperature dependence of elastic moduli versus Ga, shown in Figure 2. These data were necessary to correct for the small temperature fluctuations in the glovebox. Such corrections produced the very smooth time-dependent curves in Figure 3. Our results demonstrate that real-time aging or elastic moduli using RUS can be measured accurately. The data are marred by a glitch for the ^{239}Pu sample at 130 h caused by the glovebox being bumped; however, the slope before and after the bump is the same. Our results show that the aging rate for the ^{238}Pu-containing sample is much slower than expected, only 63% faster than the pure ^{239}Pu sample. We also observed an initial rapid transient that is puzzling. We conjecture that it is caused by radiation-damaged-induced stress relief at the sharp corners of the sample in contact with the transducers.

CP673, *Plutonium Futures — The Science,* edited by G. D. Jarvinen

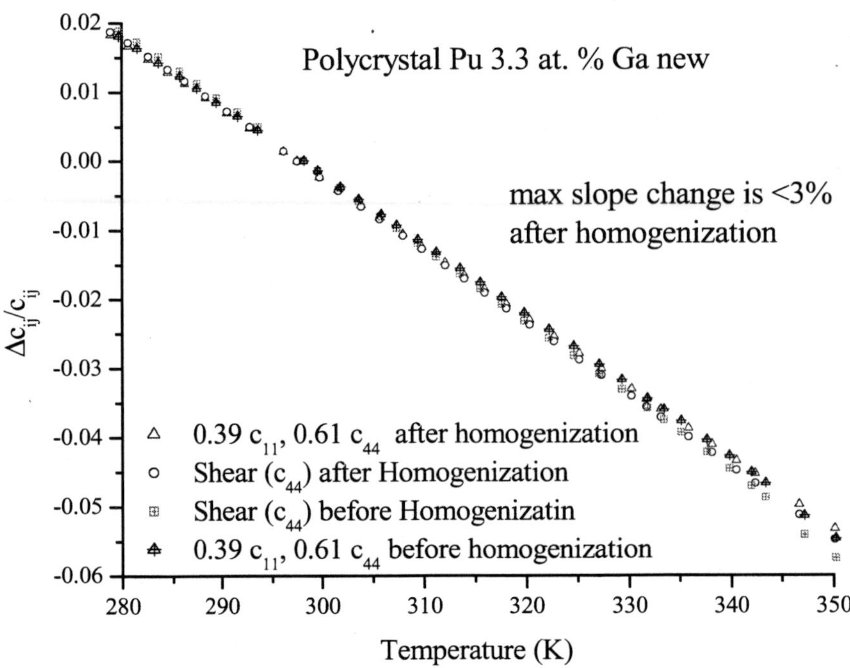

FIGURE 1. Shown is a comparison of homogenized and unhomogenized very high purity research samples of fcc Pu. The unhomogenized state was achieved by rapid solidification out of a levitation furnace. Homogenization was for 100 h at 450°C in an Ar atmosphere. No change in modulus or its temperature dependence was observed.

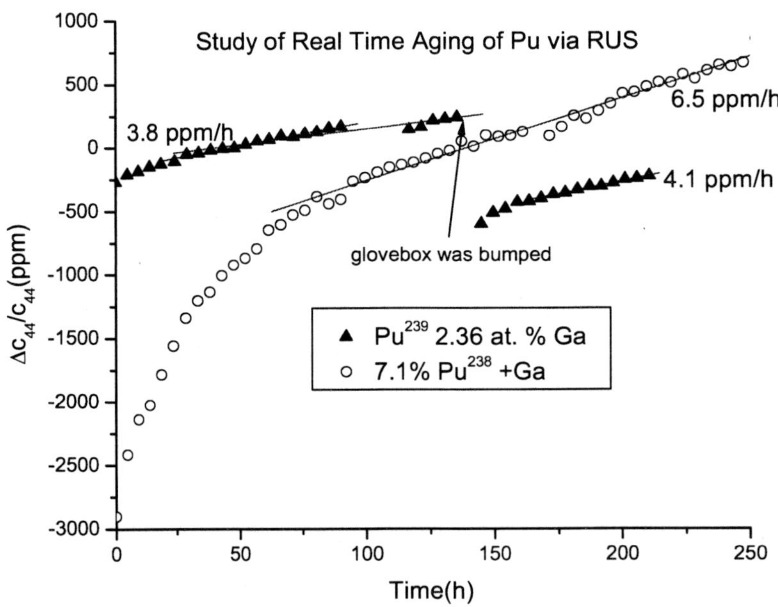

FIGURE 2. Temperature dependence of moduli for fcc Pu. Within about 3%, the measured slopes are independent of Ga.

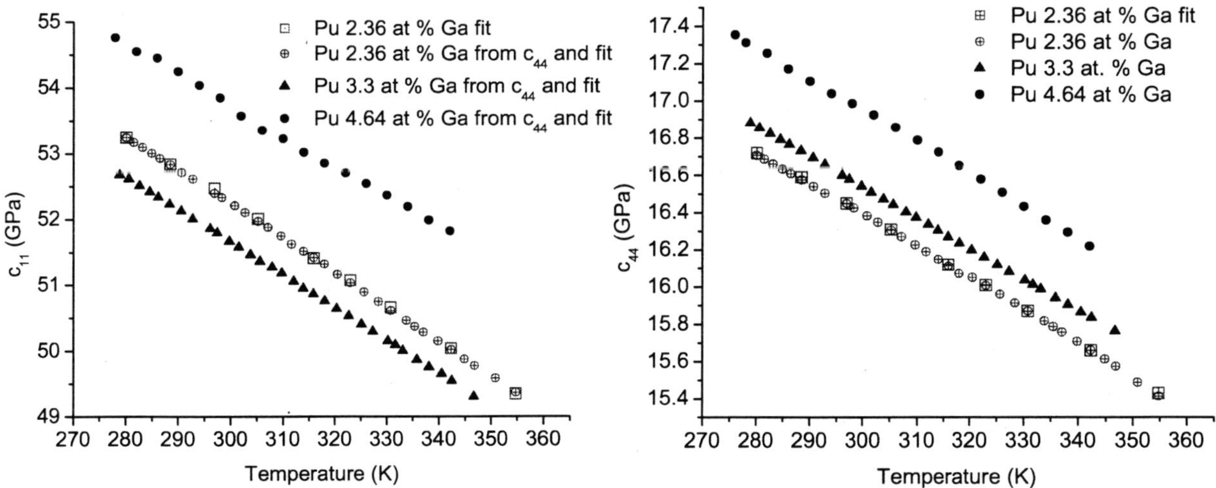

FIGURE 3. Shown is the very first observation in real time of Pu aging. We measured a pure 239 sample with 2.36 at. % Ga and the standard 7.1% Pu^{238} alloy. The time dependence was 63% stronger in the spiked alloy. We ran this at 318 K and used the data of Figure 2 to correct for the 0.05 K fluctuations in temperature over the 10-day runs. For the 239 sample, at about 130 hours, the glovebox was bumped, causing the glitch. However, the slope before and after the glitch was unchanged to our error bars of about 10%. Potential sources for this time dependence are radiation damage, phase transformation, strain relief, and He ingrowth.

SUMMARY

We have the temperature and time dependence of the elastic moduli of Ga-stabilized fcc Pu with unprecedented accuracy and precision. We find an initial transient to the time dependence, an unexpectedly low rate of change for a ^{238}Pu alloy compared to the pure ^{239}Pu sample, and the same temperature dependence near 300 K for several Ga concentrations.

REFERENCES

1. Migliori, A., and Sarrao, J. L., Resonant Ultrasound Spectroscopy, John Wiley, New York (1997).
2. Migliori, .A., Freiber, F., Lashley, J. C., Lawson, A. C., Baiardo, J. P., and Miller, D. A., J. Superconductivity: Incorporating Novel Magnetism 15, 499 (2002).

Self-Irradiation Effects in PuGa Alloys as Revealed by Positron Annihilation Spectroscopy

Benoît Oudot,[a,b] Nathalie Baclet,[a] Lionel Jolly,[a] Brice Ravat,[a] Carole Valot,[a] Pascale Julia,[a] and Manuel Grivet [b]

(a) CEA-Centre de Valduc 21120 IS-SUR-TILLE — FRANCE
Tel : 033 3 80 23 48 71, Fax : 033 3 80 23 52 17
(b) Laboratoire de Microanalyses Nucléaires, Université de Franche-Comté,
16 route de Gray 25000 BESANCON — FRANCE
Tel : 033 3 81 66 65 01, Fax : 033 3 81 66 65 22

SUMMARY

Understanding the aging of plutonium alloys today is a fundamental challenge. Positron annihilation spectroscopy has been developed to characterize the self-irradiation defects at a very fine scale. The first results on plutonium alloys, together with X-ray diffraction and high temperature dilatometry data will be presented.

INTRODUCTION

Plutonium is unstable with time and especially exhibits α decay, creating a uranium atom (E ~ 85-94 keV) and a helium atom (E ~ 5 MeV), which generate displacement cascades. Self-irradiation defects then created are susceptible to diffusion and modify the physical properties of the alloy. For instance, the swelling of plutonium alloys has been evidenced since about 25 years ago.[1] Aging effects on magnetic properties have also been detected more recently.[2]

However, the precise origin of these changes is still not understood today. This occurs for different reasons: difficulties in handling plutonium (which limits the techniques available) or the necessity to characterize the self-irradiation defects at a very fine scale.

Understanding self-irradiation in plutonium alloys then remains a fundamental challenge to be able to predict the effects of aging on the reliability and safety of the nuclear stockpile.

Aging effects are then investigated through many techniques, sensitive to different scales (X-ray diffraction, dilatometry, magnetic properties measurements, etc.) and this experimental approach is strongly coupled with multiscale modeling.[3]

Among the experimental techniques, positron annihilation spectroscopy (PAS) has been developed because this nondestructive technique allows us to characterize the size and concentration of vacancy defects that might play a fundamental role regarding the change in physical properties.

EXPERIMENTAL

For PAS experiments being performed out of a glove box, a specific confinement has been designed: the positron source inserted between two plutonium samples is first placed in an aluminum container, then placed in a sealed plastic film.[4]

CP673, *Plutonium Futures — The Science,* edited by G. D. Jarvinen

The positron source ^{22}Na is used as (maximum energy of the positron: 547 keV), which allows the probing to a maximum thickness of about 70 μm in the plutonium alloys studied. The preparation of the samples must then be strictly controlled in order to investigate the bulk of the material and to avoid artifacts from the sample preparation (defects induced by microsectioning, machining, etc.). X-ray diffraction (XRD) and PAS techniques have then been coupled to define the protocol for sample preparation. As an example, Figure 1 illustrates how the XRD diagram is affected by machining and details the electropolishing treatment that is necessary to remove the affected zone.[5]

Both lifetime and Doppler broadening spectra have been recorded; nickel was used as the standard.

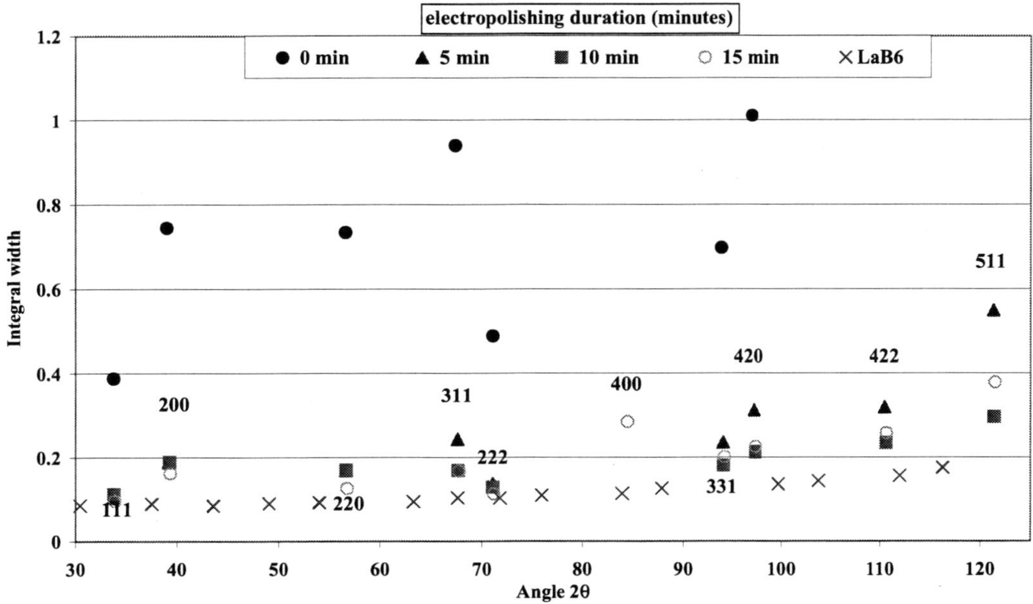

FIGURE 1. Influence of machining on the XRD diagram and effect of removing material by electropolishing.

RESULTS AND DISCUSSION

First PAS experiments have been performed on PuGa alloys with different gallium contents. Lifetime and Doppler broadening spectra have been recorded either on freshly cast alloys that are followed with time or self-irradiated alloys corresponding to different ages. As an example, Figure 2 compares the lifetime spectra of nickel (used as standard) and a plutonium alloy.

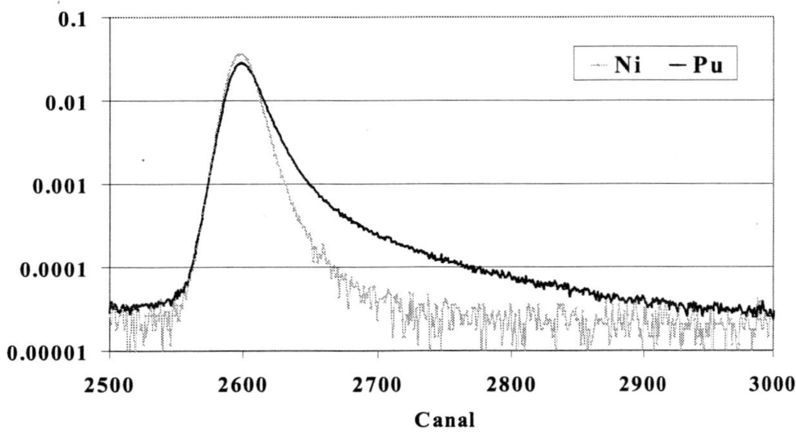

FIGURE 2. Comparison of the lifetime spectra of nickel (standard) and a plutonium alloy.

PAS has been coupled to XRD and high-temperature dilatometry, which allow both the quantification of the swelling of the plutonium alloys at a nanometric and a macroscopic scale, respectively. Swelling is indeed one of the most obvious changes that aging induces, and PAS should help to understand the origin of this phenomenon to predict if the changes observed are crucial regarding the reliability of the alloys concerned.

REFERENCES

1. Chebotarev, N. T. et al., in "Plutonium and Other Actinides," North-Holland Publishing Company (1976), p. 599.
2. Baclet, N., Dormeval, M., Pochet, P., Fournier, J. M., Wastin, F., Colineau, E., Rebizant, J., and Lander, G., "Proceedings of the "Actinides 2001" Conference, Japan, November 2001.
3. Baclet, N., Pochet, P., Oudot, B., Julia, P., Valot, C., Ravat, B., and Dormeval, M., "Matériaux 2002" Conference, Tours, France, October 2002.
4. Oudot, B., Jolly, L., and Baclet, N., "Matériaux 2002" Conference, Tours, France, October 2002.
5. Ravat, B., Jolly, L., and Valot, C., "Matériaux 2002" Conference, Tours, France, October 2002.

Phase Transformations in Delta-Stabilized Plutonium

Steven Kitching, Patrick G. Planterose, and David C. Gill

AWE, Aldermaston, Reading, RG7 4PR, UK

INTRODUCTION

The high-temperature delta phase (δ) of plutonium (Pu) can be retained at room temperature by the addition of small amounts of an alloying element such as gallium (Ga).[1] Macroscopic and microscopic variations in gallium solute concentrations will, however, affect the long-term stability of the delta phase. In low-solute regions, the delta phase is metastable and will partially transform into the alpha (α) phase at ambient or low-temperature conditions. The amount of transformation as a function of time and temperature can be represented on a time-temperature transformation (TTT) curve; see Figures 1 and 2.

Phase transformations in Pu-Ga alloys at low temperature are, however, a complex phenomenon. Alpha phase plutonium can be precipitated from low gallium δ phase alloys by cooling below room temperature. This transformation is thought to proceed by a martensitic mechanism. Under these conditions, the gallium is unable to diffuse during the shear transformation and is trapped in the alpha phase. The α phase formed in this way is expanded as it contains gallium normally insoluble in the α phase. This expanded α phase is usually designated as the alpha-prime (α′) phase[3]. The double c-curve nature of the TTT curves (Figure 1), however, indicates that the $\delta \rightarrow \alpha'$ transformation may occur by a martensitic transformation or by a massive transformation, depending on the cooling regime employed.[2,4]

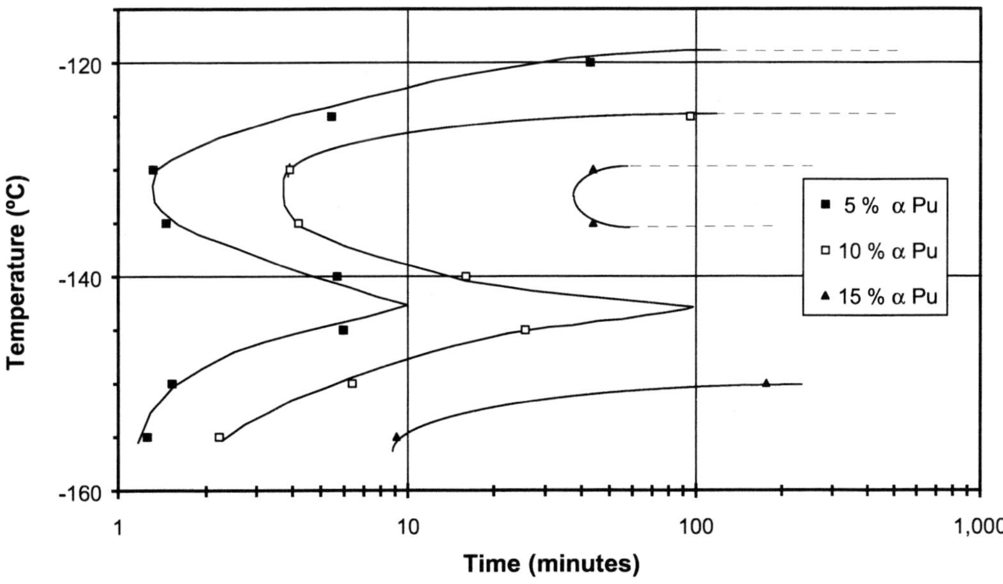

FIGURE 1. Time-temperature-transformation (TTT) curve for a 0.56 wt % plutonium-gallium alloy. Examination of time-temperature-transformation (TTT) curves for a 0.56 wt % plutonium gallium alloy demonstrates the metastability of the alloy. For example, at −125°C the plutonium will transform to contain 10% of the alpha phase in approximately 100 minutes.[2]

CP673, *Plutonium Futures — The Science,* edited by G. D. Jarvinen
2003 American Institute of Physics 0-7354-0140-3

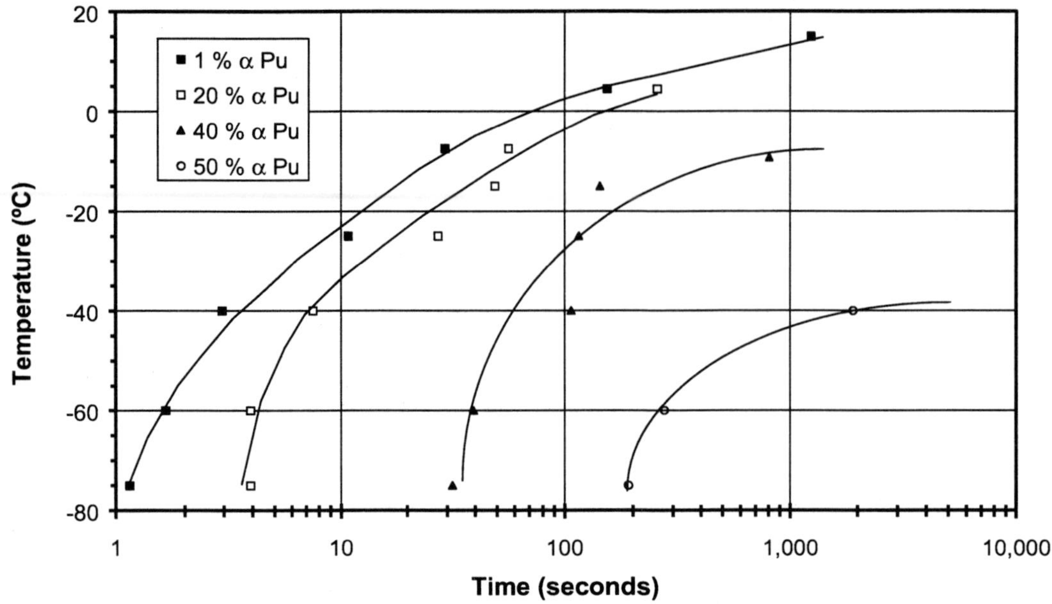

FIGURE 2. Time-temperature-transformation (TTT) curve for a 0.21 wt % plutonium-gallium alloy. A significant proportion of the delta phase can transform to the alpha phase at ambient temperature.[2]

This paper details a study of phase transformations in plutonium-gallium alloys in which samples were isothermally cold-treated in dry ice (–78ºC) for various times to induce alpha-prime formation. Differential scanning calorimetry (DSC) was used to identify and quantify the phases present on subsequent reheating of the samples.

RESULTS

A typical DSC thermogram for a 0.56 wt % Pu-Ga alloy following cold treatment in dry ice is shown in Figure 3. This clearly shows the complex nature of the post-cold-treatment phase transformation. Three peaks are resolved, the first is a 'crenellated peak' showing the α' to δ transformation that has proceeded in a 'burst'-like manner. The second peak is similar to the α to β transformation in unalloyed plutonium. The third peak is resolved as a higher temperature shoulder on the α to β peak, its cause has yet to be determined. The gallium microsegregation in the alloys was determined by electron probe microanalysis (EPMA). The phase transformations can be characterized according to the alloy content. Regions of the microstructure with a gallium content of less than 0.2 wt % are considered to be 'compositionally' unstable and are present as the alpha phase. This material transforms to delta through the beta and gamma phases on heating. Regions of the microstructure with an alloy content of less than 0.4 wt % are considered 'conditionally' stable and transform on cold treatment to an alpha-prime phase. The alpha-prime phase transforms directly to the delta phase on heating.

DISCUSSION

The results reinforce the complex metallurgical nature of plutonium alloy transformations. The nucleation and growth mechanisms of the α and α' structures are highly sensitive to the localized microstructure condition, not simply the material bulk. As well as temperature stability, the microstructure is also known to be sensitive to a wide range of impurity additions, local material defects, and mechanical work. A fundamental understanding of phase transformation mechanisms and kinetics is vital to the production of stable homogenous material, especially at low alloy concentrations.

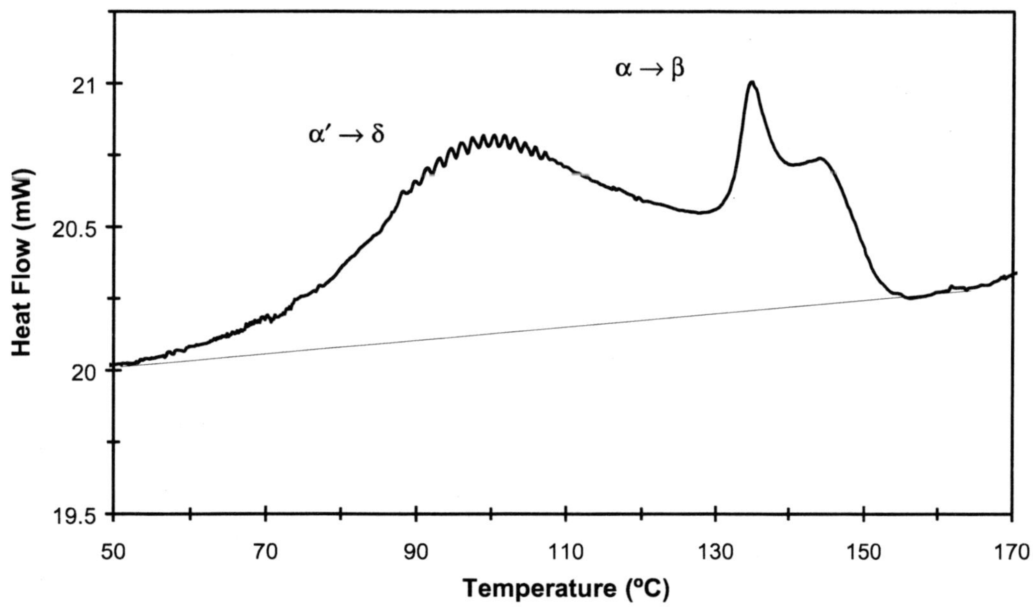

FIGURE 3. A differential scanning calorimetry (DSC) thermogram for 0.56 wt % plutonium gallium alloy. Transformations occurring on heating following a 456-hour (19-day) hold at −78°C. A proportion of the material transforms directly to the delta phase, the remainder transforms through the beta and gamma phases.

REFERENCES

1. Ellinger, F. H., Land, C. C., and Struebing, V. O., "Plutonium-Gallium Phase Diagram," J. Nucl. Mater. 12, 228 (1964).
2. Orme, J. T., Faiers, M. E., and Ward, B. J., "The Kinetics of the Delta to Alpha Transformation in Plutonium Rich Pu-Ga Alloys," in Plutonium 1975 and Other Actinides, North Holland Publishing, (1976), p. 761.
3. Hecker, S. S., Zukas, E. G., Morgan, J. R., and Pereyra, R. A., in Proc. Int. Conf. Solid-State Phase Transfer, edited by H. I. Aaronson., American Institute of Mining Engineers, New York, (1982), p. 1339.
4. Deloffre, P., Truffier, J. L., and Falanga, A., "Phase Transformation in Pu-Ga Alloys at Low Temperature and Under Pressure: Limit Stability of the δ Phase," Journal of Alloys and Compounds 271–273, 370–373 (1998).

Invar Effect, Thermal Expansion, and Elastic Softening of δ-Phase Pu

A. C. Lawson, J. A. Roberts, and B. Martinez

Los Alamos National Laboratory, Los Alamos, NM 87545, USA

INTRODUCTION

Plutonium metal is famous for its variety of crystal structures; of these, the close-packed cubic δ phase is least understood because its surprisingly large atomic volume has resisted first-principles calculation.[1] Unusual related properties—including the negative thermal expansion of the unalloyed δ phase,[2] anomalous thermal expansion behavior of the alloys,[3,4] pronounced elastic softening with increasing temperature,[5–8] and the low melting point—have remained unexplained for more than 40 years.

There has been some progress. Recently we explained[9] the anomalous thermal expansion as the consequence of the invar effect.[10,11] Figure 1 shows the energy level diagram for the invar effect.

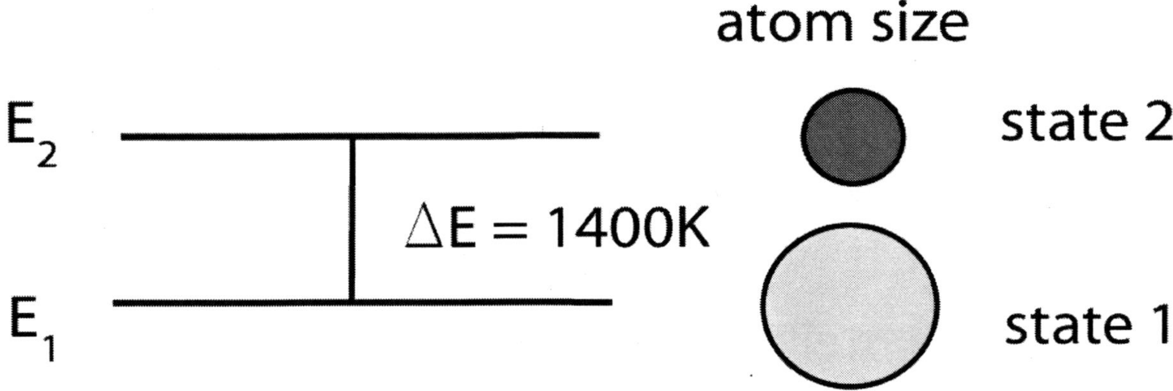

FIGURE 1. Energy level diagram for the invar effect. Two stable atomic states, 1 and 2, are distinguished by different atomic sizes and separated by energy ΔE. The thermal occupancy of state 2 is n_s, given in the text. For δ-Pu, ΔE is 1,400 K, and $R_2 < R_1$, so that the thermal expansion is negative.

Two stable atomic states, 1 and 2, are distinguished by different atomic sizes and separated by energy ΔE. The thermal occupancy of state 2 is $n_s = [1+\exp(\Delta E/kT)]^{-1}$. For δ-Pu, ΔE is 1,400 K, and $R_2 < R_1$, so that the thermal expansion is negative. In addition to the usual Grüneisen term, the thermal expansion includes the invar occupancy and the lattice constants associated with the two states; the latter are concentration dependent for the alloys, in accordance with Vegard's law. (Please consult Reference 9 for details.) Figure 2 shows the thermal expansion of Pu-Ga alloys determined by neutron diffraction. The line through the points is derived from a fit to the invar model that determines ΔE.

The low melting point can be explained by the observed elastic softening in conjunction with Lindemann rule.[7] The question of whether the invar effect can explain the elastic softening is discussed in the next section.

CP673, *Plutonium Futures — The Science,* edited by G. D. Jarvinen

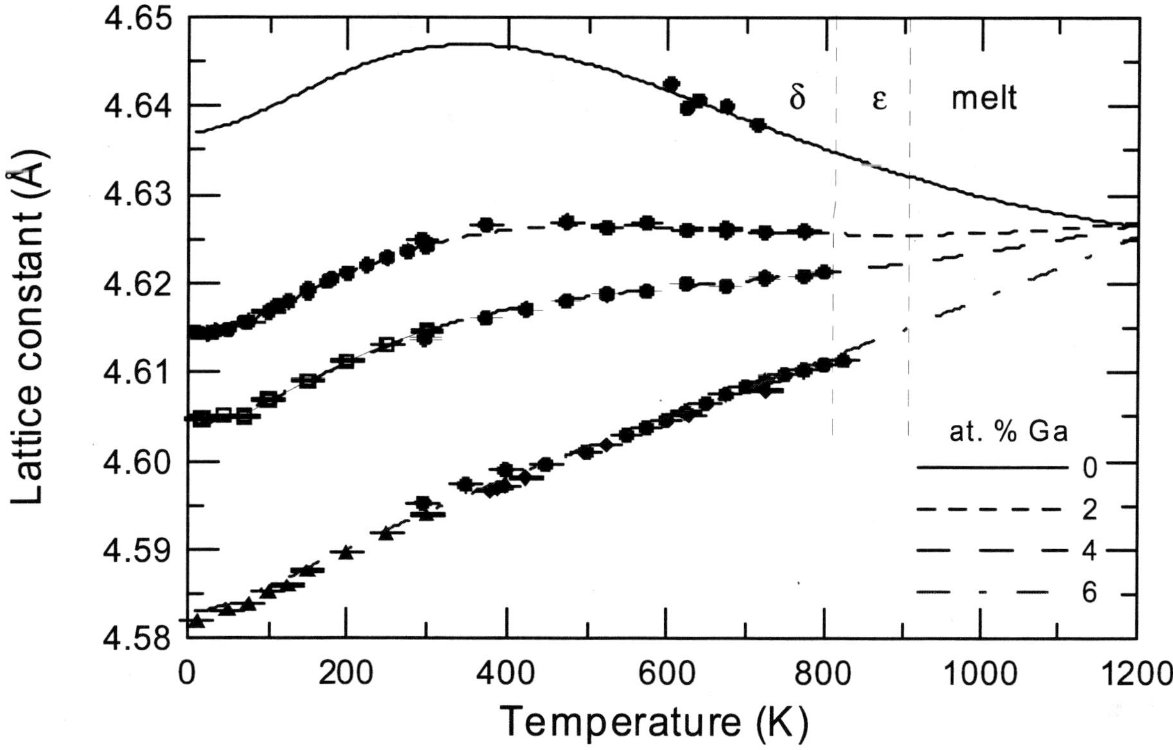

FIGURE 2. Thermal expansion of Pu-Ga alloys determined by neutron diffraction. The line through the points is derived from a fit to the invar model shown in Figure 1 that determines ΔE.

RESULTS

Figure 3 shows the temperature dependence of the Debye temperature of δ-phase Pu alloys. Two experimental curves are shown: the ultrasonic results from 1965 and the more recent neutron diffraction results. The decrease of Θ with temperature cannot be explained in terms of the usual Grüneisen effect because the lattice Grüneisen constant is quite small for Pu. But we can again use the invar occupancy to calculate $\Theta = (1 - n_S)\Theta_1 + n_S\Theta_2$, where Θ_1 and Θ_2 are the Debye temperatures of States 1 and 2, respectively. The third curve shows Θ calculated from the invar model with ΔE = 1,400 K, Θ_1 = 125 K and Θ_2 = 0, as required by comparison with the data. This means that the invar model also explains the elastic softening, but with the surprising conclusion that Θ_2 = 0. We also know from experiment[7] that Θ_1 is independent of concentration.

For the alloys, the thermal expansion is strongly dependent on concentration because the effective sizes of the atoms are as follows: Θ does not depend on concentration, because the Θ_1 associated with the lower atomic states does not. Our finding that Θ_2 is zero is surprising, but suggests an interesting (if speculative) interpretation: if we associate the higher-energy state with the liquid phase, rather than the ε phase, as was done earlier, then Θ should be taken as the shear modulus rather than the bulk modulus.

DISCUSSION

The overall conclusion is that the simple two-state invar model provides a phenomenological model of the anomalous properties of δ-phase Pu, which have remained unexplained for many years. The model enables interpolation or even extrapolation of δ-phase Pu properties through unmeasured regimes. It is desirable to obtain a first-principles explanation for the success of the invar model. In particular, we need to understand the atomic origin of the two stable atomic states—close in energy but very different in volume—that the model requires. Some efforts in this direction have been made by Cooper et al.[12] and Kotliar.[13]

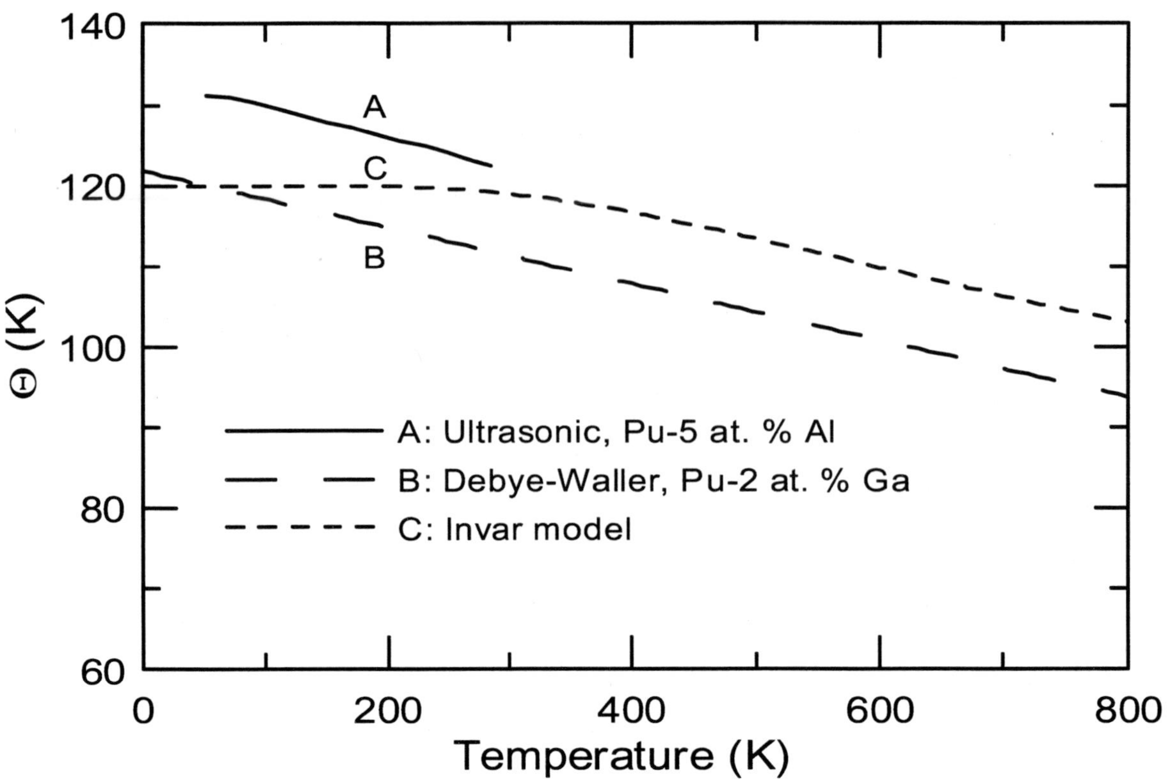

FIGURE 3. Temperature dependence of the Debye temperature of δ-phase Pu alloys. A: ultrasonic measurements.[5] B: neutron diffraction measurements.[7] C: calculated from the invar model.

REFERENCES

1. Eriksson, O., Becker, J. D., Balatsky, A. V., and Wills, J. W., J. Alloys and Compounds 287, 1 (1999).
2. Ellinger, F., J. Metals 8, 1256 (1956).
3. Elliott, R. O., Gschneidner, K. A. Jr., and Kempter, C. P., in Plutonium 1960, edited by E. Grison, W. B. H. Lord, and R. D. Fowler, Cleaver-Hume Press (1960), pp. 142–155.
4. Lee, J. A., Hall, R. O. A., King, E., and Meaden, G. T., in Plutonium 1960, edited by E. Grison, W. B. H. Lord, and R. D. Fowler, Cleaver-Hume Press (1960), pp. 39–50.
5. Taylor, J. C., Loasby, R. G., Dean, J. D., and Linford, P. F., in Plutonium 1965, edited by A. E. Kay and M. B. Waldron, Chapman and Hall (1965), pp. 162–175.
6. Lawson, A. C., Goldstone, J. A., Cort, B., Sheldon, R. I., and Foltyn, E. M., J. Alloys Comp. 213/214, 426–428 (1994).
7. Lawson, A. C., Martinez, B., Roberts, J. A., Bennett, B. I., and Richardson, J. W. Jr., Phil. Mag. B 80, 53 (2000).
8 Migliori, A., Freibert, F., Lashley, J. C., Lawson, A. C., Baiardo, J. P., and Miller, D. A., J. Superconductivity (2002), in press.
9. Lawson, A. C., Roberts, J. A., and Martinez, B., Phil. Mag. B 82, 1837 (2002).
10. Weiss, R. J., Proc. Phys. Soc. 62, 281 (1963).
11. Pepperhoff, W., and Acet, M., Constitution and Magnetism of Iron and its Alloys, Springer Verlag (2001).
12. Cooper, B. R., Vogt, O., Sheng, Q.-G., and Lin, Y. L., Phil Mag. B 79, 683 (1999).
13. Kotliar, G., J. Low-Temperature Phys. 126, 1009 (2002).

Investigations of Protactinium Metal under Pressure Provide Important Insights into Actinide Bonding Concepts

R. G. Haire,[1] S. Heathman,[2] M. Idiri,[2] T. Le Bihan,[3] A. Lindbaum,[4] and J. Rebizant[2]

[1]*Oak Ridge National Laboratory, CSD, PO Box 2008, MS-6375, Oak Ridge, TN 37831, USA*
[2]*European Commission, JRC, Institute for Transuranium Elements, Postfach 2340D-76125, Karlsruhe, Germany*
[3]*European Synchrotron Radiation Facility, B.P. 220, F-38043 Grenoble Cedex, France*
[4]*Vienna University of Technology, Institute for Solid State Physics, Wiedner Hauptstrasse 8-10/138, Wien, Austria*

INTRODUCTION

There have been several studies aimed at determining the effect of pressure on f-elements and their compounds. An important issue in these investigations is how the deceasing interatomic distances brought about by applying pressure affect the electronic nature of the materials. An especially important issue is whether pressure can force the involvement of f-electrons in bonding.

In the f-series of elements, it has been established that the metals having localized f-electrons exhibit greater atomic volumes and have more symmetrical structures than those whose f-electrons are involved in their bonding. Reviews are available covering the occurrence of f-electron delocalization in lanthanide and actinide metals that accompany the decreased interatomic distances forced by pressure.[1–3] Interestingly, after acquiring 4f-electron character in their bonding from applying pressure, lanthanide metals adopt some of the same, low-symmetry structures exhibited normally by the protactinium through plutonium elements at atmospheric pressure.

Studies of protactinium in general have been sparse, caused in part by its limited availability and radioactivity, but some of its known properties have been reviewed.[4] Its superconducting properties, high enthalpy of vaporization, and atomic volume have in part established that its bonding has 5f-electrons character at atmospheric pressure (i.e., itinerant 5f-electrons).

Protactinium, the first of the four actinide metals having itinerant 5f-electrons at atmospheric pressure, occupies an important position in the series. It is this influx of f-electron character into their bonding that generates rare polymorphism, low-symmetry crystal structures, and a smaller compressibility—facets not displayed by other metals in the Periodic Table. This itinerancy of 5f-electrons at atmospheric pressure disappears with the transplutonium elements. Our results from recent pressure studies of americium have provided important new insights into its bonding behavior under pressure,[5–7] and have altered previous conclusions reached from both experiments and theory. Structural links were established between americium and other actinides: first, between the Am-III phase and γ-Pu and then between the Am-IV phase and α-U. In the present work on protactinium, structural links were established between the high-pressure phase of protactinium and with both α-U and the high-pressure, Am-IV phase.

The experimental work reported here examined the structural behavior of protactinium to 130 GPa, using diamond anvil pressure cells (DACs). It is the first time that this metal been studied under pressure using synchrotron radiation. In addition to determining comprehensively the behavior of protactinium under pressure, we have also derived a new experimental bulk modulus and pressure derivative for it.

RESULTS

We have determined in experimental studies that the tetragonal structure of protactinium (space group, *I4/mmm*) converts to an orthorhombic, alpha-uranium structure (space group, *Cmcm*) at 77(5) GPa, where its atomic volume

CP673, *Plutonium Futures — The Science,* edited by G. D. Jarvinen

has been reduced by ~30%. This structural change (from Pa-I to Pa-II) is accompanied by a small (0.8%) decrease in volume. These changes have been interpreted as reflecting an increase in a 5f-electron contribution in protactinium's bonding over that initially present, becoming more similar to that in alpha-uranium metal at atmospheric pressure.

From the compression behavior of the Pa-I phase, the isothermal bulk modulus and its pressure derivative were obtained by fitting the experimental data to the Birch-Murnaghan equation of state. The values obtained in this work from protactinium's compression behavior were 118 (1) GPa for the modulus (B_o), and 3.4 (0.2) for the derivative (B_o'). In an earlier study of protactinium up to 53 GPa,[8] the bulk modulus and first derivative were reported as being 157(5) GPa and 1.5(0.5), respectively. Theoretical calculations have suggested a modulus of ~100 GPa.[9–11] The modulus for protactinium is considerably larger than moduli for the transplutonium metals (<50 GPa), which have localized 5f electrons.

Using a calculational approach involving Gibbs free energies, Söderlind and Eriksson[9] predicted a series of phase transitions for protactinium under pressure. The first was the transformation of protactinium's body-centered tetragonal structure to an alpha-uranium-type structure. We confirmed this transformation, but found it occurred at smaller atomic volumes and at a significantly higher pressure than predicted

We propose that pressure on protactinium forced a small increase in the 5f-contribution to its band structure, but probably less than that which exists in alpha uranium. Overall, the compression of protactinium was greater than that of uranium; at 100 GPa the relative atomic volume of protactinium was 0.64, but that of uranium was 0.72.[10]

We believe our experimental results will allow an improved understanding not only of the bonding and electronic nature of protactinium, but also for generating important systematics concerning the actinides. These new experimental data should also allow a platform from which to evaluate and fine tune theoretical predictions and/or our understanding of these actinide metals.

ACKNOWLEDGMENTS

This work was supported in part by the European Commission and by the Division of Chemical Sciences, Geoscience and Bioscience, OBES, USDOE, under contract DE-ACOR-00OR22725 with Oak Ridge National Laboratory, managed by UT-Battelle, LLC. The authors wish to acknowledge beam time provided at the ESRF in Grenoble, France, where the diffraction studies were performed, and the efforts of personnel at the ESRF. One author, A. L., wishes to thank the Austrian Academy of Sciences (APART 10739) and the Austrian Science Fund (P14932) for financial support. M. Idiri wishes to thank the Training and Mobility Program of the European Union. Special thanks are also given to G. H. Lander for useful discussions and a strong interest in these efforts

REFERENCES

1. Benedict, U., "The Effect of High Pressure on Actinide Metals," in Handbook on the Physics and Chemistry of the Actinides, edited by A. J. Freeman and G. H. Lander, Elsevier Science, Netherlands (1987), Ch. 3, pp. 227–269.
2. Benedict, U., and Holzapfel, W. B., "High-Pressure Studies—Structural Aspects," in the Handbook on the Physics and Chemistry of the Rare Earths, edited by K. A. Gschneidner, Jr., L. Eyring, G. H. Lander and G. R. Choppin, Elsevier Science, Netherlands (1993), Vol. 17, Ch. 113, pp. 245–300, and references therein.
3. Haire, R. G., "High-Pressure Behavior of f-Element Materials, Rare Earths, Resources, Science Technology and Applications," edited by R. Bautista and N. Jackson, TMS, Pennsylvania (1991), pp. 449–462, and references therein.
4. Blank, H., J. Alloys and Compounds 343, 108 (2002).
5. Heathman, S., Haire, R. G., Le Bihan, T., Lindbaum, A., Litfin, K., Méresse, Y., and Libotte, H., Phys. Rev. Lett. 85, 2961 (2000).
6. Lindbaum, A., Heathman, S., Litfin, K., Méresse, Y., Haire, R. G., Le Bihan, T., and Libotte, H., Phys. Rev. B 63, 214101 (2001).
7. Haire, R. G., Heathman, S., Le Bihan, T., and Lindbaum, A., NEA Publication, ISBN 92-64-18485-6 (2002), p.147.
8. Benedict, U., Spirlet, J. C., Dufour, C., Birkel, I., Holzapfel, W. B., and Peterson, J. R., J. Magn. and Magn. Mater. 29, 287 (1982).
9. Söderlind, P., and Eriksson, O., Phys. Rev. B 56, 10719 (1997).
10. Skriver, H. L., Andersen, O. K., and Johansson, B., Phys. Rev. Lett. 41, 42 (1978).
11. Johansson, B., and Skriver, H. L., J. Magn. and Magn. Mater. 29, 217 (1982).
12. Le Bihan, T., Heathman, S., Idiri, M., Lander, G. H., Wills, J., Lawson, A., and Lindbaum, A., Phys. Rev. B (2003), and references therein (in press).

Effects of Local Solute Ordering and Plasticity on the Delta to Alpha Transformation in Gallium-Stabilized Plutonium Alloys

C. R. Krenn, B. Sadigh, A. J. Schwartz, and W. G. Wolfer

Chemistry and Materials Science Directorate,
Lawrence Livermore National Laboratory,
University of California, Livermore, CA 94551

Abstract. The delta-to-alpha transformation in Ga-stabilized plutonium alloys has many unusual features. It is a thermally activated martensitic transformation with double-c transformation kinetics. The alpha product has an approximately 20% smaller volume than delta, but its volume can change an additional 2% depending on the transformation path. The thermally induced transformation also rarely proceeds to more than 40% completion. We use *ab-initio* techniques to look at local ordering of gallium in the alpha phase and use continuum elastoplastic energy calculations to understand the possible effects of plastic deformations during the transformations. We discuss how local ordering and plastic deformation may explain some of the unusual features of the transformation.

FIRST PRINCIPLES CALCULATIONS IN PU-GA ORDERED ALLOYS

To learn more about the structure of α′, Sadigh et al.[1] examined the eight possible substitutional Pu-Ga ordered structures in a sixteen-atom unit cell using first-principles calculations within the generalized gradient approximation and using the plane-wave pseudopotential code VASP.[2] Significantly, it was found that the relaxed cell size and total energy of the eight compounds varied a great deal. Using Zachariasen and Ellinger's nomenclature,[3] sites 6 and 8 were the lowest energy sites. Site 8 also had a significantly lower relaxed volume, which was the same as the volume of pure α. The reason that the site-8-ordered structure has such a small relaxed volume can be qualitatively understood by a crystallographic analysis of the pure α structure. The Voronoi cell associated with site 8 is significantly larger than any other site. Since gallium is oversized in the α lattice, it will cause less distortion and less volume increase in a larger site.

LATTICE CONSTANT OF α′

Because of δ's simple structure, Ga will be nearly randomly distributed in the δ lattice. When δ is rapidly quenched to form α, Ga will also be randomly distributed in the α lattice. However, there will be a large driving force for Ga trapped in energetically unfavorable sites to move to more favorable sites. This migration will occur if there is enough time and enough thermal energy for short-range diffusion to occur.

Experiments suggest that this diffusion may occur because they show that α can have a variety of densities.[4,5] As might be expected, the lower-c curve has a larger volume than the upper because ordering is more difficult at the lower temperature. The fact that two-phase material tempered at elevated temperatures for extended times has a yet still smaller volume suggests that even in the upper c-curve, only partial ordering into the site-8 variant is possible.

CP673, *Plutonium Futures — The Science,* edited by G. D. Jarvinen

SOLUTE ORDERING AND THE DOUBLE C-CURVE

Local ordering may also contribute to the appearance of the double C-curve. The difference in energy between the disordered and ordered α′ implies a separation between the two c-curves of ~90°C, which is larger that what is experimentally observed. However, Deloffre's x-ray data[4] suggest that α formed in the upper c-curve may be incompletely ordered, and this would reduce the separation. In addition, an inhomogeneous distribution of Ga in the microstructure can also contribute to the appearance of a double C-curve.[6]

PLASTICITY AND THERMAL HYSTERESIS

Without plasticity, the large volume change between δ and α (~20%) would result in elastic shear stresses exceeding several GPa. A simple analysis of an isolated spherical α particle in a δ matrix results in an elastic misfit energy that approaches 6 kJ/mole.[7] Using Adler's free-energy functions, this corresponds to a required undercooling of > 600K.[8] However, when plastic flow is allowed (assuming a 80 MPa yield strength for δ),[9] the work of transformation drops to a more reasonable value of 0.7 kJ/mole.

We have used similar estimates of the plastic work of transformation to predict the thermal hysteresis of the δ to α transformation. The upper and lower curves of the metastable phase diagram shown in Figure 1 are theoretical predictions based on a plastic dissipation model. The agreement between these curves and the experimental data points is good. We will also describe the development of models to examine the dynamic growth of martensitic plates in a plastic matrix using a 3D phase field approach.[10]

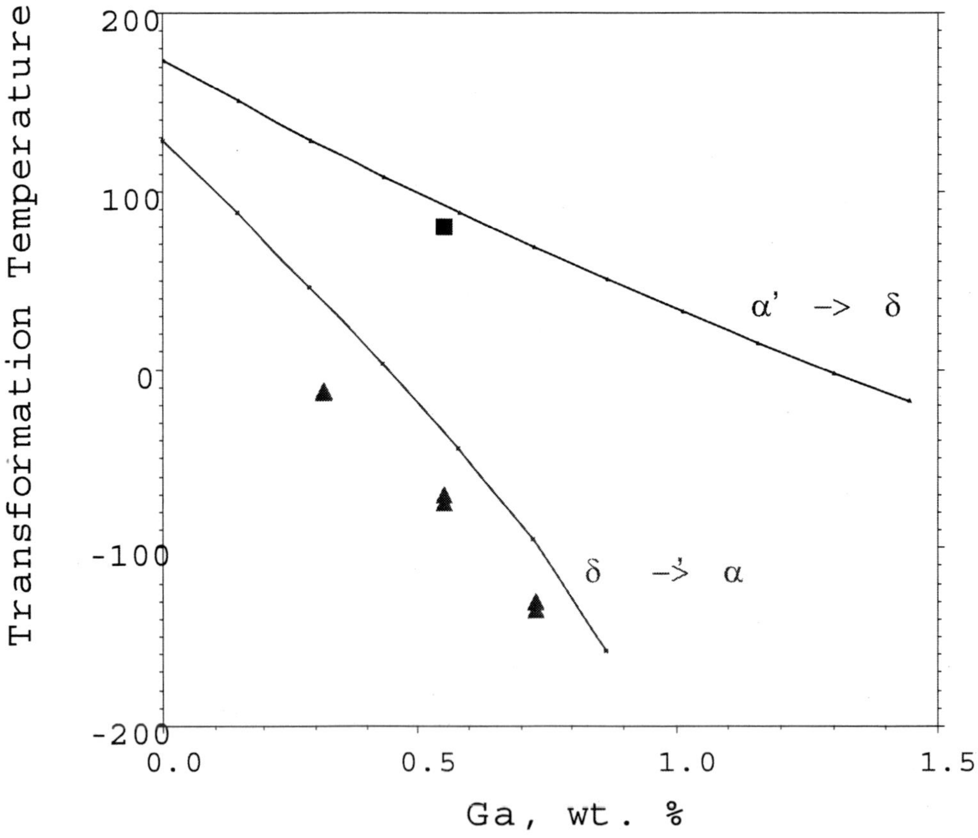

FIGURE 1. Metastable phase diagram of plutonium: The filled squares and triangles are experimental data points from Ref. 5. The solid lines are theoretical predictions based on a plastic dissipation model.

ACKNOWLEDGMENTS

This work was performed under the auspices of the U.S. Department of Energy by the University of California, Lawrence Livermore National Laboratory under Contract No. W-7405-ENG-48.

REFERENCES

1. Sadigh, B., Krenn, C. R., and Wolfer, W. G., "The Martensitic Alpha Prime Phase in the Plutonium-Gallium System," in preparation.
2. Kresse, G, and Hafner, J., "Ab initio Molecular-Dynamics Simulation of the Liquid-Metal–Amorphous-Semiconductor Transition in Germanium," Phys. Rev. B 49, 14251 (1994).
3. Zachariasen, W. H., and Ellinger, F. H., "The Crystal Structure of Alpha Plutonium Metal," Acta Cryst. 16, 777 (1963).
4. Deloffre, P., "Study of the Stability of the Delta Phase of Delta-Stabilized Plutonium Alloys," Ph.D., University of Paris, Orsay XI (1997).
5. Hecker, S. S., "Phase Stability and Phase Transformations in Pu-Ga Alloys," Progress in Mat. Sci., T B Massalski Festschrift Volume, in press (2002).
6. Hecker, S. S., private communication, Los Alamos National Laboratory (2003).
7. Lee, J. K., Earmme, Y. Y., Aaronson, H. I., and Russel, K. C., "Plastic Relaxation of the Transformation Strain Energy of a Misfitting Spherical Precipitate: Ideal Plastic Behavior," Met. Trans. A 11, 1837 (1980).
8. Adler, P. H., "Thermodynamic Equilibrium in the Low-Solute Regions of Pu-Group IIIA Metal Binary Systems," Met. Trans. A 22, 2237 (1991).
9. Gardner, H. R., "Mechanical properties," in Plutonium Handbook: A Guide to the Technology, edited by O. J. Wick, New York: Gordon and Breach (1967), pp. 59.
10. Jin, W. M., Artemev, A., and Khachaturyan, A. G., "Three-Dimensional Phase Field Model of Low-Symmetry Martensitic Transformation in Polycrystal: Simulation of ζ'_2 Martensite in AuCd Alloys," Acta Mat. 49, 2309 (2001).

SESSION SIX:
ACTINIDES IN THE
ENVIRONMENT
AND LIFE SCIENCES

Aquatic Colloids of Actinides: How Are They Generated under Natural Aquifer Conditions?

M. A. Kim,* P. J. Panak, J. I. Yun, and J. I. Kim

Institut für Nukleare Entsorgung, Forschungszentrum Karlsruhe, 76021 Karlsruhe, Germany
**Institut für Radiochemie, Technische Universität München, 85748 Garching, Germany*

INTRODUCTION

Aquatic colloids, ubiquitous in natural water, are composed of inorganic oxides, organic substances (e.g., humics), or a mixture of both.[1] An average size of the predominant number density is, in general, smaller than 50 nm and their population ranges from 10^8 to 10^{14} particles per liter of water. Actinide ions of higher charges ($z \geq 3+$) are readily absorbed into aquatic colloids and thus colloid-borne actinide can be generated. As a result, aquatic colloids may play a carrier role for the actinide migration in natural aquifer systems without substantial geochemical hindrance.[2] However, the provenance of aquatic colloids cannot be explained explicitly, and hence the generation of colloid-borne actinides (so called peudo-colloids of actinides) is not well understood.

The present paper is intended to elucidate some fundamental aspects for the provenance of aquatic colloids, in which actinides can be incorporated to become colloid-borne species. Inorganic aquatic colloids are composed primarily of hydrated aluminosilicates as primary kernels in natural aquifer systems,[3] into which water-borne trace elements are incorporated. Such heterogeneous nucleation processes are investigated in the presence of trivalent actinides (Am and Cm) in this work.

RESULTS AND DISCUSSION

Heterogeneous nucleation of Al and Si in the presence of Am or Cm results in a structural incorporation of actinides in aluminosilicates, which appear to be the kernels of most aquatic colloids. An acidic Al solution of varying concentration (10^{-3}–10^{-7} mol/L) is titrated with an alkaline Si solution of 10^{-3} mol/L to the pre-adjusted pH within 4–9, in the region where the solubility of different aluminosilicates changes drastically with a minimum at pH 6. In this process, a formation of three phases is observed: ionic phase, colloid, and precipitate. The colloid generation is ascertained by ultrafiltration with the aid of radiometry, AFM and LIBD (laser-induced breakdown detection). The composition analysis by XPS and SEM-EDXS results in a Si:Al ratio varying from 0.5 to 1.0, depending on the initial Al concentration and pH. The particle size observed by AFM and LIBD ranges from 10 nm to 50 nm, and the number density varies from 10^{11} to 10^{14} particles per liter of water, which corresponds to a colloid mass concentration of (1–5) x 10^{-5} g/l or 10–50 ppb.

Two different colloid-borne Cm species, as incorporated in hydrated aluminosilicate (HAS) colloids, are characterized by TRLFS (time-resolved laser fluorescence spectroscopy) as shown in Figure 1. The first species (Cm-HAS(I)) appears in a small amount in the whole pH range, with a maximum of 20% fraction at pH 6, and the second species (Cm-HAS(II)) is beginning to form at pH $\geq$ 5, then becomes predominant at pH $\geq$ 7. The formation of two colloid-borne Cm species is shown in Figure 2 as a function of pH. A number of hydration water molecules bound to colloid-borne Cm, as derived from the fluorescence lifetime of each Cm-HAS species, is found to be 7 for Cm-HAS(I) and 6 for Cm-HAS(II), which suggest bidentate and tridentate binding of Cm in aluminosilicate structure, respectively.

CP673, *Plutonium Futures — The Science,* edited by G. D. Jarvinen

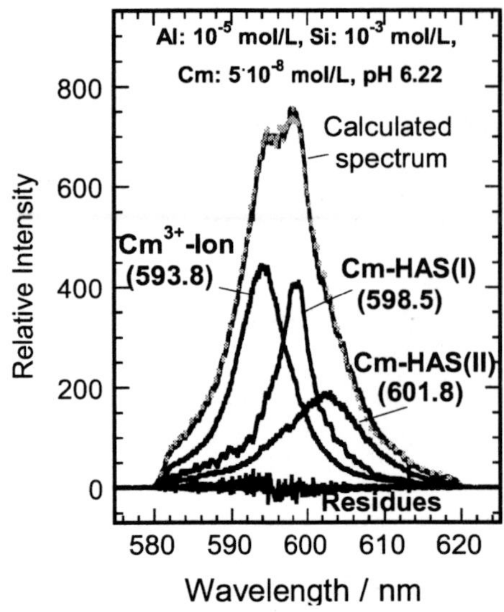

FIGURE 1. Speciation of colloid-borne Cm.

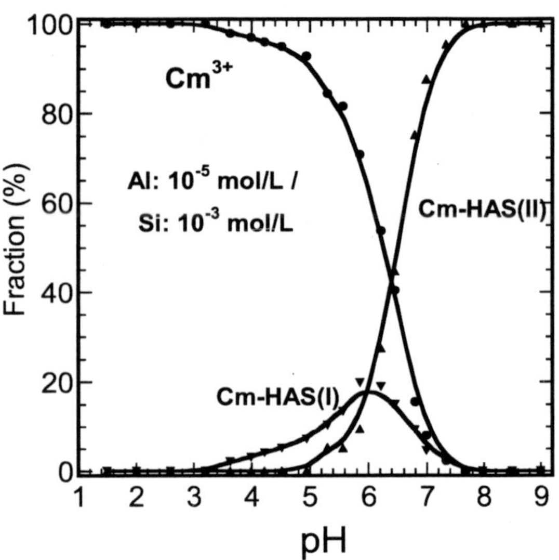

FIGURE 2. Species distribution: formation of colloid-borne Cm.

Either aging or tempering of Cm-HAS colloids leads to a nearly complete dehydration of colloid-borne Cm. This fact implies that Cm takes part as a contestant of Al(III) in the nucleation process of aluminosilicate colloids and becomes incorporated into their molecular structure. A desorption experiment indicates that Cm bound in HAS(I) and HAS(II) remains non-dissociable at pH 7 and 9, corroborating the TRLFS results as stable incorporation of Cm in HAS colloids. The present experimental results are applicable to the appraisal of how colloid-borne actinides (III) can be generated, and is helpful in better understanding the actinide migration behavior that may be primarily facilitated as colloid-borne species in natural aquifer systems.

REFERENCES

1. Kim, J. I., MRS Bull. 19, 47 (1994) and Radiochimica Acta 52, 49 (1991).
2. Artinger, R., Schuessler, W., Scherbaum, F., Schild, D., and Kim, J. I., Environmental Sciences and Technology 36, 5092 (2002).
3. Kim, M. A., Panak, P. J., Yun, J. I., Kim, J. I., Klenze, R., and Koehler, K., Colloids and Surfaces A (2002), in press.

Phytosiderophore Effects on Subsurface Actinide Contaminants: Potential for Phytostabilization and Phytoextraction

Christy Ruggiero, Scott Twary, and Elise Deladurantaye

Los Alamos National Laboratory, Chemistry and Bioscience Divisions

Phytosiderophores are produced by most grasses and many crop plants for the solubilization and uptake of Fe from soil. These small natural organic chelators, all derivatives of mugineic acid, are designed to strongly bind low-solubility Fe, and possibly other metals, in a complex and competitive soil environment and actively deliver the metal to the plant (Figure 1). Phytosiderophores form stable complexes with numerous metals.[1] Phytosiderophores have been shown to enhance soil mobility of Cu, Fe, Mn, and Zn as much as microbial siderophores and more than some synthetic chelators.[2] They are notably similar to EDTA in structure and metal-binding chemistry: Both have multi-carboxlyate chelating groups and form extremely strong complexes with Fe, and also strongly bind a number of transition metal ions (Table 1).

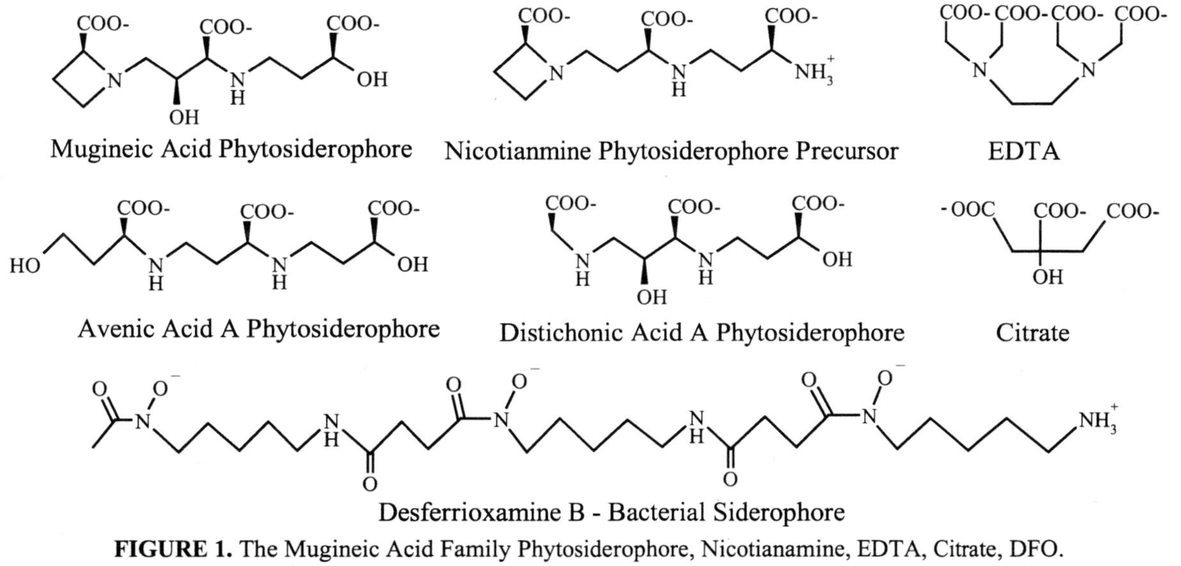

Mugineic Acid Phytosiderophore   Nicotianmine Phytosiderophore Precursor   EDTA

Avenic Acid A Phytosiderophore   Distichonic Acid A Phytosiderophore   Citrate

Desferrioxamine B - Bacterial Siderophore

FIGURE 1. The Mugineic Acid Family Phytosiderophore, Nicotianamine, EDTA, Citrate, DFO.

TABLE 1. Stability Constants (log *K*) of Phytosiderophores and Other Chelators with Various Metal Ions.

Chelator	Mn(II)	Fe(II)	Co(II)	Zn(II)	Ni(II)	Cu(II)	Fe(III)	Pu(IV)	U(VI)
Nicotianamine	8.8	~12.4	14.8	~15	16.1	18.6			
Mugineic acid	8.0	8.3		10.7	14.9	~18.4	18.1		
Desferrioxamine B		7.2	10.3	10.1	10.9	14.1	30.6	30.8	18
Citrate	3.8	4.6	5.0	5.0	5.4	5.9	11.4	~12	7.4
EDTA	13.8	14.3	16.3	18.6	16.5	18.8	25.1	25.6	7.4

CP673, *Plutonium Futures — The Science,* edited by G. D. Jarvinen

We have begun an investigation of the potential for phytoremediation (phytostabilization and/or phytoextraction) of actinide soil contaminants at DOE sites using phytosiderophore producing plants and are investigating the contribution of phytosiderophores to Pu mobility in the subsurface environment. No phyto-based technologies have been developed for stabilization or removal of Pu from soils and groundwater.

Phytosiderophore-producing plants have already been shown to accumulate U and translocate it into aboveground parts.[3] These plants may have higher Pu accumulation than other plant species.[4] Phytosiderophores and phytosiderophore-mediated uptake could be the reason for this because phytosiderophores may increase both actinide solubilization and root uptake rates. Many studies have shown that metal uptake in plants, including U and Pu, can be increased by added synthetic chelators, such as EDTA, to soils. This is because most metals have low solubility, and therefore, low bioavailability in soils that limits their uptake. Phytosiderophores will predictably strongly chelate Pu, U and Th and increase their solubility in soils, similar to EDTA and bacterial siderophores. Therefore, they will also improve plant uptake of these metals. However, when solubility of the metal is no longer limiting, plant accumulation rates are ultimately limited by the rate the plant can use or uptake a metal or metal-ligand complex. Because synthetic chelators are not necessarily directly taken up or used by the plant, they can cause increased migration of contaminant metals in the soil and actually decrease plant accumulation of metals. In contrast, phytosiderophore complexes can be directly utilized by plants; phytosiderophores are up to 100 times more efficient than anthropogenic or bacterial Fe chelators for Fe uptake into some plants.[5] Pu uptake and transport in plants appears to mimic nutrient transport,[6] and Pu most likely tracks "hard" metal nutrient-transport systems in plants, such as Fe uptake by phytosiderophores. Therefore, phytosiderophore-Pu complexes may be directly utilized by the plant, making phytosiderophores much better for increasing root uptake rates.

We plan to demonstrate potential benefits of phytosiderophore-producing plants for long-term actinide stabilization or removal. We may also show possible harm caused by these plants through increased presence of actinide chelators that could increase actinide migration in the subsurface environment (Figure 2). Here, we will present our initial research findings on the ability of applied phytosiderophores (synthesized) to chelate, solubilize, and mobilize plutonium from amorphous solids and soil samples, as compared to synthetic chelators (such as EDTA) and bacterial siderophores. We will also demonstrate the effect of phytosiderophores on the uptake rate of plutonium in plants, by comparing plutonium uptake amounts between a phytosiderophore producing plant (barley) grown hydroponically in Fe-replete and Fe-depleted conditions and a non-phytosiderophore producing plant used commonly in phytoremediation technologies (Indian Mustard), also grown in Fe-replete and Fe-depleted conditions.

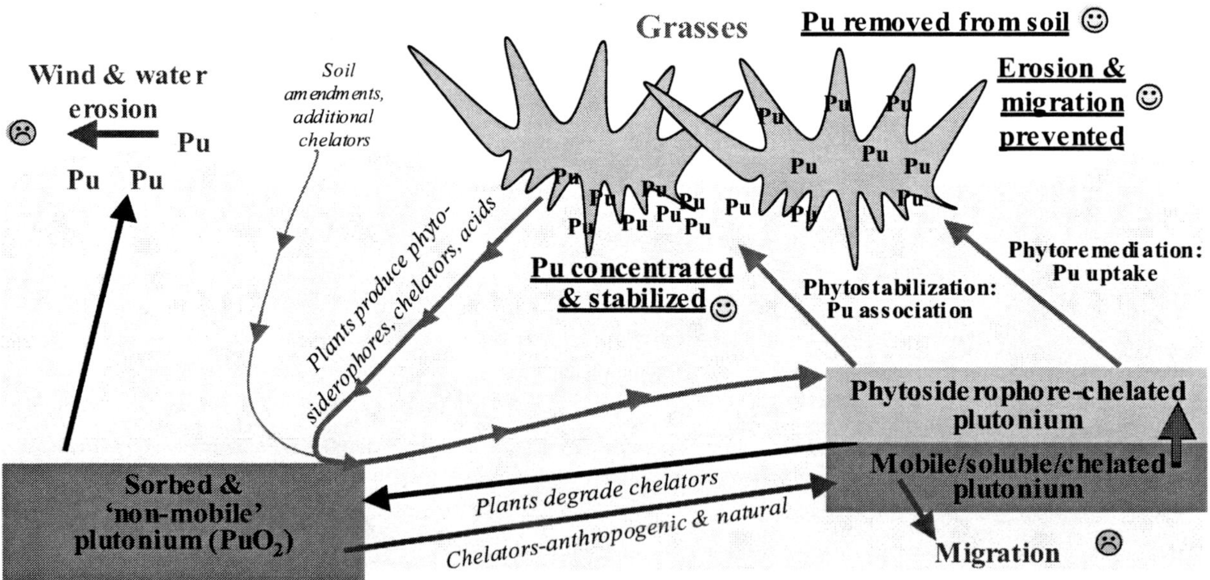

FIGURE 2. Phytosiderophore-based phytoremediation.

REFERENCES

1. (a) Matzanke, B. F., Muller, G., and Raymond, K. N., in Iron Carriers and Iron Proteins, edited by T. M. Loehr, VCH (1989), p 1; (b) Sugiura, Y., and Nomoto, K., Struct. Bonding 58, 107 (1984).
2. (a) Treeby, M., Marschner, H., and Roemheld, V., Plant Soil 114, 217 (1989); (b) Zhang, F. S., Dev. Plant Soil Sci. 54, 115 (1993); Awad, F., Roemheld, V., and Marschner, H., J. Plant Nutr. 11, 701 (1988); (c) Roemheld, V., Dev. Plant Soil Sci. 43, 159–66 (1991); (d) Stephan, U. W., Schmidke, I., Stephan, V. W., and Schloz, G., Biometals 9, 84 (1996); (e) Zhang, F. S., Plant Soil 155/156, 111 (1993).
3. Apps, M. J., Duke, M. J. M., and Stephens-Newsham, L. G., J. Radioanal. Nucl. Chem. 123, 133 (1988).
4. (a) Pimpl, M., and Schuettelkopf, H., Nuclear Safety 214, 22 (1981); (b) Vyas, B. N., and Mistry, K. B., J. Nucl. Agric. Biol. 89, 12 (1983); (c) Vyas, B. N., and Mistry, K. B., Plant Soil 345, 73 (1983); (d) Cline, J. F., and Hinds, W. T., "Uptake of Plutonium-238 by Plants Grown under Field Condition as Affected by One Year of Weathering and Aging," Issue BNWL-SA-5649 (1976).
5. (a) Ma, J. F., Kusano, G., Kimura, S., and Nomoto, K., Phytochemistry 34, 599 (1993); (b) Bar-Ness, E., Hadar, Y., Chen, Y., Roemheld, V., and Marschner, H., Plant Physiol. 100, 451 (1992); (c) Nomoto, K., Sugiura, Y., Takagi, S., Mugineic Acid Studies on Phytosiderophores, edited by G. Winkelman, D. van der Helm, and J. B. Neilands, VCH, New York, p. 400.
6. Neu, M. P., Ruggiero, C. E., and Francis, A. J., "The Biochemistry of Plutonium," book chapter in Advances in Plutonium Chemistry, 1967–2000, edited by D. Hoffman, American Nuclear Society (2002).

Biosorption of U(VI) and Pu(VI) by *Bacillus Subtilis* and a Mixture of *B. Subtilis* with Clay Mineral

T. Ohnuki, M. Samadfam, T. Yoshida, T. Ozaki, Z. Yoshida, and A. J. Francis[a]

Advanced Science Research Center, Japan Atomic Energy Research Institute, Tokai, Ibaraki 319-1195, Japan
[a]Environmental Sciences Department, Brookhaven National Laboratory, Upton, New York 11973, USA

The migration behavior of actinides in the environment as a result of microbial activity should be clarified for estimating the potential hazards of the geological disposal of high-level radioactive wastes. Actinides can exist in different oxidation states of III, IV, V, and VI in solution, and their chemical behavior depends on the oxidation states. Actinides(V) and (VI) are more mobile than actinides(III) and (IV) in ground water. The biological effect on actinide migration behavior is still being intensively investigated. We have carried out the sorption/desorption experiments of U(VI) and Pu(VI) with the microorganisms of *Bacillus subtilis* and a mixture of *B. subtilis* and clay mineral of kaolinite to clarify the role of soil microbes in actinides migration.

The sorption experiment of U and Pu by *B. subtilis* or the mixture of *B. subtilis* and kaolinite was carried out for 4×10^{-4} M U(VI) or Pu(VI) containing 0.01 M NaCl solution with its initial pH of 4.5. Concentrations of and oxidation states of U and Pu in the solutions were measured at predetermined intervals. Sorbed U and Pu with the microorganism were extracted with concentrated H_3PO_4 solution for 1 hour to determine the oxidation states. The oxidation states of U and Pu were measured by UV/VIS spectrometry. After 48 hours of contact with the U(VI) solution, the U accumulated mixture was separated from the U(VI) solution, then was washed with 0.01 M NaCl solution to provide unstained samples by drying diluted aliquots of suspension on holey carbon films (a whole mount) for TEM observation. X-ray absorption near the edge structure (XANES) analysis was carried out to determine the oxidation state of U(VI) sorbed by *B. subtilis*.

Approximately 80% and 90% of the initial concentrations of U(VI) and Pu(VI), respectively, were sorbed by *B. subtilis* from the solutions within 24 hours. The oxidation state of Pu in the solution was changed from VI to V within 2 hours. Approximately 80% of the sorbed Pu was Pu(IV) at 48 hours after the contact. On the contrary, the oxidation state of U in the solution was kept at VI. XANES analysis showed that the oxidation state of the sorbed U by *B. subtilis* was VI. These results indicate that the sorption behavior by B. subtilis differs between Pu(VI) and U(VI).

TEM observations and EDS analyses of the sorbent mixture confirmed that U is preferentially sorbed by the bacteria. The bacteria-kaolinite mixture could not be approximated as a simple noninteractive mixture of organic and inorganic sorbents and there are significant interactive effects between the bacteria and kaolinite.

It is concluded that the presence of the bacteria had a very strong effect on U(VI) and Pu(VI) sorption, even though the sorption behavior differs between them.

CP673, *Plutonium Futures — The Science*, edited by G. D. Jarvinen

Plutonium Isotope Remobilization from Natural Sediments (Gulf of Lions, Northwestern Mediterranean Sea): Estimation Based on Flume Experiments

Bruno Lansard,[1*] Sabine Charmasson,[2] Frédérique Eyrolle,[2] Mireille Arnaud,[2] and Christian Grenz[1]

[1]Centre d'Océanologie de Marseille, Université de la Méditerranée / CNRS, rue de la batterie des lions, 13007, Marseille, France. (*Corresponding author, PhD student, e-mail: lansard@com.univ-mrs.fr, telephone: 00 (33) 491041641)*

[2]Institut de Radioprotection et de Sûreté Nucléaire, Laboratoire d'Etudes Radioécologiques Continentales et de la Méditerranée, C.E.N. Cadarache, 13108, St Paul-Lez-Durance, France

SUMMARY

For the last forty years, military and industrial spent-nuclear fuels were reprocessed at the Marcoule plant located along the Rhone valley (France). Low-level plutonium isotopes contained in liquid wastes were released into the Rhone River, which represents the main fresh water input to the western Mediterranean Sea. Over these years, a reservoir of artificial radionuclides was built up both in sediments of the river and in sediments of the Rhone mouth in the Mediterranean Sea.[1] Because of the high affinity of plutonium for suspended particles, coastal sediment could be considered as one of the major sinks for plutonium but not the ultimate one.[2] In fact, this reservoir can act as a delayed source of plutonium for the water column by redissolution,[3] thermal and photochemical reactions,[4] and punctually during resuspension events. Later process occurs largely during most of the year because winds are highly frequent in this region. In 1997, the Marcoule reprocessing plant ceased its activities, leading to a significant decrease in the discharges. Therefore, in the Gulf of Lions, the contribution of the sedimentary source term could become significant. This has provided a unique opportunity to investigate the behaviour of plutonium at the sediment-water interface in a dynamic coastal environment, in order to predict with confidence the ultimate sink.[5]

A first attempt to quantify both sedimentary bed erosion and plutonium isotope remobilization from eroded particles has been carried out through flume experiments. Sediment cores were collected during the REMORA cruises held in 2001–2002 both at the Rhone estuary and in the Gulf of Lions (Figure 1). Sediments were analyzed for their plutonium content and particle size distributions. At the same time, sediment samples with undisturbed sediment-water interface were transferred within a recirculating flume for resuspension experiments. Sediment stability was determined in a 3,6 m long recirculating flume (Figure 2) filled with 300 l of seawater, representing a 10 cm high water body above the sediment interface, as described by Schaaff.[6] Flow measurements were conducted with an Acoustic Doppler Velocimeter (ADV) operating at a sampling frequency of 25 Hz and placed in the middle of the test section, 4 cm above the bottom. A turbidity sensor measured the evolution of ambient and resuspended particulate matter. Water samples were collected every 5 min at approximately mid-depth by mean of a peristaltic pump. The samples were analyzed for suspended load by filtration through Whatman GF/F filters. These experiments were used to determine the critical shear stress for incipient erosion and erosion rates of particulate matter. The erosion experiments consisted of a continuous increase during 45 minutes in the flow velocity of up to 30 cms^{-1}. After 30 min, the current was stopped, and the total body seawater was then separated by 0.5-µm filters. Plutonium isotopes were analyzed both in the particulate and dissolved phases. For several stations and different sediments, we determined the erosion fluxes of sediments under controlled hydrodynamic forcing and the resulting plutonium remobilization. The first results indicate that after a resuspension event, plutonium activity increased by a

CP673, *Plutonium Futures — The Science,* edited by G. D. Jarvinen

factor 2 to 3, both in the particulate and dissolved phases. These results underline the existence of a plutonium desorption process during resuspension events and emphasise the potential role of sediment-water interface as a secondary source of plutonium for the pelagic ecosystem of the Gulf of Lions.

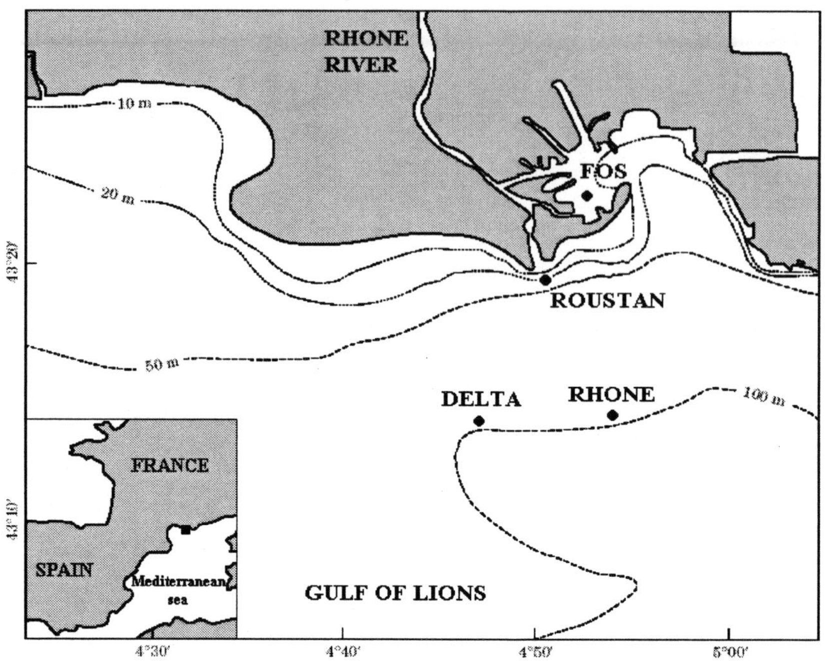

FIGURE 1. Map of the Rhone mouth with the location of the sampling sites.

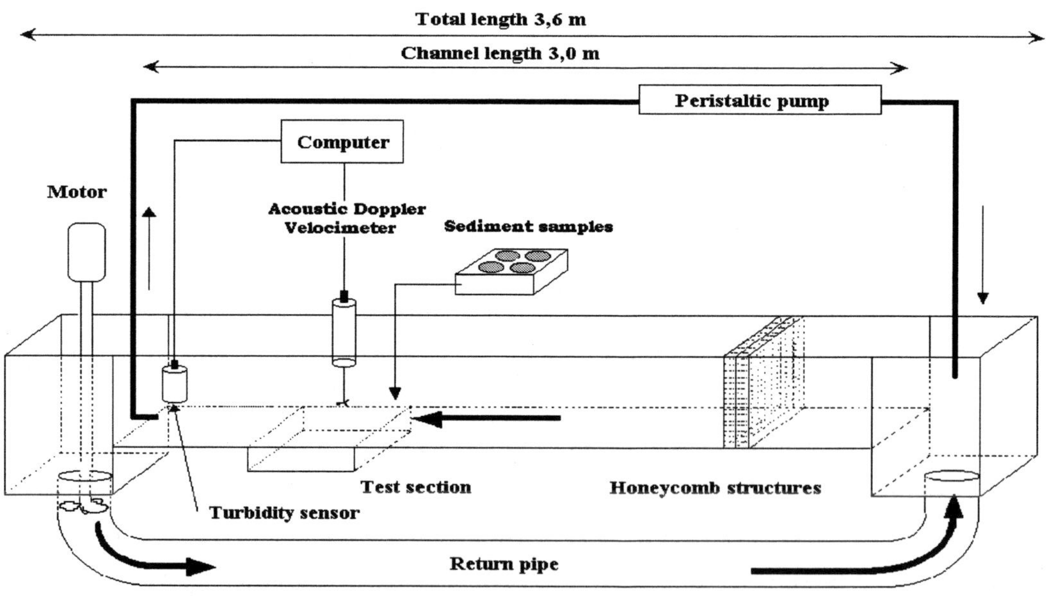

FIGURE 2. The recirculating flume used to perform erosion experiments.

REFERENCES

1. Charmasson, S., Radakovitch, O., Arnaud, M., Bouisset, P., and Pruchon, A., "Long-Core Profiles of ^{137}Cs, ^{134}Cs, ^{60}Co, and ^{210}Pb in Sediment Near the Rhône River (Northwestern Mediterranean Sea)," Estuaries 21(3), 367–378 (1998).
2. Cook, G. T., MacKenzie, A. B., McDonald, P. and Jones, S. R., "Remobilization of Sellafield-Derived Radionuclides and Transport from the North-East Irish Sea," Journal of Environmental Radioactivity 35(3), 227–241 (1997).
3. McDonald, P., Vives i Batlle, J., Bousher, A., Whittall, A., and Chambers, N, "The Availability of Plutonium and Americium in Irish Sea Sediments for Re-dissolution," The Science of The Total Environment 267(1–3), 109–123 (2001).
4. McCubbin, D., Leonard, K. S., and Emerson, H. S., "Influence of Thermal and Photochemical Reactions upon the Redox Cycling of Pu Between Solid and Solution Phases in Seawater," Marine Chemistry 80 (1), 61–77 (2002).
5. Kershaw, P. J., Woodhead, D. S., Lovett, M. B., and K. S. Leonard, "Plutonium from European Reprocessing Operations—Its Behaviour in the Marine Environment," Applied Radiation and Isotopes 46(11), 1121–1134 (1995).
6. Schaaff, E., Grenz, C., and C. Pinazo, "Erosion of Particulate Inorganic and Organic Matter in the Gulf of Lion, Comptes Rendus Geosciences 334 (15), 1071–1077 (2002).

Plutonium Colloid-Facilitated Transport in the Environment—Experimental and Transport Modeling Evidence for Plutonium Migration Mechanisms

M. Zavarin,[1] R. M. Maxwell,[1] A. B. Kersting,[2] P. Zhao,[2] E. R. Sylwester,[2] P. G. Allen,[2] and R. W. Williams[2]

[1]Energy and Environment and [2]Chemistry and Material Science Directorates, Lawrence Livermore National Laboratory, 7000 East Avenue, Livermore, CA 94551

INTRODUCTION

Natural inorganic colloids (< 1 micron particles) found in groundwater can sorb low-solubility actinides and may provide a pathway for transport in the subsurface. For example, Kerting et al.[1] found that Pu, associated with colloids fraction of the groundwater, was detected over 1 km away from the underground nuclear test at the Nevada Test Site (NTS) where it was originally deposited 28 years earlier. However, laboratory experiments have not identified the mechanisms by which Pu may sorb to colloids or exist as its own colloid and travel relatively unimpeded in the subsurface. Some data suggest that Pu sorption to colloids is a very fast process while desorption is very slow or simply does not occur.[2] Slow desorption of Pu from colloids could allow Pu sorbed to a colloid to travel much farther than if sorption were an equilibrium process. However, Pu sorption (and particularly desorption) data in the literature are scant and sometimes contradictory. In some cases, Pu desorption is rather fast, with rates dependent on colloid mineralogy.[3,4,5] Moreover, the effect of sorption and desorption kinetics (as well as other mechanisms) on colloid-facilitated transport at the field scale has not been thoroughly evaluated. This is, in part, due to limitations in colloid transport as well as sorption/desorption models.

In an effort to better understand the dominant mechanisms that control colloid-facilitated Pu transport, we have performed a series of sorption/desorption experiments using mineral colloids and Pu(IV) and Pu(V). We focused on natural colloidal minerals present in water samples collected from both saturated and vadose zone waters at the NTS. These colloid minerals include zeolites, clays, silica, Mn-oxides, Fe-oxides, and calcite. X-ray absorption fine-structure spectroscopy (XAFS) was performed to characterize the speciation of sorbed plutonium. We applied both surface complexation modeling and particle-tracking reactive transport codes to better understand the underlying processes of Pu sorption and desorption and the factors affecting Pu transport at the field scale.

RESULTS

Sorption of Pu(IV) to all mineral colloids examined occured at a very fast rate (equilibrium reached in ~24 hours). However, the affinity of Pu(IV) for the various minerals varied significantly. At pH 8, sorption to birnessite (Mn-oxide) and goethite (Fe-oxide) was much stronger than to clinoptilolite (a zeolite) and calcite. XANES data showed Pu on the mineral surfaces to be dominated by the Pu(IV) oxidation state. At pH 7–9, aqueous complexation of Pu(IV) with carbonate competes with sorption of Pu(IV) to clinoptilolite resulting in lower sorption (Figure 1).

Unlike other studies,[2] it appears from our desorption experiments on clinoptilolite that Pu(IV) desorption rates are rather fast. However, the steady state percent that remains sorbed in desorption experiments is significantly higher than in sorption experiments. For example, the free Pu in clinoptilolite desorption experiments between pH 4–10 ranged from 2% to 8% as opposed to 10% to 30% in sorption experiments. This effect is often observed in sorption/desorption experiments and is described as hysteresis.

CP673, *Plutonium Futures — The Science,* edited by G. D. Jarvinen

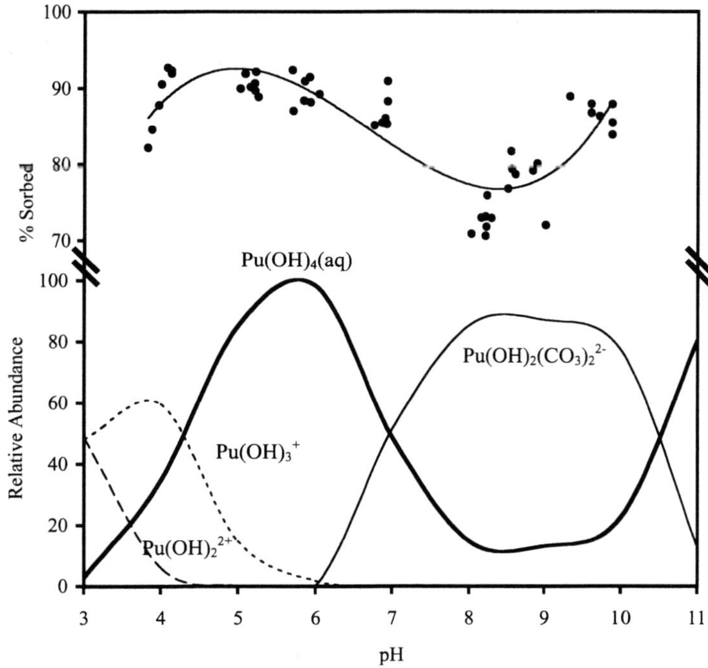

FIGURE 1. Percent Pu(IV) sorbed and aqueous speciation in clinoptilolite sorption experiments.

While Pu(V) appears to sorb to iron and manganese oxides very quickly, Pu(V) sorption rates are quite slow on aluminosilicates and calcite. Furthermore, XANES data indicate that Pu on all mineral surfaces is dominated by Pu(IV). It appears from our experiments that in the case of aluminosilicates (and possibly calcite), sorption rates are dependent on Pu(V) disproportionation and reduction in solution and subsequent sorption of Pu(IV). At pH 8, Pu(V) disproportionation and Pu sorption occurred at rates equivalent to a half-life of 0.2 years. For iron and manganese oxides, Pu(V) sorption rates are fast; mineral surfaces appear to sorb Pu(V) directly and promote Pu(V) reduction to Pu(IV).

DISCUSSION

Plutonium redox transformation rates, disproportionation rates, sorption/desorption hysteresis, and other factors may all affect colloid-facilitated Pu transport in the subsurface. Furthermore, the physical aspects of colloid transport (colloid concentration, filtration, stability, etc.) will affect Pu transport. We have applied the mechanistic Pu sorption information in conjunction with a colloid filtration to a particle transport code that can account for a number of colloid sorption and filtration mechanisms. A number of simulations were conducted using the particle transport code to interpret field-scale implications of laboratory results. Among these was a parametric sensitivity study of transport given a wide combination of forward and reverse sorption rates. This study showed several transport regimes from near-equilibrium to highly kinetic behavior. These simulations provide information regarding the importance of the various physical and chemical mechanisms on the overall transport of Pu in the subsurface. It appears that observed sorption and desorption rates may be too fast to explain the observed Pu transport at NTS. A particle model that included sorption hysteresis was also tested. Hysteresis effects may better explain the apparent colloid-facilitated Pu transport behavior.

ACKNOWLEDGMENTS

This work was performed under the auspices of the U.S. Department of Energy by University of California Lawrence Livermore National Laboratory under contract No. W-7405-ENG-48.

REFERENCES

1. Kersting, A. B., Efurd, D. W., Finnegan, D. L., Rokop, D. J., Smith, D. K., and Thompson, J. L., "Migration of Plutonium in Ground Water at the Nevada Test Site," Nature 397, 56–59 (1999).
2. Lu, N., Cotter, C. R., Kitten, H. D., Bentley, J., and Triay, I. R., "Reversibility of Sorption of Plutonium-239 onto Hematite and Goethite Colloids," Radiochimica Acta 83: 167–173 (1998).
3. Runde, W., Conradson, S. D., Efurd, D. W., Lu, N. P, VanPelt, C. E., and Tait, C. D., "Solubility and Sorption of Redox-Sensitive Radionuclides (Np, Pu) in J-13 Water from the Yucca Mountain Site: Comparison Between Experiment and Theory," Applied Geochemistry 17, 837–853 (2002).
4. Lu, N., Triay, I. R., Cotter, C. R., Kitten, H. D., and Bentley, J., "Reversibility of Sorption of Plutonium-239 onto Colloids of Hematite, Goethite, Smectite, and Silica," Los Alamos National Laboratory (1998), p. 41.
5. Lu, N. P., Triay, I. R., and Conca, J., "Uptake of Colloidal Plutonium-239(IV) onto Colloids of Hematite, Goethite, Ca-Montmorillonite, and Silica in Groundwater," Abstracts of Papers of the American Chemical Society 219, U554–U554 (2000).

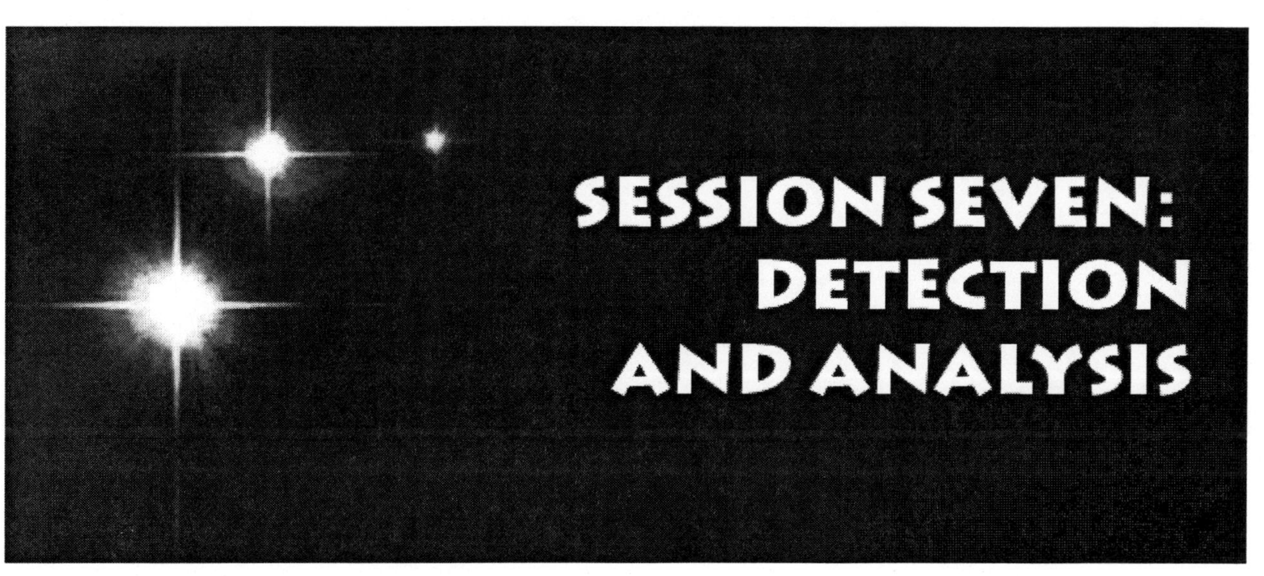
SESSION SEVEN:
DETECTION
AND ANALYSIS

The Application of Vibrational Spectroscopy to Actinide Analysis

T. J. Piper and C. D. Puxley*

Atomic Weapons Establishment, Aldermaston, Berkshire RG7 4PR, United Kingdom

INTRODUCTION

Nondestructive, in situ analysis is highly desirable for the characterization of actinide materials. The technique of vibrational spectroscopy can be applied to the nondestructive analysis of both bulk and trace actinide compounds. In particular, this technique can be used to detect and analyze typical species present on the surface of actinide materials as well as unexpected contaminants. The use of fiber-optic connections allows the remote analysis of actinide and actinide-contaminated materials to be safely carried out in gloveboxes using both Raman and Fourier Transform Infra-Red (FTIR) spectroscopies.

INTERFACING OF THE ANALYSIS INSTRUMENTS WITH THE GLOVEBOX

The interface between the Raman instrument and the probe inside the glovebox was found to be relatively straightforward to achieve by means of a pair of feedthroughs for the fiber-optic cables. These feedthroughs are incorporated into the glovebox wall. For FTIR spectroscopy, the interface through the glovebox wall between the instrument and the probe was more difficult to accomplish. Details of the spectroscopic instrumentation used and its interfacing with the glovebox will be described.

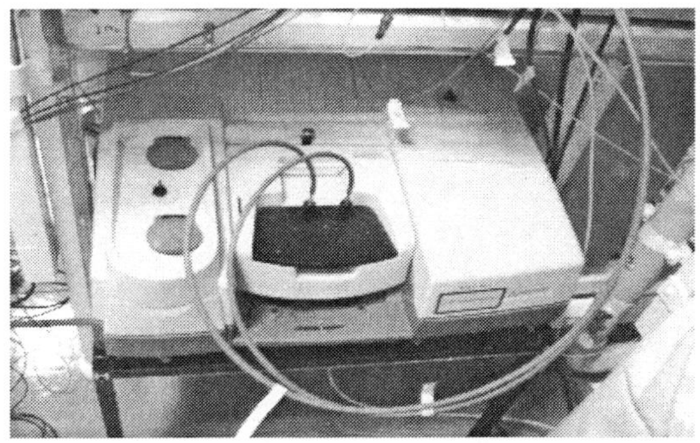

FIGURE 1. Nicolet Chalcogenide Fibre FTIR.

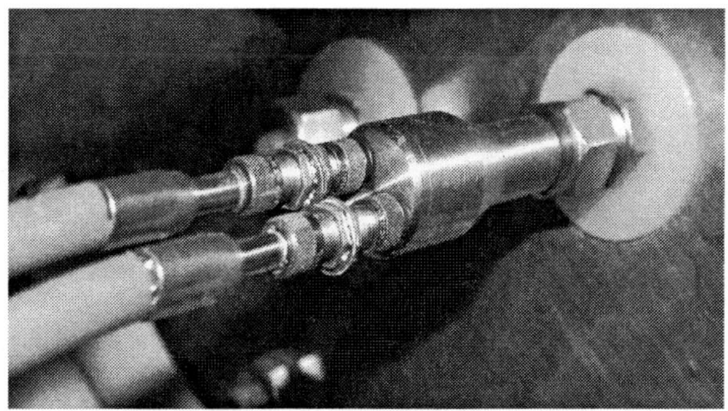

FIGURE 2. FTIR—Glovebox breakthrough.

CP673, *Plutonium Futures — The Science,* edited by G. D. Jarvinen
2003 American Institute of Physics 0-7354-0140-3

CAPABILITIES OF RAMAN AND FTIR SPECTROSCOPIES

Because of the nature of the fiber-optic connections, the applicability of FTIR spectroscopy is mainly restricted to the analysis of compounds exhibiting molecular vibrations that fall in the wave-number range of 4,000 to 1,000 cm^{-1}. However, this includes the functional groups contained within organic compounds as well as inorganic compounds making up simple oxyanions such as nitrate, carbonate, etc. The fiber-optic FTIR spectroscopy technique is complemented by the capabilities of fiber-optic Raman spectroscopy that is able to detect and analyze a wide range of inorganic compounds, including plutonium species, as well as provide information on organic compounds.

RESULTS

FTIR and Raman spectroscopy have been applied to the analysis of surface layers on actinide materials, and the impact of the actinide environment on other materials has been studied. The spectra in Figure 3 illustrate how the degradation of machine oil in contact with plutonium was followed using FTIR to monitor the changing intensity of a carbonyl band.

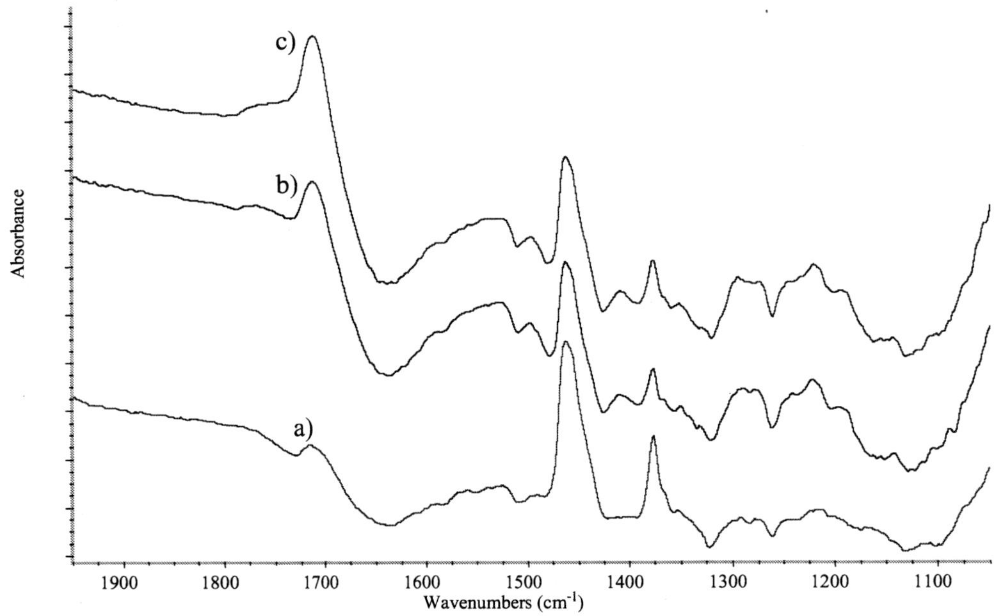

FIGURE 3. Variation in carbonyl peak intensity (ca. 1,700 cm^{-1}) after contact times of (a) 4 days, (b) 9 days, and (c) 13 days.

Optical-Fiber Bragg Grating Sensors Applied to the Study of Plutonium Alloy Aging

Pascale Julia

CEA – Centre de Valduc, 21120 Is-sur-Tille, France

In pure plutonium, the monoclinic α phase is stable at room temperature. However, the high temperature face–centered cubic δ phase can be stabilized at room temperature by adding a small amount of a so-called deltagen element such as Ga, Al, Am, or Ce. The radioactive decay of plutonium leads to the apparition of Am, U, He, and defects such as vacancies and interstitial defects. That is at the origin of some physical properties' evolution, such as swelling (elongation of material).[1] The reasons for this swelling can be distinguished in the following way:

- the defects that can be restored through an annealing treatment (vacancies, interstitials)
- the defects that can be restored with annealing at a higher temperature (He)
- the recoil atoms from the decay that cannot be restored (Am, U).

To quantify the swelling, dilatometry can be used as follows:

- Classic dilatometry allows quantification at different times (measurement after the annealing of some defects).
- Optical-fiber Bragg grating sensors[2–4] (developed by CEA – DIMRI/SIAR/LIST - 91191 Gif sur Yvette CEDEX – France). This technique allows the continuous following of the swelling.

To measure the swelling of plutonium alloys, the properties of optical-fiber Bragg grating sensors we use are sensitive to temperature, elongation, and strain. Different configurations are possible. For our application, we use three Bragg grating sensors that are carved on the core of the fiber:

- to measure the dilatation of the sample: two gratings
- to correct the elongation caused by temperature: one grating.

To have a significant answer, we must stick the grating on the sample. For this purpose, grooves have to be machined on the sample.

The operation of sticking has been optimized for a nuclear environment. Two of these gratings are used to measure the elongation of the sample and the third one is for temperature measurement. For the elongation measurement, the fiber has to be carefully stuck in a groove of the sample, but for temperature measurements, a simple contact is needed between the sample and the fiber.

The sample equipped with the optical fiber is then introduced in a furnace.

The elongation of the sample versus duration is followed under isothermal temperature conditions.

In this work, the first results obtained for small durations (< 5 months) are presented for a plutonium alloy. In the first step, the evolution of expansion is a linear function of time.

Two tests with two different samples have been performed in the same experimental conditions in order to check the reproducibility of the results.

This work presents a validation of a new technique for the continuous following of the swelling. These first results are quite interesting and many experiments have to be performed to improve first the knowledge of the technique and to get enough results to understand the phenomena inducing the swelling.

For example, the first experiments presented above are to be continued for a longer duration. Material composition and other temperatures will be studied to quantify their influence on swelling.

CP673, *Plutonium Futures — The Science,* edited by G. D. Jarvinen

REFERENCES

1. Chebotarev, N. T., and Utkina, O. N., "Relationship Between Structure and Some Properties of δ-Pu and γ-U Alloys," in Plutonium and Other Actinides, edited by H. Blank and R. Linder, North Holland Publishing Company, 1976.
2. Ferdinand, P. et al., "Optical Fiber Bragg Grating Sensors for Structure Monitoring within the Nuclear Power Plants," in Optical Fibre Sensing and System in Nuclear Environments, SCK-CEN Mol (Belgium), October 17–18, 1994.
3. Ferdinand P. et al., "Applications of Bragg Grating Sensors in Europe," International Conference on Optical Fiber Sensors OFS'97, Williamburg, Virginia, USA, October 28–31, 1997.
4. Ferdinand P. et al., "Optical Fiber Sensors Provide New Means for Measurement and Monitoring within the Nuclear Industry," International Nuclear Congress ENC'98 and World Exhibition, Nice, October 25–28, 1998.

Low-Level Detection and Quantification of Plutonium(III, IV, V, and VI) Using a Liquid Core Waveguide

Richard E. Wilson,[1,2] Yung-Jin Hu,[1,3] and Heino Nitsche[1,3]

[1]*Department of Chemistry, University of California, Berkeley, Berkeley, California*
[2]*Chemical Sciences Division, Glenn T. Seaborg Center, Lawrence Berkeley National Laboratory, Berkeley, California*
[3]*Nuclear Sciences Division, Lawrence Berkeley National Laboratory, Berkeley, California*

Understanding the aqueous chemistry of plutonium, in particular in environmental conditions, is often complicated by plutonium's complex redox chemistry. Because plutonium possesses four oxidation states, all of which can coexist in solution,[1] a reliable method for the identification of these oxidation states is needed.

The identification of plutonium oxidation states at low levels in aqueous solution is often accomplished through an indirect determination using series of liquid-liquid extraction procedures using oxidation-state specific reagents such as HDEHP and TTA.[2] Although these methods, coupled with radioactive counting techniques provide superior limits of detection, they may influence the plutonium redox equilibrium, are time consuming, waste intensive, and costly. Other analytical methods such as mass spectrometry and radioactive counting as stand-alone methods provide excellent detection limits but lack the ability to discriminate between the oxidation states of the plutonium ions in solution.

Traditionally, UV-Vis absorption spectroscopy has been used to identify oxidation state purity or mixtures in concentrated plutonium solutions. However, under the conditions of environmental studies, the plutonium concentration is often 10^{-6} M or below. The small molar absorpitivity of the plutonium ions[3] make detection by optical absorption spectroscopy at these levels impossible. To improve detection, laser photoacoustic spectroscopy has been applied to plutonium solutions with certain oxidation states in near-environmental concentrations.[4] Unfortunately, photoacoustic methods tend to be costly and time consuming.

To overcome the high detection limits associated with plutonium ions in solution, a one-meter liquid core waveguide was employed. The waveguide, consisting of a Teflon capillary with a fill volume of 250 microliters and an effective pathlength of one meter, should provide an enhancement in detection because of the one-hundred-fold increase in path length vs a conventional one-centimeter cell. This method has the advantage of nondestructively sampling the plutonium solution and also does not generate the mixed waste that is a product of the extraction protocols.

The results of this study showed that the actual increase in the detection limits was between 18 and 33 times that of the one-centimeter cell (Table 1). Calculation of the molar absorptivities from Beer's Law plots (Table 2) show good agreement with literature values.[3,4] Limits of detection for the waveguide were calculated to be 16×10^{-6} M for Pu(V), 8×10^{-6} M for Pu(III), 5×10^{-6} M for Pu(IV), and 7×10^{-7} for Pu(VI).

These detection limits allow for the identification of plutonium oxidation states in solution at concentrations relevant to selected laboratory-scale environmental studies. The fill volume of the one-meter waveguide is 250 microliters, and rapid filling and purging can be accomplished using a small peristaltic pump.

This method is currently being used in the determination of plutonium oxidation states in the supernatant of plutonium sorption experiments with iron and manganese oxides. The ability to identify oxidation states in solution in these experiments has yielded important information regarding the redox processes that occur in solution and at the mineral interface.

CP673, *Plutonium Futures — The Science,* edited by G. D. Jarvinen

TABLE 1. Comparison of Detection Limits for One-Centimeter and the One-Meter Waveguide.

Pu Ion	1-cm Detection Limit (μM)	1-m Detection Limit (μM)
III	236	7.97
IV	269	5.32
V	518	15.7
VI	12.4	0.71

TABLE 2. Calculated Molar Absorptivities and Detection Limits Using a One-Meter Waveguide. Values in parentheses are taken from Reference 3.

Pu Ion	Absorption Wavelength (nm)	Molar Absorptivity (M^{-1} cm^{-1})	Limit of Detection (μM)
III	600	33.8 (38)	7.97
IV	470	49.0 (55)	5.32
V	568	18.8 (19)	15.7
VI	830	287.9 (550)	0.71

Minimal sampling time, small volumes of waste, and the enhanced detection limits compared to those of traditional spectrophotometry make this a viable method for the identification and quantification of plutonium oxidation states in solution.

Experiments are currently under way to qualify longer path-length cells of 5 and 10 meters to further enhance the detection capabilities of these systems.

REFERENCES

1. Cleveland, J. M., The Chemistry of Plutonium, Gordon and Breach Scientific Publishers, New York (1970).
2. Nitsche, H., Lee, S., and Gatti, R., "Determination of Plutonium Oxidation States at Trace Levels Pertinent to Nuclear Waste Disposal," Journal of Radioanalytical and Nuclear Chemistry-Articles 124(1), 171–185 (1988).
3. Cohen, D. J., "The Absorption Spectra of Plutonium Ions in Perchloric Acid Solution," Journal of Inorganic and Nuclear Chemistry 18, 211–218 (1961).
4. Silva, R. J., and Nitsche, H., "Actinide Environmental Chemistry," Radiochimica Acta 70–71, 377–396 (1995).

Structural Investigations of Plutonium Zirconia-Based Materials Using the Rietveld Method with X-Ray Diffraction

R. C. Belin,[1] P. E. Raison,[2] and R. G. Haire[3]

[1] *Commissariat à l'Energie Atomique*
CEA-Cadarache DEN/DEC/SPUA/LMPC
13108 Saint-Paul-lez-Durance, FRANCE

[2] *European Commission*
Joint Research Center - Institute for Energy
Postbus Nr 2
1755 ZG Petten (N.H.), THE NETHERLANDS

[3] *Oak Ridge National Laboratory*
Chemical Sciences Division
P.O Box 2008, Oak Ridge, TN 37831-6375, USA

INTRODUCTION

In conjunction with research programs pursued in France at the French Atomic Energy Commission, which deal with nuclear waste management of long-lived radioactive elements, we are investigating the potential use of zirconia-based materials for the transmutation of americium and curium.[1,2] Of the different zirconia materials explored, pyrochlore oxides ($An_2Zr_2O_7$ where An = Am or Cm[1,2]) offer several advantages for transmutation. Some of these are a high stability under irradiation, including γ, neutrons α radiation. Regarding the storage of nuclear wastes, they provide both physical and chemical durabilities, and exhibit low-leach rates in aqueous media.[3] In the realm of basic plutonium materials science, we focus here on the study of plutonium zirconia-based compounds.

RESULTS

We are considering the plutonium pyrochlore oxide, $Pu_2Zr_2O_7$, and its related oxidized compounds, for different nuclear applications. In this regard, the pyrochlore structure was investigated in detail by the Rietveld method for refining high-quality x-ray diffraction data. Samples of the compound were prepared starting with polycrystalline PuO_2 and ZrO_2. The materials were ground/mixed in the appropriate ratios and then pressed into pellets. The samples were calcined in stabilized zirconia crucibles under reducing atmospheres (i.e., Ar/H_2) at 1,973 K for 20 hours, and the entire process repeated. Immediately before x-ray analysis, the powdered material was again treated at 1,973 K for 24 hours under Ar/H_2.

Samples for x-ray diffraction were placed into a special in-house sample holder[4] inside an inert atmosphere glovebox to avoid oxidation. Measurements were performed with a noncontaminated, high-resolution Siemens D5000 x-ray diffractometer using a curved quartz monochromator and copper radiation from a conventional tube source. The powder diffraction pattern was obtained by scanning with counting steps of: (a) 30 seconds, from 22° to 59° 2Θ; and (b) 60 seconds, from 59° to 145° 2Θ, using 0.02° step intervals. The total scan took close to 90 hours. This procedure provided excellent diffraction data for the Rietveld analysis. The room temperature patterns were refined using the Rietveld program, XND.[5] An example of the data is given in Figure 1.

CP673, *Plutonium Futures — The Science,* edited by G. D. Jarvinen

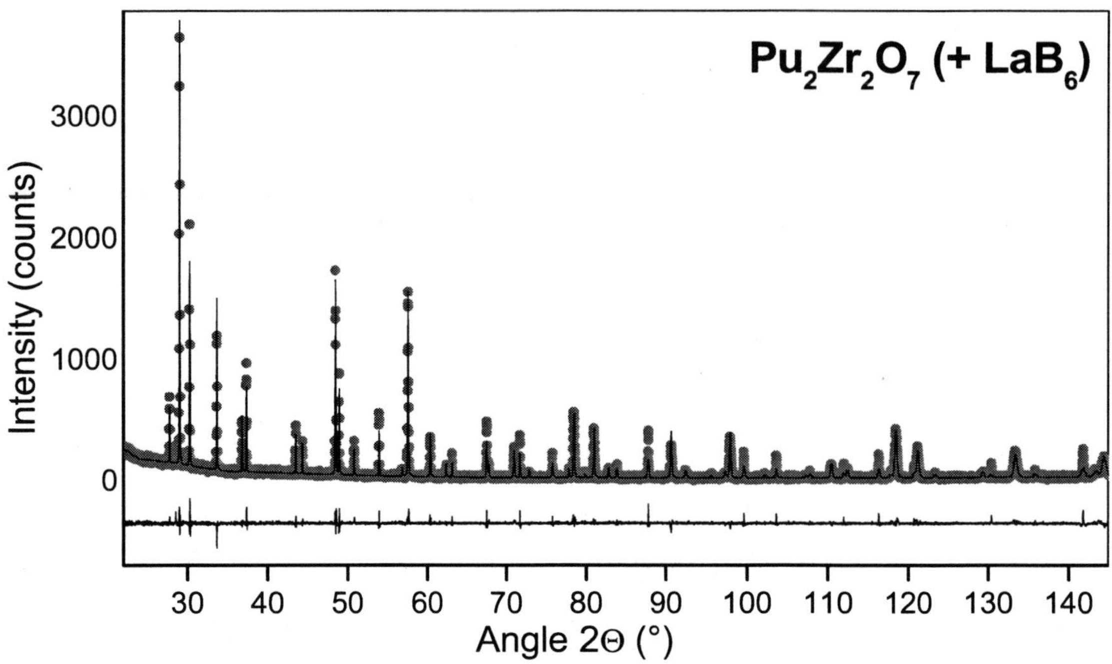

FIGURE 1. View of the Rietveld refinement of $Pu_2Zr_2O_7$, which contained LaB_6 as a standard.

The f-element pyrochlore oxides crystallize in the Fd3m, face-centered cubic system (see Figure 2), which consists of a fluorite-type structure with a double unit cell and an ordered deficiency of oxygen atoms. Positions of all atoms are fixed by symmetry, except the oxygen atoms in the 48f position. There is one unknown parameter (x), which varies from 0.309 to 0.355, its value affects the structural stability of the considered compound.

This variable parameter (x) was determined for the plutonium pyrochlore and then compared with the value calculated with the bond-valence method.[6] Because cerium is often used as a surrogate for plutonium, comparative x-ray diffraction studies were carried out with $Ce_2Zr_2O_7$ samples.

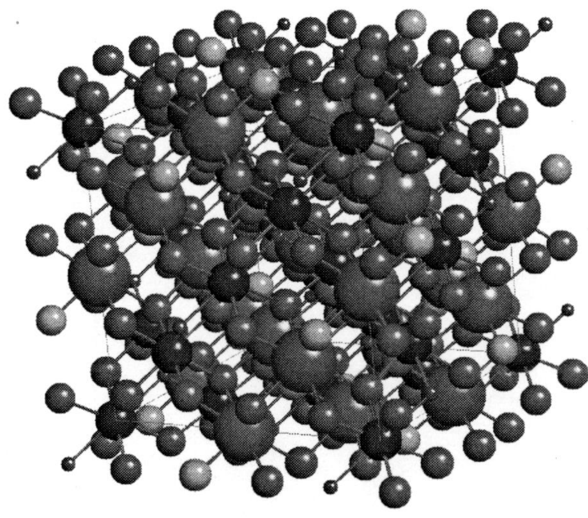

FIGURE 2. The ideal pyrochlore structure, $A_2Zr_2O_7$ (A = f-metal ion).

Both plutonium and cerium pyrochlores can oxidize in air, the temperature of initiation varying with the specific actinide or lanthanide present. During the oxidation process, the compounds can acquire sufficient oxygen to undergo phase transitions (pyrochlore structure to an oxygen-deficient or stoichiometric CaF_2 metal-dioxide lattice).[7] In an effort to study this oxidation mechanism, both cerium and plutonium pyrochlores were oxidized in air at selected temperatures (100°C steps) to determine the onset and completion temperatures of the order-disorder phase transition. The evolution of the cell's lattice parameter(s) was also followed as a function of the temperature. With the plutonium pyrochlore, the oxidation is characterized by a phase transition from the pyrochlore structure to the fluorite-type solid solution. With the cerium pyrochlore, a cubic and tetragonal biphasic system was found instead of the fluorite unit cell.

Similar work on the americium pyrochlore oxide, $Am_2Zr_2O_7$, is underway to determine experimentally the value of *(x)* and to study the potential influence of zirconium on the Am^{3+}/Am^{4+} redox couple, as has been discussed recently.[8] Given the higher radiation level of the ^{241}Am isotope used, the synthesis and handling of the samples necessitated the use of very small quantities, as compared to the other pyrochlores samples. Typically, the work was performed on some ten milligrams quantities.

The ability of zirconium to "enhance" the reduction of cations exhibiting both (III) and (IV) oxidation states has been pointed out by other authors, especially for the Ce-Zr-O system where it was shown that Zr increases significantly the ability to reduce Ce (III).[9]

In the presentation, we shall provide experimental results obtained from Rietveld analysis of the different compounds discussed above. The variation of the structural parameter *(x)* during the experimental processes will be addressed, as well as the limit of Pu^{3+}/Pu^{4+} or Ce^{3+}/Ce^{4+} ratios for which the pyrochlore structure remains viable. Finally, the influence of ZrO_2 on these redox couples will be discussed.

ACKNOWLEDGMENTS

The authors are grateful to P. Valenza and M. Rouault for their contribution in the x-ray specimens preparation and data collection and to S. Vaudez for making the plutonium pyrochlore oxide available. The effort at Oak Ridge National Laboratory was sponsored by the Division of Chemical Sciences, Geosciences and Biosciences, Office of Basic Energy Sciences, USDOE, under contract DE-ACO5-00OR22725 with ORNL, managed and operated by UT-Battelle, LLC.

REFERENCES

1. Raison, P. E., and Haire, R. G., Progress in Nuclear Energy 38, 251 (2001).
2. Raison, P. E., Haire, R. G., Sato, T., and Ogawa, T., Mat. Res. Soc. Proc. 556, 3–10 (1999).
3. Kamizono, H., Hayakawa, I., and Muraoka, S., J. Am. Ceram. Soc. 74, 863 (1991).
4. Belin, R., Valenza, P., and Raison, P., "New Hermetic Sample-Holder for X-Ray Diffraction on Radioactive Materials: Application to the Rietveld Analysis of Plutonium Compounds" (this conference).
5. Bérar, J. F., and Garnier, P., 2nd APD Conf. NIST Special Publication, Vol. 846, Washington, DC U.S. Government Printing Office, p. 212, 1992.
6. Brown, I. D., Chapter 14, Structure and Bonding in Crystals, Vol. II, Academic Press, 1981.
7. Haire, R. G., Raison, P. E., and Assefa, Z., J. Nucl. Sci. Techn. (2002), in press.
8. Raison, P. E., and Haire, R. G., "Americium and Curium in Zirconia-Based Materials: Critical Aspects of their Material Science" (this conference).
9. Fornasiero, P., Balducci, G., Di Monte, R., Kaspar, J., Sergo, V., Gubitosa, G., Ferrero, A., and Graziani, M., J. Catalysis 164, 173 (1996).

Facilitation of Trace Elemental Determination in Plutonium Oxide By Inductively Coupled Plasma Mass Spectrometry (ICP-MS)

Jeffrey Giglio,* Daniel Cummings, and John Krsul

Argonne National Laboratory–West
P.O. Box 2528
Idaho Falls, ID 83403-2528

The determination of the trace elemental content in PuO_2 is important for a variety of reasons. First, in the recycling of PuO_2, it is important to characterize the amount of additives that may hinder or limit the effectiveness of the PuO_2 in its future uses. Additionally, knowing the impurity content of the PuO_2 before use will aid in the prediction of fission and neutron activation products after use. The impurity content along with an accurate computer model is helpful in the post-use disposal of the material in a repository or in determining designated destinations for the material. This work will detail the application of a unique and effective separation scheme to remove the plutonium from a digested sample of PuO_2. This separation will facilitate the determination of impurities of the sample without exposing personnel to high doses of radiation, limit the Pu background level in the ICP-MS and lastly, allow for the determination of analytes with *m/z* s near Pu accurately and precisely.

The separation scheme developed employs a U-Teva™ resin with a novel gas-pressurized extraction cell to remove the Pu from the sample matrix. This extraction scheme uses an inert gas to move a fixed amount of sample through the resin material. Traditionally, the sample is either pumped through a resin material, or a drip column technology is used. The approach developed here limits the amount of solution needed to execute the extraction. This in turn improves the detection limits by lowering the dilution factors involved in the analysis. Additionally, because of the reduced amount of liquid used and the geometry of the column, the efficiency of the extraction is quite high. This increased efficiency allows for a more complete removal of the Pu from the matrix, reducing the radioactive fields, lowering isobaric interferences and limiting the background levels of Pu in the instrument. The extraction method described above was applied to several PuO_2 samples. Greater than 30 analytes were determined in the samples of plutonium oxide with detection levels in the low-to-mid ng/g in the solid for each analyte. Some analytes of interest determined in the samples include boron, beryllium, gallium, transition elements, and americium. In addition to the analytical results and figures of merit presented, the presentation will cover the digestion scheme employed along with the analytical work performed to verify the selectiveness and effectiveness of the extraction technique. In addition, preliminary data will be presented on additional separations developed, based on this experimental scheme discussed here. These separations will allow for the determination of trace levels of plutonium, uranium, neptunium, and americium in various matrices in one extraction.

This work is a novel alternative to the separation schemes currently applied in analytical chemistry. It is a technological step forward in that it requires less sample, is faster, has greater efficiency, produces less waste and is more reproducible than the separation schemes currently being employed. The benefits listed above lead to improved analytical figures of merit along with increased sample throughput for the laboratory.

CP673, *Plutonium Futures — The Science,* edited by G. D. Jarvinen

POSTER SESSION

MATERIAL SCIENCE/ CONDENSED MATTER PHYSICS

Evolution of Defects in Pu During Isochronal Annealing and Self-Irradiation

P. Asoka-Kumar, S. Glade, P. A. Sterne, and R. Howell

Lawrence Livermore National Laboratory, Livermore, CA 94550

We report on the evolution of defects in Pu during isochronal annealing and self-irradiation using positron annihilation spectroscopy. Positron annihilation spectroscopy is a sensitive probe (part-per-million level) for atomic-scale defects. The spectroscopic tools available at LLNL allow the determination of size, concentration, and chemical surroundings of defects in aged Pu samples.

Positron lifetime analysis was performed on eight samples aged 7 months to 42 years. All samples except the 7-month-old sample contained a high concentration of positron trapping centers. The dominant component yielded a lifetime value of ~182 ps. In aged samples, a second, longer lifetime component was observed that appears to increase in strength with the age of the sample. The observed lifetime values and their relative strengths are shown in Figure 1. The top panel corresponds to the lifetime values, and the bottom panel corresponds to the intensity of the long lifetime component.

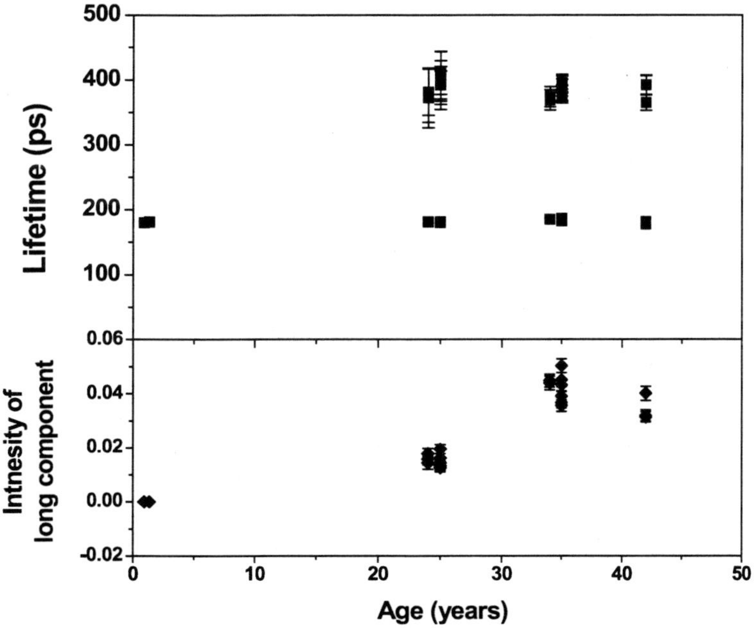

FIGURE 1. Positron lifetime values and relative intensity of the long lifetime component in aged Pu material. The intensity of the lifetime components sums to unity. The data at ~1 year is from melted and recast material.

Positron lifetime values are determined by the bubble size and He content. When He is added to a bubble, the positron lifetime is shortened as a result of the increased electron density. When the size of the bubble is known from an independent measurement, the observed positron lifetime values and the associated first-principle calculations can be used to estimate the He content of the bubble. Our positron results can be combined with the

CP673, *Plutonium Futures — The Science,* edited by G. D. Jarvinen

recent transmission electron microscopy measurements of Schwartz et al. to estimate the He content of the bubbles. The short lifetime component corresponds to 2-3 He atoms per vacancy.

A 35-year old sample was annealed to 440°C in several steps to examine the temperature stability of the two lifetime features. The isochronal anneals were performed for 30 minutes, increasing the temperature with each subsequent step. This measurement complements a previous annealing study on recast, 7-month old material. The primary objective of these studies is to establish the appropriate limits on the storage temperature of the accelerated aging samples. Figure 2 shows the intensity of the long lifetime component as a function of the annealing temperature. Also shown is the corresponding He evolution from the capsule measured using a standard He leak detector. The intensity of the long lifetime component starts to increase at about 150°C and reaches a maximum at around 250°C. A similar behavior is seen in He evolution. The storage temperature of the accelerated aging sample should be below the onset of these annealing effects.

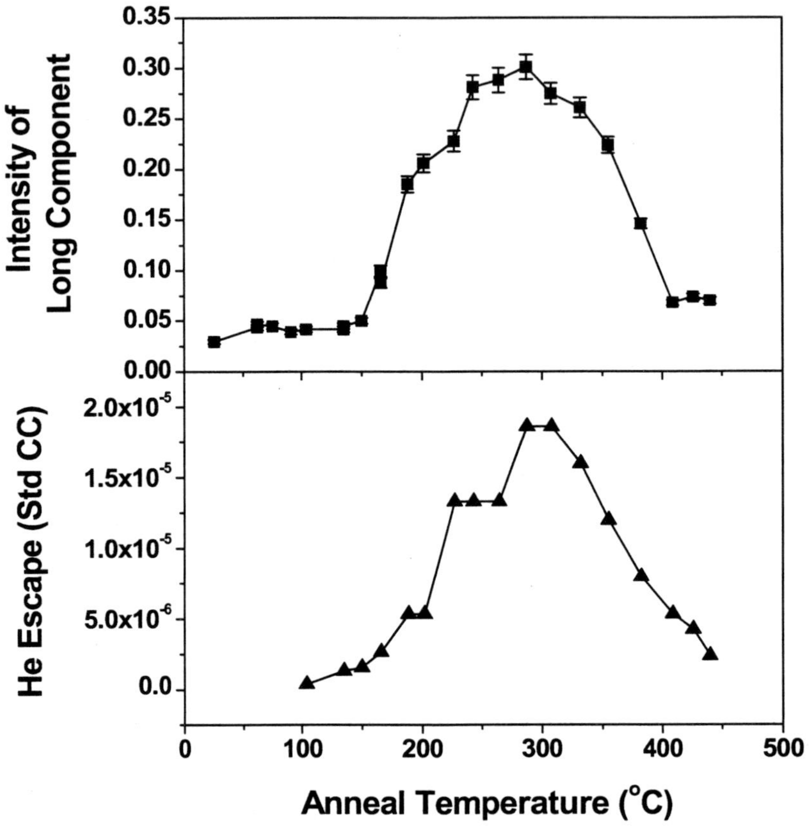

FIGURE 2. Intensity of the long lifetime component and He evolution of a 35-year old sample as a function of the isochronal annealing temperature. Each anneal lasted for 30 minutes, and the measurements are performed after cooling the sample to room temperature.

We also report results from a new spectroscopic method that examines chemical surroundings of the bubbles observed in aged Pu material. The new spectroscopy determines the momentum distribution of the electrons participating in the annihilation process. At high momentum values, these distributions are dominated by electrons bound to atoms and are therefore element specific. No significant variations are observed in aged samples, suggesting that the chemical surroundings of voids and bubbles are not evolving with age.

ACKNOWLEDGMENTS

This work was performed under the auspices of the U.S. Department of Energy by the University of California, Lawrence Livermore National Laboratory under contract No. W-7405-ENG-48.

Understanding and Predicting Plutonium Alloys Aging: A Coupled Experimental and Theoretical Approach

N. Baclet,[a] P. Pochet,[a] Ph. Faure,[a] C. Valot,[a] L. Gosmain,[b] Ch. Valot,[c] J. L. Flament,[d] and C. Berthier[d]

(a) CEA-Centre de Valduc, 21120 Is-sur-Tille, France, Tel: 033 3 80 23 51 54, Fax: 033 3 80 23 52 17
(b) CEA-Centre de Valduc, 21120 Is-sur-Tille, France, Tel: 033 3 80 23 51 54, Fax: 033 3 80 23 52 17, present adress: CEA – Centre de Saclay, 91191 Gif sur Yvette, France, Tel: 033 1 69 08 60 00
(c) LRRS , Université de Bourgogne, France, Tel: 033 3 80 39 61 83
(d) CEA – Ile de France, 91680 Bruyères le Chatel, France, Tel: 033 1 69 26 54 86, Fax: 033 1 69 26 70 24

Abstract. Understanding plutonium aging is a real challenge that requires developing very ambitious modeling and experiments. Examples of the different techniques developed and the physical values that can be reached are presented here.

INTRODUCTION

Plutonium aging is still a subject of fundamental interest today because it is necessary to better understand aging effects to be able to predict their consequences on the reliability and safety of a nuclear stockpile. Aging involves very tiny effects induced by vacancies and interstitials that are first created in the displacement cascade core and are then susceptible to diffusion in the alloy. Predicting aging effects goes through their understanding, which then requires to couple modeling and experiments to characterize changes induced by self-irradiation.

A multiscale approach, both theoretical and experimental, has then been initiated. Some dedicated techniques, such as positron annihilation spectroscopy[1] or experiments on synchrotron radiation,[2] have been developed, the challenge being to track tiny aging effects. Another challenge is also to identify which parameters are relevant for modeling and to develop the experimental techniques that are able to measure these physical parameters.

As an example, interatomic potential, which is the key for the modeling of displacement cascades by molecular dynamics, must know elastic constants and bulk modulus, data that are not available today for the PuGa alloys of interest. Specific XRD experiments are then performed to get these values. Moreover, the number of defects that are created when a decay occurs strongly depends on the threshold displacement energy that is still unknown for plutonium alloys. Migration energies of the different kinds of defects are also necessary to model the diffusion of these defects by techniques such as mesoscopic Monte-Carlo.

RESULTS AND DISCUSSION

Ideally, elastic constants are measured using single crystals that are unfortunately not available for the plutonium alloys of concern. A technique, based on X-ray diffraction performed on a polycrystalline sample submitted to a tensile test, has then been developed in the Laboratory, in order to deduce the elastic constants of the corresponding single crystal.[3] The feasibility of the method has been tested on pure copper and rhodium samples because the mechanical properties of these metals are close to those of δ-plutonium. First results are promising,[3] and the method is now being improved by taking into account the texture of the material. An encapsulated tensile test device is also being developed to perform the same x-ray diffraction experiments on plutonium alloys in the laboratory.

CP673, *Plutonium Futures — The Science,* edited by G. D. Jarvinen

Bulk modulus and its pressure derivative are fundamental parameters for molecular dynamics calculations of displacement cascades[4]. First experiments have been performed on a δ-PuGa alloy by x-ray diffraction under pressure induced by a diamond anvil cell. First results are very interesting[5] and will be presented.

Measurement of threshold displacement energy is planned by electrons irradiating a plutonium sample and simultaneously measuring the electrical resistivity at a low temperature. An increase in electrical resistivity for a given energy of the incident electrons is expected and should allow deducing the threshold displacement energy of the alloy. First tests on non nuclear materials have been performed to check the effects of confinement on the electron beam and the accuracy of the measurement.

Once they are created, defects can then diffuse, leading to a change in the physical properties of the plutonium alloy. Measurements of the migration energies of the different defects involved are planned. Holdings of plutonium samples at low temperature (to induce the defects either by electron irradiation or self-irradiation) followed by isochronal annealings should allow to identify, by electrical resistivity measurement, the stages for the migration of the different defects and then deduce the corresponding migration energies.

REFERENCES

1. Oudot, B., submitted to Plutonium Futures Conference, 2003.
2. Ravat, B., Jolly, L., and Valot, C., *Matériaux 2002* Conference, Tours, France, October 2002.
3. Gosmain, L., Valot, Ch., Baclet, N., Valot, C., and Pochet, P., *Groupe Français d'Analyse des Contraintes* Conference, Troyes, France, March 2002.
4. Pochet, P., Nuclear Instruments and Methods in Physics Research B, in press (2003).
5. Faure, Ph., private commmunication.

Computational Modeling of Uranium Hydriding and Complexes

K. Balasubramanian, Wigbert J. Siekhaus, and William McLean II

Chemistry and Material Science Directorate, Lawrence Livermore National Laboratory, University of California, Livermore CA 94550

INTRODUCTION

Uranium hydriding is an important process that has received considerable attention over many years.[1-7] Although many experimental and modeling studies have been carried out concerning the thermochemistry, diffusion kinetics and mechanisms of U-hydriding, very little is known about the electronic structure and electronic feature that governs the U-hydriding process. Yet it is the electronic structure that controls the activation barrier and thus the rate of hydriding. Moreover, the role of impurities and the role of the product UH_3 on hydriding rating are not fully understood.

An early study by Condon and Larson[1] deals with the kinetics of the U-hydrogen system and a mathematical model for the U-hydriding process. They proposed that the reaction be controlled by the diffusion of hydrogen in the reactant phase before nucleation to form the hydride phase occurs, and that the reaction be the first order for hydriding and the zero order for dehydriding. Condon[2] has also calculated and measured the reaction rates of U-hydriding and proposed a diffusion model for the U-hydriding. This model was found to be in excellent agreement with the experimental reaction rates. From the slopes of the Arrhenius plot the activation energy was calculated as 6.35 kcal/mole. In a subsequent study Kirkpatrick[3] formulated a close form for an approximate solution to Condon's equation. Bloch and Mintz[4] have proposed the kinetics and mechanism for the U-H reaction over a wide range of pressures and temperatures. They have discussed their results through two models, one, which considers hydrogen diffusion through a protective UH_3 product layer, and the second where hydride growth occurs at the hydride-metal interface. These authors obtained two-dimensional fits to experimental reaction data over a range of pressures and temperatures of experimental data to the pressure-temperature reactions. Kirkpatrick and Condon[5] have obtained a linear solution to the hydriding of uranium. These authors showed that the calculated reaction rates compared quite well with the experimental data at a hydrogen pressure of 1 atm.

Powell et al.[6] have studied U-hydriding in ultrahigh vacuum and obtained the linear rate data over a wide range of temperatures and pressures. They found reversible hydrogen sorption on the UH_3 reaction product from kinetic effects at 21°C. This demonstrates the restarting of the hydriding process in the presence of the UH_3 reaction product. DeMint and Leckey[7] have shown that Si impurities dramatically accelerate the U-hydriding rates.

We report our recent results of relativistic computations[8] that start from complete active space multiconfiguration interaction (CAS-MCSCF) followed by multireference configuration interaction (MRSDCI) computations. The cluster model computations included up to 50 million configurations for the modeling of uranium hydriding.

RESULTS

Figure 1 shows our computed potential energy surface for the insertion of a U site into H_2. As seen from Figure 1, a pure U site has to surpass a barrier of 20.9 kcal/mole for the U-hydriding. Once the barrier is surpassed, a stable product is formed that is 22.4 kcal/mole more stable than the reactants. Figure 2 shows the potential energy surface of an additional H_2 approaching UH_3 as modeled by the U^{+3} interaction with H_2. The product UH_3 is highly ionic and thus U transfers electron density to the three hydrogens resulting in a U^{+3} state. As seen from Figure 2, U^{+3}

CP673, *Plutonium Futures — The Science,* edited by G. D. Jarvinen

inserts itself into H_2 spontaneously, thus demonstrating that the U^3 site in the product UH_3 binds to H_2 spontaneously, forming a complex in which H_2 is separated far enough so as to cause the liberation of H atoms in the presence of U.

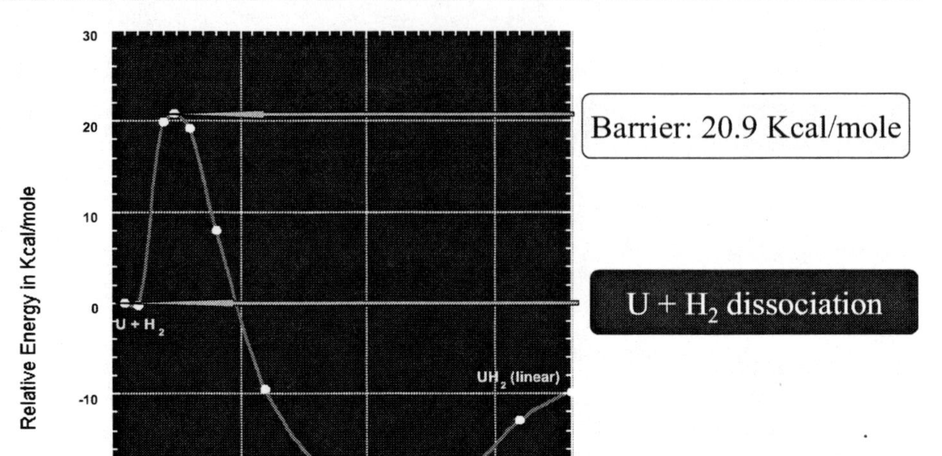

FIGURE 1. Potential energy surface for U-H_2 reaction.

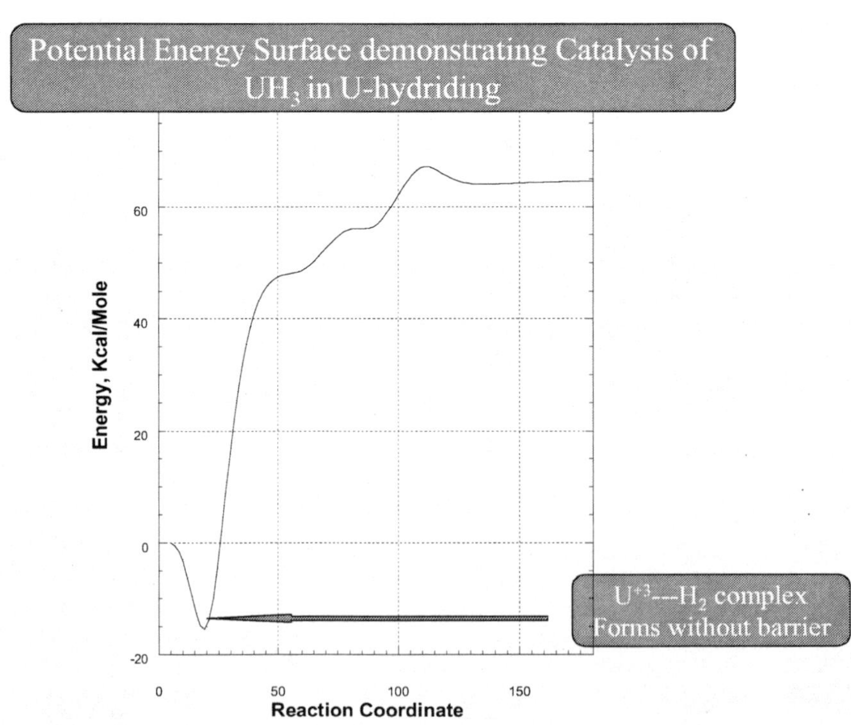

FIGURE 2. Potential energy surface for modeling UH_3-H_2 interaction.

DISCUSSION

Our computed potential energy surfaces demonstrate a 21-kcal/mol activation energy barrier for the reaction of pure U with H_2. However, the presence of the product UH_3 catalyzes the U-hydriding. We have also modeled the presence of Si impurities for the U-hydriding reaction to show that the activation barrier is lowered by the presence of Si. Our computations reveal an electron donor-acceptor model for the U-hydriding, where H_2 exchanges electronic density from its occupied $1\sigma_g$ orbital to the U(6d σ) orbital and a back donation from the U(6d π) orbital back to the H_2 $1\sigma_u$ antibonding orbital, causing the dissociation of H_2 by U. In particular the 5f or 7s orbitals of U are not involved in the dissociation of H_2. We also show that Si impurities assist the hydriding process by the spontaneous insertion of the ^1D state of Si into H_2. The UH_3 product catalyzes the hydriding process by spontaneous formation of a complex of H_2 at the U^{+3} site, which opens up the H_2 bond sufficiently to cause further U-hydriding to occur spontaneously. The bond-breaking process in the formation complex assists the formation of H atoms in the presence of U. The hydrogen atoms thus formed diffuse through the cracks to cause further U-hydriding, thus explaining the experimental observation of Powell et al.[6]

ACKNOWLEDGMENTS

This work was performed under the auspices of the U.S. Department of Energy by the University of California, Lawrence Livermore National Laboratory under Contract No. W-7405-ENG-48.

REFERENCES

1. Condon, J. B., and Larson, E. A., J. Chem. Phys. 59, 855 (1973).
2. Condon, J. B., J. Phys. Chem. 79, 392 (1975).
3. Kirkpatrick, J. R., J. Phys. Chem. 85, 3444 (1981).
4. Bloch, J., and Mintz, M. H., J. Less Common Metals 81, 301 (1981).
5. Kirkpatrick, J. R., and Condon, J. B., "Modeling Reaction Between Uranium and Hydrogen," Oak Ridge National Laboratory Internal Report, K/CSD/TM-87 (1990), pp. 1–41.
6. Powell, G. L., Harper, W. L., and Kirkpatrick, J. R., J. Less Common Metals 172, 116 (1991).
7. DeMint, A. L., and Leckey, J. H., J. Nuc. Mat. 281, 208 (2000).
8. Balasubramanian, K., Relativistic Effects in Chemistry: Part A: Theory and Techniques, Wiley Interscience, New York, NY, (1997), p. 301; K. Balasubramanian, Relativistic Effects in Chemistry Part B. Applications to Molecules and Clusters, Wiley Interscience, New York, NY, (1997), p. 527.

Phase Stability of Pu and Pu-Ga Alloys from Atomistic Calculations

M. I. Baskes, M. Stan, and K. Muralidharan

Materials Science and Technology Division
Los Alamos National Laboratory

INTRODUCTION

From a structural point of view, Pu is one of the most interesting elements in the periodic table. Its zero pressure phase diagram contains six solid phases as well as the liquid phase. Complicating our understanding of Pu is the fact that the f-electrons seem to have different character in the various crystal structures. These changes manifest themselves in the curious situation that the close-packed fcc phase has the largest volume per atom of all of the phases. Similarly, Ga has five solid equilibrium phases at various temperatures and pressures. The experimental Pu-Ga phase diagram shows eleven additional phases,[1,2] and the low-temperature, low-Ga content area is still the subject of debate.[3]

Not only are Pu and Ga scientifically challenging, they are of extreme importance to the weapons community, especially in the light of our recent commitment to being able to accurately predict stockpile aging. Aging is inherently a process that depends upon atomistic mechanisms; thus, a reliable atomic level model of the Pu-Ga system is the key to our success.

Theoreticians have been frustrated for the past few years in their inability to capture the physics of Pu using density functional methods that work well for essentially all of the elements.[4] For over the last decade it has been shown that many-body semiempirical potentials typified by the Embedded Atom Method (EAM)[5,6] have been able to represent the physical properties of a significant number of metals and alloys in excellent agreement with experiment. More recently, the EAM has been supplemented with angular forces leading to potentials that cover most of the periodic table,[7,8] including elements such as silicon[9] and tin[10] where angular forces are extremely important. Our model of tin has been a splendid success in its ability to predict the phase stability and thermodynamics in quantitative agreement with experiment.

RESULTS AND DISCUSSION

The modified EAM (MEAM) formalism has previously been applied to Pu[11] and Ga.[12] Using experimental data from the Ga-stabilized fcc δ-phase and the monoclinic α phase, a MEAM potential for Pu has been developed. This potential has been applied to all of the Pu phases. Similarly, density functional calculations using the generalized gradient approximation have been used to develop a MEAM potential for Ga. Experimental formation enthalpy[13] and the experimental lattice constant of Pu_3Ga have been used to obtain the parameters for the Pu-Ga system.

An example of the success of the model is shown in Figure 1. Here the predicted enthalpy per atom of numerous Pu-Ga alloys at 0 K relative to pure Pu and Ga in their room-temperature crystal structures is shown. With only a few exceptions, the phases that are observed in the experimental phase diagram are seen to have lower enthalpy than the phases not observed in the experiment. In general, the agreement between the predicted enthalpy with experiment is quite good. A notable exception is the experimentally observed $PuGa_4$ ($D1_b$, UAl_4 type) phase where the calculated enthalpy is significantly above the experimental value. The calculated value of the enthalpy of $PuGa_4$ ($D1_c$, $PtSn_4$ type) is predicted to lie much closer to experiment.

CP673, *Plutonium Futures — The Science,* edited by G. D. Jarvinen

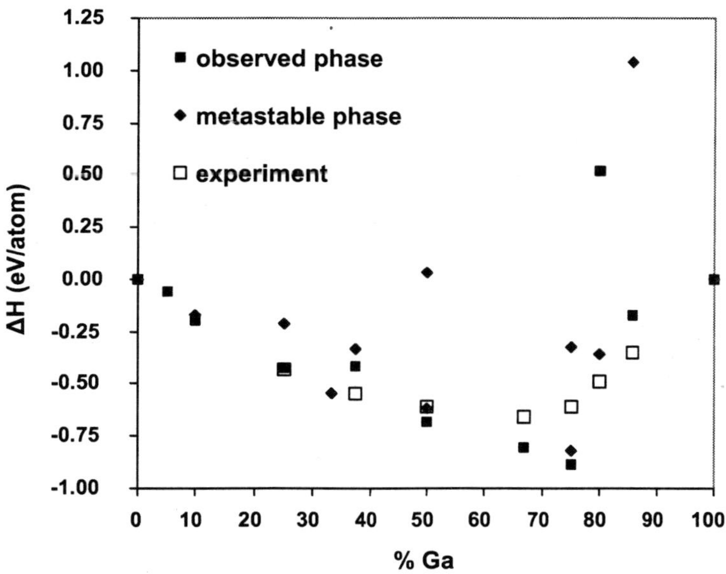

FIGURE 1. Formation enthalpy of Pu-Ga alloys relative to pure Pu and Ga vs Ga composition. The filled (open) symbols represent calculated (experimental[13]) enthalpies. Blue squares are for phases seen in the experimental phase diagram, and the red diamonds are for phases not observed.

These MEAM functions have been used to calculate the Gibbs free energy *(G)* of a number of Pu-Ga alloys with gallium content (*x*) from zero to 25% as a function of temperature *(T)* using the thermodynamic switching method[14] and thermodynamic integration of the enthalpy *(H)*,

$$\frac{G(T,x)}{T}=\frac{G^{ref}(x)}{T^{ref}(x)}-\int_{T^{ref}}^{T}\frac{H(T',x)}{T'^{2}}dT' .$$

The free energies resulting from the MD simulations have been used to calculate the chemical potential of components in all phases[15] and then to determine the equilibrium points and lines in the phase diagram.[16] The calculated phase diagram of the Pu-$Pu_{0.75}G_{0.25}$ pseudobinary system (Figure 2) shows a eutectoid point at 375 K and 27% $Pu_{0.75}Ga_{0.25}$ (that is 6.5% Ga), close to the values predicted in Reference 2. The high temperature area of the diagram requires a reevaluation of the model for the liquid phase.

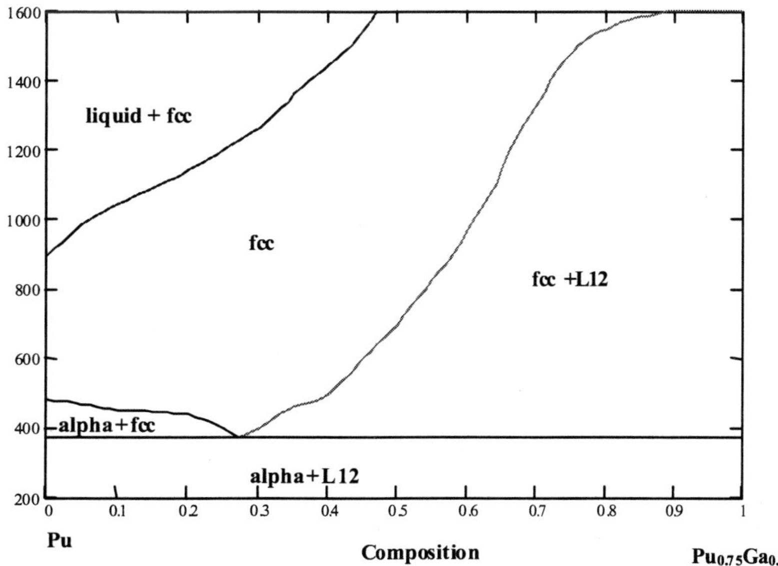

FIGURE 2. Preliminary phase diagram of the Pu-$Pu_{0.75}Ga_{0.25}$ pseudobinary systems.

REFERENCES

1. Ellinger, F. H., Land, C. C., and Strebing, V. O., J. Nucl. Mater. 12 (1964).
2. Chebotarev, N. T., Smotriskaya, E. S., Adrianov, M. A., and Kostyuk, O. E., in Plutonium 1975 and Other Actinides, Badeb Baden, Germany, North-Holland Publishing Co. (1975).
3. Hecker, S. S., and Timofeeva, L. F., in Challenges in Plutonium Science, edited by N. G. Cooper, Los Alamos National Laboratory, Los Alamos, NM (2000), p. 244.
4. Cooper, B. R., in Challenges in Plutonium Science, edited by N. G. Cooper, Los Alamos National Laboratory, Los Alamos, NM (2000), p. 154
5. Daw, M. S., and Baskes, M. I., Phys. Rev. B 29, 6443 (1984).
6. Daw, M. S., Foiles, S. M., and Baskes, M. I., Mater. Sci. Rep. 9, 251 (1993).
7. Baskes, M. I., Phys. Rev. B 46, 2727 (1992).
8. Baskes, M. I., and Johnson, R. A., Modelling Simul. Mater. Sci. Eng. 2, 147 (1994).
9. Baskes, M. I., Modelling Simul. Mater. Sci. Eng. 5, 149 (1997).
10. Ravelo, R., and Baskes, M. I., Phys. Rev. Lett. 79, 2482 (1997).
11. Baskes, M. I., Phys. Rev. B 62, 15532 (2000).
12. Baskes, M. I., Chen, S. P., and Cherne, F. J., Phys. Rev. B 66, 104107 (2002).
13. Akhachinsky, V. V., and Kopytin, L. M., "Heats of Formation of Intermetallic Compounds of Plutonium with Gallium and Tin," in Thermodynamics of Nuclear Materials, Vienna, IAEA (1968), p. 789.
14. Kirkwood, J. G., J. Chem. Phys. 3, 300 (1935).
15. Baskes, M. I., and Stan, M., Metal. and Mater. Trans., in print (2002).
16. Stan, M., in Control and Optimization in Minerals, Metals and Materials Processing (Canadian Institute of Mining, Metallurgy and Petroleum, Quebec (1999) p. 161.

Effect of Pu Valence on Acid Dissolution of Perovskite ($CaTiO_3$)

B. D. Begg, Y. Zhang, E. R. Vance, S. D. Conradson,*
and A. J. Brownscombe

ANSTO, Menai, NSW, Australia
**Los Alamos National Laboratory, Los Alamos, NM 87545, USA*

INTRODUCTION

Plutonium immobilization, whether from weapons or impure residues, is an issue of global importance and concern. Titanate mineral analogue phases have been shown to be well suited for Pu immobilization due not only to their excellent durability but also their crystal chemical flexibility to accommodate multivalent actinides.[1-3]

Perovskite, $CaTiO_3$, is one of these phases and has been shown to incorporate both trivalent and tetravalent plutonium on its Ca site.[4] Perovskite does, however, exhibit a strong preference for trivalent plutonium, which is reflected by its increased solid solubility over tetravalent plutonium. This behavior may be understood by examining the perovskite structure and the impact of substitutional dopants on the Ca site. Perovskite exhibits orthorhombic symmetry at room temperature and is stabilized when the A-site cation (in this instance Ca) is considerably larger than the B-site cation (Ti). For titanate perovskites, Ca is the smallest A-site cation for which the perovskite structure is stable. Consequently, the substitution of trivalent Pu, which is approximately equivalent in size to Ca, can be readily incorporated, but tetravalent Pu, which is considerably smaller than Ca, pushes the perovskite towards a region of structural instability.

The effect of these substitutions on the local structure around tri- and tetravalent Pu substituted into perovskite has been characterized by x-ray absorption spectroscopy. It was also important to establish whether the presence of any local disorder around the Pu site, particularly in the case of tetravalent plutonium-doped perovskite, would result in a decrease in its aqueous durability. In view of the reported water-catalyzed oxidation of PuO_2 to PuO_{2+x}, leach tests were performed under different solution redox conditions to determine the impact on the Pu-perovskite dissolution rate.

RESULTS AND DISCUSSION

Figure 1 shows Pu perovskite radial structure function plots for the reduced (Pu^{3+}) and oxidized (Pu^{4+}) perovskites containing 0.1 formula units of Pu. Significant changes have occurred to the local structure of the perovskite as a result of the change in Pu valence. Notably, a contraction of the principal Pu-O distance is associated with the presence of the smaller tetravalent Pu ion, but perhaps more significantly, a decrease in the intensity of the Pu-Ca interaction has occurred. This is consistent with the presence of local disorder associated with the substitution of Pu^{4+} on the Ca site and reflects the fact that this sample was approaching the solid solubility limit for Pu^{4+} in perovskite.

To determine whether this local disorder around the Pu^{4+} incorporated in the perovskite would impact on its aqueous durability, polished monoliths were prepared and leach tested at 90°C and pH 2 over a 21-day period. Testing was performed at pH 2 to enhance the dissolution rate. The influence of solution redox was determined by leaching the samples under both oxic (open atmosphere) and anoxic (<10 ppm oxygen) conditions. The elemental concentrations in the leachates were analyzed by ICP-MS.

CP673, *Plutonium Futures — The Science,* edited by G. D. Jarvinen

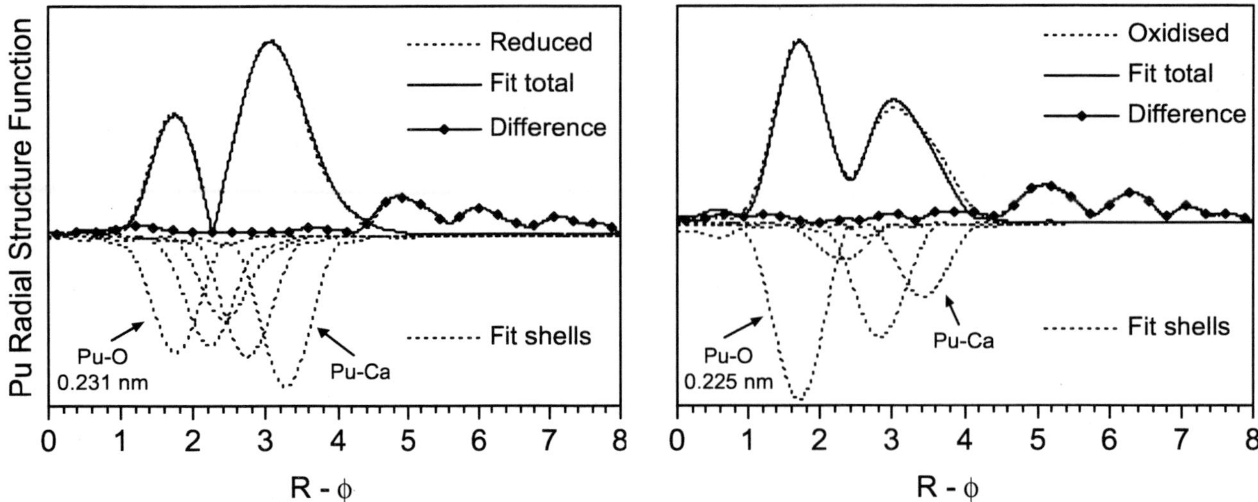

FIGURE 1. Pu radial structure function plots, showing the EXAFS fits and difference plots, for the reduced (Pu^{3+}) and oxidized (Pu^{4+}) $Ca_{0.90}Pu_{0.10}Ti_{0.80}Al_{0.20}O_3$ perovskites.

Initial results have indicated that neither the Pu valence state in the perovskite nor solution redox has any significant effect on the Pu release rate from the perovskite. So although a change in the Pu valence in the perovskite has clearly perturbed the local lattice environment, this has no discernible impact on the Pu dissolution rate in these tests. This is consistent with studies examining the effect of radiation damage, where there is significant perturbation of both the long- and short-range structure with only minimal, less than an order of magnitude, effect on dissolution rate.

The independence of the Pu dissolution rate for both the Pu^{3+}- and Pu^{4+}-perovskites with solution redox would suggest that either the dissolution mechanism does not occur through the surface oxidation of the Pu or if it does, that the inherent Pu release rates for Pu^{3+} and Pu^{4+} from perovskite are equivalent. Given reports of the water-catalyzed oxidation of PuO_2 to PuO_{2+x}, this result would suggest that should this have occurred in Pu^{4+}-perovskite leached under oxidizing conditions that the dissolution rate for Pu^{5+} is also not significantly different from Pu^{4+}. An attempt to look for evidence of a change in the Pu valence after leaching was made using diffuse reflectance spectroscopy, and although this technique is not especially surface sensitive, no evidence of oxidation of either the Pu^{3+} or Pu^{4+} was observed in the perovskites.

REFERENCES

1. Begg, B. D., Vance, E. R., Day, R. A., Hambley, M., and Conradson, S. D., in Scientific Basis for Nuclear Waste Management XX, edited by W. J. Gray and I. R. Triay. Materials Research Society, Pittsburgh, USA (1997), pp. 325–332.
2. Begg, B. D., Vance, E. R., and Lumpkin, G. R., in Scientific Basis for Nuclear Waste Management XXI, Edited by I.G. McKinley and C. McCombie, Materials Research Society, Pittsburgh, USA (1998), pp. 79–86.
3. Vance, E. R., Jostsons, A., Moricca, S., Stewart, M. W. A., Day, R. A., Begg, B. D., Hambley, M. J., Hart, K. P., and Ebbinghaus, B. B., Environmental Issues and Waste Management Technologies IV, Ceramic Transactions Volume 93, edited by J. C. Marra and G. T. Chandler, American Ceramic Society, Westerville, OH, USA (1999), pp. 323–329.
4. Begg, B. D., Vance, E. R., and Conradson, S. D., J. Alloys and Compounds 271–273, 221–226 (1998).

New Pseudophase Structure for α-Pu

J. Bouchet and R. C. Albers

Theoretical Division, Los Alamos National Laboratory, Los Alamos, NM 87545

Pure plutonium is an exotic metal that is poorly understood. The α-Pu phase (zero-temperature ground state) is believed to have mainly itinerant 5f electrons. However, δ phase (fcc) is believed to be in neither limit (neither completely itinerant nor localized), that we may describe as partially localized.[1] How to understand this phase transformation, with its associated large volume per atom expansion, has been an exciting topic of discussion within band-structure theory and related correlated-electron extensions to it for several decades now.[2–5] A correct theory should get both limits right, and is a strong constraint on the theoretical approaches that are used. The problem with including the α phase is its complicated crystal structure, which is monoclinic and involves 16 atoms per unit cell.

In this paper we propose a pseudostructure for α Pu that involves only 2 atoms per unit cell and is an orthorhombic crystal structure. We show that the real α crystal structure only involves small distortions of various atoms around this pseudostructure and that the electronic-structure energy of both systems (at least at the band-structure level) is nearly identical. By a pseudostructure we mean a simpler crystal structure where the atoms are quite close to the positions of the more complex real structure. One can view the pseudostructure as an approximation to the real structure that captures the dominant total energy of the electronic structure (heat of formation) and local bonding effects. If accurate pseudophases exist, they also make possible complicated microstructural models of the phase transformations that capture the dominant components of free-energy differences without requiring huge numbers of parameters.

In the projector augmented-wave (PAW) method, forces can be easily calculated and used to relax atoms in their instantaneous ground state. This allowed us to perform a fully relaxed calculation of the α phase (see Table 1). As in previous calculations,[6,7] the atomic volume is slightly lower compared to the experimental value. We have compared the total energies of the α structure and the γ-Pu structure, which is an orthorhombic distortion of the cubic diamond lattice. The γ phase has a larger energy and the difference is about 36 mRy. If we minimize the total energy of this structure in function of *c/a* and *b/a*, we obtain a contraction in the *z* and in the *y* direction (we will call this new structure pseudo α). The volume increases to be equal to the value obtained for the α structure, and the total energies are very close. We can see the beginning of an explanation if we compare the interatomic distances in the γ, the pseudo α, and the α structure as shown in Table 2. If the surroundings of an atom in the α structure are very complex, we can divide it into two categories: short and long bonds, well separated. An atom has four first-nearest neighbors and ten second-nearest neighbors, except the atom I that has 3 and 13 and the atom VIII that has 5 and 7. During the transformation between γ to pseudo α, the first shell of neighbors contracts around that atom while the second and the third shell expand to be very close to the fourth shell. This gives a surrounding of the four nearest neighbors and 10 second-nearest neighbors for the pseudo α, exactly what is found on the average in the α structure.

TABLE 1. Structural and Internal Parameters (A) for the Experimental and Theoretical α, Pseudo α, and γ Structures of Plutonium. The energies (mRy) are relative to the α structure.

	Pu α		γ	Pseudo α
	Exp.	**Th.**		
a	6.19	5.96	2.85	3.50
b/a	0.78	0.76	1.83	1.61
c/a	1.77	1.83	3.22	2.09
β	101.8	101.8		
Volume	20.0	18.0	17.0	18.0
Energy		0	25	3

CP673, *Plutonium Futures — The Science,* edited by G. D. Jarvinen
2003 American Institute of Physics 0-7354-0140-3

TABLE 2. Nearest Neighbors Distances (A) for the γ Pu, the Pseudo α, and the α Pu.

	Pu γ	Pseudo α	Pu α	
1 n.n	2.73 (4)	2.47 (4)	2.46-2.55 (4)	Short bonds
2 n.n	2.85 (2)	3.31 (4)		
3 n.n	2.97 (4)	3.50 (2)	3.21-3.59 (10)	Long bonds
4 n.n	3.39 (4)	3.49 (4)		

Therefore, the new phase has nearly exactly the same distribution of interatomic distances as the α phase. If we look at the [011] plane of the pseudo-α structure (see Figure 1), we can almost see the α phase. The little distortions are responsible for the very small difference in the total energy of the two structures.

Our pseudostructure is identical with the γ phase of Pu, except that the unit cell dimensions are changed. It is interesting to observe that the *b/a* and *c/a* ratios of the real γ phase of Pu have a tendency to head in the direction of the values for our new pseudophase as the temperature is lowered. Zachariasen and Ellinger measured the lattice constants as a function of temperature: the lengths of *b* and *c* both decrease with temperature, but the length of *a* increases. If the β phase did not interrupt this trend, it is likely that γ Pu would continue in this way until it would reach the phase transformation into the α phase that would then require just a tiny distortion.

In conclusion, we have discovered a very simple structure reproducing the main features of the ground state of plutonium. With this view, we were able to emphasize that the surrounding of an atom in the α structure is the crucial characteristic of the α phase. This structure belongs also to the phase diagram of plutonium and so a very simple path has been discovered between γ and α plutonium. This structure opens new possibilities for exploring the behavior of plutonium.

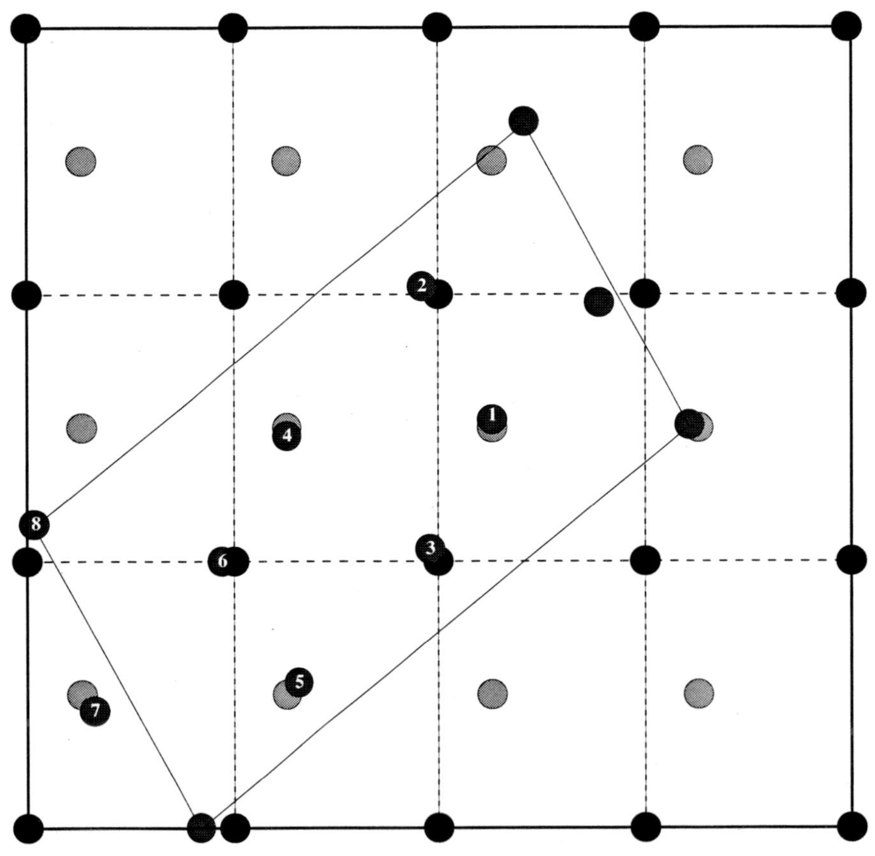

FIGURE 1. Pseudo-α structure in the (011) direction. The blue atoms are the eight positions in the α structure. The black and the grey atoms are the two different positions in the distorted diamond.

REFERENCES

1. Albers, R. C., Nature (London) 410, 759 (2001).
2. Savrasov, S. Y., Kotliar, G., and Abrahams, E., Nature (London) 410, 793 (2001).
3. Savrasov, S. Y., and Kotliar, G., Phys. Rev. Lett 84, 3670 (2000).
4. Bouchet, J., Siberchicot, B., Jolet, F., and Pasturel, A., J. Phys.: Condens. Matter 12, 1723 (2000).
5. Soderlind, P., Landa, A., and Sadigh, B., Phys. Rev. B 66, 205109 (2002).
6. Soderlind, P., Wills, J., Johansson, B., and Eriksson, O., Phys. Rev. B 55, 1997 (1997).
7. Jones, M. D., Boettger, J. C., Albers, R. C., and Singh, D. J., Phys. Rev. B 61, 4644 (2000).

The An-T-Ga Ternary System near the 1:1:5 Composition

P. Boulet, D. Bouexière, J. Rebizant, E. Colineau, and F. Wastin

European Commission, Joint Research Centre, Institute for Transuranium Elements, Postfach 2340, D-76125 Karlsruhe, Germany

Recently, LANL and the Institute for Transuranium Elements (ITU) groups reported on the discovery of superconductivity above 18K in plutonium based compounds, i.e., $PuCoGa_5$.[1,2] This new compound displays evidence for bulk superconductivity with surprising critical parameters leading to the conclusion that the superconductivity in $PuCoGa_5$ may be unconventional and possibly magnetically mediated. This discovery was accompanied with the observation of similar properties in another Pu compound of the same type, $PuRhGa_5$,[3] with lower critical parameters (T_c around 8.6 K, upper critical H_{c2} of 210 kOe) but till unusually high. As a result, the transuranium materials may represent a particularly fertile field for superconductivity of interest both for actinide research and fundamental physics.

We have then asked ourselves: Can we expect other materials displaying unconventional properties in Pu and/or other transuranium related compounds? What is the phase's relation in this class of compounds?

From the structural point of view, it is interesting to observe that ternary intermetallic 1:1:5 compounds, MTX_5 where M stands for Ce[4] or Pu,[1-2] T is a transition metal as Co or Rh, and X is In and Ga, are not unrelated to another interesting material, UGe_2 known to display superconductivity in the ferromagnetic state of the itinerant-electron system.[5] Indeed, these two systems crystallize with a very similar crystal structure in both cases made of simple cubic $AuCu_3$ building unit blocks. Moreover, analyses of as-cast (direct synthesis) of $PuCoGa_5$ samples have shown the stability of this phase to be very narrow. These two observations have led us to investigate the Pu-Ga binary and Pu-Co-Ga ternary phase diagram around the 1:1:5 stoichiometric ratio. The results are illustrated in Figure 1.

Adding a third element in the binary Pu-Ga system leads to a large variety of different compounds crystallising with quite related crystal structures. In fact, for less than a 10% change in the starting composition near 1:1:5, at least 3 distinct compounds can be obtained. On the one hand, this observation shows that all these phases are "point compounds" (i.e., do not accept any homogeneity range). On the other hand, it highlights that the pure 1:1:5 phase is the richest gallium resulting from an equilibrium with pure gallium metal.

In this study, we were also able to determine the relationship of the distinct neighbouring phases that can be found in equilibrium with each other in as-cast samples. An-T-Ga samples, with An = Np, Pu, Am, T = Fe, Co, Ni, Rh, Pd and Ir, were investigated. The unit cells for the different systems are reported in Table 1.

Similarly to that reported for cerium-based compounds, the 2:1:8 compounds crystallize in the Ho_2CoGa_8 crystal structure type. As illustrated in Figure 1, this structure is related to the $PuCoGa_5$ type and is composed of two stacked blocks of $AuCu_3$ cubic units with intercalated Co layers along the c axis. For samples with Ga deficiency, two different types of compounds crystallizing in the $YNiAl_4$ or $Ho_6Co_7Ga_{21}$ structure types are formed. The $YNiAl_4$ type adopts an orthorhombic unit cell also made of $AuCu_3$ layers, but with a different orientation as observed in the $HoCoGa_5$ type (see Figure 1). The tetragonal structure corresponding to the formula $An_6T_7Ga_{21}$ (not displayed in Figure 1) shows a very large unit cell still made up of $AuCu_3$ units.

The study of these ternary systems allowed many new ternary phases to be isolated. All crystal structures displayed are related to each other and are built of common $AuCu_3$ block types along the c axis. Physical properties are now under investigation and will be reported.

CP673, *Plutonium Futures — The Science*, edited by G. D. Jarvinen

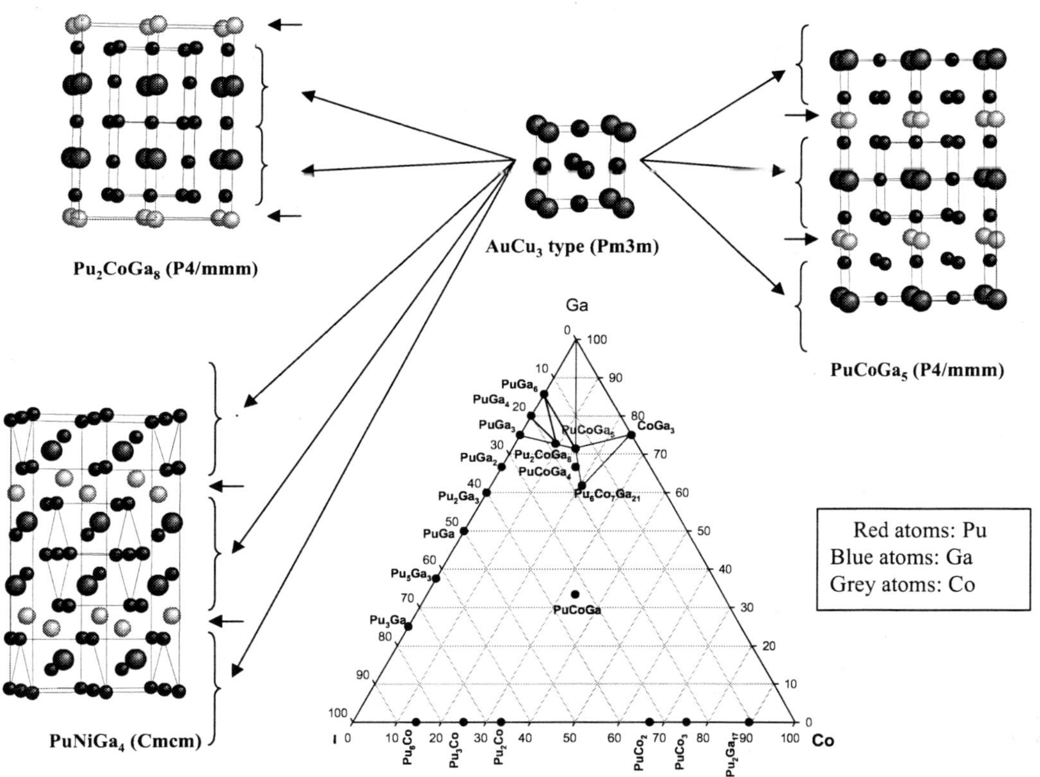

FIGURE 1. Partial isothermal section at 850°C of the ternary phase diagram Pu-Co-Ga The structures of main ternary phases observed are displayed (highlighting their common $AuCu_3$ type building blocks and the cobalt layers).

TABLE 1. Cell Parameters and Crystal Structures of the Different Compounds Obtained in the Ternary Pu-M-Ga System.

	$HoCoGa_5$ type (P4/mmm)		
	a	**c**	**zGa_2**
$PuCoGa_5$	4.235	6.793	0.308
$PuRhGa_5$	4.301	6.856	0.306
$PuIrGa_5$	4.324	6.817	0.302
$NpCoGa_5$	4.237	6.787	0.310
$PuNiGa_5$	4.245	6.796	0.307
$AmCoGa_5$	4.236	6.829	0.309
	Ho_2CoGa_8 type (P4/mmm)		
	a	**c**	
Pu_2FeGa_8	4.280	11.105	
Pu_2CoGa_8	4.269	11.063	
Pu_2RhGa_8	4.316	11.144	
Pu_2PdGa_8	4.309	11.217	
Pu_2IrGa_8	4.322	11.102	
Np_2CoGa_8	4.251	11.021	
Am_2CoGa_8	4.2555	11.165	
	$YNiAl_4$ type (Cmcm)		
	a	**b**	**c**
$PuNiGa_4$	4.121	15.477	6.491
$AmCoGa_4$	4.084	15.61	6.469
	$Ho_6Co_7Ga_{21}$ type (P4/mbm)		
	a	**c**	
$Pu_6Co_7Ga_{21}$	16.780	4.196	
$Am_6Co_7Ga_{21}$	16.828	4.197	

REFERENCES

1. Sarrao, J. L. et al., Nature 420, 297 (2002).
2. Sarrao, J. L. et al., . Phys.: Condens. Matter, in press.
3. Wastin, F. et al., J. Phys.: Condens. Matter, in press.
4. Hegger, S. S. et al., Phys. Rev. Lett. 84, 4986 (2000).
5. Saxena et al., Nature 406, 587 (2000).

Photoemission and Electronic Structure of $UCoGa_5$ and $PuCoGa_5$

M. T. Butterfield,[1] T. Durakiewicz,[1] E. Guziewicz,[1] J. J. Joyce,[1] D. P. Moore,[1] A. J. Arko,[1] L. A. Morales,[1] J. M. Wills,[1] J. L. Sarrao,[1] P. G. Pagliuso,[1] N. M. Moreno,[1] and C. G. Olson[2]

[1]Los Alamos National Laboratory, Los Alamos, NM 8754
[2]Ames Laboratory USDOE, AMES, Iowa 50011

Uranium compounds exhibit a great variety of phenomena such as Pauli paramagnetism, quadrupolar and magnetic ordering, heavy fermion, and unconventional superconductivity. It is considered that these behaviors are closely related to the crystal structure, correlation among f-electrons, and the f-electron hybridization with conduction electrons. Photoelectron spectroscopy (PES) studies have been carried out on single crystals of the compounds $UCoGa_5$ and $PuCoGa_5$ in order to determine the details of their electronic structure. While $PuCoGa_5$ is a superconductor with a T_c of 18 K,[1] and a gamma value of 77 mJ/mole K^2 the Uranium counterpart shows neither significant magnetic nor enhanced mass characteristics. We present initial angle-integrated PES measurements for both the Uranium and Plutonium based compounds taken at photon energies accessible with a Helium lamp. Both samples were cleaned by laser ablation in a vacuum of ~ $5x10^{-11}$ Torr to remove surface oxides and contamination, and measurements made down to a temperature of ~10K with an experimental resolution of ~60 meV. The Pu compound results lean towards the characteristics of its Cerium counterpart, $CeCoIn_5$ (one of the Ce based family of heavy fermion superconductors), with significant electron correlations requiring a description beyond the one-electron approximation. The overall valence band characteristics are similar to those of other Pu materials with a peak at the Fermi level and another broader peak at approximately ~1 eV, though the peak at the Fermi level is not as sharp as those observed in delta Plutonium or $PuSb_2$. A comparison of the $PuCoGa_5$ data will be made with the "mixed level" model (MLM) of Wills et al.[2] and also Density Functional Theory (DFT). The Uranium compound similarly exhibits a peak at the Fermi level and one at around ~1 eV though the ratio of the intensities of the two peaks differ from those of the Pu counterpart.

We also present angle-resolved measurements for $UCoGa_5$ taken on the PGM beamline at the SRC in Wisconsin. UCoGa5 is tetragonal and has a layered structure with units of UGa_3 separated by units of $CoGa_2$. The system cleaves well giving a nice mirror finish along the Ga plane. PES measurements were carried out on freshly cleaved samples at between 15-25 K and spectra were measured were over a wide energy range encompassing the maximum variations in Co 3d vs. U f-electron cross-sections with an overall energy resolution for the lowest photon energy of the order of ~50 meV. The photoemission spectra we will present were taken for lower photon energies (26 eV, 40 eV, and 52 eV) where good angular resolution and small lifetime broadening would allow observation of any dispersion in the U5f bands. This is a good photon energy range to achieve an adequate compromise between the high U5f photo-ionization cross section while still retaining a suitable momentum resolution. The behavior of the observable peaks within 2 eV of the Fermi energy in normal incidence data do not follow simple photoelectron cross-sectional arguments suggesting a considerable degree of hybridization of the 5f electrons. The sharpest peak observed in angle resolved data is considerably removed from the Fermi level and has a FWHM of the order of 0.5eV and shows dispersion of greater than 0.7eV. Though both materials share a very similar structure, the 5f electrons of $PuCoGa_5$ (and the 4f's of its superconducting Cerium counterpart, $CeCoIn_5$) sit on the threshold of itinerant/localized behavior whereas the 5f electrons of $UCoGa_5$ are clearly itinerant, and this leads to a rather nondescript angle resolved photoemission spectra in the case of $UCoGa_5$.

CP673, *Plutonium Futures — The Science,* edited by G. D. Jarvinen

ACKNOWLEDGMENTS

This work was supported by the U.S. DOE, Office of Basic Energy Science, Division of Materials Science and Engineering. The SRC is funded by the N.S. Foundation (grant number DMR-0084402).

REFERENCES

1. Sarrao, J. L., Morales, L. A., Thompson, J. D., Scott, B. L., Stewart, G. R., Wastin, F., Rebizant, J., Boulet, P., Colineau, E., and Lander, G. H., Nature 420, 297–298 (2002).
2. Eriksson, O., Becker, D., Balatsky, A., and Wills, J. M., J. Alloys Comp. 287, 1–5 (1999).

Thermal Modeling Experiment for Pit Storage Areas

Freddie J. Davis,[1] Eric Jensen,[1] Jonathon Ethridge,[1] Tammy French,[1] Kenneth Cary Bell,[1] and Kenneth Schwartz[2]

West Texas A&M University
BWXT Pantex

INTRODUCTION

Current pit storage is in fixed configurations that are well documented and continuously monitored to ensure that thermal loading does not exceed desired specifications. Storage configurations are currently limited in the number of pits stored, storage location, and storage container arrangement. Changes to existing storage configurations, which would modify any of the aforementioned variables, would require that the proposed configuration be created and that extensive measurements be taken to monitor heat loading. This system is very costly, time-consuming, and inflexible.

It is desired to create a method of modeling proposed storage configurations to examine potential thermal loadings before involvement of pits. This can be accomplished by constructing a computational model that can be run for any desired configuration.

This project provides experimental temperature data to support the creation and validation of a computational model. The project includes design and construction of an experimental setup of mock pit storage, design and conduct of experiments, and the acquisition of temperature data.

The objective of the development of an experimental mockup of a pit-storage area is to validate a computational model that can be used to predict the thermal characteristics of planned storage configurations. This model will allow Pantex to accurately model storage configurations to determine what thermal concerns may arise from a proposed configuration. These configurations will include location, quantity, and array orientation.

RESULTS

This work consists of (1) design and construction of experimental apparatus, and (2) conduct of experiments. The experimental apparatus consists of several items: (1) a cylindrical heat source that can be operated at various temperatures, (2) a parallel-piped enclosure to simulate a storage vault, (3) thermocouple probes,[1] (4) digital thermometers,[2] and (5) transient data recorders.[3] Items 3, 4, and 5 are commercially available and are not described within this summary.

A schematic of the cylindrical heat source is provided in Figure 1. It is an aluminum cylinder using a light bulb as the primary heat source. Thermocouples are fixed at various heights and locations on the exterior surface of the cylinder and also throughout the acrylic room to obtain a temperature profile.

This heat source has several advantages. It is economically produced, constructed from standard hardware store materials, and can be uniformly reproduced. Power levels can be easily adjusted by changing the wattage of the light bulbs. The thermocouples locations can be easily varied to the purpose of each experiment.

The simulated storage room measures 8 ft x 4 ft x 4 ft and is constructed of 1/8-in. sheets of acrylic with a wooden support superstructure. The heat sources are then placed within the room, and thermocouple probes are placed at various locations within the room. A typical experiment is illustrated in Figure 2.

CP673, *Plutonium Futures — The Science,* edited by G. D. Jarvinen

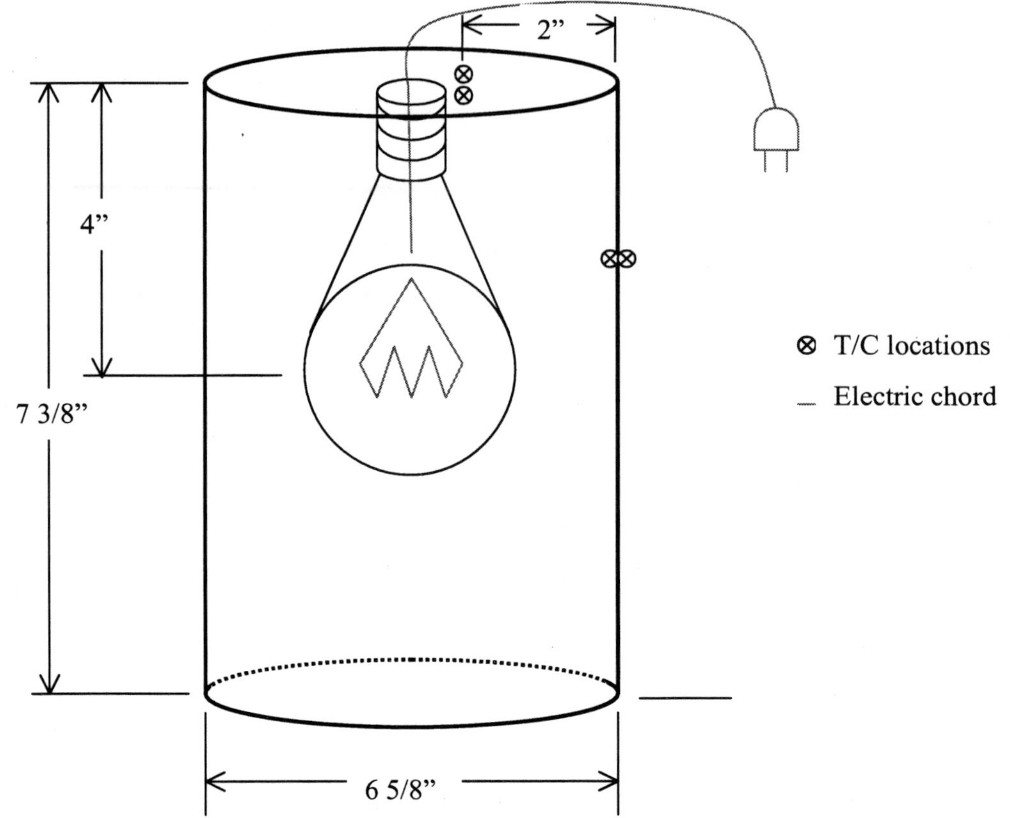

FIGURE 1. Schematic of cylindrical heat source.

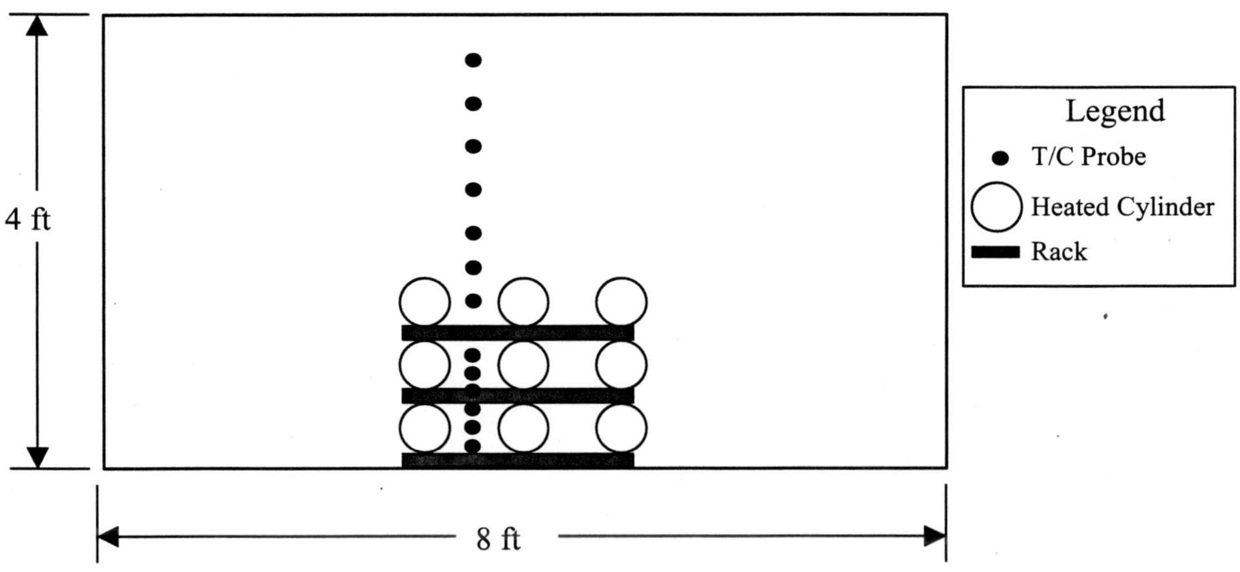

FIGURE 2. Rack configuration.

The experimental results indicate a number of variations, one of which is standing waves in the air temperature at the interstitial locations within the array of cylinders. The vertical profile of temperature is provided in Figure 3. Other experiments provide similar peculiarities and will be discussed.

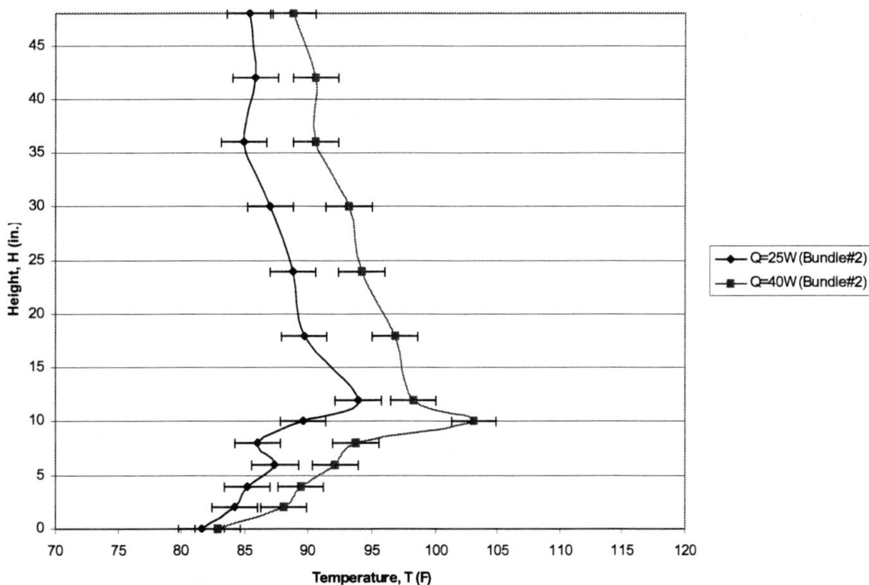

FIGURE 3. Experimental results.

DISCUSSION—SIGNIFICANCE OF THIS WORK

Several simple series of experiments have been conducted to date. Assessment of the current data indicates variation with several parameters, including: (1) bulb wattage, (2) number of heated cylinders, (3) ambient temperature, external to the acrylic room, (4) array configuration of the cylinders, (5) orientation of the cylinders, and (6) location of the array within the room. This is further complicated by the observation of a transition from laminar to turbulent flow regime that is present in many of the experiments and therefore will be difficult to model.

REFERENCES

1. Type K Mini-Thermocouple Plugs and 24 Gauge Type K Thermocouple Wire, Cat Nos., 93-840-52 and 08-530-72, Fisher Scientific Company LLC, 9999 Veterans Memorial Drive, Houston, Texas 77038, (800) 766-7000.
2. Traceable Double Thermometer with Computer Output, Cat No. 15-077-26, Fisher Scientific Company LLC, 9999 Veterans Memorial Drive, Houston, Texas 77038, (800) 766-7000.
3. DAS-4 Data Logger System, Cat No. 06-662-72, Fisher Scientific Company LLC, 9999 Veterans Memorial Drive, Houston, Texas 77038, (800) 766-7000.

Magnetism and Localization in 5f Monopnictides and Monochalcogenides Using PES

T. Durakiewicz,[1] M. T. Butterfield,[1] E. Guziewicz,[1] J. J. Joyce,[1] L. Morales,[1] A. J. Arko,[1] G. H. Lander,[2] F. Wastin,[2] and J. Rebizant[2]

MST Division, Los Alamos National Lab., Los Alamos, NM 87545, USA
European Commission, JRC, Institute for Transuranium Elements, Postfach 2340, D-76125 Karlsruhe, Germany

INTRODUCTION

The electronic structure of actinide compounds constitutes one of the unresolved mysteries of modern condensed matter physics. Existence of multipeak f-character manifolds in the valence band or degree of hybridization in actinide compounds are examples of specific questions that we are trying to address from an experimental point of view. Photoelectron spectroscopy is a tool that allows us to make a simplified yet direct determination of the electronic structure.

In our research, we performed LPLS[1] (Laser Plasma Light Source) and He lamp[2]–based photoemission experiments on single crystals of the AnSb and AnTe (An = U, Np, and Pu) series. We used the LPLS in the energy range of 40–80 eV and also the HeI (21.2 eV) and HeIIa (40.8 eV) and HeIIβ (48.4 eV) UV lines from a He discharge lamp. Sample surfaces were cleaned in UHV (5 x 10^{-11} mbar) by laser ablation to remove oxides and other surface contaminants.

RESULTS AND DISCUSSION

A three-peak manifold within 1 eV of the Fermi energy (E_f) was reported by Gouder et al. for thin films of PuSe.[3] We note excellent agreement between these in situ grown thin-film data and our results for laser-ablated single crystal PuTe. This indicates both the validity of the laser ablation process in obtaining clean surfaces, and, most importantly, similarities in the nature of 5f electronic structure between PuSe and PuTe. Throughout the entire AnSb and AnTe series, there is at least one strong peak in the photoemission spectra that can be identified from its energy dependence as associated with the main spectral weight of the 5f states. The distance of this peak from E_f (all in eV) for the compounds we have measured is USb 0.2, NpSb 0.7, PuSb 1.7, Ute 0.8, NpTe 1.3, and PuTe < 0.1. The compound showing the strongest localization is therefore PuSb (one distinct peak), and those showing the most unusual electronic structure are PuTe and PuSe (data from Reference 3), where at least three peaks exist within the first 1 eV below E_f. An energy-dependence study was performed for PuTe for energies 40.8, 48.4, 60, and 76 eV. No differences between the spectra taken for the first 4 eV below E_f were observed, other than resolution-related smoothing resulting from a lowering of the instrument resolution with increasing photon energy. This observation demonstrates that all the discussed features have a predominant 5f character, even if hybridization exists with the conduction states.

The highly symmetric simple NaCl structure of the monopnictides and monochalcogenides allows for a direct comparison of electronic structures within the AnSb and AnTe series. There exists a correlation between the binding energy of the major 5f peak and other physical properties of the compounds. Magnetic transition temperatures are diminishing with increasing binding energy of the major 5f peak within both AnSb and AnTe series. Distinct systematics are also observed in the 5f peak's binding energies being directly related to the number of 5f electrons in

CP673, *Plutonium Futures — The Science,* edited by G. D. Jarvinen

An and 5p electrons in Sb or Te, respectively. Addition of one electron (either a 5f in An or 5p in Sb, Te) consistently shifts the 5f features towards higher binding energies.

ACKNOWLEDGMENTS

This work was supported by the U.S. Department of Energy, Office of Basic Energy Sciences, Division of Materials Science.

REFERENCES

1. Joyce, J. J. et al., Surf. Sci. Interface Anal. 26, 121 (1998).
2. Durakiewicz, T. et al., Rev. Sci. Instr. 73, 3750 (2002).
3. Gouder, T. et al., Phys. Rev. Lett. 84, 3378 (2000).

Defect-Based Spin Mediation in δ-Phase Plutonium

M. J. Fluss

Lawrence Livermore National Laboratory, P.O. Box 808 East Avenue, Livermore, CA 94577, USA

INTRODUCTION

It has recently been shown, through an analysis of the magnetic susceptibility and electrical resistivity data of δ-stabilized plutonium,[1] that the hypothesis of a Kondo effect[2] ($T_K \sim 200–300$ K) in δ-plutonium is extant, thus implying that 5f electrons are localized, as in the case of other known concentrated Kondo systems. However, although explaining the origin of the anomalous resistivity in δ-stabilized PuAl, the first principal origin of the pure Pu δ phase remains as an unsolved problem. In the past, the α phase has been regarded as well understood within local density approximation (LDA). Recent theoretical work,[3] based on dynamical mean field theory (DMFT) suggests that the δ phase and the α phase of plutonium are two sides of the delocalization-localization knife edge, total energies differing by only a few tenths of an eV. In this picture, the α phase is not a weakly correlated phase: it is just slightly on the delocalized side of the localization-delocalization transition. Here, the δ phase is intrinsically stable, but the stabilized fcc binary alloys (PuAl or PuGa) are the consequence of a "destabilization" of the α phase by small amounts of impurities. In this presentation, we will suggest an alternative mechanism of defect based on spin mediation predicated on our discovery of Kondo-like impurity behavior for vacancies in δ-phase Pu(Ga).

RESULTS

Earlier, we reported on the restively measured isochronal annealing $\Delta\rho(T)/\rho_0$ and temperature dependence of the resistivity, $\rho(T)$, of annealed defect populations resulting from low-temperature (10 K to 20 K) accumulated ion damage in the stabilized delta phase of Pu(3.3 at % Ga).[4] Isochronal annealing curves from damage accumulation of self-irradiation (the alpha decay of plutonium) and from 3.8-MeV proton irradiations were measured and compared with combined molecular-dynamics and kinetic-Monte-Carlo simulations (MD-KMC).[4] As a result of these studies, we also discovered a strong inverse temperature dependence of the resistance of the defect populations produced by damage accumulation and subsequent *partial* annealing.

Damage populations resulting from low-temperature (10 K) damage accumulation were annealed at specific temperatures (30 K, 150 K, and 250 K) producing intermediate *defect populations*. The temperature dependence, $\rho(T)$, was measured by invoking Matthiesson's rule.[5] Therefore, the defect resistance was measured by subtracting the measured resistance of the defected specimen minus the resistance of the same specimen in the fully annealed state. Of course, experimental corrections were required to account for damage accumulated from self-irradiation during the measurements. The specific resistivity of the defects in the 150-K and 250-K populations were found to vary by a factor of eight over the temperature range of 150 K to 10 K. Referring to the MD-KMC modeling[4] indicates that the defect populations at 150 K and 250 K are described primarily as vacancy and small-vacancy clusters. The temperature dependence, $\rho(T)$, of these vacancy-dominated defect populations (150 K and 250 K) follows a $\rho(T) = -a\,ln(T)+b$ dependence and shares a common ln(T) intercept, all of which is suggestive of a Kondo-like magnetic impurity[2] at vacancy defect sites. A systematic deviation from a linear fit is observed in the vicinity of T = 10 K, suggesting that the energy scale or Kondo temperature is $T_K < 10$ K for vacancies and small vacancy clusters in δ-phase PuGa.

CP673, *Plutonium Futures — The Science,* edited by G. D. Jarvinen

DISCUSSION

The consequences of this observation could be profound. It suggests that the implied 5f localization in the neighborhood of vacancies and other dilations in the lattice results in localized magnetic behavior with a low Kondo temperature $T_K < 10$ K compared to that determined earlier for bulk δ-phase PuAl of $T_K \sim 200$–300 K, the latter suggesting an intermediate-valence system.[1] The question then is: Is this the origin of reduced metallic bonding? Might this be the structure-property relationship or *impurity* that precipitates a more global electronic structure change leading to the "stabilization" of the delta phase, as suggested by Cooper and his coworkers?[6]

It is interesting to note that plutonium exhibits excessive relaxation about Ga sites in δ-phase Pu(1 wt % Ga).[7] Although the Ga is found by EXAFS to be substitutional in the fcc lattice, first-neighbor Pu atoms are observed to be 3.7% contracted and the second-nearest neighbors are 0.9% contracted. An estimate for this "free volume" between first and second neighbors is about one-third the volume of a lattice vacancy. Qualitatively, this points to the possibility that the physics leading to Kondo-like behavior of vacancies may be related to the Kondo properties observed earlier in Reference 1 and is consistent with the trend in T_K as well.

Given the above, we posit the following structure-property mechanism for the stabilization of pure Pu in the δ phase. The transition to the δ phase begins at T = 593 K, but the melting point for Pu is 913 K. At two-thirds the melting point, we estimate an equilibrium vacancy population of at least $\sim 10^{-4}$. Indeed, given the low value of the EXAFS Debye-Waller factor measured for Pu-Pu bonds in stabilized δ-PuGa,[8] a higher concentration of vacancies (a lower formation enthalpy) might even be expected. Thus, it is reasonable, but remains to be proven, that a small concentration of equilibrium vacancies gives rise to the local spin sites that stabilize the delta phase of pure plutonium. What is the formalism that underlies this notion?

Such a structure-property mechanism has been described recently by Si et al.,[9] in their analysis of second-order phase transformation of the electronic structure in strongly correlated systems at T = 0 K. The mechanism is *quantum criticality*, by which it is possible to have a localized spin site propagating in time throughout the lattice and whereby such a *critical point* might manifest itself at temperatures hundreds of degrees above T = 0 K. The notion that plutonium may be an example of a new class of quantum-phase transition in which the most singular correlations between the magnetic spins at the quantum critical point have an infinite range in time, but a finite range in space, is indeed exciting. This is the opposite to what happens at a classical critical point, and leads to magnetic interactions that are almost frozen in time, yet localized in space. The low-temperature phase diagram of plutonium and its compounds may very well hold the answer to the origins of the many conflicting anomalies exhibited, provided we can manipulate the system so as to approach the implied quantum critical point.

ACKNOWLEDGMENTS

Work performed under the auspices of the U.S. Department of Energy by Lawrence Livermore National Laboratory under Contract W-7405-ENG-48.

REFERENCES

1. Méot-Reymond, J., and Fournier, J. M., J. Alloys and Compounds 232, 119 (1996).
2. Kondo, J., Prog. Theor. Phys. 32, 37 (1964); ibid, Solid State Physics 23, 184 (1969).
3. Savrasov, S. Y., Kotliar, G., and Abrahams, E., Nature 410, 793 (2001).
4. Fluss, M. J. et al., "Kondo Behavior of Defects from Ion-Irradiation in Pu(Ga)," UCRL-JC-145436 (2001)
5. Matthiesson, A., and Vogt, C., Ann. Phys., Lpzg. 122, 19 (1864).
6. Becker, J. D., Cooper, B. R., Wills, J. M., and Cox, L., J. All. Comp. 271, 367 (1998); ibid. Phys. Rev B 58, 5143 (1998).
7. Cox, L. E., Martinez, R., Nickel, J. H., Conradson, S. D., and Allen, P. G., Phys. Rev B 51, 751 (1995).
8. Allen, P. G., Henderson, A. L., Sylwester, E. R., Turchi, P. E. A., Shen, T. H., Gallegos, G. F., and Booth, C. H., Phys. Rev B 65 (2002).
9. Si, Q., Rabello, S., Ingersent, K., and Lleweilun Smith, J., Nature 413, 804 (2001).

The Properties of Actinide Nanostructures

S. C. Glade, T. W. Trelenberg, J. G. Tobin, P. A. Sterne, and A. V. Hamza

Lawrence Livermore National Laboratory, Livermore, CA, 94550

INTRODUCTION

Predicting the aging behaviors of actinide materials is a problem requiring advances in both theoretical understanding and experimental observation. This work seeks to determine the various correlation and spin contributions to the electronic structure active in actinide elements through observations of structural and electronic evolution in these materials as a function of particle size. Results from these experiments will be used to confirm the several proposed theoretical descriptions of these materials. The measurements [which will include x-ray photoelectron spectroscopy (XPS), ultraviolet photoelectron spectroscopy (UPS), auger electron spectroscopy (AES), low-energy electron diffraction (LEED), and scanning tunnel microscopy (STM)] will be carried out on a number of different uranium and plutonium systems, such as single crystalline nanoparticles and 1-D wires, produced by pulsed laser deposition (PLD).

RESULTS

The construction of the facility for the *in situ* growth of the nanoparticles and the suite of analytical instrumentation mentioned above is mostly completed. The apparatus will be required to maintain an UHV environment (10^{-11}-10^{-12} torr) while still maintaining the flexibility to carry out the objectives listed above. We plan to produce actinide thin films on silicon substrates (100 and stepped 100) through pulsed-laser deposition. The pulsed-laser deposition will be accomplished using an Nd-YAG laser (532 nm) with a 5-ns pulse length. The uranium or plutonium source material will be encased in a fused silica holder with the silicon substrate positioned to intersect the plume of actinide atoms produced when the source material is irradiated with the laser. The silicon substrates can be subsequently heated to possibly produce 1-D wires.

Pulsed laser deposition of uranium on a silicon substrate has been demonstrated with the current system. The presence of deposited uranium was confirmed by XPS (MgKα) measurements in which the 4f uranium levels were observed. Optical microscopy and AFM indicate that the uranium nanoclusters have sizes ranging from ~50 nm to ~5 μm.

Further work on uranium, including further characterization of the PLD process, will be completed before the introduction of plutonium samples into the system.

DISCUSSION

Thus far, few experiments have been preformed on actinide thin films, mostly as a result of difficulties in sample preparation. Uranium films have been produced with a number of different techniques (evaporation, magnetron sputtering, straight diode sputtering) on silicon, graphite, and metal (Pt, Pd, Mg, Al) substrates. Plutonium films have also been produced on Al and Mg substrates;[1] however, these Al and Mg substrates were polycrystalline and yielded polycrystalline films.

The goal of this study is to obtain a detailed predictive description of the structural and electronic properties of actinides materials and requires determining correlation and spin effects: Coulomb energy correlation, electron

CP673, *Plutonium Futures — The Science,* edited by G. D. Jarvinen

exchange correlation, spin and orbital polarization, spin-orbit coupling, and crystal field effects. Through photoemission electron spectroscopy, the effect of particle size on the band structure can be used to isolate the contributions from these correlation effects. This experimental data will ultimately be used to confirm the theoretical description of the structural and electronic properties of α-Pu and δ-Pu, allowing for the prediction of plutonium behavior over tens of years.

ACKNOWLEDGMENTS

This work was performed under the auspices of the U.S. Department of Energy by University of California, Lawrence Livermore National Laboratory under Contract W-7405-ENG-48.

REFERENCES

1. Gouder, T. et. al., Europhys. Lett. 55, 705 (2001).

USb_2 and $PuSb_2$ Electronic Structure: A Photoemission Study

E. Guziewicz,[1] T. Durakiewicz,[1] J. J. Joyce,[1] M. T. Butterfield,[1] L. Morales,[1] A. J. Arko,[1] J. L. Sarrao,[1] J. M. Wills,[1] and C. G. Olson[2]

[1]*Los Alamos National Laboratory, Los Alamos, NM 87545, USA*
[2]*Ames Laboratory, Iowa State University, Ames IA, USA*

The actinide pnictides comprise a very interesting group of magnetic materials. Their electronic structures have been studied on a number of occasions in recent years in order to understand the interactions that give rise to their unusual properties. Photoemission techniques are capable of providing information regarding the binding energy of the 5f electron band, as well as dispersion and hybridization with the conduction band, and therefore are an especially valuable tool for investigation of such systems.

USb_2 is an antiferromagnet below 200 K with a tetragonal layered Cu_2Sb-type structure. $PuSb_2$ has a more complicated orthorhombic $SmSb_2$-type of structure. The magnetic properties of $PuSb_2$ have not been thoroughly studied up to now, but some preliminary susceptibility study and our photoemission results indicate that $PuSb_2$ orders magnetically at low temperature.

We have studied single-crystal USb_2 by angle-resolved photoemission with synchrotron radiation in the photon energy range of 14–110 eV at the SRC in Wisconsin. The overall energy resolution for the lowest photon energy range was 24 meV, and the momentum resolution was 0.09 $Å^{-1}$. The USb_2 samples were cleaved under ultrahigh vacuum conditions, and a smooth and flat surface was obtained for angle-resolved photoemission studies.

A photoemission study of single-crystal $PuSb_2$ was performed using the LPLS (Laser Plasma Light Source) and also a He lamp in the laboratory at Los Alamos. The experimental resolution was 35 meV with a measurement temperature of 10 K. The sample surfaces were prepared by laser ablation to remove oxides and other contaminants.

The photoemission data for USb_2 and $PuSb_2$ show an extremely sharp peak near the Fermi level, as is also the case for $CeSb_2$. This narrow band near the Fermi edge would seem to be a characteristic feature of the electronic structure of the XSb_2 family (X = Ce, U, Pu). Using quantitative lineshape analysis, we determined that the natural linewidth of this peak in $PuSb_2$ is no more than 70 meV FWHM. The natural linewidth of the Fermi level peak in USb_2 was determined as not greater than 10 meV. The difference may be related not to the real lifetime broadening, but to the method of measurement. The data for USb_2 are angle-resolved, but the $PuSb_2$ data are angle-integrated. This may account for a larger linewidth in $PuSb_2$. From cross-sectional arguments, it can be confirmed that in both $PuSb_2$ and USb_2, the narrow structure near the Fermi level is dominated by 5f emissions. However, some conduction band electron hybridization is also present as is suggested by spectra taken at low photon energy. The very good energy and momentum resolution in the USb_2 study allowed observation of a 14-meV dispersion of the peak near the Fermi level, which is evidence of 5f-conduction band hybridization. We argue that the narrow peak in the $PuSb_2$ angle integrated photoemission may be a superposition of even narrower, k-resolved peaks.

The electronic structure away from the Fermi level depends on the specific actinide or lanthanide present in the XSb_2 compound. Photoemission spectra of $PuSb_2$ include a broad feature stretching from 0.25 to 4 eV. By comparison; in USb_2, the second peak observed is narrower and located between 0.25 and 0.8 eV. Cross-section and resonant photoemission measurements taken near the U5d $\rightarrow$ U5f absorption edge confirmed that this structure in USb_2 has mixed U6d-U5f character.

We present a comparison of constrained General Gradient Approximation calculations for $PuSb_2$ with the photoemission results. The calculations, broadened for instrumental resolution, temperature, and a nominal lifetime broadening show good overall agreement with the PES spectra, and the agreement near the Fermi edge is very good.

CP673, *Plutonium Futures — The Science,* edited by G. D. Jarvinen

ACKNOWLEDGMENTS

This work was supported by the U.S. DOE, Office of Basic Energy Science, Division of Materials Science and Engineering. The SRC is funded by the National Science Foundation (grant number DMR-0084402).

The Localized and Itinerant Nature of 5f Electrons in Pu and Pu Compounds

J. J. Joyce, J. M. Wills, T. Durakiewicz, M. T. Butterfield, E. Guziewicz, J. L. Sarrao, L. A. Morales, D. P. Moore, and A. J. Arko

Los Alamos National Laboratory, Los Alamos, NM 87545, USA

We report the electronic structure of Pu metal and several Pu compounds including materials with magnetic ($PuSb_2$, Pu_2RhGa_8) and superconducting ($PuCoGa_5$) transitions. The materials are experimentally determined by photoelectron spectroscopy and computationally by the mixed-level model. The photoemission measurements are generally conducted at low temperatures 10 K–80 K and experimental resolutions between 35 to 75 meV. The electronic structure calculations are a mixed-level model that is an extension of the generalized gradient approximation within density functional theory. As determined both experimentally and computationally, the 5f electrons of plutonium exhibit two different configurations, one localized and the other hybridized or itinerant. The dual nature of the Pu 5f electrons clearly indicates the boundary between localized and itinerant in the actinide series. Interestingly, the Pu 5f electrons exhibit this dual behavior over a wide ligand and crystal structure range. Limiting cases are observed on the localized side of 5f characteristics and some attribute the characteristics of alpha phase Pu metal to nearly itinerant 5f electrons. Results of controlled oxygen and hydrogen dosing for the clean Pu metal surface indicate that oxide formation disrupts the dual 5f electron configuration, but hydrogen exposure seems not to destroy the balance between localized and itinerant 5f electrons.

The photoemission measurements were carried out using the Laser Plasma Light Source (LPLS) at LANL providing a tunable photon energy range between 27 and 140 eV combined with a discharge lamp for high-resolution work providing energies of 21.2, 40.8, and 48.4 eV. The samples are cleaned by laser ablation at low temperature. This method of surface preparation compares well with other reliable methods of surface preparation including *in situ* grown thin films. Six different Pu compounds as well as α- and δ-phase Pu metal have been measured. As the number of compounds has increased, we observe the development of a common electronic structure based on the dual nature of the Pu 5f electrons, which fits many but not all of the Pu materials. This common electronic structure is observed in δ-Pu (nonmagnetic), $PuSb_2$ (magnetic transitions below 100 K believed to be antiferromagetic), $PuCoGa_5$ (superconducting below 18 K), $PuIn_3$ ($\gamma \sim 100$ mJ/mole K^2), and $PuSn_3$.

Although this broad range of Pu materials exhibits a common electronic structure with the 5f electrons in a dual capacity, the details of the electronic structure, particularly in the vicinity of the Fermi energy are distinct for each material. In particular, of the five Pu materials detailed above, $PuSb_2$ has the narrowest quasi-particle peak near the Fermi level. This peak is ~70 meV wide as measured in our angle-integrated photoelectron spectroscopy, but the other materials, including δ-Pu, show peaks near the Fermi level of 100 meV or more. Interestingly, the narrow peak in $PuSb_2$ is very close to the Fermi energy but not at the Fermi energy, possibly a hallmark for magnetic vs enhanced-mass ground states in f-electron materials. Moreover, $PuSb_2$ fits into a broader category of f-electron diantimonides including $CeSb_2$ and USb_2, which both also exhibit magnetic transitions below room temperature and peaks near but not at the Fermi energy.

The photoemission data for δ-Pu, $PuSb_2$, $PuIn_3$, and $PuCoGa_5$ are compared against electronic structure calculations based on the mixed-level model. The mixed-level model is a framework for treating the dual nature of the Pu 5f electrons where a portion of the 5f electrons are localized and do not participate in the bonding while another portion of the 5f manifold is allowed to hybridize with the conduction electrons and participate in the bonding. Within the mixed-level model, calculations show that for these materials, one obtains a good volume and a total energy minimum with 4 of 5 Pu 5f electrons localized. Furthermore, the calculations with 4 of 5 5f electrons localized agree well with the photoemission spectra. The broad range of Pu materials now being investigated at

CP673, *Plutonium Futures — The Science*, edited by G. D. Jarvinen

LANL combined with the successes of the mixed level model to describe the electronic structure are leading to a new paradigm for the understanding of Pu electronic properties.

ACKNOWLEDGMENTS

This work was supported by the U.S. Department of Energy.

Characterization of As-Cast Transmutation Alloys Containing Pu, Zr, Am, and Np*

Dennis D. Keiser, Jr. and J. Rory Kennedy

Argonne National Laboratory-West
P. O. Box 2528
Idaho Falls, ID 83403-2528

INTRODUCTION

Argonne National Laboratory is participating in a national program to develop fuels for use in transmuting long-lived transuranic actinide isotopes in nuclear reactors.[1] This program is part of the Advanced Fuel Cycle Initiative (AFCI). Argonne's focus is to develop metal fuels for transmutation. Pu-Zr-based alloys have been selected for use with additions of Am and Np. Five different alloy rods were cast using an arc-melting process, and their compositions are listed as follows (in wt %): ^{40}Pu-^{60}Zr, ^{60}Pu-^{40}Zr, ^{48}Pu-^{12}Am-^{40}Zr, ^{50}Pu-^{10}Np-^{40}Zr, and ^{40}Pu-^{10}Np-^{10}Am-^{40}Zr. Microstructural characterization was performed using SEM/EDS/WDS. This paper describes the observed alloy microstructures and the distribution of the various alloy constituents.

EXPERIMENTAL PROCEDURE

The charge materials for the alloys were melted using an arc-melting process, and the final rods were cast using quartz molds. The samples were transversely sliced from the as-cast rods using a slow-speed Buehler saw and then polished using grinding papers through 1200 grit. Microstructual analysis was performed using a Zeiss 960A scanning electron microscope. Both secondary and backscattered electron images were taken to evaluate the alloy microstructures. Elemental compositions and spatial distributions were obtained with an Oxford EDS and WDS spectrometer equipped with ISIS LINK software.

RESULTS AND DISCUSSION

Alloys Without Np

SEM analyses of the ^{40}Pu-^{60}Zr, ^{60}Pu-^{40}Zr, and ^{48}Pu-^{12}Am-^{40}Zr alloys suggest the presence of single-phase microstructures. All alloy compositions are given in weight percent. Based on the X-ray maps generated using SEM/WDS, it does appear that there is an even distribution of Pu and Zr in the binary alloys and of Pu, Am, and Zr in the ternary alloy.

For the ^{40}Pu-^{60}Zr alloy, zirconium-oxide phases appeared on the phase boundaries when oxygen ingress from the outer surface of the sample was high enough. Otherwise, a single-phase structure was observed. The grain boundaries and Widmanstätten structure observed for the ^{40}Pu-^{60}Zr alloy suggest the presence of prior β-Zr phase

* Argonne National Laboratory's work was supported by the U.S. Department of Energy, Office of Nuclear Energy, Science and Technology, under contract W-31-109-ENG-38.

CP673, *Plutonium Futures — The Science,* edited by G. D. Jarvinen

boundaries and the precipitation of α-Zr phases in the alloy microstructure. Results from X-ray diffraction analyses on this alloy will be used to confirm this observation.

Looking at the ^{60}Pu-^{40}Zr and ^{48}Pu-^{12}Am-^{40}Zr alloys, it does not appear that adding 12 wt % Am to a Pu-40 wt % Zr alloy affects the morphology of the alloy in a way that is recognizable using SEM/EDS/WDS analysis. No secondary phases seem to develop, suggesting that Am is soluble in the single phase that comprises the microstructurc of a ^{60}Pu-^{40}Zr alloy.

Alloys That Contain Np

SEM analyses of the ^{50}Pu-^{10}Np-^{40}Zr and ^{40}Pu-^{10}Np-^{10}Am-^{10}Zr alloys suggest the presence of multiphase microstructures. X-ray maps show a heterogeneous distribution of the alloy constituents. In the ^{50}Pu-^{10}Np-^{40}Zr alloy, areas enriched in Np and Pu are observed along with areas enriched in Zr. Similarly, the ^{40}Pu-^{10}Np-^{10}Am-^{10}Zr alloy exhibited regions enriched in actinides, viz., Pu, Am, and Np, and regions enriched in Zr. This distribution of alloy constituents suggests that the addition of Np to a Pu-^{40}Zr or a Pu-^{10}Am-^{40}Zr alloy results in a microstructure that contains multiple phases, which is different than what is observed when only Am is added to a Pu-^{40}Zr alloy.

CONCLUSIONS

Based on the results of the alloy microstructural characterization described above, it appears that arc melting techniques can be employed to cast homogeneous alloy rods from Pu, Am, Np, and Zr. The as-cast alloys are single phase for the ^{40}Pu-^{60}Zr, ^{60}Pu-^{40}Zr, and ^{48}Pu-^{12}Am-^{40}Zr compositions and are multiphase for the ^{50}Pu-^{10}Np-^{40}Zr and ^{40}Pu-^{10}Np-^{10}Am-^{40}Zr alloy compositions. The uniform distribution of alloy constituents observed for the various alloys reflects favorably on the possibility of employing these alloys as fuels for the transmutation of long-lived transuranic actinide isotopes.

ACKNOWLEDGMENTS

Support for this work was provided by the U.S. Department of Energy, Office of Nuclear Energy, Science, and Technology, under Contract No. W-31-109-ENG-38.

REFERENCES

1. Hayes, S. L. et al., in Proc. Accelerator Applications/Accelerator Driven Transmutation Technology and Applications '01, Reno, NV, November 12–15, 2001, American Nuclear Society (2001).

Investigation of the Thermal Characteristics of an Americium Bearing Pu-^{40}Zr-Based Alloy

J. Rory Kennedy

Argonne National Laboratory

INTRODUCTION

Nonfertile fuel types under consideration for use in the United States Department of Energy (U.S.-DOE) Advanced Accelerator Applications (AAA) actinide transmutation program included inert matrix metallic alloys. Because the properties of the multicomponent alloys eventually required for use as actinide-burning fuels are to a very large extent unknown, it is a rising imperative that the physical characteristics of these materials, as well as the processing requirements to final fuel fabrication, be determined. Some of the binary metal systems have not been studied; few of the ternary metal systems and virtually none of the quaternary metal systems have been studied.

Because americium (Am) is known to have an unusually high vapor pressure, it was anticipated that long duration sintering processes would result in significant losses of Am during fuel fabrication. These losses would be difficult to quantify and ultimately prevent fabrication of a compositionally consistent product. A laboratory-scale modified arc-melting technique was developed to fabricate metal-alloy fuel pin slugs with minimal loss of Am. These results are documented below.

At least the principal features of the Pu-Am and Pu-Zr binary metal-alloy phase diagrams are established. That of the Am-Zr system is not known. Reported here are results from thermal analysis studies on a single composition in the ternary Pu-Am-Zr system: Pu-^{12}Am-^{40}Zr.

RESULTS

A number of arc-melting casts were made of the Pu-^{12}Am-^{40}Zr target composition, each produced from a minimum of three homogenization melts followed by a casting melt. The feedstock charge materials were a 77.86 (±0.39)Pu-21.2 (±4.2)Am weight percent alloy and nuclear grade zirconium metal (errors at the 2 sigma level in parentheses). Chemical analysis on five different cast samples gave an average weight percent composition of 48.52 (±0.24)Pu-12.0 (±2.4)Am-40.7 (±4.1)Zr. Average losses of <1% of the Am per melting operation were observed.

X-ray diffraction patterns obtained from samples cut from the as-cast pins could be indexed to an fcc phase of the δ-Pu, β-Am type (a = 45.7 pm), most likely representing a δ-$(Pu_{1-x}Am_x)$Zr phase in the ternary system.

Samples were investigated for their thermal characteristics by thermogravimetric-differential thermal analysis (TG-DTA) and thermomechanical analysis (TMA). The heating curves on the as-cast materials are shown in Figure 1. The TMA experiments indicate an apparent single phase at lower temperatures with a linear expansion coefficient of about 8×10^{-6} K^{-1} up to about 710°C by extrapolation method. The phase transition appears as a contraction to a higher density phase with a linear expansion coefficient of about 5×10^{-6} K^{-1}. This same phase transition is observed in the DSC/DTA curve showing an endotherm with an onset temperature of about 700°C. During the course of the DSC/DTA heating, some slight mass gain was observed (~0.16 wt %) that is speculated to be oxidation but represents less than 2 molar percent of the material.

The subsequent cooling curves obtained for the Pu-^{12}Am-^{40}Zr alloy samples are shown in Figure 2. The phase transition around 700°C is reversible with onset temperatures of about 730°C and 710°C for the TMA and DSC/DTA experiments, respectively. Both sets of curves indicate the possibility of one or more broader unclear transitions at lower temperatures. A mass gain of 0.09 wt % was also observed during cooling.

CP673, *Plutonium Futures — The Science,* edited by G. D. Jarvinen

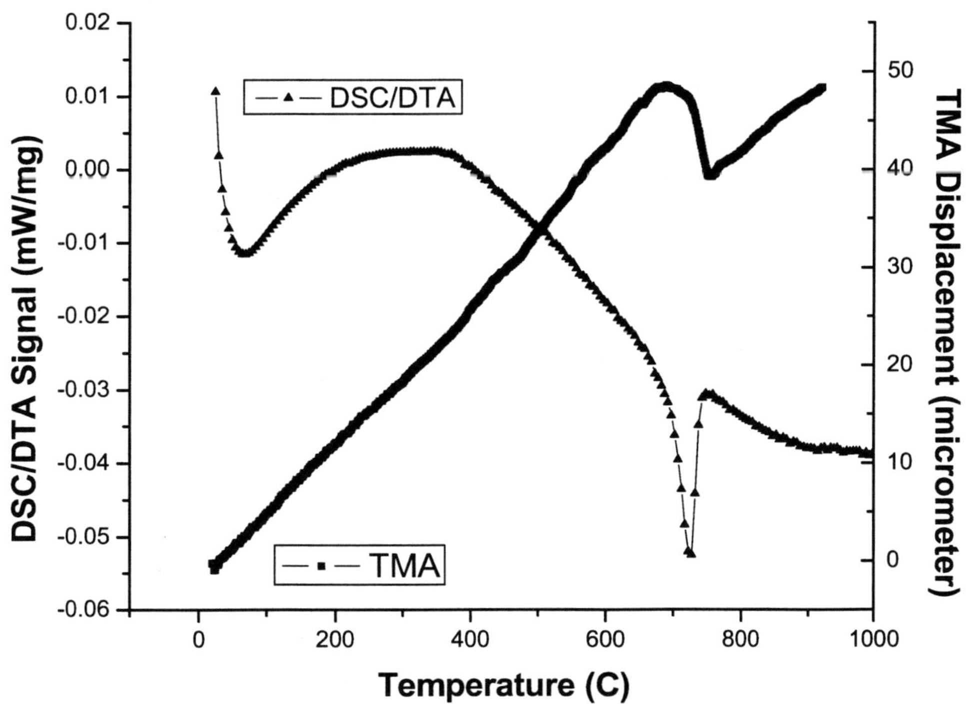

FIGURE 1. DSC/DTA and TMA heating curves for as-cast Pu-12Am-40Zr alloy. Exotherm to positive values for DSC/DTA plot.

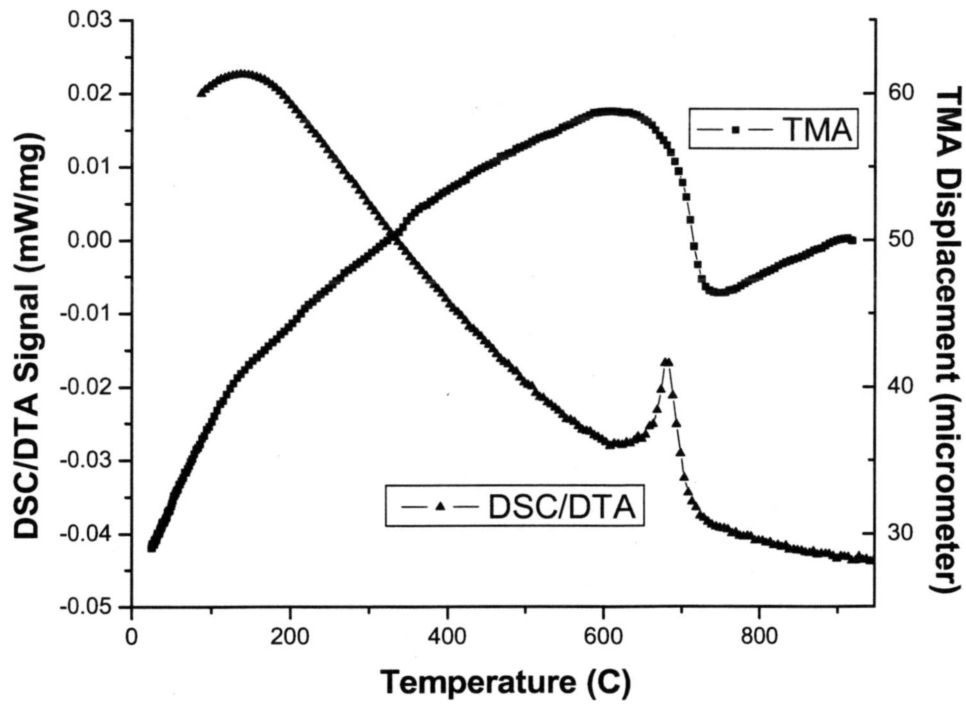

FIGURE 2. DSC/DTA and TMA cooling curves for Pu-12Am-40Zr alloy samples following heating curves in Figure 1. Exotherm-to-positive values for DSC/DTA plot.

DISCUSSION

Americium has an unusually high vapor pressure considering its consensus melting point (1,176°C)[1] and the low vapor pressure and melting points of its preceding elements in the actinide period. In order to produce a consistent metallic product containing Am, a process must be employed that minimizes Am loss. An early report[2] suggested that arc-melting techniques can incur significantly less Am losses than classical powder-processing techniques. Studies conducted in the course of the work reported here concur with the earlier report. Unfortunately, Am could only be analytically quantified to the ±20% level in both the feedstock material and the product alloys. Although processes are currently being developed to better quantify Am in these type alloys, the arc-casting technique used in this study produces an excellent product with very little Am loss.

According to the Pu-Am phase diagram,[2] Pu and Am have a broad mutual solubility range in the δ-Pu, β-Am, fcc, phase field. The phase field extends from room temperature up to the solidus. Even at its narrowest, the phase field extends from Pu-^{5}Am to Pu-^{75}Am. Should this mutual solubility carry over to the ternary Pu-Am-Zr system, the phase behavior of Pu-^{12}Am-^{40}Zr might be expected to mimic that of Pu-^{40}Zr. The study reported on here confirms this expectation. According to the Pu-Zr phase diagram,[2] Pu-^{40}Zr exists as two thermodynamically stable phases below about 270°C and as a single fcc phase above about 350°C. The Pu-Zr fcc phase is a stabilized δ-Pu phase having a much expanded phase field compared to Pu metal. In the as-cast Pu-^{12}Am-^{40}Zr alloys, only a single fcc phase was observed at room temperature, although minor amounts of other phases cannot be ruled out. The very broad fcc phase stabilization in the Pu-Am system would explain this. In Pu-^{40}Zr, a phase transition from an fcc (δ-PuZr) to a bcc structure occurs at 640°C. Both Pu (ε-Pu) and Zr (β-Zr) have a bcc structure at higher temperatures and are mutually soluble. The phase transition from fcc to bcc in Pu is known to be accompanied by a contraction of the lattice to a denser phase. A similar transition is observed from the TMA experiments on Pu-^{12}Am-^{40}Zr (Figures 1 and 2) only at elevated temperatures (700ºC–710ºC). It is suggested here that this transition also represents an fcc-to-bcc transition with the bcc phase being denser than the fcc phase. In Pu metal, the fcc δ-Pu phase is reported to exhibit an unusual negative linear coefficient of expansion.[3] The δ-Pu-^{12}Am-^{40}Zr phase produced here shows a more normal positive linear expansion coefficient, probably due to the influence of both the Am and Zr.

REFERENCES

1. Oetting, F. L., Rand, M. H., and Ackermann, R. J., The Chemical Thermodynamics of Actinide Elements and Compounds, Part 1. The Actinide Elements, IAEA, Vienna (1976).
2. Ellinger, F. H., Johnson, K. A., and Struebing, V. O., J. Nucl. Mater. 20, 83 (1966).
3. See Plutonium Handbook—A Guide to the Technology, edited by O. J. Wick, Amer. Nucl. Soc. Pub. (1980).

Theory for δ-Pu and δ-Pu Based Alloys

Alex Landa,[1] Per Söderlind,[1] and Andrei Ruban[2]

[1]*Lawrence Livermore National Laboratory, University of California, P.O. Box 808, Livermore, CA 94550, USA*
[2]*Center for Atomic-Scale Materials Physics, Technical University of Denmark, DK-2800 Lyngby, Denmark*

INTRODUCTION

At atmospheric pressure, plutonium metal exhibits six crystal structures upon heating from room temperature to its melting point of 913 K. Among these phases, δ-Pu has received significant interest in the metallurgical community because of its high ductility that makes it easy to machine and form. The pure δ-Pu phase is stable at temperatures 593–736 K, but small amounts of a so-called "δ-stabilizer," for example Al, Ga, In, Tl, Sc, Zr, Zn, Am, or Ce, can help retain this phase down to ambient temperature. It is difficult to identify any common properties among the δ-stabilizers. The δ phase can be stabilized by an impurity with the atomic volume larger than δ-Pu (Sc, In, Ce, Tl, and Am) as well as with smaller impurities (Al, Ga, Zn, and Zr). Most of the δ stabilizers (Al, Ga, In, Tl, Sc, Am, and Ce) are trivalent, Zr is tetravalent, and Zn is divalent. Here we explain this stabilizing effect in terms of the balance between ordered and disordered magnetism and how this balance is offset by the addition of an alloying component.

Recently, Söderlind[1] found that an AF (CuAuI) structure is the zero-temperature, ground-state magnetic configuration for δ-Pu. At higher temperatures, δ-Pu is stabilized by magnetic interactions driving a disordered magnetic state. Söderlind et al.[2] proposed that δ-Pu is a disordered magnet that upon cooling undergoes transformation to an AF (CuAuI) structure, with a mechanical destabilization and phase transition to a lower symmetry phase as a result. Considering that the ground-state AF structure is closely followed by the mechanically stable disordered magnetic state, about 2.5 mRy higher in energy, these authors came to the conclusion that the spin entropy could favor the disordered moment state at higher temperatures. Here, we apply these new ideas[1,2] to directly calculate the transition temperature (T_c) of δ-Pu. For this purpose, we have applied density functional theory, implemented in the scalar-relativistic, spin-polarized Korringa-Kohn-Rostoker method within the Green's function formalism (the technical details of the present calculations are very similar to those published[2]). The effect of magnetic moment disorder as well as compositional disorder was treated by means of the coherent potential approximation, implemented in the former case, in conjunction with the disordered local moment (DLM). In order to calculate the DLM state, one uses a random mixture of two distinct magnetic states, namely the spin up (Pu^u) and spin down (Pu^d). The configurational energy of such a binary $Pu^u_{50}Pu^d_{50}$ alloy was mapped into the Ising type Hamiltonian and effective cluster interactions (ECI) were defined by the structure inverse method. These ECI have been applied in Monte Carlo simulations.

RESULTS AND DISCUSSION

The MC simulations were performed by using the Metropolis algorithm for a 13,824-site box (24 x 24 x 24) with periodic boundary conditions. The simulation box was initially filled with randomly distributed Pu^u and Pu^d atoms. The temperature was subsequently lowered by 2 K after 5000 MC steps per atom (MCS/A) and first 1000 MCS/A were discarded before calculation of the thermodynamic averages. In Figure 1, we show the total energy per atom and its temperature derivative in the MC simulations. The first order phase transition DLM→AF occurs at $T_c \approx$ 548 K, which is in fair agreement with the experimental temperature of stabilization of δ-Pu phase (593 K). Although an account of spin-orbit coupling is not essential for the qualitative behavior of δ-Pu, our preliminary

CP673, *Plutonium Futures — The Science,* edited by G. D. Jarvinen

estimation of the ECI-SIM calculated within the FP-LMTO formalism indicates that implementation of relativistic effects within currently used spin-polarized Green's function technique should increase calculated T_c by 10%–15% and improve agreement with experimental data.

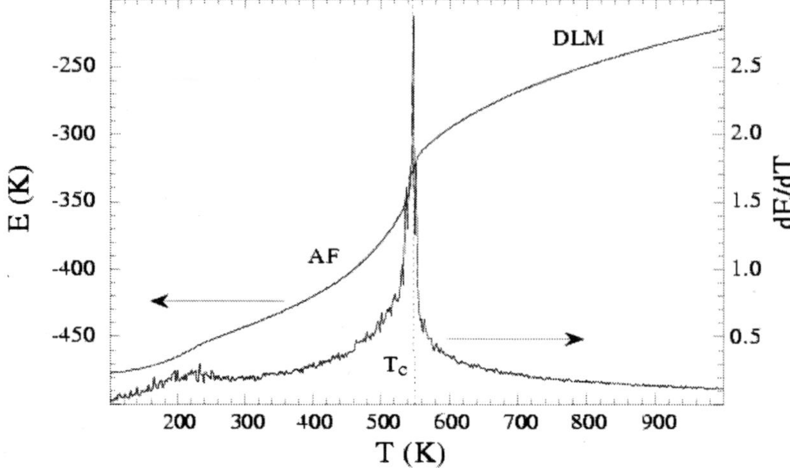

FIGURE 1. The configurational energy per atom E (K) and its first temperature derivative (dE/dT) as a function of temperature in the Monte Carlo simulations.

Figure 2 shows the energy difference between equilibrium DLM and AF spin configurations of $Pu_{90}X_{10}$ disordered fcc alloys, where X (= Sc, In, Ce, Tl, Am, Cm, Th, and Ac) represents the elements with a size exceeding that of δ-Pu (group I), and X (= Ni, Co, Fe, Mn, Zn, Ga, Al, and Zr) represents elements with a size smaller that of δ-Pu (group II). One can see how doping Pu with a large solute atom could lower the total energy of the DLM phase with respect to the AF phase (~2.5 mRy for pure δ-Pu) and thereby stabilize δ-Pu to lower temperatures. It is interesting to note that the nonmagnetic metals from group II (Zn, Ga, Al, and Zr) also promote stabilization of δ-Pu, although this effect is very weak. On the other hand, we found that the magnetic 3d transition metals from group II (Mn, Fe, and Co) strongly destabilize δ-Pu, in agreement with their experimental phase diagram.

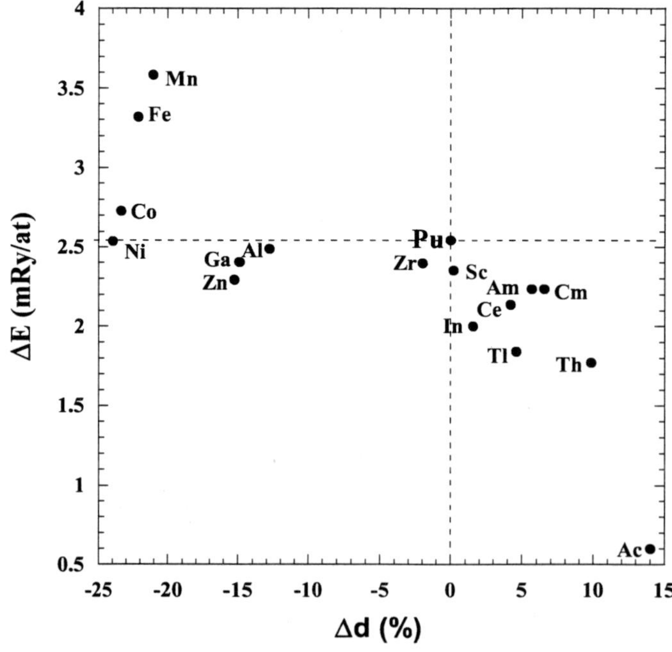

FIGURE 2. The energy difference between equilibrium DLM and AF configurations of $Pu_{90}X_{10}$ disordered fcc alloys as a function of the difference in Wigner-Seitz radius (Δd) between the components (X and δ Pu).

Table 1 shows calculated equilibrium atomic volume (V), bulk modulus (B), and formation energy (ΔE^f) of DLM $Pu_{75}X_{25}$ fcc disordered solid solutions as well as DLM Pu_3X ($L1_2$) compounds (X = Al, Ga, In, and Tl). For comparison, we also show experimental data available[3–6] as well as results of previous spin-restricted DFT calculations.[7] Calculated atomic volumes of Pu_3X compounds agree with the experimental data. In addition, we predict relatively soft bulk moduli, in the range of 42–52 GPa for these compounds (experimental bulk moduli are not available). Our calculations further show that for every Pu_3X compound under consideration, $\Delta E^f < 0$ (indicating that this compound is stable) and also that the formation of this compound is more favorable than the formation of the disordered fcc $Pu_{75}X_{25}$ solid solutions.

TABLE 1. Atomic Volumes (in Å^3), Bulk Modulus (in GPa), and Formation Energies (in mRy/at) for Pu Compounds and Alloys.

System	Volume				
	Experiment	Present Theory	Published Theory[7]	B_{theory}	ΔE^f
$Pu_{75}Al_{25}$		23.3		46.9	7.22
Pu_3Al	23.0[3]	23.2	17.6	50.7	- 4.16
$Pu_{75}Ga_{25}$		23.5		45.5	5.51
Pu_3Ga	22.6[4]	23.0	17.6	52.6	- 10.93
$Pu_{75}In_{25}$		25.6		44.4	- 0.25
Pu_3In	26.0[5]	25.7	20.8	48.7	- 7.43
$Pu_{75}Tl_{25}$		26.4		42.4	5.36
Pu_3Tl	26.3[6]	26.4	21.2	47.7	- 1.63

These efforts could be viewed as the first step towards an *ab initio* treatment of the Pu-X phase diagrams.

ACKNOWLEDGMENTS

This work was performed under the auspices of the U.S. Department of Energy by the University of California Lawrence Livermore National Laboratory under contract W-7405-ENG-48.

REFERENCES

1. Söderlind, P., EuroPhys. Lett. 55, 525 (2001).
2. Söderlind, P., Landa, A., and Sadigh, B., Phys. Rev. B 66, 205109 (2002).
3. Ellinger, F. H., Land, C. C., and Miner, W. N., J. Nucl. Mater. 5, 165 (1962).
4. Ellinger, F. H., Land, C. C., and Struebing, V. O., J. Nucl. Mater. 12, 226 (1964).
5. Ellinger, F. H., Land, C. C., and Johnson, K. A., Trans. Metall. Soc. AIME 233, 1252 (1965).
6. Bochvar, A. A., Konobeevsky, S. T., Kutaitsev, V. I., Menshikova, T. S., and Chebotarev, N. T., in Proceedings of the United Nations, International Conference on the Peaceful Uses of Atomic Energy, United Nations, Geneva (1955), p. 184.
7. Becker, J. D., Wills, J. M., Cox, L., and Cooper, B. R., Phys. Rev. B 54, 17265 (1996); ibid. 58, 5143 (1998); Becker, J. D., Cooper, B. R., Wills, J. M., and Cox, L., J. Alloys Comp. 271–273, 367 (1998).

A Comparison of the Design of Russian and U.S. Containers for Plutonium Oxide Storage

Caroline F. V. Mason,[1] Stanley J. Zygmunt,[1] Douglas E. Wedman,[1] P. Gary Eller,[1] Randall M. Erickson,[1] Walter J. Hansen,[1] and Gary D. Roberson[2]

[1]*Los Alamos National Laboratory, Los Alamos, NM 87545*
[2]*U.S. Department of Energy*

INTRODUCTION

The safe storage of plutonium in the form of plutonium oxide (PuO_2) is a major concern in countries with significant plutonium inventories. The goal is to stabilize and package oxide in such a way that the possibility of leaks and failures are unlikely.

Currently in Russia, PuO_2 is stored[1] at the Mining and Chemical Combine (MCC, Zheleznogorsk) and at the Siberian Chemical Combine (SCC, former Tomsk-7). (Plutonium metal is stored at PA "Mayak" and is not addressed here). Current storage containers for Russian PuO_2 do not meet modern safety requirements. Further, every three years the gaskets have to be replaced. The containers can become overpressurized due to radiation processes, and this results in possible container failures.[1]

In the U.S., PuO_2 is present at several Department of Energy (DOE) sites.[2] U.S. reports of long-time storage of PuO_2 show a few cases of storage container failures [2] among thousands of intact cases. Major causes of malfunction are metal oxidation in nonairtight packages and gas pressurization from inadequately stabilized oxide. Because of these failures, the U.S. DOE adopted a standard[3] for stabilization, packaging and storage of plutonium-bearing material that addresses these vulnerabilities.

Research programs both in Russia[4] and the U.S.[5] continue to evaluate metal corrosion, gas generation (e.g., Reference 6), and interaction of PuO_2 with residual water (e.g., Reference 7) that may contribute to package failure.

CRITERIA

The stabilization requirements of the DOE standard are intended to accomplish the following objectives:

- Eliminate reactive materials such as substoichiometric plutonium oxides;
- Eliminate organic materials;
- Reduce water content to less than 0.5 wt % and reduce equivalent quantities of species that might produce water;
- Minimize potential for readsorption of water above a 0.5 wt % threshold; and
- Stabilize any other potential gas-producing constituents.

The standard also addresses the perceived benefit of calcining PuO_2 to reduce the respirable fraction of the powder.

Although there are Russian regulations regarding the packaging[8] of PuO_2, there are no analogous regulations to the U.S. 3013 standard[3] regarding the stabilization of PuO_2 powder before storage.

CP673, *Plutonium Futures — The Science,* edited by G. D. Jarvinen

RUSSIAN STORAGE CONTAINER DESIGN[8–11]

Figure 1 shows the inner container and convenience can to store oxide[10] that is being developed at the A. A. Bochvar All Russia Research Institute of Inorganic Materials (VNIINM), Moscow. The inner container is sealed with a bolt and gasket and is configured to hold the inner container to prevent motion during handling. In some variants of design, the outer container is welded rather than bolted.

FIGURE 1. Photograph of inner Russian container for PuO_2 storage[10] with convenience can.

The system is designed to hold about 4 kg of PuO_2. The dimensions of the inner container are ~10 in. tall with a diameter of ~10 in. The welded or bolted cover contains an O-ring compression in the top center that allows for gas to be sampled or vented through a built-in filter. The exterior of the can is electropolished. Concern centers around the ability to decontaminate the outer surface of the inner container lid. This may[11] be decontaminated electrolytically or by the use of polymeric coatings.

Figure 2 shows a schematic of the Russian design[8] with the inner and outer containers and a convenience can. There is an alternate version with the convenience can having a wider mouth.

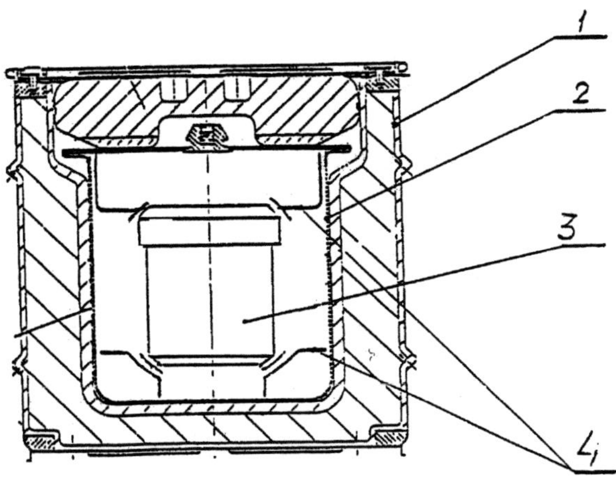

FIGURE 2. Schematic of Russian container design[8] showing: 1—outer container, 2—inner container, 3—convenience can, 4—design elements.

U.S. STORAGE CONTAINER DESIGN[12]

Figure 3 shows the British Nuclear Fuels Ltd. (BNFL) Boundary (Outer) Container and Figure 4 a schematic of the U.S. designed system.

FIGURE 3. British Nuclear Fuels Limited (BNFL) Boundary Container (Outer Container).

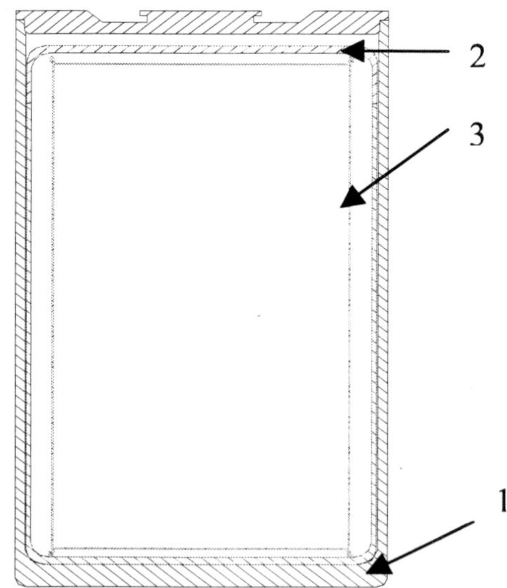

FIGURE 4. Schematic of U.S. can design showing: 1—outer container, 2—inner container, 3—convenience can.

The schematic shows the outer container (the BNFL Boundary Container) with the lifting fixture on the top. Both outer and inner containers are made of welded stainless steel and are designed to the 3013 standard. The dimensions of the whole system are ~10 in. tall with a diameter of ~5 in. The inner convenience can is not required by the 3013 standard but has been made of tin-plated rolled steel as in a food pack can. After the oxide is placed in the convenience can, the can is placed in the inner container that is welded shut. Its outer surface is decontaminated before being placed in the outer container that is also welded[13] shut and can then be placed in a standard shipping container.

The specifications[3] for storage for PuO_2 require the oxide to be thermally stabilized in an oxidizing atmosphere for two hours at 950°C before storage. This ensures that the water content is less than 0.5 wt %, as required by the standard and accomplishes other safety-related goals. The containers are designed to withstand pressures in excess of 700 psig—the theoretical maximum pressure that can be obtained under "worst-case" conditions of 0.5 wt % moisture, 19W heat generation, and 211°C.

COMPARISON

Both Russian and U.S. systems hold about 4 kg of PuO_2. The Russian containers are roughly twice as wide as the U.S. design.

For primary containment, the Russian design uses bolts and gaskets, but the U.S. uses a hermetically sealed welded container.[13] One resulting concern of the Russian design is the alpha radiolytic degradation of gaskets and covers, made of rubber, polypropylene, and fluoroplastics when in direct contact with plutonium.

The Russian system is designed to be attached directly with the plutonium-handling line both for loading and unloading, but in the U.S. system, the containers are inserted into the process line (glovebox line).

In the U.S., containers have to meet the 3013 standards, but in Russia no formal equivalent to the 3013 standard exists.

ACKNOWLEDGMENTS

The authors acknowledge the support of John Baker and Sotirios Thomas, Program Managers for the Materials Disposition Program, U.S. NNSA, and all scientists in Russia working on this program. In addition this work was supported by the DOE Nuclear materials stewardship Project Office, Albuquerque Operations Office. The authors gratefully acknowledge discussions with many subject matter experts over the course of the past several years.

REFERENCES

1. Glagovsky, E. M., communication to S. J. Zygmunt (January 2002).
2. Eller, P. G., Szempruch, R. W., and McClard, J. W., "Summary of Plutonium Oxide and Metal Storage Package Failures," Los Alamos National Laboratory document LA-UR-99-2896 (1999).
3. "DOE STANDARD, Stabilization, Packaging, and Storage of Plutonium Bearing Materials," DOE-STD-3013-2000 (September 2000).
4. LANL-VNIINM Agreement C14610017-35/M9 (August 25,1999).
5. Mason, R., Allen, T., Morales, L., Rink, N., Hagan, R., Fry, D., Foster, L., Wilson, E., Martinez, C., Valdez, M., Hempel, F., and Peterson, O., "Materials Identification and Surveillance: June 1999. Characterization Status Report," Los Alamos National Laboratory document LA-UR-99-3053 (1999).
6. Livingston, R. R., and Duffey, J. M., "Gas Generation Testing of Plutonium Dioxide," ANS Fifth Topical Meeting on Spent Nuclear Fuel and Fissile Materials Management, Charleston, SC, September 2002.
7. Vladimirova, M. V., "Mathematical Modeling of Radiation-Chemical Processing PuO_2 Sorbed Water System," ANS Fifth Topical Meeting on Spent Nuclear Fuel and Fissile Materials Management, Charleston, SC, September 2002.
8. Glagovsky, E. M., Project Manager-Coordinator, "Preliminary Requirements for a Packaging Set (PS) for Storage and Transportation of Weapons Origin PuO_2," Report V 3.4.4, RF Ministry of Atomic Energy, A. A. Bochvar All-Russia Research Institute of Inorganic Materials (VNIINM), Moscow (1998).
9. Glagovsky, E. M., Project Manager-Coordinator, "Development of Technically Viable Options for the Inner Container and Package Set Technical Requirements," Report V 3.2, RF Ministry of Atomic Energy, A. A. Bochvar All-Russia Research Institute of Inorganic Materials (VNIINM), Moscow (1999).

10. Glagovsky, E. M., Project Manager-Coordinator, "Description of the Technological Process of the Fabrication of the Inner Container: Description of Equipment Needed for Manufacturing Inner Container and Loading Can," Report X 3.5, RF Ministry of Atomic Energy, A. A. Bochvar All-Russia Research Institute of Inorganic Materials (VNIINM), Moscow (2001).
11. Glagovsky, E. M., Project Manager-Coordinator, "Upgraded Design of Inner Container. Schematic and Possible Decontamination Procedures," Report X 3.4, RF Ministry of Atomic Energy, A. A. Bochvar All-Russia Research Institute of Inorganic Materials (VNIINM), Moscow (2001).
12. Rivera, Y. M., American Glove Box Society, Orlando, FL, August 2002, Los Alamos National Laboratory document LA-UR-02-4175 (2002).
13. Fernandez, R., Horrell, D. R., Hoth, C. W., Pierce, S. W., Rink, N. A., Rivera, Y. M., and Sandoval, V. D., "Plutonium Metal and Oxide Container Weld Development and Qualification," Los Alamos National Laboratory report LA-13029 (January 1996).

Plutonium Hydriding Research Facility

Gordon W. McGillivray, Ian M. Findlay, Robert M. Harker, and Ian D. Trask

Atomic Weapons Establishment, Tadley, Reading RG7 4PR

In order to be able to consign massive plutonium to long-term storage with any confidence, it is necessary to understand, and thereby be able to model predictively, plutonium's corrosion behavior in a wide range of storage environments.

One of the unique problems associated with plutonium is its extremely high reactivity towards hydrogen, with linear penetration rates of 55 μm/s quoted for a hydrogen pressure of 1 bar.[1] Even when stored in a nominally hydrogen-free environment, plutonium has the ability to generate free molecular hydrogen by a number of routes. A reaction between plutonium and water can lead to the generation of plutonium dioxide and hydrogen/plutonium hydride. Additionally, plutonium can generate hydrogen through the radiolysis of hydrogenous organic materials that may be present either as packing materials or as trace contaminants. Apart from the fact that hydrogen is a corrosive agent in its own right, plutonium hydride can also act to catalyze the oxidation of plutonium; enhancing the oxidation rate in pure oxygen by a factor of 10^{13}.[2] The conversion of massive plutonium into dispersed solids such as hydride and oxide enhances greatly the risk of plutonium migration into the wider environment. As a result, a thorough understanding of the thermodynamics, kinetics, and mechanism of the plutonium hydriding reaction is a necessary precursor to the consignment of massive plutonium to long-term storage.

The year 2003 will see the installation, commissioning, and setting to work of a dedicated plutonium hydriding research facility at AWE. The apparatus consists of two independently operated gas-handling lines, each with its own unique reaction cells. Between them, the two lines cover an operating temperature range of 20°C–700°C and an operating pressure range of 0–30 bar A. The general design of both lines is similar. Each has a bank of storage volumes that can be used to store and mix reactant and processing gases. These gases can then be delivered to the reaction cell through a bank of pressure controllers and flow controllers that will allow experiments to be conducted under a wide range of constant pressure and variable pressure conditions.

One unique feature is a "visual" reaction cell with a hot stage that can operate in the temperature range 20°C–450°C. Coupons are mounted on the hot stage below an optical window. A microscope/camera/image capture system allows the initiation and radial growth of hydriding corrosion sites to be followed in real time. The camera has a 1,200 x 1,600 pixel resolution, allowing the identification of features as small as 8.3 x 6.3 microns within a 1-cm^2 field (the size of a typical test coupon). A x10 magnification capability will allow for the identification of submicron sized features within a 1-mm^2 field, and a computer controlled x–y axis tracking capability will allow the mapping of the entire surface of a 1-cm^2 test coupon at the maximum resolution.

The apparatus will have independent hydrogen and deuterium supplies, both generated electrolytically at a purity of >99.9999%. The ability to use both hydrogen and deuterium will allow the effect of hydrogen isotopes on reaction parameters such as hydride spot initiation and growth rates to be determined, the aim being to use such information to gain an insight into the reaction mechanism. In addition to hydrogen and deuterium, the apparatus will incorporate an oxygen delivery line that can be used both to generate controlled oxide surfaces on the plutonium metal and effect the disposal of postexperiment hydride reaction product as oxide. Additional gas storage bottles will deliver a range of gases, representative of those that are likely to be present within a plutonium storage environment.

REFERENCES

1. Haschke, J. M., and Martz, J. C., J. Alloys Comp. 266, 81–89 (1998).
2. Haschke, J. M., Actinide Hydrides. "Topics in Element Chemistry: Synthesis of Lanthanide and Actinide Compounds," Kluwer Academic Publishers (1991), pp. 1–53.

CP673, *Plutonium Futures — The Science,* edited by G. D. Jarvinen
2003 American Institute of Physics 0-7354-0140-3

Thermodynamic and Spectral Properties of Compressed Ce Calculated by the Combination of the Local Density Approximation and Dynamical Mean Field Theory

A. K. McMahan,[1] K. Held,[2] and R. T. Scalettar[3]

[1]*Lawrence Livermore National Laboratory, University of California, Livermore, CA 94550*
[2]*Max-Planck-Institut für Festörperforschung, D-70569 Stuttgart, Germany*
[3]*Physics Department, University of California, Davis, CA 95616*

INTRODUCTION

It has long been accepted that the evolution between localized and itinerant character in the actinides is caused by 5*f* electron correlation effects.[1-3] Similar behavior occurs in the lanthanides, and in particular Ce,[4] Pr, [5] and Gd.[6] All exhibit pressure-induced phase transitions with unusually large volume changes. A thorough understanding of this general phenomenon should therefore also benefit from a study of the lanthanides as well. To this end, the present work reports calculations for the Ce transition using the merger of the local density approximation (LDA) and Dynamical Mean Field Theory (DMFT). This new approach marries the material specificity and richness of the LDA with a truly correlated solution from the DMFT. It has been used recently in the study of Pu by Savrasov, Kotliar, and Abrahams.[7]

RESULTS

We have calculated thermodynamic and spectral properties of Ce metal over a wide range of volume and temperature, including the effects of 4*f* electron correlations, by the merger of the local density approximation and dynamical mean field theory (DMFT). The DMFT equations are solved using the quantum Monte Carlo technique supplemented by the more approximate Hubbard I and Hartree Fock methods. At large volume, we find Hubbard split spectra, the associated local moment, and an entropy consistent with degeneracy in the moment direction. On compression through the volume range of the observed γ–α transition, an Abrikosov-Suhl resonance begins to grow rapidly in the 4*f* spectra at the Fermi level, a corresponding peak develops in the specific heat, and the entropy drops rapidly in the presence of a persistent, although somewhat reduced local moment. Our parameter-free spectra agree well with experiments at the α- and γ-Ce volumes, and a region of negative curvature in the correlation energy leads to a shallowness in the low-temperature total energy over this volume range that is consistent with the γ–α transition. As measured by the double occupancy, we find a noticeable decrease in correlation on compression across the transition; however, even at the smallest volumes considered, Ce remains strongly correlated with residual Hubbard bands to either side of a dominant Fermi-level structure. These characteristics are discussed in light of current theories for the volume collapse transition in Ce. This work has appeared in print elsewhere.[8]

DISCUSSION

This work presents a thorough exposition of thermodynamic and spectral properties of compressed Ce using the new LDA+DMFT technique, and the results shed light on the γ–α transition and on the pressure-induced evolution

CP673, *Plutonium Futures — The Science*, edited by G. D. Jarvinen

from localized towards itinerant behavior in this material, which may have broader implications for other *f*-electron metals.

ACKNOWLEDGMENTS

Work by AKM was performed under the auspices of the U.S. Department of Energy by the University of California, Lawrence Livermore National Laboratory under contract No. W-7405-ENG-48. K. Held acknowledges support by the Alexander von Humboldt Foundation, and R. T. Scalettar from NSF-DMR-9985978.

REFERENCES

1. Benedict, U., J. Alloys Comp. 193, 88 (1993); Holzapfel, W. B., J. Alloys Comp. 223, 170 (1995).
2. Brooks, M. S. S., Johansson, B., and Skriver, H. L., in Handbook on the Physics and Chemistry of the Actinides, edited by A. J. Freeman and G. H. Lander, North-Holland, Amsterdam, Vol. 1 (1984), p. 153.
3. Boring, A. M., and Smith, J. L., Los Alamos Science 26, 90 (2000).
4. Koskimaki, D. G., and Gschneidner, K. A. Jr., in Handbook on the Physics and Chemistry of Rare Earths, edited by K. A. Gschneidner Jr. and L. R. Eyring, North-Holland, Amsterdam (1978), p. 337; Olsen, J. S., Gerward, L., Benedict, U., and Itié, J. P., Physica 133B, 129 (1985).
5. Mao, H. K. et al., J. Appl. Phys. 52, 4572 (1981); Smith, G. S., and Akella, J., J. Appl. Phys. 53, 9212 (1982); Grosshans, W. A., and Holzapfel, W. B., J. Phys. (Paris) 45, C8 (1984); Zhao, Y. C. et al., Phys. Rev. B 52, 134 (1995); Baer, B. J. et al. (in press).
6. Hua, H., Vohra, Y. K., Akella, J., Weir, S. T., Ahuja, R., and Johansson, B., Rev. High Pres. Sci. & Tech. 7, 233 (1998).
7. Savrasov, S. Y., Kotliar, G., and Abrahams, E., Nature 410, 793 (2001); preprint, cond-mat/0106308.
8. Held, K., McMahan, A. K., and Scalettar, R. T., Phys. Rev. Lett. 87, 276404 (2001); McMahan, A. K., Held, K., and Scalettar, R. T., Phys. Rev. B (in press), cond-mat/0208443.

The Failure of Russell-Saunders Coupling in the 5f States of Plutonium

K. T. Moore,[1] M. A. Wall,[1] A. J. Schwartz,[1] B. W. Chung,[1]
D. K. Shuh,[2] R. K. Schulze,[3] and J. G. Tobin[1]

[1]*Lawrence Livermore National Laboratory, Chemistry and Materials Science Directorate, L-350, Livermore, CA 94550, U.S.A.*
[2]*Lawrence Berkeley National Laboratory, Department of Chemical Sciences, Berkeley, CA 94720, U.S.A.*
[3]*Los Alamos National Laboratory, MST-6, MS G755, Los Alamos, NM 87545, U.S.A.*

The nature of the Pu-5f electronic structure is still under debate.[1–6] Many of the complications are derived from the necessity of explaining the phase-specific behavior of Pu and Pu alloys, particularly the low-symmetry monoclinic α phase and the high-symmetry fcc δ phase. Experimentally, there are severe impediments, such as the present inability to grow large single crystals and the radioactive and chemical hazards of the materials. Theoretically, no single model has gained universal acceptance because of the limitations of each approach. Recent advances include the application of dynamical mean-field theory (DMFT) to δ-Pu,[1] generalized gradient approximation with a Hubbard U (GGA-U) to δ-Pu[2] and density functional theory with gradient density corrections and spin-orbit polarization to α- and δ-Pu.[3] However, many of the same questions remain from earlier formulations.[4,5] Several of these key questions revolve around the interaction of the spin and orbital angular momenta. In fact, until now, it was even unclear which momentum-coupling scheme should be used with the Pu-5f states.

In the foundation of each of the above approaches to modeling[1–5] there are two limiting cases for the coupling of angular momenta in multielectronic systems: Russell-Saunders (LS) coupling (Figure 1) and JJ coupling. For a simple, two-electron system, these can be expressed as shown in Table 1. For atoms where the spin-orbit coupling is weak, the interactions between the orbital angular momenta of individual electrons is stronger than the spin-orbit coupling between the spin and orbital angular momenta. It can be assumed that the orbital angular momenta of the individual electrons add to form a resultant orbital angular momentum L, and that the individual spin angular momenta add to form a resultant spin angular momentum S. L and S are then summed to form the total angular momentum J. For heavier elements with larger nuclear charge, the spin-orbit interactions become as strong as the interactions between individual spins or orbital angular momenta. Therefore, the spin and orbital angular momenta of each electron couples to form individual electron angular momenta. These individual electron angular momenta j_1, j_2, j_3... are then summed to give the total electron angular momenta J. It has been shown that while the two schemes produce similar trends in derived quantities, there is an important discrepancy between the results of the two schemes, which generates very different results on an element-by-element basis. Historically, the Russell-Saunders approach has been shown to be generally highly successful with the rare earths, and because of this early modeling of actinides was based on a nonrelativistic approach, i.e., neglecting spin-orbit splitting.[4] More recent calculations have either explicitly included the spin-orbit splitting[3] or generated results that are consistent with a JJ scheme.[2] This JJ coupling should become more relevant as the atomic number increases, however, until now there remained significant uncertainty about which coupling scheme is appropriate. In this document, we will demonstrate that the disappearance of prepeak structure at energies below the main continuum peak of the $O_{4,5}$ absorption edge of Pu metals is direct proof of the occurrence of JJ coupling in Pu.

High-energy, electron-energy loss spectroscopy (HE-EELS) results acquired using a transmission electron microscope (TEM) are shown for α-Th, α-U and α- and δ-Pu in Figures 2 and 3. The salient feature of the $O_{4,5}$ HE-EELS edges in Figure 2 is the disappearance of the small prepeak below the main continuum peak in both α- and δ-Pu. This is direct evidence of the presence of JJ coupling in Pu because the absence of a prepeak correlates to the filling of the $5f^{5/2}$ sublevel in the $5d^{10}5f^5 \rightarrow 5d^9 5f^6$ transition.

CP673, *Plutonium Futures — The Science,* edited by G. D. Jarvinen

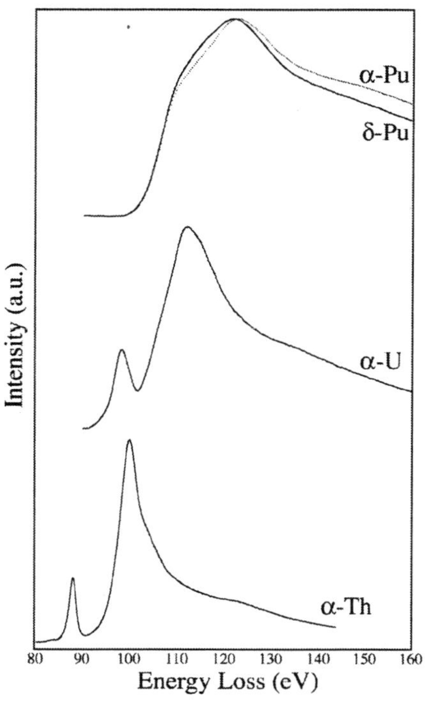

FIGURE 1. The failure of Russell-Saunders coupling in the 5f states of plutonium.

TABLE 1. RS vs JJ Coupling for Two Particles (1,2) and Four Angular Momenta (l_1,l_2,s_1,s_2).

RS	JJ
$L = l_1 + l_2$	$j_1 = l_1 + s_1$
$S = s_1 + s_2$	$j_2 = l_2 + s_2$

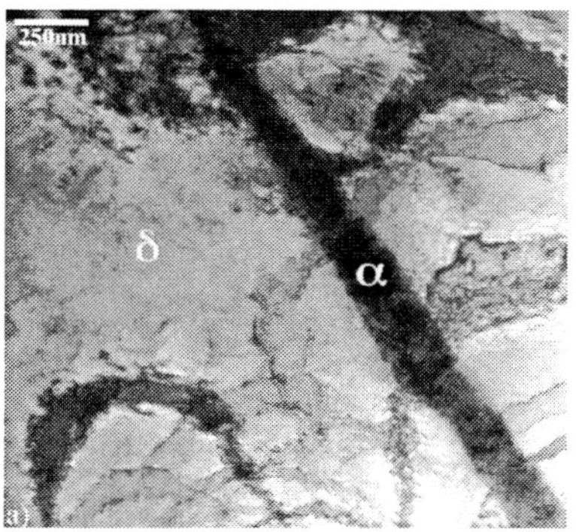

FIGURE 2. The $O_{4,5}$ (5d → 5f) absorption edges from α-Th, α-U, α-Pu, and δ-Pu acquired by HE-EELS in a TEM.

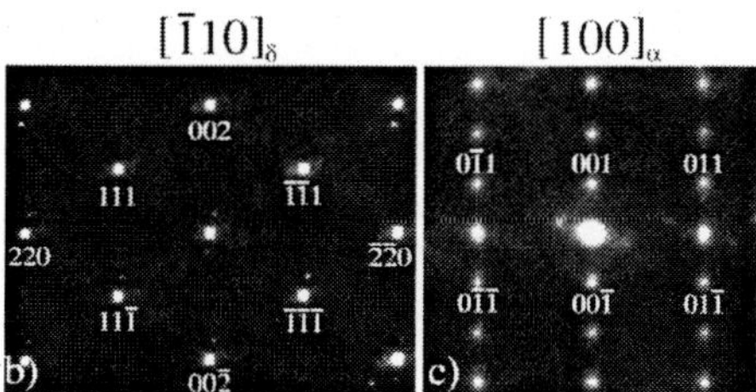

FIGURE 3. (a) A bright-field TEM image acquired near a $[\bar{1}10]_\delta \| [100]_\alpha$ zone axis showing an α plate in a δ matrix, (b) a $[\bar{1}10]$ diffraction pattern from δ, and (c) a $[100]$ diffraction pattern from α, each with a number of reflections indexed.

Synchrotron radiation-based x-ray absorption spectroscopy (XAS) and photoelectron spectroscopy measurements were performed at the Advanced Light Source to support these results. The Pu $O_{4,5}$ XAS absorption edges (5d → 5f) for α- and δ-Pu in Figure 4(a) are quite similar to the HE-EELS edges reported in Figure 1. The similarity between HE-EELS and XAS can be further confirmed by comparing our U and Th HE-EELS to previously published absorption results. In both the HE-EELS and XAS $O_{4,5}$ absorption edge of Th and U the pre-peak is present. In the case of Pu, both the HE-EELS and XAS of the Pu 5d → 5f transition exhibit only the large continuum structure with no prepeak; a small shift of approximately 1 eV between α- and δ-Pu; a peak leading edge in the region of 110 to 120 eV; and a decrease on the higher energy side as the photon energy goes toward 160 eV.

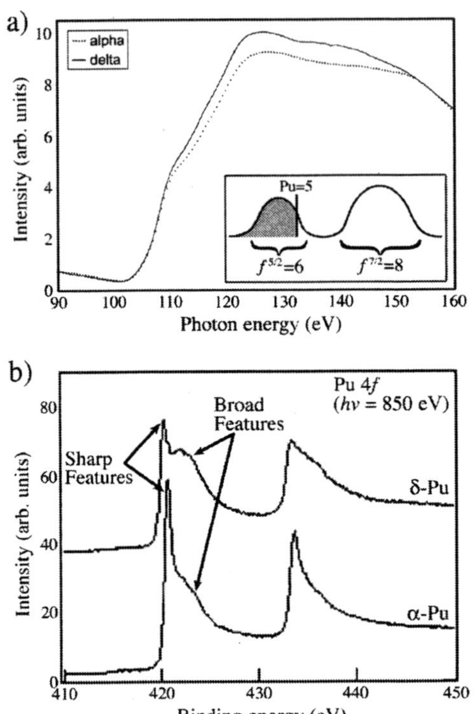

FIGURE 4. (a) The $O_{4,5}$ (5d → 5f) absorption edges from α-Pu and δ-Pu acquired by XAS using synchrotron radiation. (b) Core-level photoemission spectra from a large polycrystalline δ-Pu sample and a polycrystalline α-Pu sample.

In the case of the XAS samples, in situ phase determination was made through the photoelectron spectroscopy of the Pu 4f levels, as shown in Figure 4(b). The double component structure of each spin-orbit split peak, with a shoulder or broad maximum following the sharper line, has been shown previously by Havela and Gouder[7] to be a reliable means of differentiating the α- and δ-Pu phases.

In the 4f levels of the lanthanides, where Russell-Saunders coupling occurs, the only filling effects that are observed are at the end of the series. In the actinides, the filling can be seen to occur midseries, when the $(5f^{5/2})$ occupation goes from 5 to 6 in the process of XAS or HE-EELS. There is evidence that the filled $(5f^{5/2})^6$ exists in the Am ground state and affects its properties.[8,9] Because we have found direct evidence of spin-orbit splitting in the 5f states of Pu, Russell-Sunders coupling schemes must fail for the 5f states of Pu and the other actinides and only the use JJ coupling is appropriate.

REFERENCES

1. Savrasov, S. Y., Kotliar, G., and Abrahams, E., Nature 410, 793 (2001).
2. Savrasov, S. Y., and Kotliar, G., Phys. Rev. Lett. 84, 3670 (2000).
3. Söderlind, P., Europhys. Lett. 55, 525 (2001).
4. Skrivers, H. L., Andersen, O. K., and Johansson, B., Phys. Rev. Lett. 41, 42 (1978).
5. Johansson, B., Phys. Rev. B 11, 2740 (1975).
6. Hecker, S. S., Harbur, D. R., and Zocco, T. G., Prog. Mat. Sci., in press; "Challenges in Plutonium Science," Los Alamos Science 26 (2002), Vols. I and II.
7. Havela, L. et al., Phys. Rev. B 65, 235118 (2002).
8. Naegele, J. R. et al., Phys. Rev. Lett. 52, 1834 (1984).
9. Pénicaud, M., J. Phys: Condens. Matter 9, 6341 (1997).

Aging and Phase Stability in δ-Stabilized Pu

L. A. Morales, A. C. Lawson, S. Conradson, E. N. Butler, D. P. Moore, M. Ramos, J. A. Roberts, and B. Martinez

Los Alamos National Laboratory

Optimal extraction of information from powder diffraction patterns requires an accurate description of the diffraction line shape. Most refinement techniques, such as Rietveld, treat the diffraction line width as a smooth function of d-spacing or diffraction angle. Anisotropic line-shape broadening in which the peak width is not a smooth function of d spacing, is sometimes observed in powder patterns and has been difficult to model with whole pattern fitting or Rietveld techniques. Recently this technical problem has been overcome by incorporating the anisotropic broadening model of Stephens into the GSAS Rietveld refinement package (1, 2). If the influence of stacking faults can be neglected, the broadening of the diffraction line shape can be attributed mainly to the strain fields surrounding dislocations and the domain size or ordering length. In particular, this dislocation model of the mean square strain, which is related to the peak width, employs a knowledge of the individual contrast factors, C, of dislocations related to particular Burgers vectors as well as to the elastic constants of the material of interest (3, 4).

Such an approach has been applied to the understanding of the anisotropic line broadening and microstrain effects now discovered in various δ-stabilized Pu-Ga alloys. X-ray diffraction powder patterns were obtained with a Scintag X2 diffractometer, the high intensity powder diffractometer (HIPD) at the Los Alamos Neutron Science Center, and on beamline 7-2 at the Stanford Synchrotron Radiation Laboratory (SSRL). For a variety of aged samples, line profiles of the first seven reflections were measured with negligible instrumental line broadening. High resolution diffraction data were collected at SSRL.

The peakwidths obtained by neutron diffraction for $^{242}Pu_{0.98}Ga_{0.02}$ were found to be temperature dependent. During the first cooling cycle, the microstrain broadening could be distinguished from instrumental broadening by 150 K; the peak broadening increased as the sample was cooled to 15 K. The microstrain relaxed but was not removed as the sample was warmed to room temperature. The line broadening was reproduced during the second and third cooling of the sample. The microstrain was completely relieved only when the sample was heated to 650 K.

For each research alloy sample examined by x-ray diffraction, anisotropic line-shape broadening and microstrain effects were observed during each variety of diffraction experiment. In the x-ray diffraction experiments, the overall microstrain scaled with the age of the specimen, however, texturing of some samples interfered with this correlation. The dislocation density and domain size were calculated for each sample studied. The dislocation density was found to increase with time, while the domain size tended to decrease with sample age. However, high-resolution diffraction studies at SSRL clearly showed that in some samples the peaks were split into multiple reflections. Further experiments showed that the degree of peak splitting scaled with the Ga composition and distribution. Well-homogenized samples showed no peak splitting and could be fit with a cubic (FCC) model. Materials in which the Ga composition was modulated in order to induce microstrain, showed multiple reflections.

The microstrain broadening was assumed to be a measure of dislocations induced by processing and aging. The evolution of microstrain broadening for naturally aged samples less than 10 years old is attributed to the formation of point defects such as Frenkel pairs (vacancies and interstitials) resulting from radioactive decay. These point defects can coalesce with time to yield dislocations and even dislocation loops. The observed anisotropic line broadening is attributed to the anisotropic elastic constants for δ-stabilized plutonium. The peakwidths for the family of 001 reflections are exagerated because that is the "soft" crystallographic direction in δ-stabilized plutonium. As the strain fields evolve, the δ-stabilized plutonium lattice yields in the 001 direction.

Recent high-resolution x-ray diffraction data on freshly prepared alloys with an induced Ga gradient as well as older, naturally aged samples clearly show peak splittings. These patterns were indexed to a tetragonal cell. As the

CP673, *Plutonium Futures — The Science,* edited by G. D. Jarvinen

stain fields around various sites evolve, the lattice distorts to form a meta-stable phase, new to the Pu-Ga system, which may be an intermediate between δ-stabilized plutonium and the two-phase mixture of alpha plutonium and Pu_3Ga. The formation of this meta-stable phase associated with a modulated Ga composition is consistent with spinodal decomposition.

REFERENCES:

1. Stephens, P. W., 1999, J. Appl. Cryst. **32** 281.
2. Larson, A. C., and Von Dreele, R. B., 1986, Los Alamos National Laboratory Report LAUR 86-748.
3. Ungár, T., Leoni, M. and Scardi, P., 1999, J. Appl. Cryst. **32** 290.
4. Ungár, T., and Tichy, G., 1999, Phys. Stat. Sol. **(a) 171** 425.

Measurement of Mechanical Properties of Delta Plutonium Metal Using Spherical Microindentation

Roberta N. Mulford[1] and Robert Asaro[2]

[1]*NMT-15, Los Alamos National Laboratory*
[2]*Department of Structural Engineering, UCSD*

Spherical microindentation is a simple adaptation of the instrumented microhardness technique for evaluating the mechanical properties of a material from small samples. This method was applied to plutonium to determine whether changes in constitutive properties result from self-induced radiation damage that accumulates over many years.

Radiation damage by autoradiolysis in delta plutonium metal is hypothesized to be the cause of slight density changes observed in the metal at long times.[1] Voids, vacancies, and dislocations remain after dissipation of the recoil energy of the daughter uranium atom produced during the decay. Accumulation of such damage structures results in a reduction in bulk density in the metal. The helium atom produced in the decay is captured in the lattice, eventually forming helium bubbles, which have been seen in TEM measurements.[2]

Both radiation damage and accumulation of helium may influence the phase stability and mechanical properties of the metal as it ages. Mechanical properties, in turn, may govern the evolution of helium bubbles into voids, and an understanding of changing mechanical properties may facilitate prediction of the onset of void swelling, which produces radical changes in the metal. To investigate the variation of the constitutive behavior of delta plutonium metal during aging, a method is required that will track constitutive properties from examination of small samples, permitting multiple measurements to be performed on an individual sample over time to track any evolution of constitutive properties.

Spherical microindentation is an integrated test that can be used to measure hardness and other mechanical properties near the surface of a metallic specimen. Methods for analyzing the data have evolved from the earliest relation of Brinnell hardness to work-hardening coefficient and to flow stress,[3,4] to sophisticated analysis of the strain fields resulting from various pressure distributions under the spherical indenter.[5] The varying shape of the contact between the indenter and the metal surface ("sinking in" or "piling up.") can be described as a function of work hardening in the specimen. This nonideal contact has been addressed by a variety of adjustment methods to the radius used in the geometrical analyses, in an attempt to produce an accurate and unique description of flow stress and work hardening from a set of indentation records.

Modern instrumentation allows continuous recording of load and displacement during indentation, which supports a more sophisticated analysis than was previously possible and allows existing methods of analysis to be applied in a more thorough way to obtain mechanical properties directly.

We have performed sets of spherical indentation experiments on plutonium and on tantalum, copper, and aluminum. These other metals allowed development of the analysis method and confirmation of the method against known mechanical threshold stress (MTS) constitutive model[6] parameterizations and measured stress-strain curves for materials with a variety of behaviors, without reference to plutonium. Indents on plutonium yielded aging data, but were not readily calibrated against a set of known properties for a standard sample.

Finite element (FE) modeling of the systems was performed independently of the experiments, using the MTS constitutive model[6] with parameters determined from the material of the same pedigree as that used for the indentation experiments (Figure 1). The FE modeling of the anticipated behavior of the plutonium used a simplified version of the MTS model in which temperature dependence was neglected.

CP673, *Plutonium Futures — The Science,* edited by G. D. Jarvinen

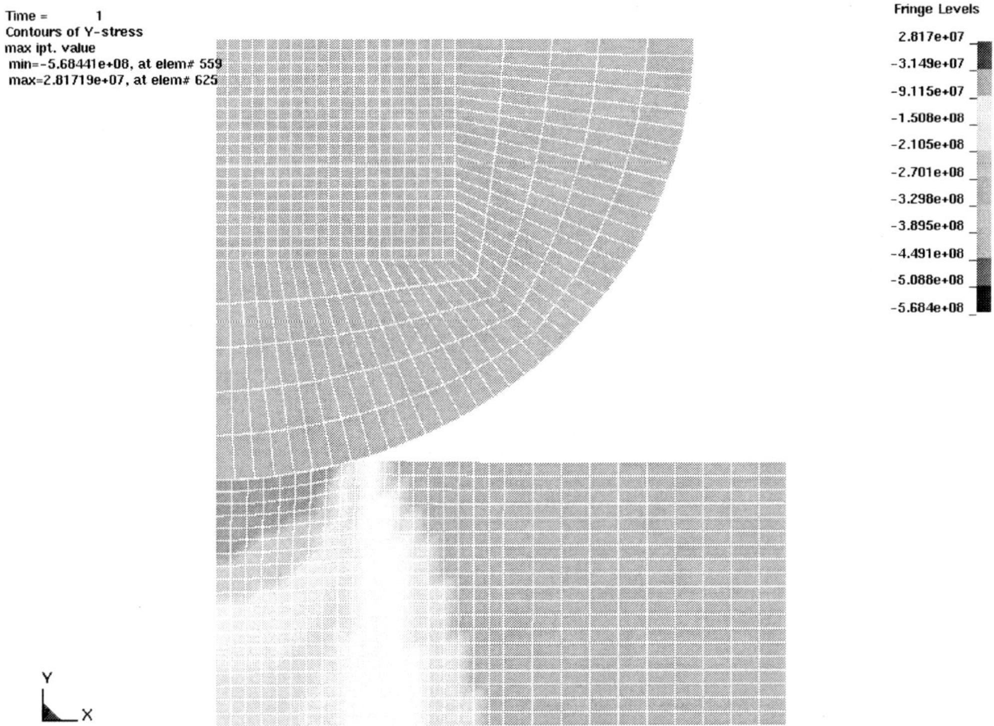

FIGURE 1. FE model output showing stresses resulting from loading by spherical indenter.

Two methods of analysis were applied to the data; a geometrical method closely following that of Hill,[2] and a 2-D simulation using the FE code. Comparison of the two methods demonstrated the validity of the geometrical method and allowed its accuracy to be investigated. Specifically, prediction of the stress distribution using the FE model allowed the selection of a functional form to describe the loading used in the geometrical model. Any assumptions regarding the stress distribution affect the slight correction required to accurately describe the radius of contact of the indenter in the geometrical model.

The geometrical analysis of the indentation data permitted the determination of the work-hardening coefficient n by two methods and the determination of flow stress. Integrating the plastic work done during indentation indicated the deviation from the ideally plastic case and allowed the evaluation of the hardening coefficient n. (Figure 2.) Alternately, the stress and strain could be calculated from load and area at each point in the indentation, and fit to a power law hardening expression to give values for the work-hardening coefficient n and flow-stress κ from each indentation record. From n, the correction to the contact radius was calculated, and a revised analysis is done in terms of the actual radius of contact to obtain a final value for flow stress. Comparing the results with known properties of the standard materials indicated the accuracy of the analysis schemes. Some scatter was observed between different indentation records, probably related to the grain size in the metal. The influence of creep on the progress of the indent was examined.

This new method may allow rapid and relatively easy estimation of the mechanical properties of materials, using small samples. If the sensitivity of the method is adequate, application of this method to plutonium will allow measurement of the mechanical properties of a large number of modern and aged plutonium samples to determine the effect of aging on mechanical properties of the plutonium metal and to determine which other parameters affect variation in the mechanical properties of the aging metal.

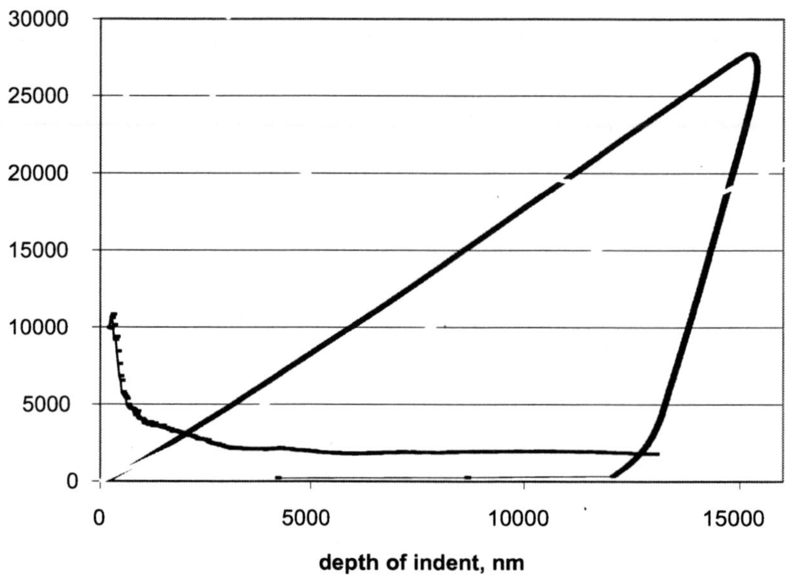

FIGURE 2. The value of the work-hardening coefficient n (scaled by 10^4) is determined by a method based on integrating the load-displacement curve to obtain work done during the indent. The region in which the indentation is predominanatly elastic can be seen as n approaching 1 early in the indent.

REFERENCES

1. Mulford, R., and Trujillo, E., "Radiation Damage in Delta Plutonium Metal, Measured as Change in Bulk Density," this meeting.
2. Zocco, T. G., Los Alamos Science 26, 286 (2000).
3. Norbury, A. L., and Samuel, T., J. Iron Steel Inst. 117, 673 (1928).
4. Tabor, D., The Hardness of Metals, Clarendon Press, Oxford (1951).
5. Hill, R., Storakers, B., and Zdunek, A. B., Proc. Roy. Soc London A 423, 301 (1989).
6. Follensbee, P., and Kocks, F., Acta Metall. 36, 81–93 (1988).

Formation of Plutonium Hydride PuH_2: Description of the Reaction Rate Surface as a Function of Pressure and Temperature

Roberta N. Mulford and Damian C. Swift

Los Alamos National Laboratory, Los Alamos, NM 87544

The hydriding reaction of plutonium metal is used increasingly in the recovery and processing of plutonium. There is thus an increased need for an understanding not only of the thermodynamic parameters governing the reaction, but also of the kinetic behavior to be expected with variations in process parameters such as pressure and temperature. A mathematical description of the behavior of the hydriding process in the entire reaction space in P, T, and rate provides a predictive capability, and enables the reaction rate to be optimized. In addition, given a well-defined description of this rate surface as a function of pressure and temperature, process parameters may be selected to optimize other important parameters such as particle size. These parameters may in some cases be mapped as regions or separate functions onto the surface described here.

We examined published data on the hydriding rate for Pu and developed a functional form for the rate in terms of the pressure and temperature state of the system. A function describing the rate as a function of both pressure and temperature simultaneously was required to fit all the available data. Treatment of data as a function of temperature alone, i.e., a projection of the data on an isobaric surface, indicates a dependence on temperature that exhibits a large amount of scatter and does not indicate any particular functional dependence. See Figure 1.

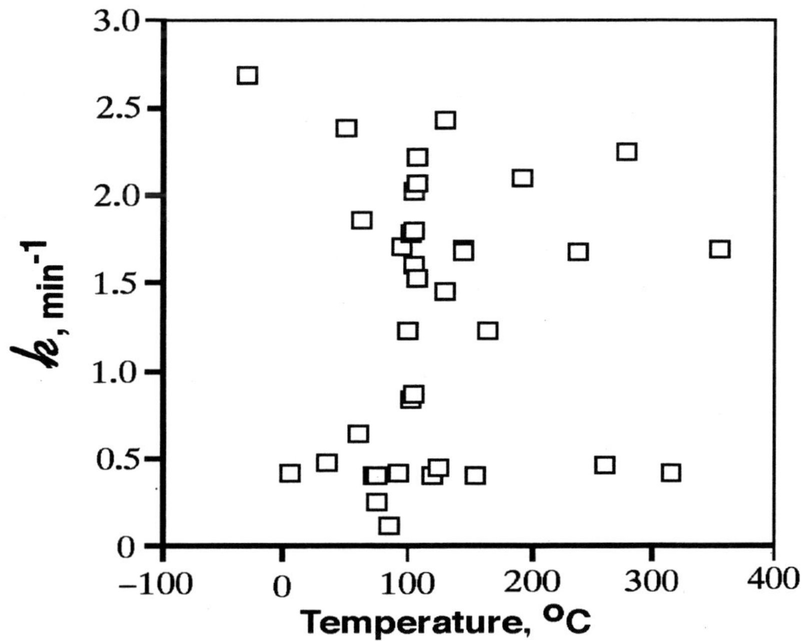

FIGURE 1. Dependence of rate k on temperature in degrees C.

CP673, *Plutonium Futures — The Science,* edited by G. D. Jarvinen

The function used to describe the data is:

$$k_{hyd}(p,T) = \alpha \frac{\alpha}{\beta p^{-1/2} + \gamma} \left(1 - \frac{T}{T_{eqm}}\right)^{1/m} , \quad (1)$$

where

$$T_{eqm} = \frac{8165}{10 - \log_{10} p} - 273, {}^oC , \quad (2)$$

and $\alpha = 1.1$, $\beta = 1/k_2K_{1/2} = 0.7608476$ (min.torr$^{1/2}$), $\gamma = 1/k_2 = 1/2.69$ (min), m = 2.5.

The function selected was provided with an adjustable curvature in temperature by choosing an algebraic form with a fitting parameter *1/m*. This approach provides a practical empirical fit to the data: no information on the activation energy of the reaction was inferred from this functional form. The function was bounded in the P vs T plane and in the rate vs P plane by known functions published previously.[1,2] The function describing rate as a function of pressure given in Reference 2 is seen to require a slight correction, about 10%, as it was determined from data at various temperatures projected onto an isothermal plane. See Figure 2.

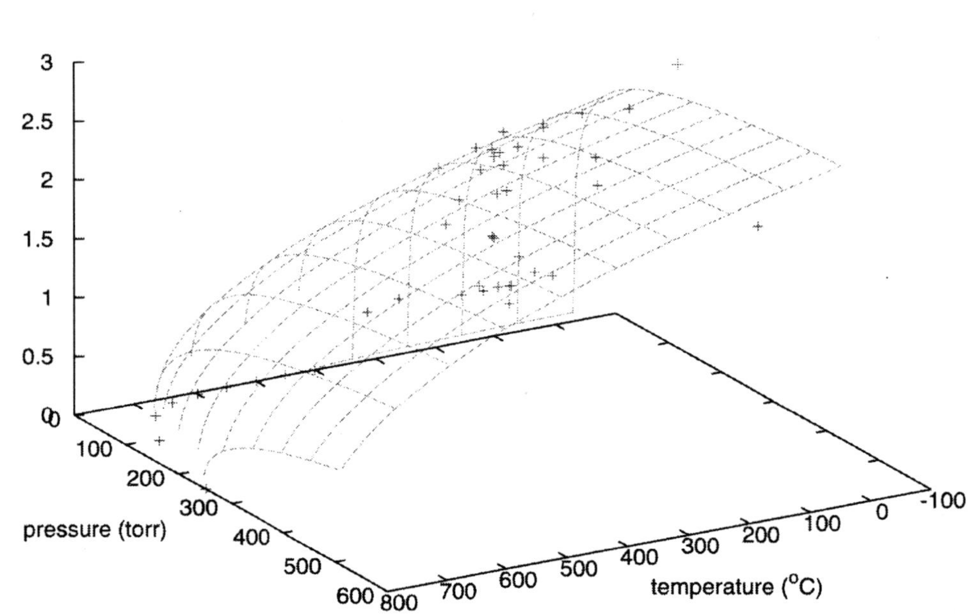

FIGURE 2. A view of the surface described by Equation (1) in P, T, and rate (k) space.

The actual dependence of the rate on temperature is expected to have an Arrhenius-like form, $\exp(-E_a/RT)$ rather than the power-law function chosen. The feasibility of this form is discussed, and the quality of fit to known data described. Participation of a second species, PuH_3, in the reaction was examined and determined to require a minor extension to the model presented.

The reaction is seen to exhibit a rather weak (*1/m* = ~1/2) dependence on temperature. Near the equilibrium line where k [forward] = ~ k [reverse], the rate obviously decreases rapidly toward zero. The variation of rate with pressure is seen to be very strong in the low-pressure region where surface coverage by hydrogen dominates the reaction rate, a behavior well described in the literature.[2] The thermal behavior probably indicates a mechanism in which transport plays a major role.

The application of this function to a hydride-dehydride recovery process provides a useful description of the behavior of the system during the onset of hydriding, a transient period during which the reaction heats the system and temperature may fluctuate. This portion of any process is difficult to control or monitor, because of the relatively rapid changes occurring in the system.

The function described provides a useful tool for the practical application and manipulation of the hydriding reaction. Description of the system in terms of Arrhenius rates yields some information on the activation energies of the system.

REFERENCES

1. Mulford, R. N. R., and Sturdy, G. E., J. Am. Chem. Soc. 77, 3449 (1954).
2. Stakebake, J. L., J. Electrochem. Soc. 128, 2383 (1981).

Radiation Damage in Delta Plutonium Metal, Measured as Change in Bulk Density

Roberta N. Mulford and E. Ann Trujillo

NMT-15, Los Alamos National Laboratory

Autoradiolysis in delta plutonium metal produces radiation damage, which is evident as a noticeable change in lattice constant over several years after casting, and as a slight density change in the metal over long times. Recoil energy of the daughter uranium atom produced in the decay is the primary source of the energy deposited in the lattice. The consequent formation of voids, vacancies, and dislocations results in a reduction in the bulk density in the metal. Dislocation density increases with the time of irradiation,[1] although thermal annealing following the initial damage cascade removes most of the decay-generated dislocations[2] and may remove dislocations and other structures remaining from previous decay cascades. This "overwriting" of new damage cascades on previously accumulated damage accounts for a reduction in the rate of density decrease after the first 2 to 3 years as saturation of the nascent lattice with damage structures proceeds.[3,4] Accrued radiation damage structures, dislocations, vacancies, and voids have implications for mechanical properties and ultimately for the physical integrity of the metal.[1,5] Measurement of the magnitude of volume expansion, besides providing useful indications of time scales for material performance, may establish limits for some of the time constants for dislocation formation and annihilation, vacancy mobility, or helium migration.

The helium atom produced in the decay is captured in the lattice, eventually forming helium bubbles, which have been seen in TEM measurements.[6] Presence of the helium is assumed to be a driving factor in the density decrease and provides some potential for void swelling. Void swelling depends on vacancy and dislocation formation up to a critical defect population, and on helium mobility. Void swelling in response to radiation damage and helium ingrowth has been seen in other metals[1] and exhibits an induction time that depends on the metal and on the rate of helium ingrowth.

To examine the progress of radiation damage and the formation of helium bubbles within the metal, the density of delta plutonium has been measured as a function of age in years. These data complement other measurements, such as x-ray diffraction determination of lattice constants that measure an increase in the lattice constant with time. The amount of helium can be estimated from the known radioactive decay rate and age of the metal because very little helium is evolved at the surface of the metal over time.

A density change is seen in the bulk density of delta plutonium metal in samples having ages between 15 and 42 years. Densities of a large number of samples of old plutonium metal have been measured to provide some statistical evaluation of the magnitude of density change as a function of metal age. The density change appears to be constant with age over 2–4 years. Several alloys with gallium are examined. The density change appears to be a function of the concentration of alloying agent in the metal. The rate of change is –0.029 %/annum for a 1% Ga alloy, and the variation with alloy concentration appears to be 0.91 for each percent of gallium added to the metal.

The uncertainty in the rate is about ±0.006%/annum, reflecting the large scatter in the behavior of individual samples. The uncertainty varies with the alloy. The error in each measurement is about 0.0004 g/cm^3 and varies with the size and thermal output of the samples. Rates of density change are determined both from the examination of trends in density in an ensemble of samples of different ages, but also by comparing the densities of individual samples from year to year over several years.

The rate of decrease in density measured at these advanced ages is compared with the measured rate for density decrease in short-term radiation damage observed in newly cast samples.[3] Although rates of initial rapid and slower long-term density change differ, the dependence of these two rates on gallium concentration have similar

CP673, *Plutonium Futures — The Science,* edited by G. D. Jarvinen

magnitudes: the rate of the transient effect changes by 0.85 for each percent of gallium added to the metal, and the rate of the bulk density changes by 0.91 for each percent of gallium added.

Where initial strain in rolled or pressed samples can be estimated, the change in bulk density varies slightly with initial strain, as expected for a mechanism involving an increase in dislocation density.

Data presented here may serve to indicate the degree of dimensional and structural integrity that can be expected in delta plutonium metal as it ages. Radiation damage structures and bubbles may be expected to influence somewhat the response of plutonium metal to mechanical insult or to thermal fluctuation.[6] The data presented here may indicate what processes proceed in the Pu as it ages and what effect these will have on the properties or thermal stability of the metal.

The time scale for the onset of void swelling in plutonium has been estimated, but never measured. Estimates of an induction time shorter than 40 to 45 years are not borne out by the data presented here. Estimates may be improved by inclusion of experimental data for the number of helium atoms per vacancy, the dislocation density and mobility, and other data that may be derived by comparison of this data with other studies.

The observed rate of density decrease with time exceeds that expected solely from the influence of helium buildup. The number of vacancies per decay, that is, vacancies per helium atom formed, is apparently in the range of 5–7, rather than on the order of 0.5 to 2, as predicted by theory.[7] Radiation damage from the uranium recoil atom, appears to dominate density change in the 15–45-year time frame. The dependence on alloy concentration also suggests that retained damage is the governing factor in the density change because helium production is nearly independent of alloy. A variation in the stabilization of damage structures as a function of alloy concentration is an interesting possibility suggested by the data presented here.

REFERENCES

1. Weir, J. R., "The Effect of High Temperature Reactor Irradiation on Some Physical and Mechanical Properties of Beryllium," in The Metallurgy of Beryllium, Institute of Metals, London (1963).
2. Kapinos, V. G., and Bacon, D. J., Phys. Rev. B 50(18), 13194–13203 (1994).
3. Chebotarev, N. T., and Utkina, O. N., in Plutonium and Other Actinides, edited by H. Blank and R. Lindner, North-Holland Publishing Co., Amsterdam (1976), pp. 559–565.
4. Nelson, R. D., Land, C. C., and Ellinger, F. H., in The Plutonium Handbook, edited by O. J. Wick, American Nuclear Society (1980), p. 323.
5. Tappin, D. K., Bacon, D. J., English, C. A., and Phythian, W. J., J. Nucl. Matl. 205, 92–97 (1993).
6. Zocco, T. G., Los Alamos Science 26, 286 (2000.)
7. Wolfer, W. G., Philosophical Mag. A 59, 87 (1989).

Analysis of Actinide Compressibility and Structure at High Pressures in Comparison with Lanthanide and Transition Metal Behaviour

B. A. Nadykto and O. B. Nadykto

RFNC-VNIIEF, Arzamas-16 (Sarov), Nizhni Novgorod region, 607190,
E-mail:nadykto@vniief.ru, Fax: 83130 45772

The analysis of P(ρ) curves in the experiments on static compression of materials in diamond anvils[1–8] and shock experiments[9,10] indicate the presence of kinks that are an evidence of a change in material behaviors under pressure. When the material crystalline structure remains unchanged, the change in electron structure of solid atoms (i.e., electron redistribution between inner and outer shells) can only account for the kinks. The data on static compression of thorium, uranium, americium, and lanthanide series testifies that the material electron structure changes under pressure.

A method for calculating the energy of a compressed-atom state was proposed in References 11 and 12. This model provides approximate analytical expressions describing elastic energy and pressure in compression and allows estimation of outer electron energy states such as $E_n = 9AB_0/2N_A\rho_n$.

The computational curves and experimental data for Th,[3] U, and Am[9] are compared in Figure 1. The computational curves for uranium have been obtained from processing of the data of shock experiments.[9] As seen from Figure 1, the experimental data suggest the existence of phases in Th, U, and Am at pressures higher than 50–150 GPa, in which bulk moduli are close, B_0 = 340–400 GPa.

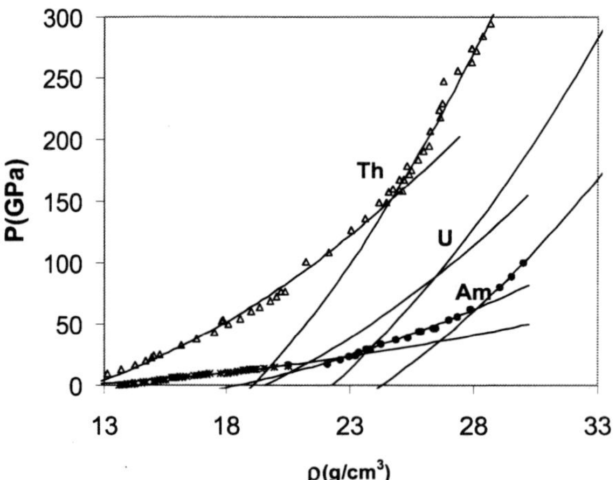

FIGURE 1. Pressure P(ρ) on normal isotherms Th, U, and Am. Curves are calculational data. Experimental points are from References 1 and 3.

The experimental data on americium compression by a high pressure of up to P = 100 GPa is reported in Reference 1. The transition from phase AmI to phases AmII and AmIII changes the slope of the P(ρ) curve in accordance with the dramatic decrease in the bulk modulus (more than by a factor of 2) and an increase in the equilibrium volume for these phases. Despite the different crystalline structures of the AmII and AmIII phases, P(ρ)

CP673, *Plutonium Futures — The Science,* edited by G. D. Jarvinen

for them can be described by a single curve with the same values of bulk modulus, $B_0 = 13.0$ GPa, and equilibrium density, $\rho_n = 11.8$ g/cm^3.

At a pressure higher than 150 GPa in thorium, there is a transition to the phase of 19.1 g/cm^3 equilibrium density, 400 GPa bulk modulus, and 227 eV outer electron energy per atom. This energy corresponds to six electrons in the outer shell, and this means that completed shells are involved in the electron rearrangement.

The recent measurements of the crystalline structure of neodymium,[4] praseodymium,[5] and samarium[6] show that low-symmetry structures, i.e., monoclinic and orthorhombic, are dominating in them at high pressure. In so doing, the high-pressure phases of the elements corresponding to the highest measured pressures have an equilibrium density about two times higher than that of the parent phase, a high bulk modulus (320–430 GPa), and a high atomic cell energy (160–200 eV) corresponding to five or even six electrons in the outer atomic shell.

The run of the curve of actinide atomic radii versus the atomic number of the element appears in Figure 2 along with similar data for lanthanides and 3d, 4d, and 5d transition metals. For light actinides, the dependence resembles the transition metal behavior.

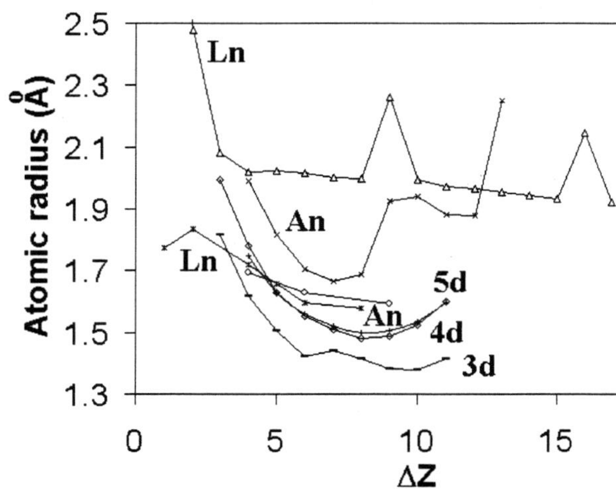

FIGURE 2. Atomic radius of actinides, lanthanides, and transition metals versus atomic number. Circles denote atomic radii of high-pressure phases of actinides, crosses those of lanthanides.

The drastic difference between the behaviors of light actinides and 4f elements is typically accounted for by the different behavior of 5f electrons in actinides (till plutonium inclusive) and in lanthanides. In light actinides 5f electrons are believed to make a contribution to the bond in solid, whereas 4f electrons are core-level electrons located deep in the inner shell; in americium 5f electron localization occurs and configuration of the following actinides is similar to that of lanthanides.

This explanation does not seem undeniable. In this description, americium must be a bivalent metal (an analog of europium) containing seven electrons in 5f state. The atomic radius of a bivalent metal is significantly longer than that of the neighboring trivalent elements. The atomic radii of the neighboring elements are close to those of americium and, like americium, they are trivalent metals. According to the data,[13] in the actinide series, einsteinium is a bivalent metal, while californium that precedes it exhibits properties of bivalent metal in thin films (like samarium that precedes europium in the lanthanide series).

So it would be appropriate to suggest that an electronic configuration with a half-filled 5f shell takes place in einsteinium. But in that case, the electronic configuration in americium is $5f^2 6d^3 7(sp)^3$, i.e., it is closest to praseodymium in the electronic configuration. Next, it should be presumed that light actinides are transition 6d metals and the actinides following the americium fill of the 5f shell with partially filled 6d shell. The plutonium behavior features may relate to the transition from the 6d shell filling to the 5f state filling. An argument in support of the 6d nature of light actinides (before plutonium) can be the recent indication[14] to the existence of oxide PuO_4, which is similar to OsO_4.

After einsteinium, the shells can be filled by analogy with heavy lanthanides. In this case, the following members of the series must be trivalent metals up to the appearance of configuration $5f^{14}$ where the transition to the bivalent state, similar to ytterbium, is possible. However, 6d electrons can compete, and the continuation of the 6d state filling with the retention of seven 5f electrons and a bivalent metal state should not be excluded. The metal state of heavy actinides is very hard to study because the quantity of these materials is very small and their radioactivity is extremely high. Nevertheless, reasoning from the study of the temperature dependence of element vapor volatility, it is concluded in References 15 and 16 that Es, Fm, Md are bivalent metals.

The density of transition metal oxides is, as a rule, significantly less (by a factor of 1.5–2) than that of metal. The same behavior is characteristic of light actinides. For light actinides, oxide density is higher than metal density and for heavy lanthanides it makes (0.9–1) of the metal density. For bivalent europium and ytterbium, the oxide density is higher by a factor of 1.4 and 1.3, respectively, than the metal density and close to the trivalent lanthanide density. The americium oxide density is close to the americium metal density.

Figure 2 also presents the equilibrium atomic radii of the high-pressure phases for lanthanides and actinides; their behaviors are close to each other.

Reasoning from the analysis of data for many elements, we can suggest that the hybridization of electrons is possible only when their principal quantum numbers are the same and, hence, the filling of inner d and f shells does not change the element valence (the number of outer electrons). The valence can change due to the transition of d or f electrons to the outer shell, but in the form of s, p, and even d or f electrons with the principal quantum number higher than that of the inner d or f shell. The electron energy in the inner d or f shell is significantly larger than the outer electron energy because of the larger magnitude of the inner-core charge. However, the energy of transition of the d(f) electron to the outer shell can be small because in the transition, the energy of electrons located higher than the d(f) shell increases because of the increase in the electric charge acting on them by one.

ACKNOWLEDGEMENTS

The work was carried out with the financial support by the ISTC under Project #1662.

REFERENCES

1. Lindbaum, A., Heathman, S., Litfin, K. at al., Phys. Rev. B 63, 214101 (2001).
2. Vohra, Y. K., Beaver, S. L., Akella, J. et al., J. Appl. Phys. 85(4), 2451 (1999).
3. Vohra, Y. K., and Akella, J., Phys. Rev. Lett. 67, 3563 (1991).
4. Chesnut, G. N., and Vohra, Y. K., Phys. Rev B. 61(6) R3768–3771 (2000).
5. Chesnut, G. N., and Vohra, Y. K., Phys. Rev.B 62(5), 2965–2968 (2000).
6. Chesnut, G. N., and Vohra, Y. K., Proceedings of International Conference on High Pressure Science and Technology (AIRAPT-17), Honolulu, Hawaii, July 25–30, 1999, edited by M. H. Manghnani, W. J. Nellis, and M. F. Nicol, pp. 483–486.
7. Hu, J. Z., Mao, H. K., Guo, Q. Z., and Hemley R. J., Proceedings of International Conference on High Pressure Science and Technology (AIRAPT-17), Honolulu, Hawaii, July 25–30, 1999, edited by M. H. Manghnani, W. J. Nellis, and M. F. Nicol, pp. 1039–1042.
8. Takemura, K., Phys. Rev. B 50(22), 16238–16246 (1994).
9. Marsh, S. P., LASL Shock Hugoniot Data, University of California, Berkley, 1980.
10. Trunin, R.F. (editor), "Experimental Data on Shock Compression and Adiabatic Expansion of Condensed Materials," RFNC-VNIIEF, Sarov, 2001.
11. Nadykto, B. A., Physics-Uspekhi, 36(9) 794–827 (1993).
12. Nadykto, B. A., VANT. Ser. Teor. i Prikl. Fizika, 3, 58–73 (1996).
13. Chemistry of Actinides, Vol. 2, edited. by J. Cats, G. Sieborg, and L. Morss, Moscow, Mir Publishers (1997), pp. 558–621.
14. Domanov, V. P., Buklanov, G. V., and Lobanov, Yu.V., Abstracts of International Conference Actinides 2001, Hayama, Japan, November 4–9, 2001, p. 166.
15. Zvara, I., Belov, V. Z., Domanov, V. P. et al., Joint Inst. Nucl. Res. Rep. JINR P6-10334, Dubna (1976).
16. Hubener, S., Radiochem. Radioanal. Lett. 44. P. 79–86 (1980).

Local Structure and Vibrational Properties of α'-Pu Martensite and Ga-Stabilized δ–Pu

Erik J. Nelson,[1] Kerri J. M. Blobaum,[2] Mark A. Wall,[2] Patrick G. Allen,[1] Adam J. Schwartz,[2] and Corwin H. Booth[3]

[1]Seaborg Institute for Transactinium Science, Lawrence Livermore National Laboratory, P.O. Box 808, Livermore, CA 94551
[2]Materials Science and Technology Division, Lawrence Livermore National Laboratory, P.O. Box 808, Livermore, CA 94551
[3]Chemical Sciences Division, Lawrence Berkeley National Laboratory Berkeley, CA 94720

Small additions of Ga can stabilize the fcc δ phase in Pu down to room temperature. Upon further cooling of Ga-doped Pu, a diffusionless martensitic partial transformation of the alloy occurs. In the transformed regions, the δ structure, in which Ga atoms are soluble and stable, changes to the monoclinic α structure, in which Ga atoms are insoluble but effectively trapped because of the very slow diffusion rates of Ga in the α matrix. This metastable phase consisting of Ga dopant atoms trapped within a monoclinic α-Pu matrix is termed the α' phase.

The transition from the δ to the α' phase results in a 25% volume contraction. This large effect is believed to be due to a change in the nature of the Pu 5f electrons from delocalized (α) to localized (δ), and marks a transition in the electronic behavior of the actinides from the delocalized light actinides (Ac-Np) to the localized, lanthanide-like behavior of the heavier actinides. Because of the large changes in mechanical and electronic properties between the two phases, this complex δ–> α' martensitic phase transformation in Pu-Ga alloys has been the subject of many experimental and theoretical studies.[1] The role of Ga in the transformation, its stabilization of the δ-Pu phase at room temperature, and its location in the resulting mixed α'-δ material are several important issues that still need to be resolved in the understanding of this system and that are addressed by the results of this study. These results are important in the fundamental understanding of this phase transition, and the nature of the stability of the δ phase of Pu-Ga alloys is an important issue in stockpile stewardship and Pu metallurgy.

In this study, extended x-ray absorption fine-structure spectroscopy (EXAFS) is used to investigate the local atomic environment and vibrational properties of plutonium and gallium atoms in the α' and δ phases of a mixed-phase 1.9 at. % Ga-doped Pu alloy. EXAFS results measured at low temperature compare the structure of the mixed-phase sample with a pure δ-Pu sample. EXAFS spectral components attributed to both α'-Pu and δ-Pu were observed in the mixed-phase sample. Ga K-edge EXAFS spectra indicate local atomic environments similar to the Pu L_{III}-edge EXAFS results, which suggests that Ga can be substituted for Pu atoms in both the monoclinic α'-Pu and fcc δ-Pu structures. In δ-Pu, we measure a Ga-Pu bond length contraction of 0.11 Å with respect to the Pu-Pu bond length. The corresponding bond-length contraction around Ga in α'-Pu is only 0.03 Å. Results from temperature-dependent Pu L_{III}-edge EXAFS measurements are fit to a correlated Debye model, and a large difference in the Pu-Pu bond Debye temperature is observed for the α' and δ phases: $\theta_{cD}(\alpha') = 159.1 \pm 12.5$ K versus $\theta_{cD}(\delta) = 120.4 \pm 2.6$ K. These Debye temperatures are consistent with those from earlier studies on α-Pu and δ-Pu phases determined by using other methods.[2-3] A similar analysis is performed on a single-phase 3.3 at. % Ga-doped δ-Pu sample. The results of temperature-dependent Pu L_{III}-edge and Ga K EXAFS determines pair-specific correlated-Debye temperatures, $\theta_{cD}(\delta)$, of 110.7 ± 1.7 K and 202.6 ± 3.7 K, for the Pu-Pu and Ga-Pu pairs, respectively, in the δ-Pu phase.[3] Because the Debye temperature can be related to a measure of the lattice stiffness, these results indicate the Ga-Pu bonds are significantly stronger than the Pu-Pu bonds in the δ-Pu phase. In addition, the difference in the Debye temperatures for the α' and δ phases are related to the observed stability of Ga in δ-Pu and metastability of Ga in α'-Pu.

CP673, *Plutonium Futures — The Science*, edited by G. D. Jarvinen

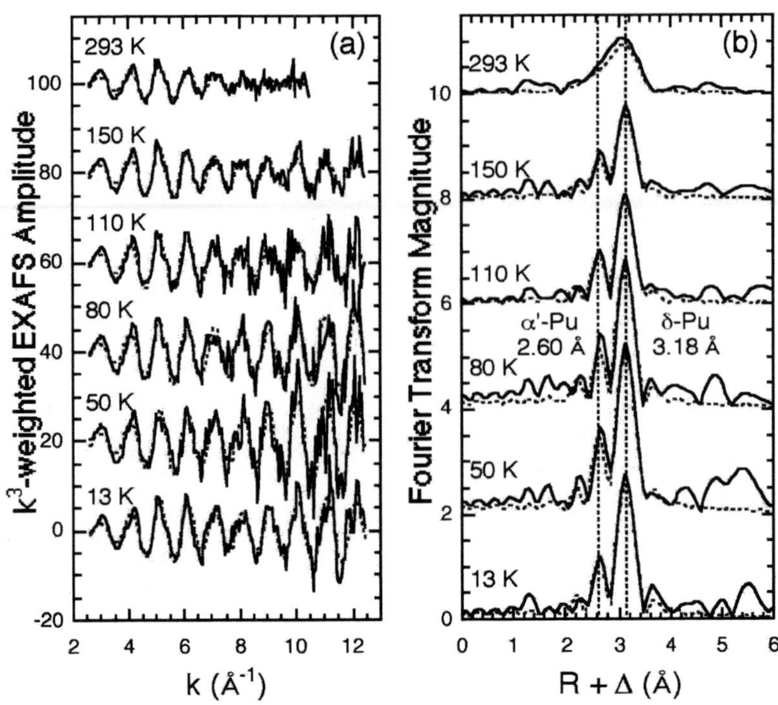

FIGURE 1. (a) Pu L_{III}-edge k^3-weighted transmission EXAFS data and (b) Fourier transforms (FTs) of the EXAFS for the partially transformed 70% δ-Pu / 30% α'-Pu alloy. For each plot, the solid line is the data, and the dashed line is the fit to the data. Note the differences between the thermal damping of the δ and α' component peaks at R+Δ ~ 3.0 Å and 2.4 Å, respectively, as a function of temperature.

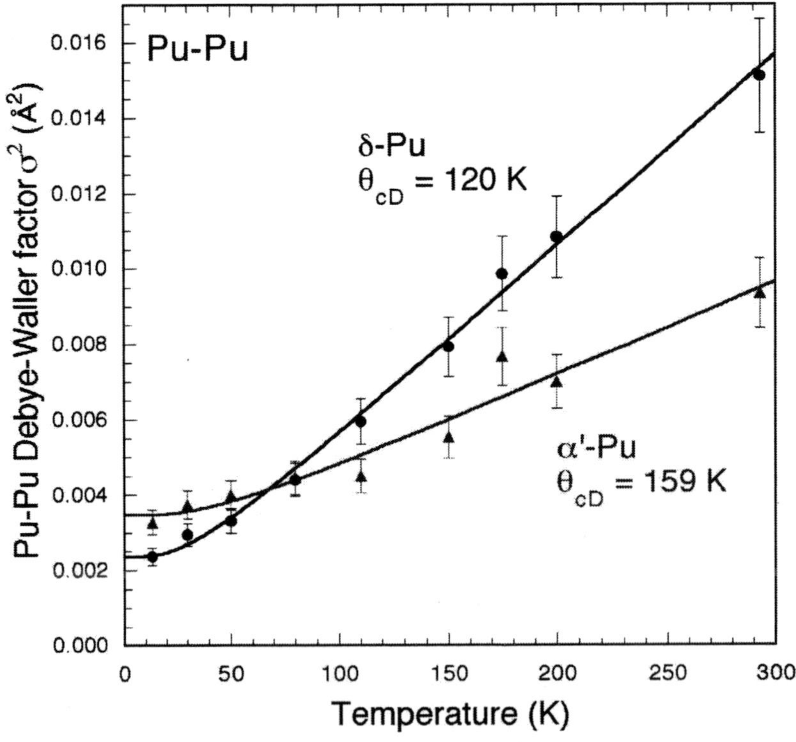

FIGURE 2. Plot of Debye-Waller factor vs temperature for first-shell Pu-Pu bonds in α'-Pu and δ-Pu as determined from fits to the Pu L_{III}-edge EXAFS. The data are plotted along with fits generated using the correlated Debye model.

This work was performed under the auspices of the U.S. Department of Energy by University of California Lawrence Livermore National Laboratory under contract No. W-7405-Eng-48, under LDRD funding. UCRL-JC-151541

REFERENCES

1. See articles within Los Alamos Science, "Challenges in Plutonium Science," 26, Vols. I–II (2000).
2. Lee, J. A., and Waldron, M. B., "The Actinide Metals," in Inorganic Chemistry, Series Two (Baltimore, University Park Press, Vol. 7, Chap. 7 (1975), p. 250.
3. Allen, P. G., Henderson, A. L., Sylwester, E. R., Turchi, P. E. A., Shen, T. H., Gallegos, G. F., and Booth, C. H., "Vibrational Properties of Ga-Stabilized δ-Pu by Extended X-Ray Absorption Fine Structure," Phys. Rev. B 65, 214107 (2002).

Sintering of Plutonium Oxide Powder to Near Theoretical Density

Thomas P. O'Holleran, Kenneth J. Bateman, and Dennis L. Wahlquist

Argonne National Laboratory—West
P. O. Box 2528
Idaho Falls, ID 83403-2528

INTRODUCTION

Plutonia bodies approximately one centimeter in size proved difficult to fabricate by conventional sintering at densities greater than 95% of theoretical. Hot isostatic pressing (HIPing) is a technique used in industry to fabricate dense ceramic and metal parts, sometimes to near-net shape. In this technique, high pressure is applied during the heating cycle to assist the sintering process. Pressure is applied isostatically to the load by pressurizing the gas (in this case argon) inside the HIP heating chamber. Pressure is transferred to the load, which is sealed in an evacuated container.

HIPing has been used to fabricate a glass-bonded ceramic waste form to immobilize radioactive salt waste from electrometallurgical treatment of spent nuclear fuel.[1] Ceramic waste forms ranging in size from 2.54 cm in diameter and 7.5 cm in length to 45.7 cm diameter and 76 cm long were successfully consolidated in stainless-steel containers.[2]

In order to evaluate the HIP process for densifying PuO_2 powder, material interactions were first investigated using CeO_2 as a surrogate for plutonia. Also, in order to avoid potential distortion and density gradients resulting from "end effects" at the top and bottom of the container, alumina- and yttria-stabilized zirconia filler materials were tested. The filler materials were placed above and below the plutonia charge. Nickel and titanium spacers to prevent interaction of the plutonia with the filler materials were also evaluated. This experiment was conducted at 1,623 K and at atmospheric pressure. The results of these tests were used to select materials for HIPing plutonia.

RESULTS

In the materials compatibility experiment, yttria-stabilized zirconia showed greater shrinkage than the alumina or the ceria. This result was desirable in order to ensure that full pressure would be transferred to the plutonia. The titanium spacers oxidized badly and crumbled, but the nickel spacers showed little evidence of interaction with any of the materials that they contacted. Based on these results, yttria-stabilized zirconia and nickel spacers were selected for HIPing plutonia.

Plutonium oxide powder was HIPed to a maximum temperature of 1,673 K and a maximum pressure of 200 MPa. The HIP schedule is shown graphically in Figure 1. Figure 2 shows a photograph of a container after HIPing.

The container was cut open with a low-speed abrasive saw, and the immersion density of the plutonia was measured. The resulting density of 11.45 g/cm^3 is 99.9% of theoretical density. Scanning electron microscopy revealed a rather coarse microstructure with grain sizes about 25–50 μm, and occasional triple point pores (see Figure 3).

CP673, *Plutonium Futures — The Science,* edited by G. D. Jarvinen

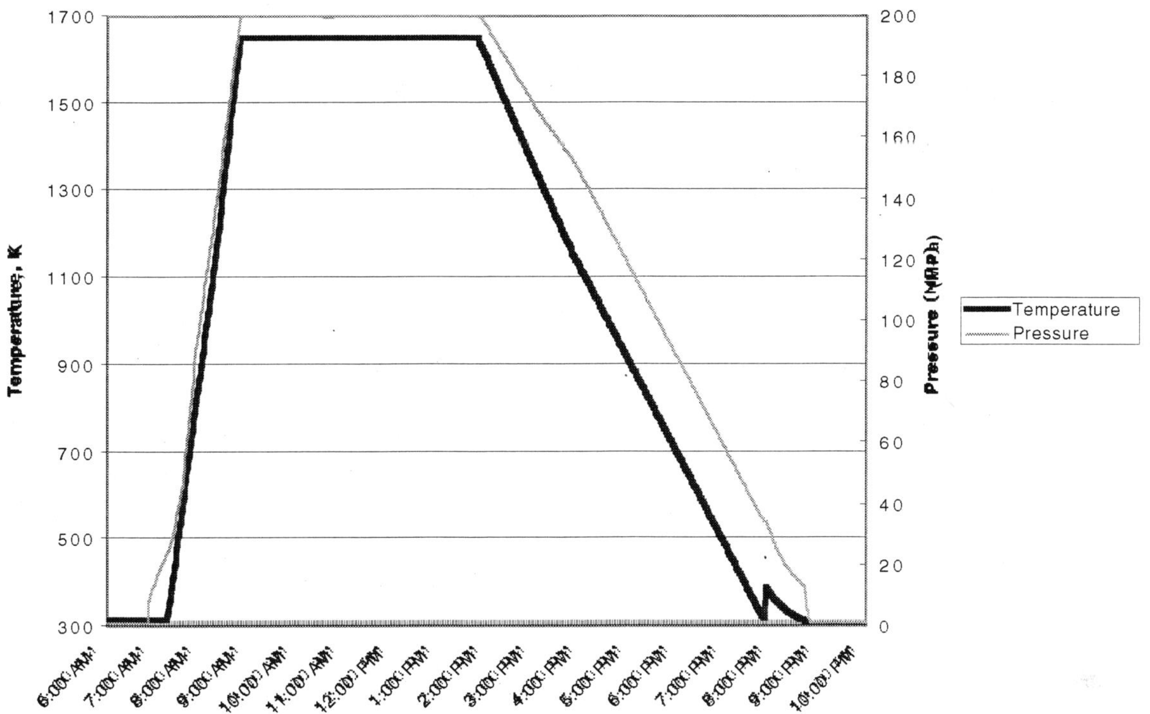

FIGURE 1. HIP cycle used to sinter plutonia showing the temperature and pressure as a function of time.

FIGURE 2. Photograph of a plutonia-containing HIP container after HIPing. The plutonia is in the central bulge, with zirconia filler above and below. Maximum diameter at the top of the container is 2.54 cm.

FIGURE 3. Secondary electron image of HIPed plutonia. Pull-outs in this image reveal grain size and occasional triple point pores. Surface roughness is an artifact of sample preparation.

DISCUSSION

The results reported here represent a novel application of conventional technology (HIPing) to sinter plutonium oxide. Near-theoretical density was achieved without the use of sintering aids or special atmospheres to control redox conditions. The established ability of this technology to produce near-net shape parts suggests potential applications such as nuclear fuel or radioisotope heat-source fabrication.

REFERENCES

1. Simpson, M. F., Goff, K. M., Johnson, S. G., Bateman, K. J., Battisti, T. J., Toews, K. L., Frank, S. M., Moschetti, T. L., and O'Holleran, T. P., "A Description of the Ceramic Waste Form Production Process from the Demonstration Phase of the Electrometallurgical Treatment of EBR-II Spent Fuel," Nuclear Technology 134 (2001).
2. Bateman, K. J., Moricca, S., Eddowes, T., and Rigg, R. H., "Metal Canister Designs for Hot Isostatic Pressing of Ceramic and Glass/Ceramic Matrices for Immobilizing Radioactive Elements," Proceedings of PM2TEC 2003, International Conference on Powder Metallurgy and Particulate Materials, Las Vegas, Nevada, June 8–12, 2003 (to be published).

An Alternative Interpretation of the Existence and Importance of the $PuO_{2+X}.H_2O$ Solid Solution

Mark T. Paffett,[1] Doug Farr,[2] and Dan Kelly[1]

[1]Chemistry Division
[2]Nuclear Materials Division
Los Alamos National Laboratory

The existence and molecular form of the hyperstoichiometric oxide PuO_{2+x} is reexamined and evaluated from previous reports and recent, additional, unpublished data. The physical data used in this reexamination includes x-ray photoelectron spectroscopy (XPS),[1] x-ray absorption near-edge structure (XANES),[2] and extended x-ray absorption fine structure (EXAFS)[2,3] data. Based upon a critical look at the strengths and weaknesses inherent to each technique and the sum of the evidence, an alternative description of the physical entity, $PuO_{2+x}.H_2O$ is postulated. The XANES data indicate that the hyperstoichiometric PuO_{2+x} oxide resulting from corrosion by water is best described as a solid solution with the Pu metal center in a valence state between +4 and +6 (most likely +V).[2] The EXAFS data indicate that lattice expansion and accommodation of the oxygen from the corrosion agent (water) is predominately in the form of lattice hydroxyls. XPS data indicate a hydroxylated surface whose thickness depends on exposure conditions (temperature, P_{H_2O}). Finally, the original XRD data[4,5] are reexamined and found to be insufficient to unequivocally ascribe observed behavior to that of an ordered material with a well-established and confirmed structure containing strictly interstitial oxygen anions. The hyperstoichiometric oxide $PuO_{2+x}.H_2O$ interfacial system is best described by a variety of oxyhydroxide species atop an expanded PuO_2 lattice that has varying degrees of oxygen containing species (that may include coordinated hydroxyl or water molecules). Furthermore, the complex nature of the actual hyperstoichiometric oxide PuO_{2+x} has several pragmatic implications regarding gas generation and the evolution of molecular species during thermal treatments or excursions that will be discussed.

INTRODUCTION AND BACKGROUND

The existence and importance of the hyperstoichiometric oxide PuO_{2+x} has recently been the center of much concern in the context of long-term storage and disposition options for much of the excess Pu in the DOE inventory.[6,7] Clearly, for safe storage and transportation, an adequate understanding of its existence, properties, and behavior are warranted. Most of the physical data substantiating the existence and nature of the hyperstoichiometric oxide PuO_{2+x} is attributed to the works of Haschke, et al. and is detailed in a series of articles and reports.[4,5,8–16] Where relevant (and available), the central physical data substantiating this previous molecular level description of the $PuO_{2+x}.H_2O$ entity is briefly reexamined. Many of these previous reports and published articles ascribing molecular-level detail to this entity are reexamined in the light of an alternative plausible interpretation(s) concerning the molecular composition and identity of the hyperstoichiometric oxide PuO_{2+x}. New data are included in this reevaluation that substantiates this alternative view. Finally, the significance and importance of the interfacial molecular transformations and gas-liberation behavior in long-term storage scenarios will be discussed within the construct of this alternative interpretation of the physical nature of the material.

CP673, *Plutonium Futures — The Science,* edited by G. D. Jarvinen

SUMMARY OF XPS RESULTS: $PUO_{2+X}.H_2O$

In recent data obtained by Farr et al.[1] the PuO_yX_z (X= H, C) model compounds, hydroxylated PuO_2 powders, and PuO_2 solids exposed to water vapor were examined by XPS. In addition, other oxides and hydroxides were examined with particular emphasis on both the Pu 4f and O 1s regions. In this recent work, the Pu 4f, O1s, and C1s line shapes were fit with multicomponent spectra to unequivocally demonstrate the presence of a variety of species. The XPS results of Farr et al.[1] contained the following significant findings:

(1) Surface hydroxyls were ubiquitous in all PuO_2 samples exposed to water vapor or ambient air;
(2) Higher valent Pu species were observed consistent with Pu in a valence state greater than +4;
(3) Surface hydroxyls were persistent to temperatures as high as 590°C; and
(4) Surface hydroxylation occurs in proportion to temperature and water-vapor exposure.

In conclusion, for PuO_2 samples exposed at room temperature to water vapor, an escape depth analysis of the attenuation and line-shape fits of either the Pu 4f or the O 1s spectra estimates that the hydrated (or hydroxylated layer) thickness is on the order of 3.2 ± 0.5 nm atop the PuO_2 substrate.

SUMMARY OF XANES AND EXAFS RESULTS: $PUO_{2+X}.H_2O$

Examination of the $PuO_{2+x}.H_2O$ entity by XANES and EXAFS has been repetitively performed over the last several years, and the complete, unabridged, technical papers have recently appeared.[2,3] The most significant points in these recent works are:

(1) The excess charge in PuO_{2+x} is localized on Pu ions, resulting in a mixed valence compound with discrete Pu(IV) and (most likely) (V).
(2) The O occurs as oxo ions with a somewhat expanded bond length relative to mononuclear coordination compounds, up to 1.9 Å. The coordination geometries therefore appear fairly typical. How these fit into the lattice to occupy 20% or more of the Pu sites is an open question. If they occur as serpentine channels, they need not diffract.
(3) The O anions are disordered, and this includes materials that do not have excess O and therefore do not show oxo ions. It is also true that all samples of PuO_2 examined show such disorder, including most high-temperature fired ones.

In summarizing this characterization work on materials prepared following a reaction of PuO_2 with water, the following view holds. With respect to bond length and other considerations, the most accurate conceptual model for PuO_{2+x} is actually PuO_2-H_2O, where the PuO_2 host incorporates water, which can and does hydrolyze, in such a way that it forms highly stable adducts. Rapid chemical equilibrium with H_2O also occurs involving hydrolysis, addition, and substitution as well as oxidation, and these reactions occur throughout the bulk material.

DISCUSSION

The potential for radiolytic gas generation leading to container pressurization is a major issue affecting the ability of the Department of Energy Environmental Management Program to transport and secure nuclear materials for long-term storage or for possible further disposition. The current ability to predict sealed-container pressures and gas-phase compositions is restricted by a limited, finite understanding of the fundamental processes that contribute to gas generation over Pu-bearing solids. In previous works we have examined the relative roles of interfacial reactions that potentially lead to gas evolution.[17–18] Embodied in these mathematical models are the thermodynamics and kinetics of water adsorbing and desorbing from PuO_2, radiolysis of adsorbed water in a variety of physically plausible scenarios, and hydrogen and oxygen recombination reactions and additional reaction rates for the suggested water and oxygen-assisted corrosion of PuO_2 to form hyper-stoichiometric PuO_{2+x} can also be added. The incorporation of this alternative view of the hyper-stoichiometric PuO_{2+x} and reactions leading to gas evolution will be presented with an emphasis on plausible storage scenarios.

ACKNOWLEDGMENTS

Funding for this investigation was made possible by the Nuclear Materials Stabilization Program Office, United States Department of Energy, Albuquerque Operations and Headquarters Offices (under the auspices of the DNFSB 94-1 Research and Development Project). The authors specifically acknowledge the assistance and permission of several cited authors (References 2 and 3) in allowing access to preprints of works and permission to cite results from these works.

REFERENCES

1. Farr, J. D., Morales, L., Neu, M., and Schulze, R. K., "XPS Characterization of the Surfaces of PuO_2 following Corrosion by Water Vapor" (2003), manuscript in preparation.
2. Begg, Bruce D., Clark, David L., Conradson, Steven D., den Auwer, Christophe, Ding, Mei, Espinosa-Faller, Francisco, Gordon, Pamela L., Hess, Nancy J., Hess, Ryan, Keogh, D. Webster, Morales, Luis, Neu, Mary, Paviet-Hartmann, Patricia, Runde, Wolfgang, Tait, C. Drew, Veirs, Kirk, and Villella, Phillip M., "Speciation and Unusual Reactivity in PuO_{2+x}," Inorg. Chem. (2003), in press.
3. Conradson, Steven D., "Application of X-Ray Absortion Fine Structure Spectroscopy to Materials and Environmental Science," Applied Spectroscopy 52, 252A (1998).
4. Stakebake, Jerry L., Larson, D. T., and Haschke, John M., "Characterization of the Plutonium-Water Reaction II: Formation of a Binary Oxide Containing Pu(VI)," J. Alloys and Compounds 202(1-2), 251–263 (1993).
5. Haschke, John M.. "Hydrolysis of Plutonium. Plutonium-Oxygen Phase Diagram," Los Alamos National Laboratory, Los Alamos, New Mexico, USA, edited by Lester R. Morss and J. Fuger, Transuranium Elem. Symp., Meeting Date 1990, American Physical Society, Washington, DC (1992), pp. 416–425.
6. "Stabilization, Packaging, and Storage of Plutonium-Bearing Materials," Department of Energy, DOE-STD-3013-2000 (September 2000).
7. "Assessment of Plutonium Storage Safety Issues at Department of Energy Facilities," U.S. DOE Report DOE/DP/0123T, U.S. Department of Energy, Washington, DC (1994).
8. Haschke, John M., Allen, Thomas H., and Morales, Luis A., "Reaction of Plutonium Dioxide with Water: Formation and Properties of PuO2+x," Science (Washington, DC) 287(5451), 285–287 (2000).
9. Haschke, John M., and Allen, Thomas H., "Equilibrium and Thermodynamic Properties of the PuO2+x Solid Solution," J. Alloys and Compounds 336(1–2), 124–131 (2002).
10. Haschke, J. M., Allen, T. H., and Morales, L., "Reactions of Plutonium Dioxide with Water and Hydrogen-Oxygen Mixtures: Mechanisms for Corrosion of Uranium and Plutonium," J. Alloys and Compounds 314(1–2), 78–91 (2001).
11. Morales, L., Allen, T., and Haschke, J., "Kinetics of the Reaction between Plutonium Dioxide and Water from 25°C to 350°C: Formation and Properties of the Phase PuO2+x," AIP Conference Proceedings 532 (Plutonium Futures—The Science), 114–116 (2000).
12. Morales, Luis A., and Haschke, John M., "Kinetics and Thermodynamics of the Reaction between Plutonium Dioxide and Water from 100°C to 350°C," Book of Abstracts, 219th ACS National Meeting, San Francisco, California, March 26–30, 2000, NUCL-107, American Chemical Society, Washington, DC (2000).
13. Haschke, John M., and Ricketts, Thomas E., "Adsorption of Water on Plutonium Dioxide," J. Alloys and Compounds 252(1–2), 148–156 (1997).
14. Haschke, John M., Allen, Thomas H., and Stakebake, Jerry L., "Reaction Kinetics of Plutonium with Oxygen, Water, and Humid Air: Moisture Enhancement of the Corrosion Rate," J. Alloys and Compounds 243(1–2), 23–35 (1996).
15. Stakebake, Jerry L., Larson, D. T., and Haschke, John M., "Characterization of the Plutonium-Water Reaction II: Formation of a Binary Oxide Containing Pu(VI)," J. Alloys and Compounds 202(1–2), 251–263 (1993).
16. Haschke, J. M., Hodges, A. E. III, Bixby, G. E.; and Lucas, R. L., "Reaction of Plutonium with Water. Kinetic and Equilibrium Behavior of Binary and Ternary Phases in the Plutonium + Oxygen + Hydrogen System," Avail. NTIS Report (1983), (RFP-3416; Order No. DE83006752), 24 pp., from Energy Res. Abstr. 8(8), Abstr. No. 17327 (1983).
17. Paffett, M., and Kelly, D., "The Essential Elements of Modeling Gas Generation From Well Defined Plutonium Materials," American Nuclear Society Topical Conference on Spent Nuclear Fuel and Fissle Materials, Charleston, South Carolina, September 2002, American Nuclear Society Publications, Los Alamos National Laboratory document LA-UR-02-4349.
18. Kelly, D., and Paffett, M. T., "Advanced Modeling and Experimental Validation of Complex Nuclear Material Waste Forms of Potential Transportation Concern," Proceedings of Waste Management 2002, Tucson Arizona, February 2002, American Nuclear Society Publications.

Americium and Curium in Zirconia-Based Materials: Critical Aspects of Their Structural Properties

P. E. Raison[1*] and R. G. Haire[2]

[1] *European Commission*
Joint Research Center - Institute for Energy – HFR unit
P.O. Box 2, 1755 ZG Petten – The Netherlands
**on a leave of absence from the Commissariat à l'Energie Atomique*
CEA Cadarache DEN/DEC/SPUA, France

[2] *Oak Ridge National Laboratory*
Chemical Sciences Division
P.O. Box 2008, Oak Ridge, TN 37831-6375, USA

INTRODUCTION

Mixed oxide phases of actinides with the Group IV transition metals, titanium, zirconium and hafnium, have become increasingly attractive for a variety of nuclear-related applications. These include uses as matrices for transmuting americium and curium as new conceptual reactor fuels for customized waste forms, as well as for other applications. A number of chemical combinations are possible and many of these have been discussed in the literature over the past years.[1-4]

We have investigated zirconium-based binary or ternary oxide systems that appear promising for these applications because they can be composed to yield cubic phases. Specifically, the AmO_2-ZrO_2-Y_2O_3, Cm_2O_3-ZrO_2, and AmO_2-Cm_2O_3-ZrO_2 phase diagrams have been investigated, primarily by X-ray diffraction, using Am-243 and Cm-248 isotopes. We have also extended our studies to include selected lanthanides and other actinides from plutonium through californium in order to develop a systematic screening of these oxide-based ceramics and to explore the self-irradiation effect on their structural properties.

RESULTS

In the AmO_2-ZrO_2-Y_2O_3 system, Y_2O_3 was selected to stabilize the cubic form of zirconia. The goal was to form very stable fluorite-type, cubic solid solutions. We found that the final products exhibited cell parameters varying linearly with the americium content. The stability of these cubic solutions heated in different atmospheres was also investigated. It was determined that some compositions underwent a cubic-to-monoclinic phase transition under reducing atmospheres, while other compositions remained stable under similar experimental conditions.

We pursued the use of curium as a stabilizing agent in place of yttrium, as the latter has a nonnegligible neutron capture cross section that represents a "neutron cost" in transmutation schemes. This incorporation of curium for phase stabilization also avoids the need for partitioning americium and curium because both actinides could be treated simultaneously in transmutation schemes. In this regard, investigations of the binary Cm_2O_3-ZrO_2 system was undertaken to establish the concentrations of curium necessary for stabilizing zirconia-based matrices in a cubic form. It was determined that Cm^{3+} does stabilize cubic ZrO_2-based forms for limited compositions. Specifically, only $(Cm_{0.25}Zr_{0.75})O_{1.875}$, and $(Cm_{0.5}Zr_{0.5})O_{1.75}$ were found as single-phased materials, existing as a fluorite structure or as a pyrochlore structure[5-6] when the two metallic ions were at a 1:1 ratio. Other compositions resulted in diphasic

CP673, *Plutonium Futures — The Science*, edited by G. D. Jarvinen

systems. These results are in agreement with findings for various Ln_2O_3-ZrO_2 systems and could be correlated as a function of ionic radii of the metal ions.[7]

After studying the selective incorporation of either americium or curium in zirconia, we have investigated the global Am-Cm-Zr-O system for compositions close to those encountered in reprocessing schemes. It was found that for many actinide compositions, single-phased materials (either fluorite- or pyrochlore-type phases) could be formed. It was also observed here that Zr could influence the redox behavior of the Am^{4+}/Am^{3+} redox couple by extending its reduction domain.

As part of this effort, attention was paid to the stability of the materials produced with regard to self-irradiation with time. The results were compared to information reported in the literature for related Group IV oxide materials.

Plans exist for irradiating some of such zirconia-based actinide materials to be irradiated in France at the PHENIX reactor and in the High Flux Reactor (HFR) in the Netherlands, as part of European projects on nuclear waste management. In this context, the use of such materials for these technological applications will be presented as well.

The presentation will provide the structural data obtained for the various systems mentioned and compare their behavior with known data for other reported actinide-zirconia materials and lanthanide-zirconia systems.

ACKNOWLEDGMENTS

The French contribution of this work was co-funded by the "Commissariat à l'Energie Atomique" and "Electricité de France." The first author is grateful to both institutions for their continuous support and interest in that project, especially to the "Fuel Research Department" at the CEA Cadarache. The effort at Oak Ridge National Laboratory was sponsored by the Division of Chemical Sciences, Geosciences and Biosciences, Office of Basic Energy Sciences, USDOE, under contract DE-AC05-00OR22725 with ORNL, managed and operated by UT-Battelle, LLC. The authors are indebted to the Office of Basic Energy Science, U. S. Department of Energy, for the use of the particular actinide isotopes, made available through the transplutonium production facilities at ORNL.

REFERENCES

1. Raison, P. E., and Haire, R. G., Progress in Nuclear Energy 38, 251 (2001).
2. Sickafus, K. E., Matzke, H., Hartmann, Th., Yasuda, K., Valdez, J. A., Chodak III, P., Nastasi, M., and Verrall, R. A., J. Nucl. Mat. 274 NO1-2, 66 (1999).
3. Yu, N., Sickafus, K. E., Kodali, P., and Nastasi, M., J. Nucl. Mat. 244, 266 (1997).
4. Nitani, N., Yamashita, T., Matsuda, T., Kobayashi, S., and Ochmichi, T., J. Nucl. Mat. 274, 15 (1999).
5. Raison, P. E., Haire, R. G., Sato, T., and Ogawa, T., Mat. Res. Soc. Proc., 556, 3–10 (1999).
6. Haire, R. G., Raison, P. E., and Assefa, Z., J. Nucl. Sci. Techn. (in press, 2002).
7. Assefa, Z., Haire, R. G., and Raison, P. E., J. Nucl. Sci. Techn. (in press, 2002).

Quantum Size Effects in Hexagonal Plutonium Layers

A. K. Ray[1] and J. C. Boettger[2]

[1]Physics Department, University of Texas at Arlington, Arlington, Texas 76019
[2]Applied Physics Division, Los Alamos National Laboratory, Los Alamos, New Mexico 87545

INTRODUCTION

During the past two decades, considerable theoretical efforts have been devoted to studying the electronic and geometric structures and related properties of surfaces to high accuracy. One of the many motivations for this burgeoning effort has been a desire to understand the detailed mechanisms that lead to surface corrosion in the presence of environmental gases; a problem that is not only scientifically and technologically challenging but also environmentally important. Such efforts are particularly important for systems like the actinides for which experimental work is relatively difficult to perform as a result of material problems and toxicity.

Among the actinides, plutonium is particularly interesting in two respects.[1] First, Pu has, at least, six stable allotropes between room temperature and melting at atmospheric pressure, indicating that the valence electrons can hybridize into a number of complex bonding arrangements. Second, plutonium represents the boundary between the light actinides, Th to Pu, characterized by itinerate 5f electron behavior, and the heavy actinides, Am and beyond, characterized by localized 5f electron behavior. In fact, the high-temperature fcc δ phase of plutonium exhibits properties that are intermediate between the properties expected for the light and heavy actinides. These unusual aspects of the bonding in bulk Pu are apt to be enhanced at a surface or in a thin layer of Pu adsorbed on a substrate, as a result of the reduced atomic coordination of a surface atom and the narrow bandwidth of surface states. For this reason, Pu surfaces and films may provide a valuable source of information about the bonding in Pu. Also, studies of the work functions of thin layers of Pu should provide us with detailed knowledge about quantum oscillations as a function of the film thickness in Pu layers, i.e., the well-known quantum size effect in work functions.

This investigation has thus concentrated on hexagonal Pu layers that correspond to the (111) surface of δ-Pu. Although the monoclinic α phase of Pu is more stable under ambient conditions, there are advantages to studying δ-like monolayers. First, small amounts of impurities can be used to stabilize δ-Pu at room temperature. Second, grazing-incidence photoemission studies combined with the calculations of Eriksson, et al.,[2] suggest the existence of a small-moment δ-like surface on α-Pu.

RESULTS AND DISCUSSIONS

The properties of the hexagonal Pu layers have been determined here with a *new, fully relativistic* version of the linear combinations of a Gaussian-type orbitals fitting-function (LCGTO-FF) method, as embodied in the program GTOFF. This technique is distinguished from other variants of the LCGTO method by its use of two auxiliary GTO basis sets to expand the charge density and the exchange-correlation (XC) integral kernels. The charge fitting function coefficients are determined variationally, by minimizing the error in the Coulomb energy, but the XC coefficients are obtained with a least-squares fit. In its nonrelativistic form, the LCGTO-FF method is known to yield results that are comparable to results produced by other all-electron, full-potential DFT methods. The relativistic implementation of the LCGTO-FF method has progressed through several stages over the years. Scalar relativity was initially implemented in GTOFF using a nuclear-only Douglas-Kroll-Hess (DKH) transformation that neglected relativistic terms involving cross products of the momentum operator. The implementation of relativity in GTOFF was subsequently extended to include all of the scalar-relativistic cross-product terms and spin-orbit

CP673, *Plutonium Futures — The Science,* edited by G. D. Jarvinen

coupling terms that are produced by the nuclear-only DKH transformation. The most recent version of GTOFF has progressed beyond this by the development of the screened nuclear spin-orbit approximation, which approximately incorporates two-electron spin-orbit coupling effects. This *fully relativistic* LCGTO-FF method has recently been applied by us to study hydroxylated actinide oxide surfaces.[3]

The overall precision of any LCGTO-FF calculation is, to a large extent, determined by the selection of the three basis sets. In this work, the orbital basis set used for Pu started with a 23s20p15d11f uncontracted basis set derived from an atomic basis set. This basis set was contracted into a 17s14p11d7f basis with coefficients taken from scalar relativistic atom calculations using the same DFT models as were used in the film calculations. The basis set was then augmented with an s-type function with an exponent of 0.07, a p_Z-type function with an exponent of 0.08, and a d-type function with an exponent of 0.12. The charge and XC basis sets were 25s2d and 21s2d, respectively. The two-dimensional Brillouin zones for the hexagonal layers were sampled on uniform meshes with 19 irreducible points. The SCF cycle was iterated until the total energy was stable to within 0.01 mRy.

For the hexagonal Pu layers, we have studied the effects of scalar relativity vs full relativity (i.e., spin-orbit coupling included) at the generalized gradient approximation (GGA) level of modern density functional theory. Because of severe demands on computational resources, the lattice constant was not optimized, and all the computations have been carried out at the experimental lattice constant. The effects of spin polarization have been studied only for the Pu mono- and dilayers and the results indicate that spin-orbit coupling has a much stronger effect on the layer properties compared to the effects of spin polarization. We found this to be true in a detailed study of Pu monolayer earlier, using an earlier version of GTOFF without simultaneous inclusions of spin polarization and spin-orbit coupling.[4] Thus, spin polarization has not been included in the three-layer calculations. For each combination of approximations, total energies, cohesive energies, surface energies, work functions, and spin moments have been calculated. The cohesive energies for the n layers have been calculated with respect to n monolayers as also (n-1) layers and a monolayer. Table 1 shows the results of the total energies and the binding energies. In Figure 1, we have plotted the surface energies and the work functions of the Pu layers at both the scalar relativistic and fully relativistic levels. The surface energies are calculated from,

$$E(N) = E_B N + 2 E_s , \tag{1}$$

where E_B is the bulk energy calculated, in a first approximation, here from,

$$E_B = E_3 - E_2 , \tag{2}$$

and the work functions are calculated as the negative of the Fermi energy. It is clear, from the graph, that both the surface energies and the work functions show quantum oscillations.

TABLE 1. Total Energies and Cohesive Energies for Pu_n (n = 1-3) Layers. Total energies are in atomic units, and cohesive energies, indicated in brackets, are in eV. The first number represents cohesive energies with respect to n monolayers and the second number, with respect to (n-1) layers and the monolayer.

System	NSP-SR	SP-SR	NSP-SO	SP-SO
Monolayer	–29512.690243	–29512.777988	–29616.726443	–29616.774180
Dilayer	–59025.469338 (2.42)	–59025.596318 (1.10)	–59233.535959 (2.26)	–59233.596054 (1.30)
Trilayer	–88538.241698 (4.65, 2.23)		–88850.335182 (4.24, 1.98)	

In conclusion, we have presented results on electronic and geometric structure properties of Pu layers, using the LCGTO-FF method. Results clearly indicate the existence of quantum size effects in Pu layers. The work of AKR was supported by the Welch Foundation, Houston, Texas under Grant Y-1525. The work of JCB was supported by the U.S. Department of Energy under contract W-7405-ENG-36 and the LDRD program at Los Alamos National Laboratory.

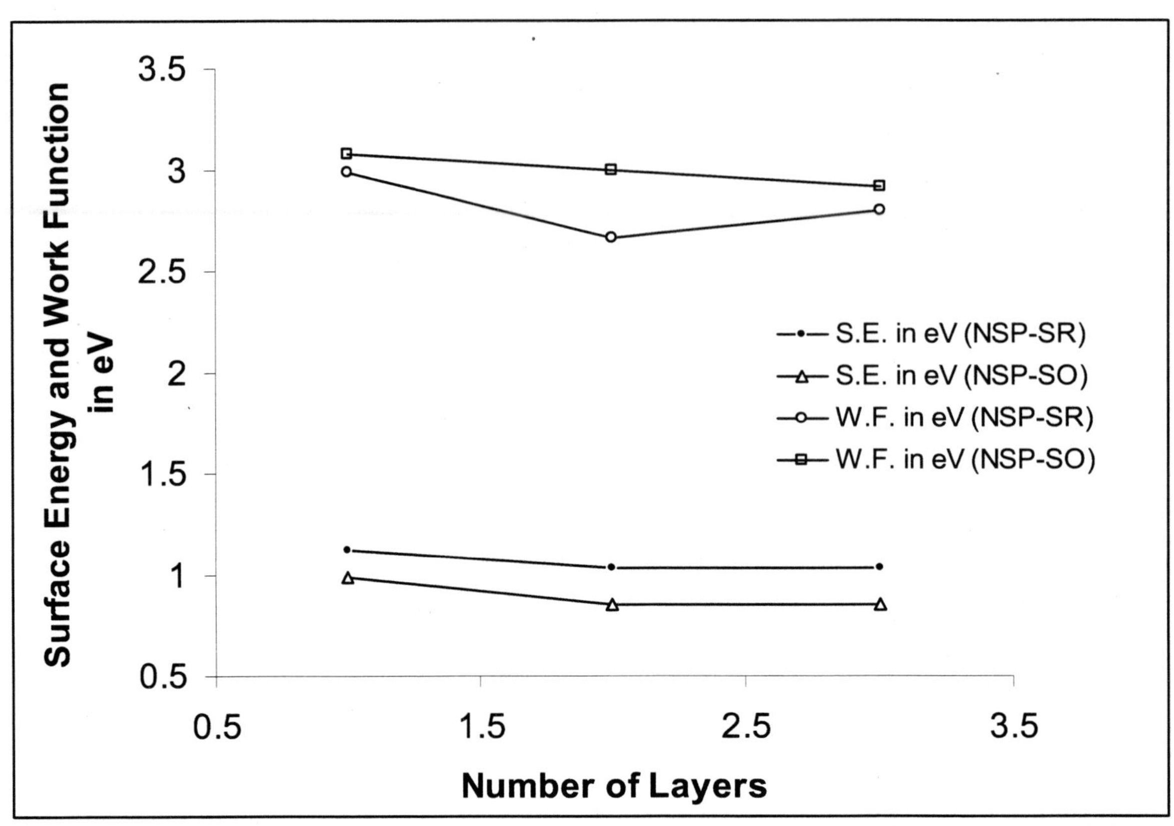

FIGURE 1. A plot of the surface energies and the work functions of the Pu layers at both the scalar relativistic and fully relativistic levels.

REFERENCES

1. Plutonium Futures—The Science, edited by K. K. S. Pillay and K. C. Kim, American Institute of Physics, Melville, New York (2000).
2. Eriksson, O., Cox, L. E., Cooper, B. R., Wills, J. M., Fernando, G. W., Hao, Y. G., and Boring, A. M., Phys. Rev. B 46, 13576 (1992).
3. Boettger, J. C., Phys. Rev. B 57, 8743 (1998); ibid, 62, 7809 (2000); Boettger, J. C., and Ray, A. K., Int. J. Quant. Chem. 90, 1470 (2002).
4. Ray A. K., and Boettger, J. C., Eur. Phys. J. B 27, 429 (2002).

Theory for α- and δ-Pu

Per Söderlind, Alex Landa, and Babak Sadigh

Lawrence Livermore National Laboratory, University of California, P.O. Box 808, Livermore, CA 94550, USA

INTRODUCTION

During the last few years, there has been intense research, both theoretically and experimentally[1] aimed to understand the fundamental properties of Pu and the anomalous, but technologically important, δ phase. Pu undergoes several phase transitions as a function of temperature at ambient pressure, and δ-Pu, stable between about 600 K and 740 K, is the least dense phase. This phase is quite different from the ground-state α phase, for instance: (i) δ-Pu has an atomic volume about 25% greater than α-Pu. (ii) The crystal symmetry of δ-Pu is cubic, but α-Pu condenses in a very complex low-symmetry monoclinic phase. (iii) The thermal expansion of δ-Pu is small and negative, but it is very large and positive for α-Pu. These remarkable inconsistencies occur for two phases of Pu separated by only a few hundred degrees and naturally suggest that the electronic structure in Pu is rapidly changing with temperature. The ambient pressure phase diagram of Pu[2] is shown in Figure 1.

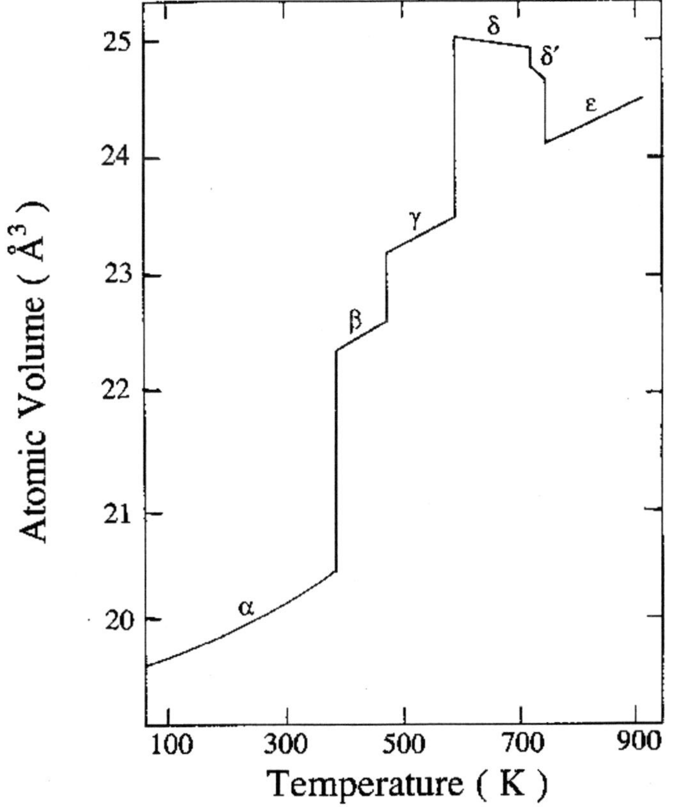

FIGURE 1. The experimental (Reference 2) phase diagram of plutonium.

CP673, *Plutonium Futures — The Science,* edited by G. D. Jarvinen

In the following paper we are trying to understand both α-Pu and δ-Pu from first-principles density-functional theory (DFT) expanding the work by Söderlind[3,4] and coworkers. The earlier study[3] of α-Pu showed that standard DFT, including gradient corrections to the electron exchange/correlation approximation, gave fair to good agreement with experiments for properties such as density, bulk modulus, and ground-state crystal structure. It was clear from this and other publications, however, that this theory was incomplete and unable to describe δ-Pu. For this reason, many recent semiempirical models[5–8] have been presented to better describe this phase. Several authors[4,9–11] have suggested that the δ phase might involve magnetism because allowing for the formation of spin moments in DFT makes it reproduce many of the critical properties of δ-Pu. This idea is controversial because magnetic susceptibility measurements[12] show only a weak temperature dependence for plutonium. Here we present the most recent calculations for δ-Pu and try to shed some light on the controversy. In addition to our study of δ-Pu we have made an effort to relax the monoclinic α phase of plutonium to confirm the accurate description of α-Pu provided by DFT.

We have combined three computational methods in the present investigation of Pu. First, a full-potential muffin-tin orbitals (FPLMTO) method, which includes spin-orbit interaction as well as orbital polarization, is an all-electron method with no geometrical approximations to the one electron potential or charge density. The advantage with this method is very accurate total energies but at the cost of great computational effort. Next, we applied a Korring-Kohn-Rostoker (KKR) method within a Green's function formalism. This method has the advantage of treating disorder, magnetic or compositional, within the coherent potential approximation (CPA) while at the same time being computationally very efficient. Last, we used so-called VASP code[13] with the projector augmented wave (PAW) technique. This approach is a fast and accurate way to calculate total energies as well as interatomic forces. In the present version, however, it does not include spin-orbit coupling. Magnetic disorder was treated, in the FPLMTO and the PAW, by super-cell calculations (special quasi-random structures), but in the KKR the so-called disordered local moment (DLM) model was applied. More details of these methods as applied for Pu were given in the paper by Söderlind et al.[4]

RESULTS AND DISCUSSION

Let us first focus on our present results for δ-Pu, in Figure 2. This figure shows nonmagnetic (NM), ferromagnetic (FM), antiferromagnetic (AF), and disordered magnetic (D or DLM) total energies for δ-Pu as a function of atomic volume. Notice that the AF configuration has the lowest energy regardless of computational method. This energy is, however, only slightly lower than that of the disordered magnetic calculation. It is also clear from this plot that the nonmagnetic treatment gives much too small atomic volume and much too high total energy to be appropriate. It was found that the AF configuration gives rise to a mechanical instability[4] in δ-Pu, but the disordered state was mechanically stable. This, in conjunction with the expected higher entropy for the disordered state, makes it plausible that δ-Pu is a disordered magnet. This would then help to explain the stabilization of δ-Pu at elevated temperatures. Upon cooling, pure δ-Pu tends to order in an antiferromagnetic fashion, which in turn promotes a structural phase transformation to a lower symmetry structure because of the mechanical instability of the AF phase. Therefore, the energy difference between the disordered and ordered magnetic phases becomes fundamental to the transition temperature in δ-Pu, and is currently[14] under investigation.

We can conclude from the calculations, presented in Figure 2, that a magnetic exchange interaction is needed to reproduce the experimental lattice constant for δ-Pu, at least within DFT. A magnetic treatment also gives a good bulk modulus and a total energy that is only a few mRy (a few 100 K) from the ground-state α-Pu[11] total energy. To make further contact with experiments, we show the calculated spectra in Figure 3. Here the magnetic order was AF, but calculations for other magnetic configurations gave similar results. Notice the well-defined narrow peak close below the Fermi level in δ-Pu. This behavior is in good accordance with experimental[4,15] photoemission. A nonmagnetic calculation, however, is not able to reproduce this behavior at all.

Next we turn to our detailed calculations for α-Pu. This phase solidifies in a very complex monoclinic structure of eight distinct atoms and a total of 16 atoms in the unit cell. Using the PAW technique, we calculated ab initio forces and were able to relax the numerous structural parameters of α-Pu. As in the case of δ-Pu, we allowed for a spontaneous formation of magnetic moments also in these calculations. Similar to δ-Pu, an AF ordering[16] was preferred in α-Pu. There is generally a good agreement between PAW and experiments[17] with the exception of the atomic equilibrium volume and the *b/a* axial ratio, which were about 8% too small and 4% too large, respectively. For this reason, we performed FPLMTO calculations; first the atomic volume was calculated for the fixed experimental geometry, and then the *b/a* axial ratio was relaxed, keeping all other structural parameters frozen.

These FPLMTO calculations were nonmagnetic but included spin-orbit coupling. Both the atomic volume and the *b/a* axial ratio are improved considerably; the atomic volume is about 3% too small, but the *b/a* ratio is in perfect agreement with the experiment. Our conclusions from these calculations are that α-Pu has a tendency to form magnetic moments (aligned antiparallel), but that a nonmagnetic treatment gives a very good description of the details of the crystal structure when spin-orbit coupling is included. It is also interesting to compare the spectra from α-Pu and δ-Pu in Figure 3. Quantitatively, both spectra compare rather well with photoemission,[15,17] and the satellite below the Fermi level is much reduced in α-Pu compared to δ-Pu. In the spectra[17] by Havela et al., this satellite essentially disappeared in α-Pu but was substantial in δ-Pu. It should be noted, however, that the experimental photoemission for α-Pu has shown discrepancies between measurements and may not be very well established.

In conclusion, both α-Pu and δ-Pu are well described within DFT, and in the case of δ-Pu, the formation of magnetic moments are absolutely necessary. Magnetic susceptibility measurements have not been able to confirm this picture for δ-Pu, and we have some speculations why this is so. First, spin and orbital moments are close to canceling each other so that the total moment in δ-Pu is small. For a complete cancellation, the total magnetic moment is zero, and no contribution to the susceptibility is expected. Next, the moments are disordered in δ-Pu with a zero net total moment, again not visible in the magnetic susceptibility measurements. These claims are substantiated by FPLMTO calculations[18] that do include magnetic fields in excess of 10 T. Finally, there could exist screening of the magnetic moments (not accounted for in the DFT) that may explain the measurements.

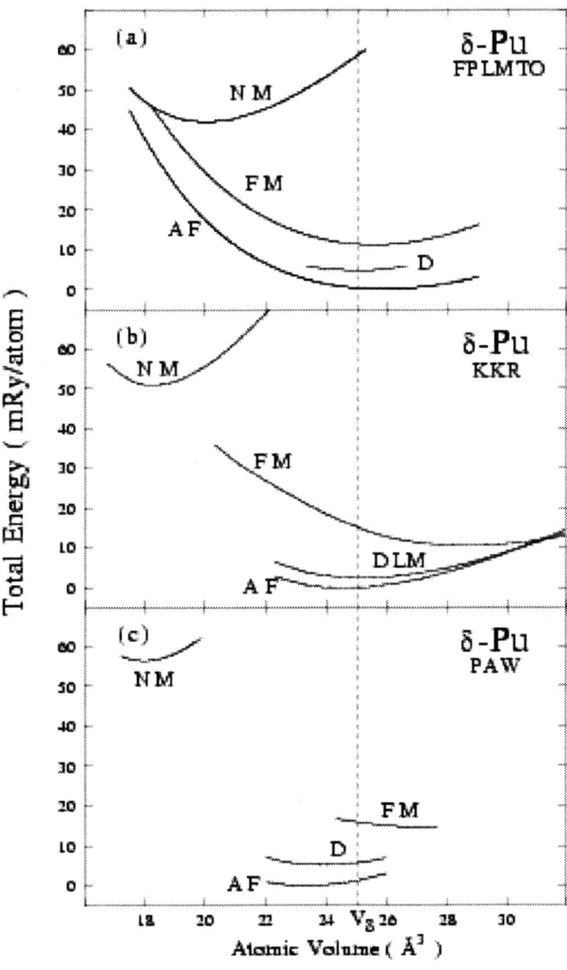

FIGURE 2. FPLMTO (a), KKR (b), and PAW (c) total energies for fcc Pu in nonmagnetic (NM), ferromagnetic (FM), antiferromagnetic (AF), and disordered (D) configurations. Dashed line indicates experimental equilibrium volume.

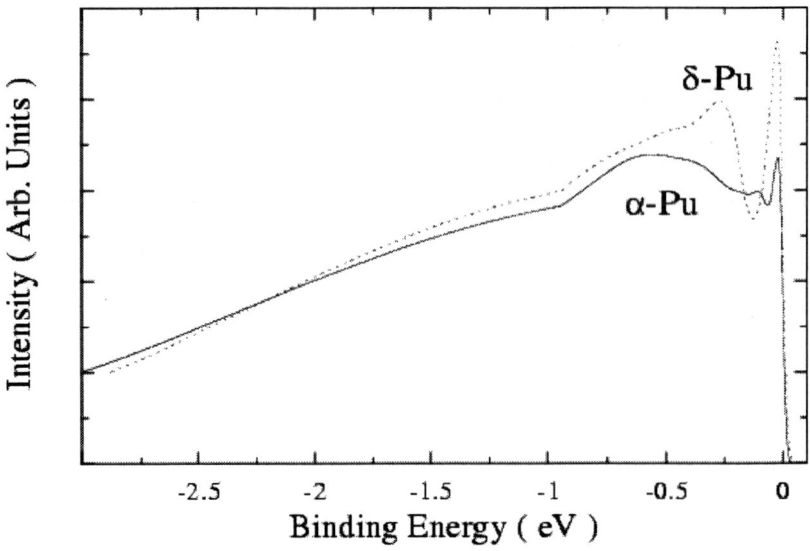

FIGURE 3. FPLMTO total electronic density of states for α-Pu and δ-Pu (dashed). The spectra have been convoluted with lifetime broadening as described by Arko et al. (Reference 15).

ACKNOWLEDGMENTS

We thank O. Eriksson, J.M. Wills, B. Johansson, J. Tobin, and A. Hjelm, for their help and discussions. This work was performed under the auspices of the U.S. Department of Energy by the University of California Lawrence Livermore National Laboratory under contract W-7405-ENG-48.

REFERENCES

1. Hecker, S. S., MRS Bulletin 26, 672 (2001).
2. Young, D. A., Phase Diagrams of the Elements, University of California Press, Berkeley (1991).
3. Söderlind, P., Eriksson, O., Johansson, B., and Wills, J. M., Phys. Rev. B 55, 1997 (1997).
4. Söderlind, P., Landa, A., and Sadigh, B., Phys. Rev. B 66, 205109 (2002).
5. Penicaud, M., J. Phys.: Condens. Matter 9, 6341 (1997).
6. Eriksson, O., Becker, J. D., Balatsky, A. V., and Wills, J. M., J. Alloys Compd. 287, 1 (1999).
7. Bouchet, J., Siberchicot, B., Jollet, F., and Pasturel, A., J. Phys.: Condens. Matter 12, 1723 (2000).
8. Savrasov, S. Y., and Kotliar, G., Phys. Rev. Lett. 84, L31 (2000).
9. Postnikov, A. V., and Antropov, V. P., Comput. Mater Sci. 17, 438 (2000).
10. Wang, Y., and Sun, Y., J. Phys.: Condens. Matter 12, 1723 (2000).
11. Söderlind, P., EuroPhys. Lett. 55, 525 (2001).
12. Meot-Reymond, S., and Fournier, J. M., J. Alloys Comp. 232, 119 (1996).
13. Kresse, G., and Hafner, J., Phys. Rev. B 47, 558 (1993); Kresse, G., and Furthmüller, J., ibid. 54, 11169 (1996); Kresse, G., and Joubert, D., ibid. 59, 1758 (1999).
14. Landa, A., Söderlind, P., and Ruban, A. V. (to be published).
15. Arko, A. J., Joyce, J. J., Morales, L., Wills, J., Lashley, J., Wastin, F., and Rebizant, J., Phys. Rev. B 62, 1773 (2000).
16. Sadigh, B., Söderlind, P., and Wolfer, W. G. (to be published).
17. Havela, L., Gouder, T., Wastin, F., and Rebizant, J., Phys. Rev. B 65, 235118 (2002).
18. Söderlind, P. (to be published).

Theory of Positron Annihilation in Helium-Filled Bubbles in Plutonium

P. A. Sterne and J. E. Pask

Lawrence Livermore National Laboratory, Livermore, CA 94550, USA

Positron annihilation lifetime spectroscopy is a sensitive probe of vacancies and voids in materials.[1] This nondestructive measurement technique can identify the presence of specific defects in materials at the part-per-million level. Recent experiments by Asoka-Kumar et al.[2] have identified two lifetime components in aged plutonium samples—a dominant lifetime component of around 182 ps and a longer lifetime component of around 350–400 ps. This second component appears to increase with the age of the sample and accounts for only about 5 percent of the total intensity in 35-year-old plutonium samples.

First-principles calculations of positron lifetimes are now used extensively to guide the interpretation of positron lifetime data.[3] At Livermore, we have developed a first-principles finite-element-based method for calculating positron lifetimes for defects in metals.[4] This method is capable of treating system cell sizes of several thousand atoms, allowing us to model defects in plutonium ranging in size from a mono-vacancy to helium-filled bubbles of over one nm in diameter.

In order to identify the defects that account for the observed lifetime values, we have performed positron lifetime calculations for a set of vacancies, vacancy clusters, and helium-filled vacancy clusters in delta-plutonium. The calculations produced values of 143 ps for defect-free delta-Pu and 255 ps for a mono-vacancy in Pu, both of which are inconsistent with the dominant experimental lifetime component of 182 ps. Larger vacancy clusters have even longer lifetimes. The observed positron lifetime is significantly shorter than the calculated lifetimes for mono-vacancies and larger vacancy clusters, indicating that open vacancy clusters are not the dominant defect in the aged plutonium samples.

When helium atoms are introduced into the vacancy cluster, the positron lifetime is reduced due to the increased density of electrons available for annihilation. For a mono-vacancy in Pu containing one helium atom, the calculated lifetime is 190 ps, while a di-vacancy containing two helium atoms has a positron lifetime of 205 ps. In general, increasing the helium density in a vacancy cluster or He-filled bubble reduces the positron lifetime, so that the same lifetime value can arise from a range of vacancy cluster sizes with different helium densities.

In order to understand the variation of positron lifetime with vacancy cluster size and helium density in the defect, we have performed over 60 positron lifetime calculations with vacancy cluster sizes ranging from 1 to 55 vacancies and helium densities ranging from zero to five helium atoms per vacancy. The results indicate that the experimental lifetime of 182 ps is consistent with the theoretical value of 190 ps for a mono-vacancy with a single helium atom, but that slightly better agreement is obtained for larger clusters of 6 or more vacancies containing 2–3 helium atoms per vacancy. For larger vacancy clusters with diameters of about 3–5 nm or more, the annihilation with helium electrons dominates the positron annihilation rate; the observed lifetime of 180 ps is then consistent with a helium concentration in the range of 3 to 3.5 He/vacancy, setting an upper bound on the helium concentration in the vacancy clusters. In practice, the single lifetime component is most probably associated with a family of helium-filled bubbles rather than with a specific unique defect size.

The longer 350–400 ps lifetime component is consistent with a relatively narrow range of defect sizes and He concentrations. At zero He concentration, the lifetime values are matched by small vacancy clusters containing 6–12 vacancies. With increasing vacancy cluster size, a small amount of He is required to keep the lifetime in the 350–400 ps range, until the value saturates for larger helium bubbles of more than 50 vacancies (bubble diameter > 1.3 nm) at a helium concentration close to 1 He/vacancy.

CP673, *Plutonium Futures — The Science,* edited by G. D. Jarvinen

These results, taken together with the experimental data, indicate that the features observed in TEM data by Schwartz et al. are not voids, but are in fact helium-filled bubbles with a helium pressure of around 2–3 helium atoms per vacancy, depending on the bubble size. This is consistent with the conclusions of recently developed models of He-bubble growth in aged plutonium.

ACKNOWLEDGMENTS

This work was performed under the auspices of the U.S. Department of Energy by the University of California Lawrence Livermore National Laboratory under Contract No. W-7405-ENG-48

REFERENCES

1. See, e.g., Positron Spectroscopy of Solids, edited by A. Dupasquier and A. P. Mills, Jr., IOS Amsterdam (1995).
2. Asoka-Kumar, P., Glade, S., Sterne, P. A., and Howell, R., presented at this meeting.
3. Puska, M. J., and Nieminen, R. M., Rev. Mod. Phys. 66, 841 (1994).
4. Sterne, P. A., Pask, J. E., and Klein, B. M., Applied Surface Science 149, 238 (1999).

Thermal- and Radiation-Induced Interactions of Water on UO_2 Surfaces

J. Stultz, M. T. Paffett, and S. A. Joyce

Chemistry Division, Los Alamos National Laboratory, Los Alamos, NM 87545

Most plans for the disposition of surplus nuclear materials involve storage in sealed containers where the evolution of gases from reactions of adsorbed water could present both pressure and flammability hazards.[1] Despite efforts such as calcining the material to minimize the water content before packaging, both residual moisture and readsorbed water may be present in the final containers. Given the anticipated temperature excursions during transportation and storage, this water may thermally desorb, increasing the pressure, and/or thermally dissociate to produce H_2 gas, increasing flammability hazards. In addition, the radiation from the nuclear material may induce radiolysis of the water with the likely products being water vapor, H_2, O_2, and H_2O_2. In order to better understand the relative importance of the thermal- and radiation-induced chemistry, we have studied the interactions of water on single crystals of uranium dioxide.

All experiments were conducted in an ultrahigh vacuum chamber in order to clean and characterize the single crystal samples and to minimize the effects of readsorption in the interpretation of the results. The single crystal $UO_2(111)$ surface was prepared by cleaning with Ar ion sputtering, annealed at high temperatures and examined with Auger electron spectroscopy to determine surface cleanliness. The principal experimental techniques employed were temperature-programmed desorption (TPD) for studying the thermal chemistry and electron-stimulated desorption (ESD) to simulate the effects of radiolysis. In TPD, the surface is exposed to controlled quantities of water at low temperatures; the surface is then heated, and gaseous product evolution is monitored with a mass spectrometer. Evolved gas-phase products are directly determined, and binding energies and reaction kinetics can be obtained from an analysis of the desorption temperature profiles. An electron gun is employed to simulate the effects of radiation because the inherent radiation flux at the surface from these planar samples is too low to induce significant chemistry in reasonable time scales. Ionization events and the production of low-energy secondary electron emission from 5-MeV alpha particles accounts for over 99% of its energy loss. Therefore, a controlled external electron source acts as an appropriate simulant for inducing the majority of the interfacial radiation chemistry. Both positive ion and neutral species desorption are monitored with a mass spectrometer. $D_2{}^{16}O$ was used for all exposures to reduce background effects from the small amounts of residual H_2O in the vacuum system.

The TPD results for the clean UO_2 surface are relatively straightforward. The only desorbing species is water. No D_2, O_2, or other species are detected. Two distinct states for adsorbed water are observed. A low temperature, nonsaturating state is observed. This is ascribed to the formation of condensed multilayers of water on the surface by comparison with previous work.[2] A saturating, high-temperature state is assigned to the direct interaction between water and the UO_2 surface. A simple kinetic analysis of this monolayer state yields a binding energy of approximately 13–15 kcal/mole.

In contrast to the thermal chemistry, the electron-induced chemistry is more complex. D_2 and O_2 gas are the primary ESD products, along with small amounts of water. No D_2O_2, an important product of the radiolysis of bulk water, is observed. The absolute and relative amounts of D_2 and O_2 produced are functions of the temperature and coverage of water. The highest yields occur for multilayer water. D_2 is produced in hyperstoichiometric amounts (e.g., the ESD D_2:O_2 ratio is at least 5; ideally it is 2 in D_2O). The cause of the oxygen deficiency in the neutral products is currently unknown. For monolayers, and lower-coverage water films, the overall yields decrease, but the D_2:O_2 ratio increases (exceeding 10). In addition to neutral species, positive ions were monitored as well. For the clean UO_2 surface, ESD only observed O, H, and F ions. The H comes from the small adsorption of water from the background gases in the system, and the F is attributed to impurities in the fluxes used to produce the UO_2 crystal. This result highlights the potential sensitivity of ion-desorption to minority low concentration species. The D ion is

CP673, *Plutonium Futures — The Science,* edited by G. D. Jarvinen

the only species observed from multilayers of water, consistent with previous studies.[3] For submonolayer amounts of water, D and OD ions are the primary products. The ESD D and OD signals persist to surface temperatures up to 600 K. From the thermal studies, all molecular water was observed to desorb below 300 K. In order to determine the origin of the OD species, isotopically labeled ^{18}O was incorporated into the surface by ion sputtering. Upon exposure to $D_2{}^{16}O$, ^{18}OD ions were observed, indicating that at least some of the oxygen comes from the substrate as the result of dissociation at minority defect sites.

The TPD results indicate that the majority of water interacts molecularly on a clean UO_2 (111) surface with a binding energy only slightly greater than water with itself. However, this small change may have important implications in "real world" environments. A simple steady-state kinetic model can be used to determine the vapor pressure of water needed to form the direct, monolayer film. Based of the TPD-derived binding energy of 13–15 kcal/mole, a water pressure on the order of a few millitorr is enough to saturate the monolayer film. This pressure is below that used in most gloveboxes; therefore, most processed material will, except in the driest boxes, readsorb one layer of water. Most of the standards developed for nuclear materials storage require accounting for at least this level of water loading in designing the containers. The electron-irradiation results indicate that over prolonged periods, significant radiolytic production of H_2 and O_2 can occur. In all cases, H_2 was produced in excess of O_2, which is consistent with previous work on gas generation from PuO_2 powders. As the absolute yields decrease and relative H_2:O_2 yields increase with decreasing water loading, these results suggest that minimizing moisture content would be beneficial not only in reducing the pressure hazards, but also in ensuring that the gas composition is outside the flammability limits for hydrogen/oxygen mixtures.

ACKNOWLEDGMENTS

Funding for this investigation was made possible by the Los Alamos National Laboratory LDRD program and by the Nuclear Materials Stabilization Program Office, United States Department of Energy, Albuquerque Operations and Headquarters Offices, under the auspices of the DNFSB 94-1 Research and Development Project.

REFERENCES

1. Department of Energy (DOE), DOE-STD-3013-2000, Stabilization, Packaging, and Storage of Plutonium-Bearing Materials, U.S. Department of Energy (September 2000).
2. Henderson, M. A., "The Interactions of Water with Solid Surfaces: Fundamental Aspects Revisited," Surface Science Reports 46, 1 (2002).
3. Sieger, M. T., Simpson, W. C., and Orlando, T. M., "Electron-Stimulated Desorption of D{sup +} from D{sub 2}O Ice: Surface Structure and Electronic Excitations," Phys. Rev. B. 56, 4925 (1997).

Laser-Driven Materials Experiments: New Capabilities for Probing Dynamic Behavior

Damian C. Swift, Thomas E. Tierney, Dennis L. Paisley, George A. Kyrala, Randall P. Johnson (P-24), and Allan Hauer (P-DO)

Los Alamos National Laboratory, Los Alamos, NM 87545

Accurate models of the response of materials to dynamic loading are important for simulations of the performance and safety of weapons. These models are devised using experiments in which samples are subjected to high pressures, usually from shock waves. New classes of models are being developed using the detailed underlying physics, such as dislocation dynamics, rather than *ad hoc* empirical relations to describe the response. Laser-driven materials experiments offer promising possibilities to contribute to model development, including convenient new diagnostics such as transient x-ray diffraction to measure response at the microstructural level. These experiments provide a bridge between single crystal and polycrystalline behavior.

We have used laser-based experiments to investigate the dynamic response of a range of materials, using the TRIDENT facility at Los Alamos. Loading is generated either by the impact of a flyer plate (accelerated with laser energy) or by direct irradiation of the surface of a sample, in which case material pressure is induced by the reaction to ablation of the surface ablation. Laser-driven flyer experiments have already been performed on actinides; here we discuss the possibilities of direct drive. One attraction of the TRIDENT facility is the high degree of flexibility in controlling the laser. For instance, the intensity history can be varied to induce a shock wave, or quasi-isentropic compression (pressure increasing smoothly). By altering the laser energy and the size of the focal spot on the sample, the pressure can be varied in a controllable way between about 1 GPa and 1 TPa. Unlike the more conventional gas gun and high-explosive methods—and also pulsed-power discharge such as the Z facility—there is less stray energy or momentum around the experiment from the sabot, detonation products, etc. This makes it easier to recover and analyze the sample. Furthermore, because the time at which pressure is generated is well defined by the arrival of the laser pulse, it is straightforward to synchronize diagnostic measurements such as surface velocity history to ~50 ps accuracy. Velocities are measured by the laser Doppler VISAR technique, of a point or a line on the sample. A particular attraction of TRIDENT is the capability of obtaining transient x-ray diffraction (TXD) records in situ as a sample is loaded. Essentially monochromatic x-rays are generated conveniently from a hot laser-induced plasma, produced by focusing another beam onto a thin foil. In this way, we have obtained simultaneous records of surface velocity history and deformation of the crystal lattice (Figure 1). We have also obtained time-resolved records of electrical conductivity from the polarization-dependent reflectivity (ellipsometry), and we are developing diagnostics to measure temperature at the surface of the sample with subnanosecond time resolution.

We have obtained useful VISAR and TXD records from a variety of materials; TXD data mainly from single crystals but also with some preliminary data from polycrystalline foils. Because of constraints on the number and position of x-ray streak cameras, we are limited in the range of angles that can be covered during a single experiment. However, we have distinguished shock compression, which is uniaxial at the atomic level, from plastic flow, and we have also observed—though not uniquely identified—phase transformations in silicon[1] (Figure 2) and beryllium.[2] The combination of these diagnostics on laser-driven experiments is a particularly good complement for molecular dynamics simulations, as it becomes possible to probe the behavior of solids at the level of the crystal lattice and to compare experimental data directly against atomistic simulations. An advantage of small sample sizes is of course that less material is involved: a convenience when dealing with toxic, radioactive, or expensive materials.

CP673, *Plutonium Futures — The Science,* edited by G. D. Jarvinen

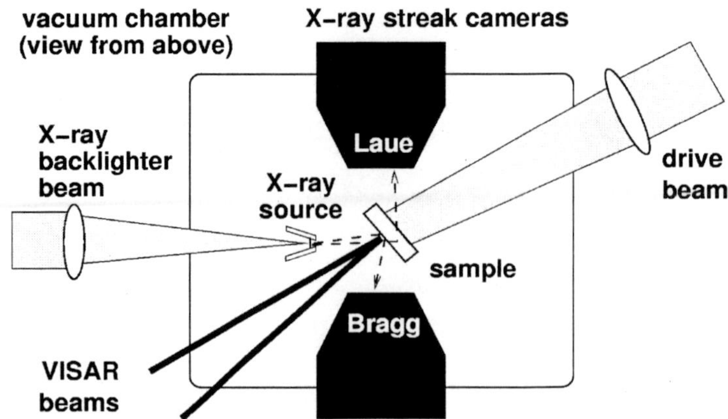

FIGURE 1. Schematic layout of target chamber for direct-drive materials experiments at TRIDENT, with velocimetry and transient x-ray diffraction.

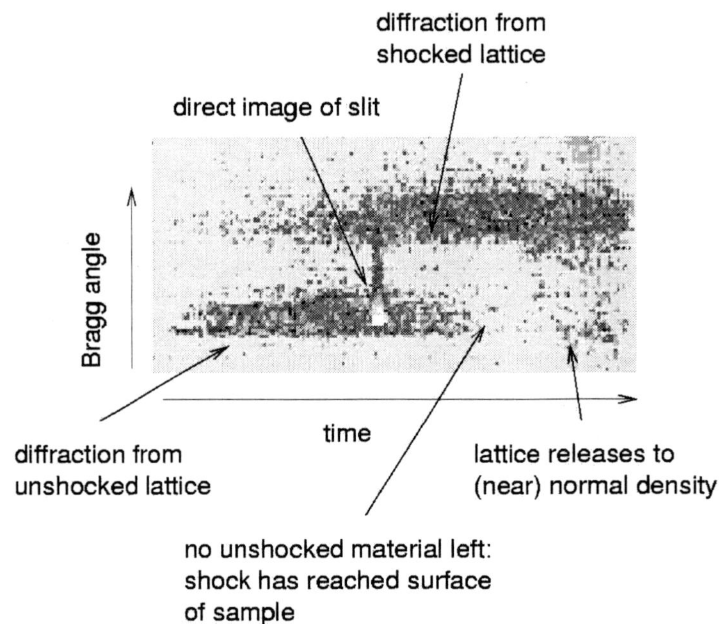

FIGURE 2. Example transient x-ray diffraction record from a shock-loaded single crystal of silicon.

There are significant challenges—administrative as well as technical—in applying these techniques to actinides, particularly plutonium. Modified types of target chambers are needed to cope with contamination. More importantly, compared with our previous experiments on elements of lower atomic number, x-rays of the energy normally used for TXD would not penetrate far into actinide samples, so it would be difficult to synchronize the x-ray source with the arrival of the shock wave; the diffracted signal could also be much smaller. These problems can be addressed by altering the experiment somewhat, e.g., by using a short-pulse (~picosecond) backlighter to generate higher-energy, more penetrating x-rays. We believe that only incremental improvements, of fairly low risk, would be needed to apply these new techniques to plutonium.

REFERENCES

1. Swift, D. C., Ackland, G. J., Hauer, A., and Kyrala, G. A., Phys. Rev. B 64, 214107 (2001).
2. Swift, D. C., Paisley, D. L., Kyrala, G. A., and Hauer, A., Proc. 2001 APS Topical Conference on Shock Compression of Condensed Matter, AIP CP620, American Institute of Physics (2002).

Acoustic Resonance Spectroscopy for Structural Evaluation

Margaret T. Trujillo, Claudette C. Trujillo,
David A. Miller, and Joseph Baiardo

Los Alamos National Laboratory, Los Alamos, NM 87545

Acoustic Resonance Spectroscopy (ARS) is a nondestructive testing technique first applied to the analysis of weapons components by personnel in Los Alamos in the early 1990s. ARS uses ultrasonic transducers to excite and measure the resonant response of a structure. Experimentally, a structure is rested on an array of transducers, one of which loads the pit with a sinusoidally varying load and another that measures the structural response to the load. A correlation is then drawn between the driving frequency of the load and the measured response to determine the resonant frequencies of the structure. The resonant frequencies of the structure are a function of both the material physical properties and the geometry, and therefore give a fingerprint of the component. This fact makes ARS a powerful technique to monitor changes over time in a single part, or to compare manufactured parts to verify consistency in the manufacturing process.

The current primary application of the ARS technique is to measure the resonant frequencies of pits within the Pit Surveillance Project. This presentation will describe the current experimental setup of the ARS technique and the process for implementing the technique. Also, the usefulness of this technique to identify structural and material changes will be demonstrated using common structural components.

CP673, *Plutonium Futures — The Science,* edited by G. D. Jarvinen
2003 American Institute of Physics 0-7354-0140-3

Thermodynamics of Pu-Based Alloys

P. E. A. Turchi,[1] P. G. Allen,[1] L. Kaufman,[2] Shihuai Zhou,[3] and Zi-Kui Liu[3]

[1]Lawrence Livermore National Laboratory (L-353), P. O. Box 808, Livermore, CA 94551
[2]Dept. of Materials Science and Engineering, MIT, Cambridge, MA 02139
[3]Dept. of Mater. Sci. and Eng., The Pennsylvania State University, University Park, PA 16803

The study of the equilibrium thermodynamic properties of Pu-based alloys (Pu-X, where X = Al, Fe, Ga, Ni) in the framework of the CALPHAD approach. Assessment of the equilibrium phase diagrams in the whole range of alloy composition has been performed. Predictions are made on the low-temperature phase diagrams of Pu-Ga and Pu-Al for which controversy has been amply discussed in the past. The calculated phase diagrams of Pu-Ga and Pu-Al confirm the experimental results, and in particular, a low-lying eutectoid decomposition temperature for the $\delta \rightarrow \alpha + Pu_3M$ (M = Ga, Al) transition is predicted for Pu-Ga and Pu-Al. An overall picture for the stability properties of Pu-Ga and Pu-Al that reconciles the results of past studies carried out on these alloys is proposed. Additional assessment will be presented for the phase diagram of the binary Pu-Ga and Pu-Ni alloys. Finally, preliminary results will be presented on the kinetics of some of the reactions occurring in Pu-Ga alloys, and suggestions are made for further studies.

ACKNOWLEDGMENTS

This work was performed under the auspices of the U.S. Department of Energy by the University of California Lawrence Livermore National Laboratory under Contract W-7405-ENG-48.

CP673, *Plutonium Futures — The Science,* edited by G. D. Jarvinen

Accelerated Molecular Dynamics Study of Vacancies in Pu

Blas Pedro Uberuaga,[1] Steven M. Valone,[2] M. I. Baskes,[2] and Arthur F. Voter[1]

[1]*Theoretical Division and* [2]*Materials Science and Technology Division*
Los Alamos National Laboratory
Los Alamos, NM 87545 USA

From its central place in the nuclear weapons program to its role in modern fission reactors, the importance of Pu cannot be overstated. A complete understanding of the properties of Pu begins at the atomic level, where point defects—vacancies and interstitials—dominate the behavior. These defects govern the self-diffusivity of Pu, control the motion of impurities or dopants that might be present, and play a key role in the relaxation after a radiation damage event. The goal of this work is to study the dynamical properties of Pu vacancies as a first step towards deepening our understanding of the fundamental defect dynamics in Pu.

Molecular dynamics (MD) simulation is an important tool for understanding materials at the atomic scale. MD allows us to do what experiments cannot: focus on a small piece of material where we can follow the motion of individual atoms. In doing so, we can directly probe the behavior of vacancies and interstitials, as well as other defects, providing input to higher-level models. One limitation of MD, however, is that the time scales accessible are relatively short: ns to μs. To overcome this time scale problem, we employ recently developed accelerated dynamics techniques,[1] in particular parallel-replica dynamics[2] and temperature-accelerated dynamics.[3] These methods allow us to reach longer time scales than regular MD. In contrast to a kinetic Monte Carlo approach, they make no previous assumptions about the atomistic mechanisms available for dynamical evolution.

Parallel-replica dynamics achieves parallelization of time. If the rare events that govern motion from state to state in the material are first-order processes, then parallel-replica dynamics is exact, correctly describing even correlated dynamical events. By parallelizing on M processors, one can achieve a maximum speedup of M compared to a single-processor simulation. Reaching this theoretical maximum efficiency depends on how infrequent the events are because there is some overhead involved each time an event is observed.

Temperature-accelerated dynamics (TAD) are potentially much more powerful but also involve more assumptions about the nature of the system. In particular, if the system obeys the harmonic transition state theory, TAD gives exact state-to-state dynamics. The TAD algorithm involves running MD at a high temperature T_{high} higher than the temperature of interest, T_{low}. From each basin (or state), the times of attempted escape events at T_{high} are then extrapolated to T_{low}, allowing an appropriate low-temperature event to be selected. Using this algorithm, substantial boost factors (relative to direct MD) are achievable for some systems.[4]

In this work, we use a modified embedded atom method (MEAM)[5] potential fit to the properties of Pu. To simplify the analysis somewhat and to simulate the stabilizing effects of Ga-doped Pu, we have used both the parameters published by Baskes,[6] as well as a modified parameter set in which $t_3 = 0$. We will refer to these as Pu(1) and Pu(2), respectively. Pu(2) stabilizes the FCC phase of Pu, relative to the other phases, much as Ga does in real Pu.

We have applied the parallel-replica method to study the motion of a single vacancy in both Pu(1) and Pu(2) at 900 K. Running on 38 processors, we achieved a speedup of 36 times. This simulation, taking one week of wall-clock time, would have taken 36 weeks on a single processor.

The motion of the Pu crystal at this temperature (900 K) is characterized by global partial distortions of the crystal into phases other than FCC. Thus, with the vacancy in one lattice site, there are a multitude of local minima decorating the potential energy basin belonging to the vacancy. At this temperature, vacancy hops are relatively rare: in a simulation of Pu(1) for 1.1 μs, we saw only five diffusive events. Three of these events were simple vacancy hops, in which an adjacent Pu atom moved directly into the vacancy site, resulting in the vacancy moving one nearest-neighbor distance. The energy profile of the minimum energy path for such an event has been calculated

CP673, *Plutonium Futures — The Science,* edited by G. D. Jarvinen

with the nudged elastic band (NEB) method[7] and is illustrated in Figure 1 for Pu(2).[8] The barrier for a single hop is about 1.55 eV. Even for this simple event, the complexity of the potential can be seen. The analogous curve for Pu(1) has a barrier of 1 eV.[9]

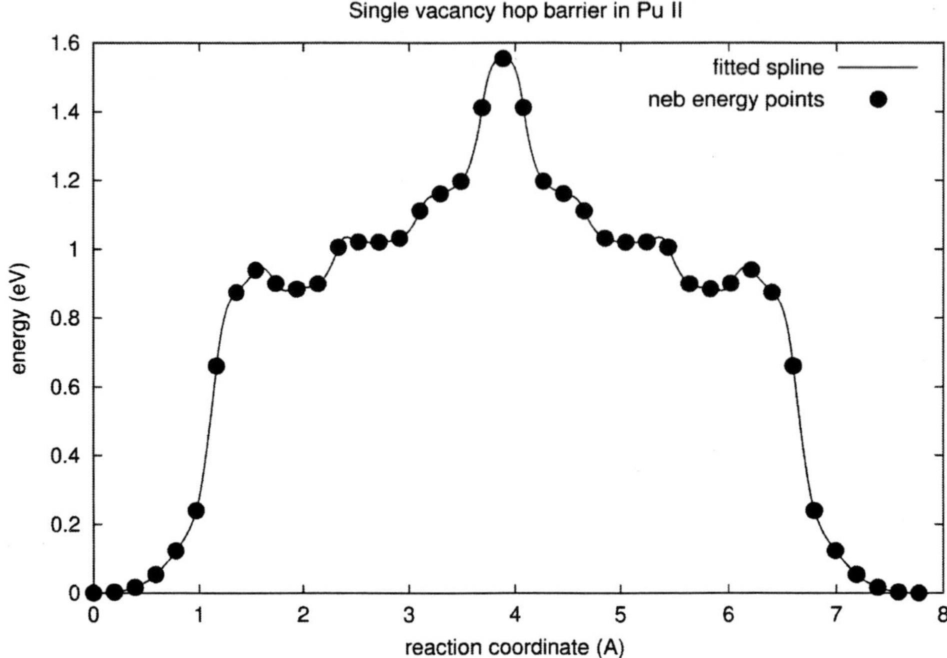

FIGURE 1. Energy profile of minimum energy path (from nudged elastic band) for a single vacancy hop in Pu(2).

Interestingly, we also saw two events in which the vacancy executes a double hop. Two Pu atoms move in a concerted motion so that the vacancy moves two nearest-neighbor spacings during one event. An example of this type of event is illustrated in Figure 2. Although we have not converged a nudged elastic band for this process, initial calculations suggest that the barrier for this mechanism is on the order of 1 eV for the Pu(2) potential, though more analysis needs to be done to truly characterize this event and compare it to the single vacancy hop mechanism.

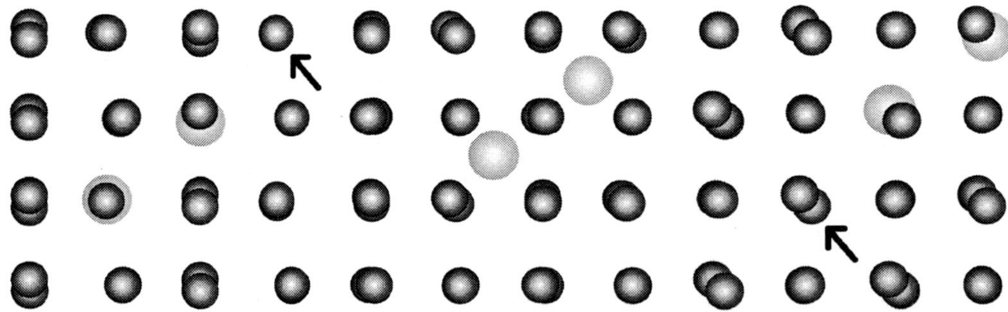

FIGURE 2. Double vacancy hop in Pu(1). The two atoms that move from one lattice site to another are highlighted (lighter color and larger size) for clarity (only a subset of the 255-atom system cell is shown). Arrows indicate the position of the vacancy initially (left frame) and after the event (right frame). The slight distortion away from the perfect FCC structure can be seen.

We have also performed a parallel-replica simulation of a vacancy in Pu(2). As mentioned, Pu(2) differs from Pu(1) in that the FCC phase is stabilized somewhat. The crystal does not distort globally into other phases but does display small local distortions. In one simulation of 1.0 μs at 900K, we saw one event, which was a single hop. Although the statistics from this preliminary study are extremely limited, it appears that the diffusion rate at T = 900 K for the Pu(2) model (1 event/1.0 μs) is lower than for Pu(1) (5 events/1.1 μs).

To summarize, we are beginning to apply accelerated dynamics methods to study defect dynamics in Pu. In our initial study using parallel-replica dynamics, we have identified two different diffusion mechanisms for a single vacancy. We are in the process of extending this work to small clusters of vacancies, and we are adapting the TAD method for this system to achieve even longer time scales.

REFERENCES

1. Voter, A. F., Montalenti, F., and Germann, T. C., Annu. Rev. Mater. Res. 32, 321 (2002).
2. Voter, A. F., Phys. Rev. B 57, R13985 (1998).
3. Sorensen, M. R., and Voter, A. F., J. Chem. Phys. 112, 9599 (2000).
4. Sprague, J. A., Montalenti, F., Uberuaga, B. P., Kress, J. D., and Voter, A. F., Phys. Rev. B 66, 205415 (2002); Montalenti, F., Sorensen, M. R., and Voter, A. F., Phys. Rev. Lett. 87, 126101 (2001).
5. Baskes, M. I., Nelson, J. S., and Wright, A. F., Phys. Rev. B 40, 6085 (1989); Baskes, M. I., Phys. Rev. B 46, 2727 (1992).
6. Baskes, M. I., Phys. Rev. B 62, 15532 (2000).
7. Henkelman, G., Uberuaga, B. P., and Jonsson, H., J. Chem. Phys. 113, 9901 (2000); J. Chem. Phys. 113, 9978 (2000).
8. Because of the metastable distortions that exist in Pu(1), there are many local minima along the minimum energy path from one vacancy site to the next. These minima cause convergence problems for NEB, so we opted to calculate the minimum energy path for the vacancy motion in Pu(2).
9. Valone, S. M., Baskes, M. I., Stan, M., Mitchell, Terence E., Lawson, Andrew C., and Sickafus, K. E., J. Nucl. Mater. (in press).

Atomistic Models of Point Defects in Plutonium Metal

Steven M. Valone,[‡1] Michael I. Baskes,[1] and Blas P. Uberuaga[2]

Materials Science and Technology[1] and Theoretical[2] Divisions, Los Alamos National Laboratory, Los Alamos, NM, 87545 USA
[‡]Corresponding author: smv@lanl.gov

The aging properties of plutonium (Pu) metal and alloys are driven by a combination of materials composition, processing history, and self-irradiation effects. Understanding these driving forces requires a knowledge of both thermodynamic and defect properties of the material. The multiplicity of phases and the small changes in temperature, pressure, and/or stress that can induce phase changes lie at the heart of these properties. In terms of radiation damage, Pu metal represents a unique situation because of the large volume changes that accompany the phase changes. The most workable form of the metal is the fcc (δ-) phase, which in practice is stabilized by the addition of alloying elements such as Ga or Al. The thermodynamically stable phase at ambient conditions is the monoclinic (α-) phase, which, however, is 20% lower in volume than the δ phase. In stabilized Pu metal, there is an interplay between the natural swelling tendencies of fcc metals and the volume-contraction tendency of the underlying thermodynamically stable phase. This study explores the point defect properties that are necessary to model the long-term outcome of this interplay.

Point-defect properties are atomistic in nature. To study point-defect production and migration, it is necessary to construct an atomistic model of the interactions among Pu atoms. Recently, progress has been achieved in the form of a modified embedded atom (MEAM)[1,2] potential for pure Pu.[3] The MEAM potential was able to capture the most salient features of atomic volume and enthalpy of Pu metal and liquid metal as a function of temperature at zero pressure (Figures 1a and 1b). Most significantly, the atomic volume difference between the α- and δ-phases was captured nearly quantitatively.

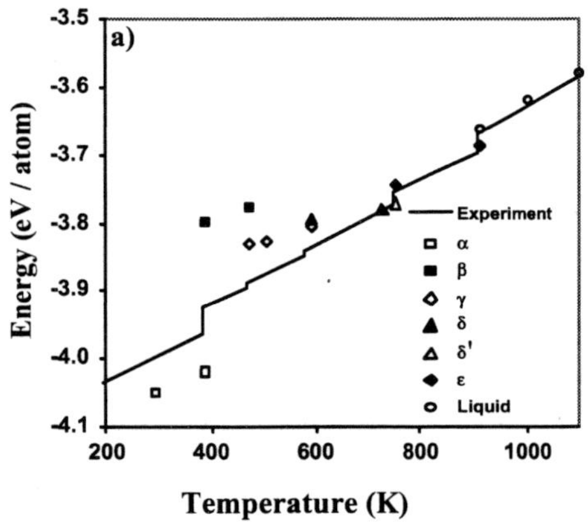

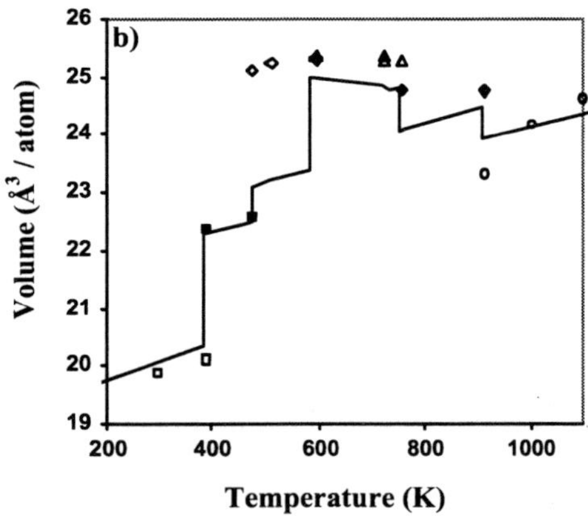

FIGURE 1. MEAM potential model of Pu metal at low pressures as a function of temperature. The differences in atomic energy and volume between α and δ Pu are quite close to the experimental values.

Here we use this potential to simulate the point defects that form as the result of a self-irradiation cascade in Pu metal at room temperature (RT) and ambient pressure. The first property of interest is the minimum displacement threshold energy (DTE). The DTE represents the minimum energy in a cascade required to cause the formation of a Frenkel pair. The potential represented in Figure 1 (model I) has a very low DTE, 10 eV, compared to other fcc

CP673, *Plutonium Futures — The Science,* edited by G. D. Jarvinen

metals.[4] See Table 1. For this reason, we consider a second set of parameters (model II) in which the fcc phase is stable all the way down to 0 K. The DTE for model II is 18 eV.

TABLE 1. Comparison of DTEs (E_d) for the Two Models of Pu and Some fcc Metals. The experimental values are for polycrystalline samples using electron beam irradiation, except for Pu.

Metal	Pu	Ag	Al	Au	Cu	Ni	Pt	Pb
Experiment E_d (eV)	33[a] 14[b]	25[e]	16[e]	36[e]	19[e]	23[e]	24[e]	12.5-15[f]
MEAM E_d (eV)	10[c] 18[d]			20.9[c]	17.7[c]	23.7[c]	24.1[c]	15[c]

[a]Reference 6 value based on sublimation energies in fcc metals.
[b]Reference 7 value based on melting temperature of metals.
[c]Reference 4.
[d]Reference 4 (Model II).
[e]Reference 5.
[f]Reference 8.

The reason for the low DTE in model I can be directly related to the low-temperature properties of the model. Below 100 K, the fcc lattice of model I undergoes a rhombohedral distortion.[4] In real Pu metal, the fcc lattice is unstable as the temperature is decreased (Figure 1a). So model I is realistic in this sense. The fcc lattice in model II has no such low-temperature distortions. Consequently, it behaves more fcc-like, as one might expect for Ga-stabilized Pu. Currently, there is no way to validate either model.

Next, we are concerned with interstitial defects. Each cascade produced by the fission of a Pu atom produces about 2,500 Frenkel pairs. The structure of the interstitial for model I is a (100) split dumbbell, but the interstitial for model II remains in an octahedral site. The migration barrier for the interstitial in model I is 0.055 eV.[4] This value is typical for fcc metals.

Vacancies migrate much more slowing, leading to an imbalance between interstitials and vacancies. In model I, the barrier is at most 1 eV, but the value for model II is 1.55 eV. These values appear to be high relative to the experimental value of 0.68 eV measured by isochronal annealing.[9] Parallel replica dynamics simulations[10] shows that mono-vacancy migration proceeds by two-atom hops as often as by one-atom hops.

Finally, the bias volume, resulting from the difference between interstitial and vacancy volumes, provides a simple measure of the driving force for void swelling. Model I has no measurable bias volume. Figure 2 shows bias volume for Model II Pu as well as for Ni and Pb. The Ni and Pb values are close to the experimental values.

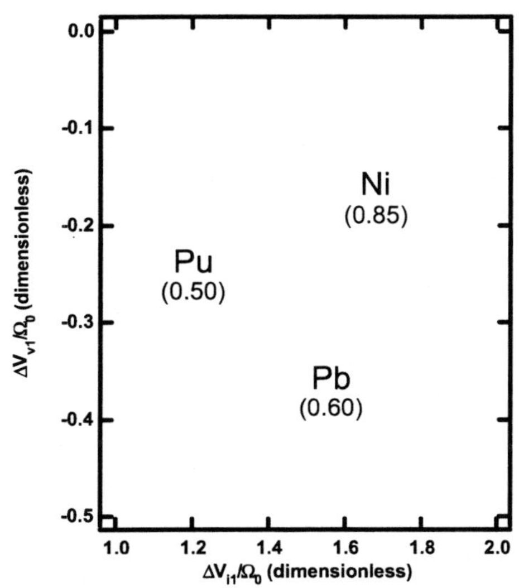

FIGURE 2. Bias volumes from MEAM potentials. Pu value is for model II.

REFERENCES

1. Daw, M. S., and Baskes, M. I., Phys. Rev. Lett. 50, 1285 (1983); Phys. Rev. B 29, 6443 (1984).
2. Baskes, M. I., Phys. Rev. Lett. 59, 2666 (1987); Phys. Rev. B 46, 2727 (1992).
3. Baskes, M. I. Phys. Rev. B 62, 15532 (2000).
4. Valone, S. M., Baskes, M. I., Stan, M., Mitchell, Terence E., Lawson, Andrew C., and Sickafus, K. E., J. Nucl. Mater., accepted.
5. Averbach, R. S., and Diaz de la Rubia, T., Solid State Physics 51, 281 (1998).
6. Stevens, M. F., Zocco, T. G., Albers, R. C., Becker, J. D., Walter, K. C., Cort, B., Paisley, D. L., and Nastasi, M. A., "Final Report: Fundamental and Applied Studies of Helium Ingrowth and Aging in Plutonium," Los Alamos National Laboratory document LA-UR-98-2606 (1998).
7. Wade, W. Z., J. Nucl. Mater. 38, 292 (1971); Wade, W. Z., Short, D. W., Walden, J. C., and Magana, J. W., Metall. Trans. A 9A, 965 (1978).
8. Diaz de la Rubia, T., Soneda, N., Caturla, M. J., and Alonso, E., J. Nucl. Mater. 251, 13 (1997).
9. Wirth, B. D., Schwartz, A. J., Fluss, M. J., Caturla, M. J., Wall, M. A., and Wolfer, W. G., MRS Bulletin 26, 679 (2001).
10. Voter, A., F., Phys. Rev. B 48, 13985 (1998).

Local Atomic Structure in Uranium-Niobium Shape Memory Alloys

Carole Valot,[a] Steve Conradson,[b] David Teter,[b] Daniel Thoma,[b] and Eric Peterson[b]

[a] *CEA/Valduc/Département de Recherches sur les Matériaux Nucléaires, 21120 Is sur Tille, France*
[b] *Material Science and Technology Division, MST-8, Los Alamos National Laboratory, Los Alamos, NM 87544, USA*

Uranium is used as a reactor fuel and in many other applications because of its high density and nuclear features. However, to improve its corrosion resistance and ductility, it has been necessary to alloy it with other metallic elements such as Mo, Nb, Ti, Zr.[1] This work concerns UNb alloys that exhibit the shape memory effect (SME), which occurs for Nb concentrations close to the monotectoid concentration. The SME seems to be strongly correlated to the martensitic transformations occurring in these alloys.[2] There is an extensive literature describing the different phases in the UNb system, the structural relationships between them, and their stages of appearance. Moreover, it is apparent that the solute plays an important role in defining the type of stabilized phase as a function of solute concentration. Thus, further investigations using different approaches should be pursued to better understand the structure of these alloys, including the short-range as well as long-range order and the associations between them. An important objective in such studies would be to elucidate the role of the Nb solute in the phase stability and SME.

Actually, even if many solid solutions are considered as substitutional, recent local structure experiments often show much more complex behavior.[3,4] A more detailed description of the structure is therefore likely to be required to understand the complex physical properties of the material. Even when the long-range order (usually characterized by the Bragg peaks in a diffraction pattern) implies a substitutional solid solution, probing finer length scales through the local structure reveals an important role of the solute that seems to be the key of phase stability or particular physical properties such as the SME.

We have studied UNb alloys using the powerful methods accessed with the high brightness, tunable x-rays available from synchrotron radiation, including structure refinement from X-Ray Diffraction (XRD) pattern (Rietveld), Radial Distribution Function (RDF) from EXAFS measurements, and resonant Pair Distribution Function from Wide Angles X-ray Scattering (WAXS-PDF). The aim is to describe more precisely the arrangement of the atoms in these alloys from the very local scale to the long-range order. This description on multiple length scales, enhancing the long-range average structure provided by conventional crystallographic analysis, is to be one of the keys to explain complex macroscopic properties such as SME in UNb alloys. The energy-dependent measurements provide element-specific information that differentiates between the U and Nb atoms, complementing the conventional Rietveld analysis that assumes a homogeneous, substitutional structure. In some cases, especially delta Pu alloys, it has been shown that a nanostructure, extremely difficult to observe with classical XRD, can be detected by much more local techniques such as EXAFS and WAXS-PDF. The WAXS-PDF technique already used for amorphous materials is now also in widespread use to explain complex physical properties linked to the complex local structure in crystalline materials.[5]

The synchrotron experiments were performed at the Stanford Synchrotron Radiation Laboratory (SSRL) on the 11.2 and 7.2 beam lines. UNb samples of different contents (from 5 at% to 20 at%) were prepared at the Los Alamos National Laboratory.

EXAFS experiments were performed at the uranium L_3 edge (17,178 eV) and the niobium K edge (19,005 eV). These experiments show that the local environment of the Nb atoms is drastically different from the one of the U atoms (Figure 1). Considering the fact that the Nb atoms' neighbors are essentially U atoms (Nb-U), these results

CP673, *Plutonium Futures — The Science,* edited by G. D. Jarvinen

imply the presence of two kinds of U atoms, those whose environments are principally other U atoms and those that link or connect sets of Nb atoms that are likely to differ significantly from the environment posited by the crystal structure. The Nb atoms interact primarily with this second set of U atoms, indicating that the Nb atoms may interact and perturb the U atoms that link them.

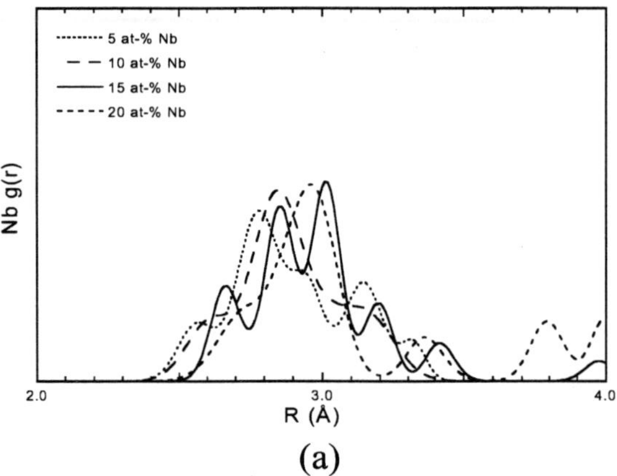

(a)

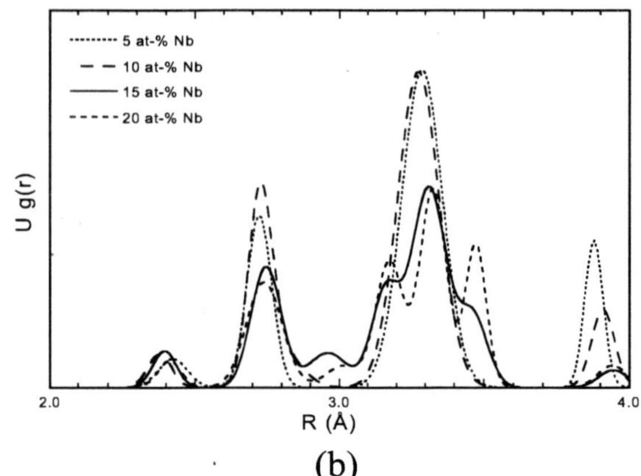

(b)

FIGURE 1. UNb alloys radial distribution function as a function of Nb content. (a) Uranium L_3 edge, (b) Nb K edge.

To complement the EXAFS results at a longer length scale, a WAXS-PDF study was performed on one UNb alloy (14%), which exhibits the SME effect. These WAXS-PDF experiments were not standard because the choice to work in the anomalous way was done, this in order to get a selectivity between the U and Nb as in EXAFS. Thus, three XRD patterns were recorded respectively close to the U L_3 edge, the Nb K edge, and below these two edges. Even if these anomalous experiments have to be improved by measurements at a larger number of energies, they already give interesting results. A small but significant evolution of the pair-distribution function as a function of energy indicates the presence of distinct U and Nb local environments: that of the U closer to the crystal structure and that of the Nb displaying noncrystallographic distances that suggest cooperativity between the strained Nb sites. Moreover, the poor correspondence between the measured PDFs and those calculated from the crystal shows the importance of the local structure in the complex properties of these materials.

REFERENCES

1. Vandermeer, R. A., "Phase Transformations in a Uranium 14 at % Niobium Alloy," Acta Metall. 28, 383 (1980).
2. Field, R. D., Thomas, D. J., Dunn, P. S., Brown, D. W., and Cady, C. M., "Martensitic Structures and Deformation Twinning in the U-Nb Shape Memory Alloys," Phil. Mag. A 81(7), 1691 (2001).
3. Dormeval, M., Baclet, N., Valot, C., Rofidal, P., and Fournier, J. M., "Crystalline and Electronic Structure of Pu-Ce and Pu-Ce-Ga Alloys Stabilized in the □ Phase," J. Alloys and Compounds, accepted, to be published.
4. Conradson, S. D., "Where is the Gallium? Searching the Plutonium Lattice with XAFS," Los Alamos Science 26 (2000).
5. Egami, T., "PDF Analysis Applied to Crystalline Materials," in Local Structure from Diffraction, edited by S. J. L. Billinge and M. F. Thorpe, Plenum Press, New York (1998).

Intensity Patterns of Diffuse X-Ray Scattering from Thermally Populated Phonons in fcc δ-Pu-Ga

Joe Wong,[a] M. Holt,[b,c] H. Hong,[b] M. Wall,[a] A. J. Schwartz,[a] P. Zschack,[b] and T.-C. Chiang[b,c]

[a]Lawrence Livermore National Laboratory, University of California, PO Box 808, Livermore CA 94551
[b]Frederick Seitz Materials Research Laboratory, University of Illinois at Urbana-Champaign, 104 South Goodwin Avenue, Urbana, IL 61801-2902
[c]Department of Physics, University of Illinois at Urbana-Champaign, 1110 West Green Street, Urbana, IL 61801-3080

Abstract. Thermal diffuse scattering (TDS) x-ray intensity patterns from phonons in an fcc δ-Pu-Ga alloy have been recorded for the first time. A small-diameter (25-μm) monochromatic 18-keV x-ray beam was used to obtain good-quality TDS images from a polycrystalline δ-Pu-Ga specimen. The recorded TDS images agree qualitatively well with theoretical images predicted by a generalized Morse potential for an fcc lattice with parameters determined directly from the known elastic constants of the alloy. These results illustrate the power of TDS analysis for phonon studies in otherwise inaccessible systems and are the target basis for a future determination of actual phonon dispersion curves in fcc-Pu materials using a combination of both TDS experiments and high-resolution inelastic x-ray scattering (HRIXS).

Taking advantage of the high brightness of undulator synchrotron beams and a recently developed technique for fabricating large-grain Pu materials,[1] we have designed a TDS[2–5] experiment and successfully measured patterns of x-ray scattering by thermal phonons in an fcc Pu-Ga alloy. The experimental results compared well qualitatively with those calculated using an ionic generalized Morse potential[6] with parameters determined directly from known elastic constants.[7]

Figure 1(a) is a transmission x-ray scattering image from a [111] grain in the microstructure of the Pu-Ga alloy. This experimental TDS image is a summation of sixty exposures, each of which was recorded in 30 s using a 25-μm beam at 18.037 keV. The data clearly exhibit a three-fold symmetry characteristic of that along the [111] direction in the face-centered cube lattice. The intensity pattern is due to x-ray scattering from thermally populated acoustic phonons at room temperature on the Ewald sphere that are closest to the neighboring reciprocal lattice points. These data represent the *first* phonon data recorded for Pu materials. The experimental TDS pattern compared qualitatively well with the simulated [111] image for fcc δ-Pu shown in Figure 1(b). The spotted Debye rings in the experimental image [Figure 1(a)] were indexed and were attributed to δ-Pu-Ga (major phase) and a PuO impurity (minor phase). TDS images from (100) grains have also been measured and showed the expected four-fold symmetry.

Simulations of x-ray scattering images were performed using an ionic Morse potential (MP) plus a long-range Coulumb attraction to describe interatomic forces, which has been applied with quantitative success in several fcc metals by Mohammed et al.[6] In combination with elastic constants of a similar fcc Pu-Ga alloy determined by Ledbetter and Moment,[7] the room temperature TDS patterns and dispersion curves of δ-Pu along the three principal directions were calculated following a recently established procedure.[5–7] The simulated patterns for [111] are shown in Figure 1(b). The dispersion curves of δ-Pu along with those of Cu[8] and Pb[9] were plotted in Figure 2. The Pb curves show pronounced anomalies as a result of electron-phonon interaction (dips at the zone boundaries) that are completely ignored by this simple model, and likely Pu will be complicated as well.

CP673, *Plutonium Futures — The Science,* edited by G. D. Jarvinen

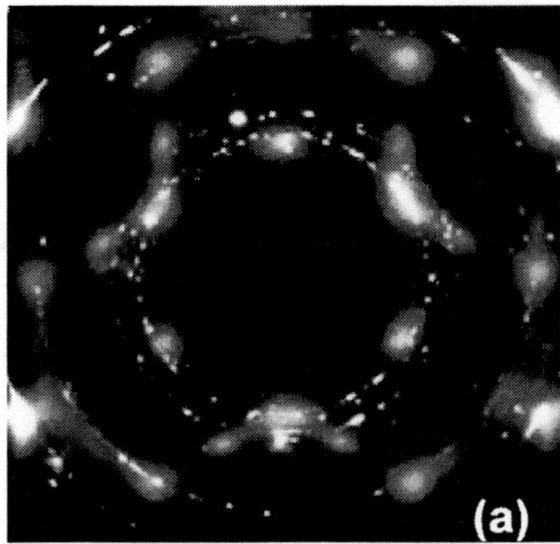

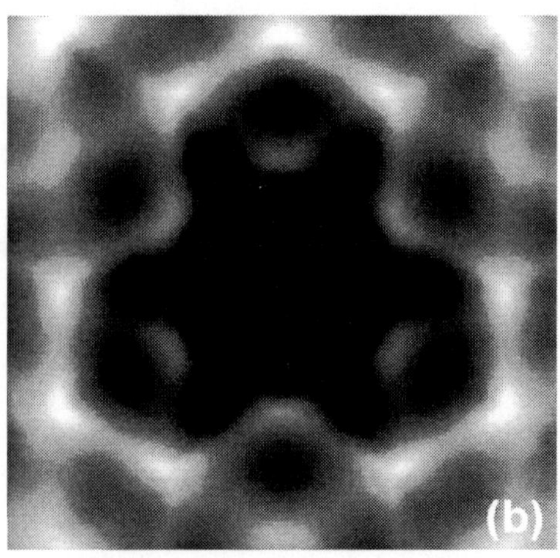

FIGURE 1. Transmission x-ray scattering images from an fcc δ Pu-Ga alloy along the [111] direction. (a) Experimental TDS image (in red) is a summation of sixty exposures, each 30 s long and recorded using a 25-μm beam at 18.030 keV. The data clearly show a three-fold symmetry of high-density acoustic phonons on the Ewald sphere that are closest to the neighboring reciprocal lattice points and compare quite well with the simulated [111] TDS image for fcc Pu shown in (b).

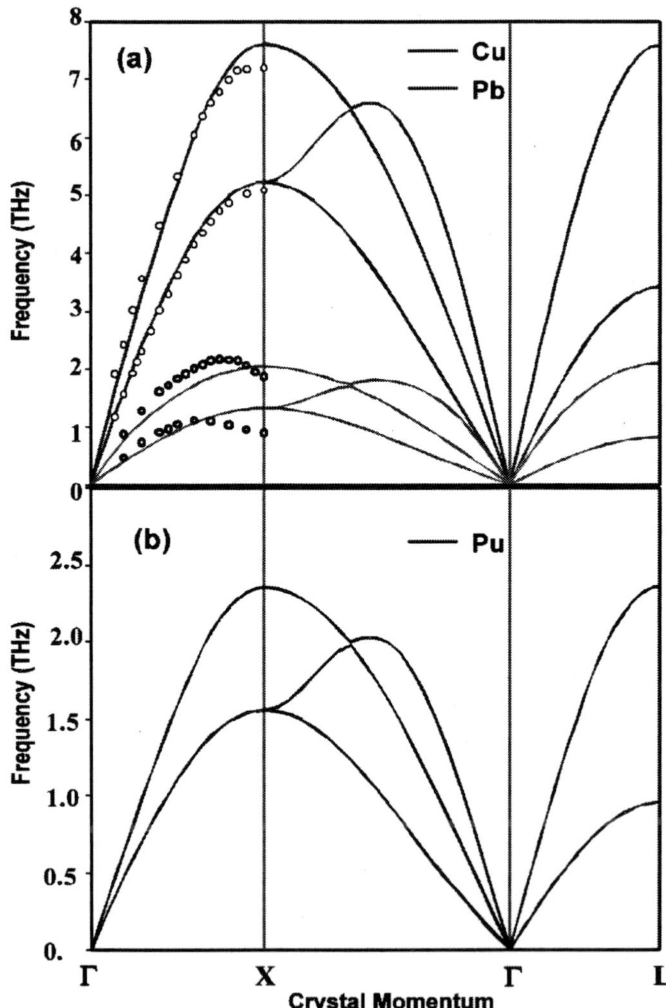

FIGURE 2. Calculated phonon dispersion curves for (a) Cu and Pb, and (b) δ-Pu using generalized Morse potentials. The open circules in (a) for the dispersion curves of Cu[8] and Pb[9] are experimental data from inelastic neutron scattering.

The present TDS results and preliminary simulation of the phonon dispersions of δ-fcc-Pu alloys demonstrate the feasibility of using modern x-ray scattering methods with intense undulator synchrotron beams to derive phonon dispersions and lattice dynamical data for materials not readily possible with conventional inelastic neutron scattering as a result of a high neutron absorption cross section and/or nonavailability of large-enough single crystals. These results also serve as an experimental guide for quantitative determination of phonon dispersion curves in Pu and other 5f actinide systems using a powerful combination of high-Q-resolution TDS experiments and high-energy resolution inelastic x-ray scattering.

REFERENCES

1. Lashley, J., Stout, M. G., Pereyra, R. A., Blau, M. S., and Embury, J. D., Scripta Mater. 44, 2815 (2001).
2. Wu, Z., Hong, Hawoong, Aburano, R., Zschack, P., Jemian, P., Tischler, J., Chen, Haydn, Luh, D.-A., and Chiang, T.-C., Phys. Rev. B59, 3283 (1999).
3. Holt, M., Wu, Z., Hong, Hawoong, Zschack, P., Jemian, P., Tischler, J., Chen, Haydn, and Chiang, T.-C., Phys. Rev. Lett. 83, 3317 (1999).
4. Holt, M., Zschack, P., Hong, Hawoong, Chou, M. Y., and Chiang, T.-C., Phys. Rev. Lett. 86, 3799 (2001).
5. Holt, M., Czoschke, P., Hong, H., Zschack, P., Birnbaum, H. K., and Chiang, T.-C., Phys. Rev. B66, 64303 (2002).
6. Mohammed, K., Shulka, M. M., Milstein, F., and Merz, J. L., Phys. Rev. B29, 3117 (1984).
7. Ledbetter, H. M., and Moment, R. L., Acta Metallurgica 24, 891 (1976).
8. Svensson, E. C., Brockhouse, B. N., and Rose, J. M., Phys. Rev. 155, 619 (1967).
9. Brockhouse, B. N., Arase, T., Gaglioto, G., Roa, K. R., and Woods, A. D. B., Phys. Rev. 128 (1962).

Intergrowth Structure in U- and Hf-Bearing Pyrochlore and Zirconolite: Analytical Electron Microscopy Investigation

Huifang Xu[1] and Yifeng Wang[1, 2]

[1]*Department of Earth and Planetary Sciences, The University of New Mexico, Albuquerque, NM 87131*
[2]*Sandia National Laboratories, Carlsbad, NM 88220*

The concept of Synroc was originally proposed by Ringwood et al., and the first Synroc fabrication technology was developed by Dosch et al. During the last two decades, Synroc has been subjected to extensive studies. Synroc immobilizes radionuclides by incorporating them into appropriate phases and forming solid solutions. With large polyhedra (with coordination numbers ranging from 7 to 8) in the structures, Synroc is able to accommodate a wide range of radionuclides (e.g., actinides, Pu, U, Ba, Sr, Cs, Rb, Tc, etc.) as well as neutron absorbers (e.g., Gd and Hf). U- and Pu-loaded Synroc generally contains phases of pyrochlore and zirconolite. Various Synroc formulations (e.g., Synroc-C, Synroc-D, Synroc-E, Synroc-F etc.) have been developed for specific high-level wastes. Homogeneity of sintered ceramics is an important factor to long-term chemical durability. This article describes chemistry and phase relation of Ce- and U-bearing pyrochlore and zirconolite. Ce^{3+}, Ce^{4+}, U^{4+}, and U^{6+} in the studied crystalline phases may be used as chemical analogues for Pu^{3+}, Pu^{4+}, and Pu^{6+}, respectively.

SAMPLES AND EXPERIMENTS

The starting material with stoichiometry of $Ca(U_{0.5}Ce_{0.25}Hf_{0.25})Ti_2O_7$ was prepared by mixing and grinding powders of TiO_2 (anatase), UO_2, $Ca(OH)_2$, and CeO_2 with deionized water. The mixed material was dried overnight in air at about 90 °C. The dried material was pressed and then sintered at 1350°C for 5 hours in an environment of slowly flowing Ar gas at one atmosphere. Part of the sintered sample was crushed in acetone. Drops of acetone suspension containing the crushed particles were dropped on perforated C-coated nylon grids for transmission electron microscopy (TEM) observation. All TEM and energy-dispersive spectroscopy (EDS) results were carried out with a JEOL 2010 high-resolution TEM and Oxford Link ISIS EDS system. Point-to-point resolution of the high-resolution TEM is 0.19 nm. Mineral standards were used for quantification of collected EDS data. We used a theoretical k_{Ce} value, because no Ce-silicate standard was available for Ce. Electron energy-loss spectroscopy (EELS) studies of Ce were carried out with a JEOL 2010F HRTEM with a GIF system.

RESULTS AND DISCUSSION

TEM and EDS results show that the sample contains both pyrochlore and zirconolite phases. Figure 1 shows a selected-area electron diffraction (SAED) pattern and high-resolution TEM image of the pyrochlore. EDS spectra show that the pyrochlore is relatively rich in U, Ce, and Ca with respect to zirconolite. EELS data indicate that Ce in zirconolite is in the form of Ce^{3+}, and Ce in the pyrochlore is in the forms of Ce^{3+} and Ce^{4+}; and the ratio of $Ce^{4+}/(Ce^{4+} + Ce^{3+})$ is about 0.6 according to the intensity ratio of M_4 and M_5 edges. Ce^{4+} is sensitive to the high-energy electron beam. Spread beam was used for collecting the EELS spectra to avoid a reduction reaction of Ce^{4+}. The composition of the pyrochlore and the zirconolite are listed in Table 1 according to the obtained EDS spectra and EELS data. The chances for finding pyrochlore and zirconolite grains is about 10 to 1.

CP673, *Plutonium Futures — The Science,* edited by G. D. Jarvinen

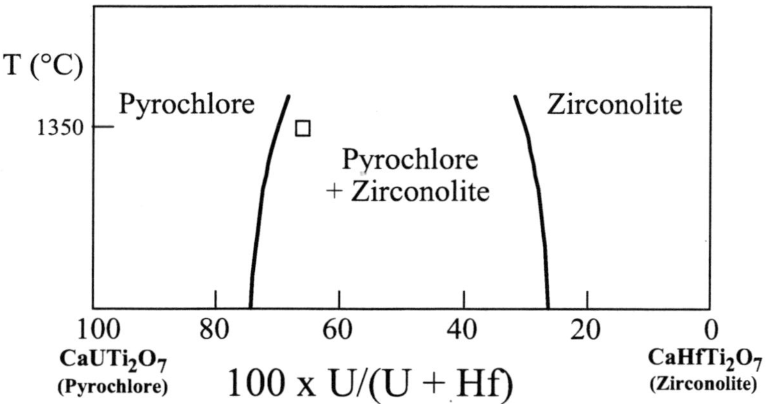

FIGURE 1. A proposed immiscibility gap between pyrochlore and zirconolite in the binary system of $CaUTi_2O_7$–$CaHfTi_2O_7$. The upper part of the figure illustrates the composition of the pyrochlore and zirconolite in a pseudo-ternary system of UO_2–HfO_2–TiO_2.

In some pyrochlore grains, there are zirconolite lamellae within the pyrochlore. The (001) of zirconolite is parallel to the (111) of the pyrochlore. A similar kind of phenomenon was also reported by Buck et al. The cell parameters measured from the SAED patterns are a = 10.5 Å for the pyrochlore (cubic) and a = 12.5 Å, b = 7.4 Å, c = 11.6 Å, β = 100° for the zirconolite (monoclinic), respectively. A high-resolution TEM image shows that the interface between pyrochlore and zirconlite is straight and coherent. The value of d_{111} (6.1 Å) of the pyrochlore is slightly larger than d_{002} (5.7 Å) of the zirconolite, because the pyrochlore contains more U (large ionic radius) than the zirconolite does. The compositions of the zirconolite lamellae and its neighboring pyrochlore are also listed in Table 1. Such kinds of lamellae result from epitaxial growth during the crystallization of both the pyrochlore and the zirconolite. The (001) plane of the zirconolite is very similar to the (111) plane of the pyrochlore. In some areas, the zirconolite lamellae display disordered structure that results from nonperiodic stacking of 2-layer and 3-layer zirconolite polytypes such as those that occur in Zr-zirconolite.

TABLE 1. Chemical Formulae of the Pyrochlore and Zirconolite.

Pyrochlore	Zirconolite
$Ca_{0.99}(Ce^{3+}_{0.12}Ce^{4+}_{0.18}U_{0.52}Hf_{0.21})(Ti_{1.95}Hf_{0.05})O_7$	$(Ca_{0.91}Ce_{0.09})(Ce^{3+}_{0.10}U_{0.25}Hf_{0.66})(Ti_{1.99}Hf_{0.01})O_7$
$Ca_{1.02}(Ce^{3+}_{0.13}Ce^{4+}_{0.19}U_{0.54}Hf_{0.16})(Ti_{1.96}Hf_{0.04})O_7$*	$(Ca_{0.94}Ce_{0.06})(Ce^{3+}_{0.09}U_{0.25}Hf_{0.65}Ti_{0.01})Ti_{2.00}O_7$*
$Ca_{1.01}(Ce^{3+}_{0.13}Ce^{4+}_{0.19}U_{0.55}Hf_{0.15})(Ti_{1.95}Hf_{0.05})O_7$	$(Ca_{0.90}Ce_{0.10})(Ce^{3+}_{0.06}U_{0.25}Hf_{0.67})Ti_{2.00}O_7$
	$Ca_{1.00}(Ce^{3+}_{0.13}Ce^{4+}_{0.18}U_{0.52}Hf_{0.19})(Ti_{1.96}Hf_{0.04})O_7$
$(Ca_{0.92}Ce_{0.08})(Ce^{3+}_{0.09}U_{0.26}Hf_{0.65})Ti_{2.00}O_7$	$(Ca_{0.93}Ce_{0.07})(Ce^{3+}_{0.11}U_{0.26}Hf_{0.64})Ti_{2.00}O_7$
$Ca_{1.00}(Ce^{3+}_{0.13}Ce^{4+}_{0.19}U_{0.49}Hf_{0.21})(Ti_{1.94}Hf_{0.06})O_7$	
$Ca_{1.02}(Ce^{3+}_{0.12}Ce^{4+}_{0.18}U_{0.54}Hf_{0.18})(Ti_{1.94}Hf_{0.06})O_7$*	$(Ca_{0.88}Ce_{0.12})(Ce^{3+}_{0.05}U_{0.26}Hf_{0.66}Ti_{0.02})Ti_{2.00}O_7$*
Average	
$Ca_{1.01}(Ce^{3+}_{0.13}Ce^{4+}_{0.19}U_{0.52}Hf_{0.18})(Ti_{1.95}Hf_{0.05})O_7$	$(Ca_{0.91}Ce_{0.09})(Ce^{3+}_{0.08}U_{0.26}Hf_{0.66}Ti_{0.01})Ti_{2.00}O_7$
U/(U+ Hf) = 0.72	U/(U + Hf) = 0.28

Note: The chemical formulae are normalized to 7 oxygen. The formulas marked with an asterisk (*) indicate lamellae composed of intergrown pyrochlore and zirconolite. The ratio of U/(U + Hf) only considers U and Hf in site B positions of zirconolite and pyrochlore with stoichiometry of $ABTi_2O_7$. The U/(U+ Hf) ratio for the pyrochlore is about 0.69 if all Hf atoms are at B sites.

According to the obtained EDS and TEM data, it can be proposed that there is an immiscibility gap in a solid solution between U-pyrochlore ($CaUTi_2O_7$) and Hf-zirconolite ($CaHfTi_2O_7$) at 1,350°C. The ratios of U/(U+Hf) in the B site of $ABTi_2O_7$ are 0.72 for the pyrochlore and 0.28 for the zirconolite (Table 1). If all Hf atoms are at the B site, then the ratio of U/(U+Hf) is 0.69 for the pyrochlore. It can be estimated that there are about 11 mol. % of zirconolite (including isolated zirconolite grains and zirconolite lamellae within the pyrochlore) in the mixture of pyrochlore and zirconolite, with bulk stoichiometry of $Ca(U_{0.5}Ce_{0.25}Hf_{0.25})Ti_2O_7$. If all Hf atoms are at the B site, there will be about 5 mol. % of pyrochlore (as a low limit) in the mixture system. It is proposed that there are about 5 to 11 mol. % of zirconolite in the system. It is proposed that a single-phase pyrochlore ($Ca(U,Hf)Ti_2O_7$) may be synthesized at 1,350°C if the ratio of U/(U+Hf) is larger than 0.72, and a single-phase zirconolite ($Ca(Hf,U)Ti_2O_7$)

may be synthesized at 1,350°C if the ratio of U/(U+Hf) is smaller than 0.28. Any mixtures with the U/(U+Hf) ratio between 0.72 and 0.28 will form a mixture of pyrochlore and zirconolite. The ionic radius difference between Pu^{4+} and Hf^{4+} is smaller than that between U^{4+} and Hf^{4+}. However, the chemical (Gibbs free energy of formation) difference between Pu^{4+} and Hf^{4+} is larger than that between U^{4+} and Hf^{4+}. It can be inferred that an immiscibility gap between $CaPuTi_2O_7$ and $CaHfTi_2O_7$ will be similar to that between $CaUTi_2O_7$ and $CaHfTi_2O_7$. The phase diagram may provide guides for preparing homogeneous Hf-bearing pyrochlore for hosting U and Pu.

ACKNOWLEDGMENTS

This work is based upon research conducted at the Transmission Electron Microscopy Laboratory in the Department of Earth and Planetary Sciences of the University of New Mexico, which is partially supported by NSF and State of New Mexico. The authors also thank NSF of China for partial support of this study.

Microstructure and Composition of a Ce-pyrochlore: A Chemical Analog for Pu-pyrochlore

Huifang Xu,[a] Yifeng Wang,[a,b] Laurence A. J. Garvie,[c] Robert L. Putnam,[d] and Alexandra Navrotsky[e]

[a]*Department of Earth and Planetary Sciences, The University of New Mexico, Albuquerque, NM 87131, e-mail: hfxu@unm.edu*
[b]*Sandia National Laboratories, 115 North Main Street, Carlsbad, NM 88220*
[c]*Department of Geological Sciences, Arizona State University, Tempe, AZ 85287-1404*
[d]*Los Alamos National Laboratory, Los Alamos, NM 87545*
[e]*Department of Chemical Engineering and Materials Science, University of California at Davis, One Shields Avenue, Davis, CA 95616-8779*

Abstract. Ce-pyrochlore ($CaCeTi_2O_7$), is a chemical analogue for $CaPuTi_2O_7$, which is a proposed ceramic waste form for deposition of excess weapon-usable Pu in geological repositories. Ce-pyrochlore was synthesized by firing and annealing a mixture of $Ce(NO_3)_4$, TiO_2, and $Ca(OH)_2$ with a stoichiometry of $CaCeTi_2O_7$ at 1,300°C for 50 hours. The annealed product contains Ce-pyrochlore, Ce-bearing perovskite, CeO_2 (cerianite), and minor CaO. Electron energy-loss spectroscopy (EELS) was used to determine the valence of Ce in the synthesized materials using the shape of the Ce $M_{4,5}$ edge. Cerium in the perovskite is dominated by Ce^{3+}. The $Ce^{4+}/\Sigma Ce$ in the pyrochlore is 0.8, giving $(Ca_{0.87}Ce^{3+}_{0.20}Ce^{4+}_{0.86}Ti_{0.05})Ti_2O_7$. High-resolution TEM images show that the boundary between pyrochlore and perovskite is semicoherently bonded. The orientational relationship between the neighboring pyrochlore and perovskite is not random. There are no glassy phases observed at the grain boundaries between pyrochlore and perovskite, and between CeO_2 and pyrochlore. It is postulated, based on the presence of trivalent Ce in the Ce-pyrochlore, that the neutron poisons such as Gd can be incorporated into the $CaPuTi_2O_7$ phase.

INTRODUCTION

Pyrochlore-like phases are considered durable crystalline hosts for weapon-usable Pu waste. The Ce-pyrochlore phase, $CaCeTi_2O_7$, is a chemical analogue for the real waste form, $CaPuTi_2O_7$. Pyrochlore and zirconolite phases, with the stoichiometry of $CaMTi_2O_7$ (or $MCaTi_2O_7$) can be considered as a distorted derivative structure of fluorite, where M represents tetravalent cations such as Zr, Hf, U, Pu, and other actinides. The large 7- or 8-coordinated polyhedra allow pyrochlore and zirconolite phases to accommodate a range of radionuclides such as Pu, U, and Ba, as well as neutron poisons such as Hf and Gd. Most previous analogue studies focused on the zirconolite and Hf-zirconolite phases. However, the Pu-loaded waste form will have the pyrochlore structure because of the relatively large size of Pu. In this paper, we describe the microstructure and chemistry of a Ce-pyrochlore phase.

RESULTS AND DISCUSSIONS

The Ce-pyrochlore samples were synthesized from a mixture of $Ca(OH)_2$, $Ce(NO_3)_4$, and TiO_2 (anatase) fired at 1,300°C for 50 hours. The composition of the mixture was targeted to the stoichiometry of $CaCeTi_2O_7$, i.e., Ca:Ce:Ti = 1:1:2. The $Ce^{4+}/\Sigma Ce$ ratios were determined using the shape of the Ce $M_{4,5}$ edge with a multiple, linear, least-squares fitting method with Ce^{4+} and Ce^{3+} spectra as the end members. The mixed-valence spectra exhibit characteristics of the two end-member single-valence spectra and can be simulated by linear combinations of Ce

CP673, *Plutonium Futures — The Science,* edited by G. D. Jarvinen

$M_{4,5}$ spectra from CeO_2 and Ce_2O_3. The Ce^{4+} and Ce^{3+} $M_{4,5}$ reference spectra were scaled to each other by normalizing their edges to the continuum above the edge. In this way, the integrated areas of the $M_{4,5}$ edges are quantifiably related and reflect the ratio of f-holes of Ce^{4+} and Ce^{3+}.

Backscattered electron images of the reactant show a heterogeneous mixture of ceramic phases. The samples contain Synroc phases of Ce-pyrochlore, cerianite (CeO_2), and perovskite. The Ce-pyrochlore and perovskite contain Ce^{3+}. However, cerianite in the sample is almost pure CeO_2. Based on electron microprobe analyses, Ca:Ce:Ti ratios for the pyrochlore and perovskite formed at 1,300°C are 1:1.23:2.36, and 1:0.53:1.71, respectively. The composition of the pyrochlore formed at 1,300°C deviates from the ideal stoichiometry of $CaCeTi_2O_7$.

The TEM reveals phases of Ce-pyrochlore, Ce-bearing perovskite, and cerianite. There are inclusion-like perovskite, pyrochlore, and minor CaO crystals within large crystals of cerianite. There are no open cracks or glassy phases at the grain boundaries. TEM images show that the boundary between cerianite and Ce-bearing perovskite (Figure 1) is semicoherently bonded. There are edge dislocations at the grain boundaries (Figure 1), and preferred orientations between the phases. The (001) lattice fringes of cerianite are roughly parallel to the (001) of the pyrochlore, and the (001) lattice fringes of cerianite are parallel to the (001) lattice fringes of perovskite. The Ce-bearing perovskite formed at 1,300°C has orthorhombic symmetry. There are no glassy phases at the grain boundaries between pyrochlore and cerianite nor the boundary between CeO_2 and perovskite. In addition, there is no intermediate $CeTiO_4$ phase between cerianite and Ce-pyrochlore. This is different from a $CaZrTi_2O_4$ system, in which a $ZrTiO_4$ phase is observed between zirconolite and zirconia.

The Ce $M_{4,5}$ edges, acquired by EELS, of CeO_2 and Ce-perovskite are consistent in shape with Ce^{4+} and Ce^{3+}, respectively (Figure 2). The Ce $M_{4,5}$ edge of the pyrochlore is similar in shape to the Ce^{4+} spectrum but with a low energy tail indicative of Ce^{3+}. The $Ce^{4+}/\Sigma Ce$ ratio determined for the pyrochlore is 0.8, with little variation from grain to grain. According to the atomic ratios of Ca:Ce:Ti for pyrochlore and perovskite and Ce^{4+} contents in pyrochlore, the structural formula of the Ce-pyrochlore formed at 1,300°C can be represented as $(Ca_{0.87}Ce^{3+}{}_{0.21}Ce^{4+}{}_{0.86}Ti_{0.05})Ti_2O_7$. The results indicate that high-temperature annealing will result in a small amount of Ce^{3+} in the pyrochlore. This indicates that both Ce- and Pu-pyrochlore can incorporate trivalent Gd and tetravalent Hf into the crystal structure as neutron poisons. The structural formula of Gd- and Hf-bearing Pu-pyrochlore can be written as $(Ca_{1-x}Gd_{2x}Pu_{1-x-y}Hf_y)Ti_2O_7$.

It is also reported that Pu-pyrochlore is thermodynamically stable with respect to TiO_2 and perovskite in an environment that is depleted in aqueous silica and carbon dioxide.[1] However, Pu-zircon is thermodynamically unstable. The corrosion products of Pu-pyrochlore in an oxidizing environment are PuO_2, TiO_2, and other stable phases. The corrosion rate for Pu-pyrochlore is extremely low. Therefore, Pu-pyrochlore is a thermodynamically feasible and stable waste form to host Pu in the environment of WIPP, NM.

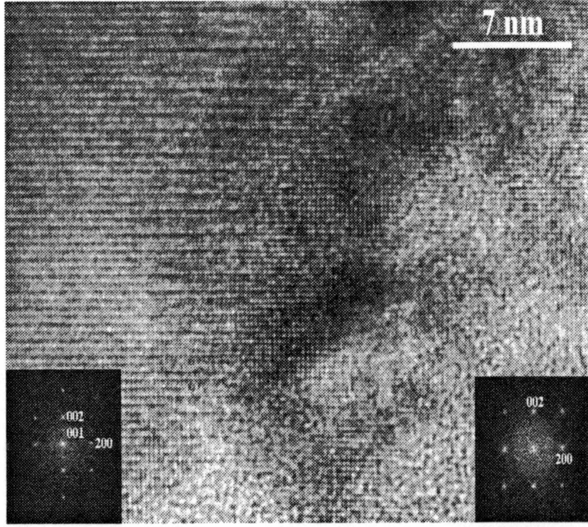

FIGURE 1. High-resolution TEM image of the Ce-pyrochlore-dominated Synroc sample illustrating a grain boundary between cerianite (right) and Ce-bearing perovskite (left). Dark areas at the boundary are the result of stress from dislocation cores.

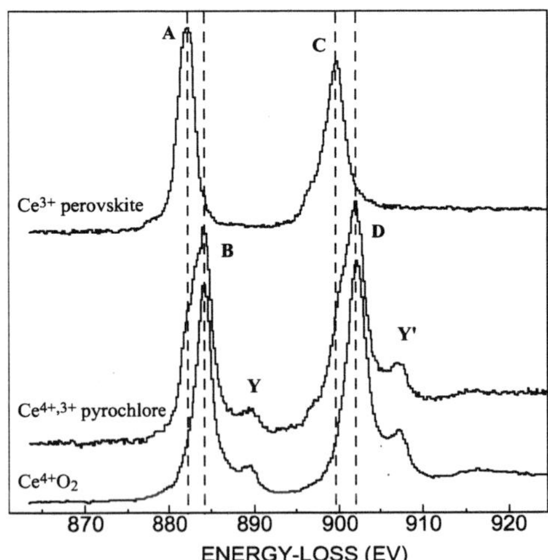

FIGURE 2. EELS spectrum from Ce-pyrochlore and Ce-bearing perovskite. The Ce-pyrochlore is dominated by Ce^{4+}, and the perovskite by Ce^{3+}.

ACKNOWLEDGMENTS

This work is based upon research conducted at the Transmission Electron Microscopy Laboratory in the Department of Earth and Planetary Sciences of the University of New Mexico, which is partially supported by NSF (CTS98-71292) and the State of New Mexico. This work was supported by grant EAR-0087714 (for LAJG) from the National Science Foundation.

REFERENCES

1. J. Nuclear Mater. 275, 216–220 (1999).

ACTINIDE COMPOUNDS AND COMPLEXES

Structural Trends and Bonding of the *5f*-Elements (U-Am) with the Oxoligand IO_3^-

Amanda C. Bean,[*] Brian L. Scott,[†] Thomas Albrecht-Schmitt,[‡] and Wolfgang Runde[†]

** Nuclear Materials & Technology, Los Alamos National Laboratory, Los Alamos, NM 87545*
[†]Chemistry Division, Los Alamos National Laboratory, Los Alamos, NM 87545
[‡]Department of Chemistry, Auburn University, Auburn, AL 36849

INTRODUCTION

The solid-state chemistry of transuranium compounds has received considerably less attention than their uranium analogs, owing to decreased availability and the highly specialized facilities needed to safely study long-lived α-emitters. However, understanding the behavior of the early transuranium elements is critical for assessing their environmental impact as long-term contributors to radioactive dose in nuclear waste repositories.[1,2] Of particular interest is how these elements might react with fission-product radionuclides, such as ^{129}I, or their derivatives.[3,4] In fact, iodine can exist in solution in both oxidized and reduced forms, i.e., as IO_3^- and I^-, and studies on the nature of ^{129}I in nuclear waste suggested the existence of iodate, IO_3^-.[2]

Although the potential importance of low-solubility transuranium iodates for lanthanide/actinide separation was recognized about fifty years ago, we still lack a thorough understanding of their structural and spectroscopic properties. Few Np(V) and Np(VI) iodates, such as $NpO_2IO_3 \cdot 0.5KIO_3 \cdot 2H_2O$ and $[Co(NH_3)_6][(NpO_2)_2(IO_3)_7] \cdot 7H_2O$, have been reported; however, these compounds were characterized only by elemental analyses and IR spectroscopy.[5,6] A brownish-yellow dense crystalline compound, stated as being $NpO_2(IO_3)_2 \cdot 2H_2O$, was reported to form upon evaporation.[5] In the past, the constitution of plutonium iodates was even more ambiguous, with solubility data reported for a compound of suggested formula $Pu(IO_3)_4$.[7]

Here we report the synthesis and structural characterization of transuranium (Np, Pu, Am) iodates. The structural differences are correlated with the compounds' spectroscopic properties within the series of actinyl [U(VI),N$_p$(VI),Pu(VI)] iodates. The properties of the first Am(III) iodates will be compared to those of the chemically analogous lanthanide series.

RESULTS

Hydrothermal synthesis was used to produce single crystals (Figure 1), which were analyzed to obtain structures and Raman spectra of five novel transuranium iodates. The transuranic iodates, $NpO_2(IO_3)_2(H_2O)$ (**2**), $NpO_2(IO_3)_2 \cdot H_2O$ (**3**), $PuO_2(IO_3)_2 \cdot H_2O$ (**4**),[8] and two forms of $Am(IO_3)_3$ (**5,6**) have been prepared from the aqueous reactions of Np(V), Pu(IV or VI), or Am(III) with KIO_4 or H_5IO_6 at 180°C. The actinyl(VI) compounds contain pentagonal bipyramidal $[AnO_7]$ (An = Np, Pu) polyhedra with the axial trans-dioxocation $[O = An = O]^{2+}$. The pyramidal iodate $[IO_3]$ anions link the $[AnO_7]$ units into sheets that interact with one another through intermolecular $IO_3^- IO_3^-$ interactions (Figure 2). In compound **2**, the neptunium atom is bound to four iodate oxygens and one terminal water molecule in the equatorial plane. Compounds **3** and **4** are isostructural and oxygen atoms from bridging iodate anions occupy the five equatorial sites around the neptunyl and plutonyl moieties. The iodate anions occur as both doubly and triply bridging anions and link the actinyl cations into sheets as shown in Figure 3. Both types of iodate anions have their stereochemically active lone-pair of electrons aligned on one side of each layer

CP673, *Plutonium Futures — The Science,* edited by G. D. Jarvinen

creating a polar structure. Raman spectra of these compounds and $UO_2(IO_3)_2(H_2O)$ show a sequential shift of the ν_1 AnO_2^{2+} stretch to lower wavenumbers as the atomic number of the actinide is increased. The structure of α-$Am(IO_3)_3$ (**5**) is isostructural with $Nd(IO_3)_3$, both of which contain nine-coordinate metal centers made up from the oxygens of iodate ligands. The three-dimensional extended structure is built from doubly and triply bridging iodate ligands. In contrast to the chemical analogue Nd(III), we also found a β-form (**6**) of $Am(IO_3)_3$ that is unique among the *f*-elements. Crystallographic data: **2**, orthorhombic, space group *Pcan*, a = 7.684(2) Å, b = 8.450(2) Å, c = 12.493(3) Å, Z = 4; **3**, orthorhombic, space group $Pna2_1$, a = 7.314(1) Å, b = 11.631(2) Å, c = 9.449(2) Å, Z = 4; **4**, orthorhombic, space group $Pna2_1$, a = 7.320(1) Å, b = 11.636(2) Å, c = 9.473 (2) Å, Z = 4; **5**, monoclinic, space group $P2_1/c$, a = 7.243(2) Å, b = 8.538(3) Å, c = 13.513(4) Å, β = 100.12, Z = 4; **6**, monoclinic, space group $P2_1/n$, a = 8.871(3) Å, b = 5.932(2) Å, c = 15.315(4) Å, β = 96.95, Z = 4.

0.12 x 0.12 x 0.21 mm^3

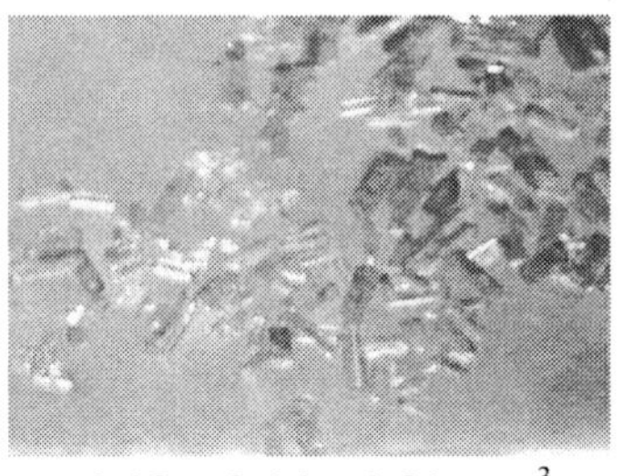

0.12 x 0.16 x 0.21 mm^3

FIGURE 1. Single crystals of $UO_2(IO_3)_2H_2O$ (**1**), $NpO_2(IO_3)_2H_2O$ (**2**), and $PuO_2(IO_3)_2{\cdot}H_2O$ (**4**), respectively.

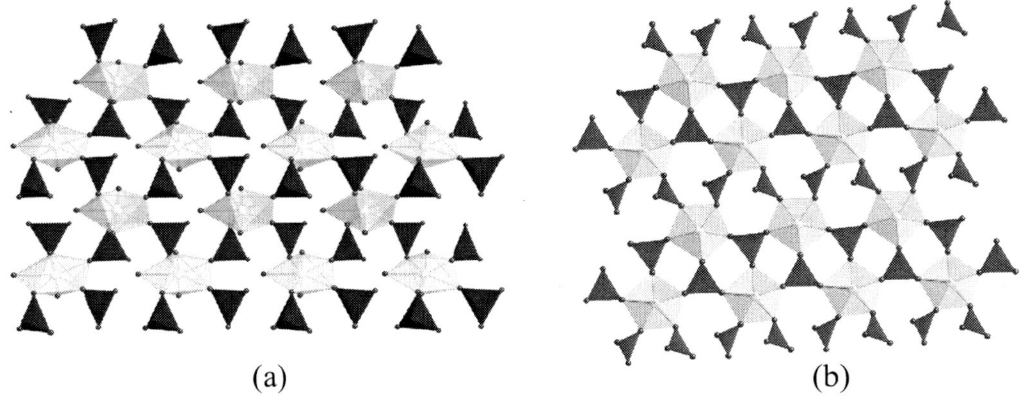

FIGURE 2. Comparison of the two-dimensional sheets found in (a) $UO_2(IO_3)_2H_2O$ (**1**), $NpO_2(IO_3)_2H_2O$ (**2**) and (b) $NpO_2(IO_3)_2{\cdot}H_2O$ (**3**) $PuO_2(IO_3)_2{\cdot}H_2O$ (**4**).

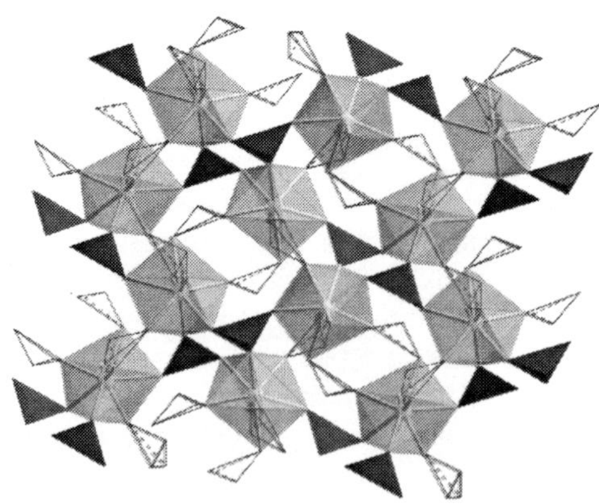

FIGURE 3. The three-dimensional network of $Am(IO_3)_3$ (**5**) demonstrating the doubly and triply bridging coordinations of iodate.

DISCUSSION

Expanding the structural knowledge and spectroscopic characterization of uranium and transuranic compounds is a challenging and important task that needs to be approached in a strategic manner. This research is the beginning of a successful synthetic approach, other than traditional solid-state synthesis, in producing novel compounds and/or unraveling the structural properties of important materials containing the series of the light actinide elements by utilizing mild hydrothermal conditions. The structural trends from $UO_2(IO_3)_2H_2O$ (**1**) to the intermediate species of $NpO_2(IO_3)_2H_2O$ (**2**) and $NpO_2(IO_3)_2 \cdot H_2O$ (**3**) and finally, to one of the few Pu(VI) structures, $PuO_2(IO_3)_2 \cdot H_2O$ (**4**), have shown a remarkable diversity of the ability of the actinide atom and its participation in bonding to build such extended structures. In addition, the heavier element, americium, poses interesting structural chemistry to further explore in the future, in hopes of stabilizing the AmO_2^{2+} unit in a single crystal that could be characterized by single-crystal X-ray crystallography.

REFERENCES

1. Kaszuba, J., and Runde, W. H., Environ. Sci. Technol. 33, 4427–4433 (1999).
2. Efurd, D. W., Runde, W. H., Banar, J. C., Janecky, D. R., Kaszuba, J., Palmer, P. D., Roensch, F. R., and Tait, C. D., Environ. Sci. Technol. 32, 3893–3900 (1998).
3. Doshi, G. R., Joshi, S. N., and Pillai, K. C., Journal of Radioanalytical and Nuclear Chemistry 155, 115–127 (1991).
4. Robertson, D. E., Smith, M. R., Koppenaal, D. W., Kiddy, R. A., Strebin, R. S., Brauer, F. P., Ross, G. A., Baldwin, D. L., and Hornibrook, C., Waste Management (Tucson, Arizona) 2, 287–294 (1991).
5. Blokhin, V. I., Bukhtiyarova, T. N., Krot, N. N., and Gel'man, A. D., Russ. J. Inorg. Chem. 17, 1742–1746 (1972).
6. Tsivadze, A. Y., Muchnik, B. I., and Krot, N. N., Russ. J. Inorg. Chem. 17, 1746–1749 (1972).
7. O'Connor, P. R., data quoted in USAEC Report CN-1764 (1944).
8. Runde, W. H., Bean, A. C., Albrecht-Schmitt, T. E., and Scott, B. L. Chem. Comm. (to be published, 2003).

Characterization of Multiple Pu(IV) Chloride Complexes in Aqueous Solution by Coordinated Visible Absorption, EXAFS and Crystal Structure Studies

John M. Berg, Steven D. Conradson, John H. Matonic, Mary P. Neu, and Sean D. Reilly

Los Alamos National Laboratory, Los Alamos, NM 87545, USA

The complexation of Pu(IV) by chloride in aqueous solutions has long played a prominent role in plutonium-separations chemistry, yet the chloride coordination numbers of the dominant complexes and their stability ranges remain in question. The recent OECD review of plutonium solution thermodynamics concludes that, although there are reported observations of solution complexes up to $PuCl_4$(aq), only $PuCl^{3+}$(aq) has had its formation constant reliably established.[1] Multiple reports of formation constants for $PuCl_2^{2+}$(aq) are in significant disagreement, and reported observations of $PuCl_3^{+}$(aq) are sparse.

Visible absorption spectra of actinide coordination complexes with partially filled metal ion 5f shells are characterized by numerous overlapping peaks of moderate width and intensity as a result of intra-5f transitions. The energies and intensities of the peaks are reasonably sensitive to changes in the inner coordination sphere. Indeed, modeling of spectral changes with solution conditions has long been a supplemental tool in complexation studies of these metal ions. However, the bulk of spectroscopic modeling efforts on Pu complexation equilibria in particular were conducted before multivariate analysis techniques became widely employed or indeed technically feasible in most laboratories. Changes in absorbance at, typically, one to three wavelengths were quantified and modeled. Even the comparatively recent study by Giffaut of Pu(IV) chloride complexation, the only spectroscopic study of this system referenced in the OECD review, modeled quantitative measurements of absorbance at two wavelengths to extract formation constants.[2] The potential bias resulting from wavelength selection and the considerable uncertainty caused by noise in the data limited the reliability of the results, particularly because three or more absorbing species are the norm rather than the exception for these chemical systems. The potential improvements offered by multivariate analyses are significant for these types of information-rich spectral data.

We have conducted several spectrophotometric studies of Pu(IV) complexation in aqueous HCl and in several series of $HCl/HClO_4$ mixtures at formally constant ionic strength. Series of spectra at different total ionic strengths were merged into a single data matrix for simultaneous analysis under the hypothesis that a common set of pure species spectra should account for all of the observed absorbance. Multivariate curve resolution (MCR) with constraints was applied to discern the total number of spectroscopically distinct chemical species and their approximate stability fields. The MCR results were then used to provide initial parameter guesses and constraints for multivariate classical equilibrium models in which the species concentrations were constrained to obey chemical equilibrium equations. Using this approach, we have determined that five distinct complexes of differing chloride coordination number are necessary to explain Pu(IV) spectra in solutions up to a 9-molal HCl concentration.

Because the optical data do not contain directly accessible information about the absolute coordination numbers of the absorbing species, we used data from other techniques to constrain the analysis of the optical spectra. We conducted EXAFS studies under a few, carefully selected sets of solution conditions to get average chloride coordination numbers under a range of conditions. We also prepared $PuCl_6^{2-}$ in crystalline powders and in nonaqueous solvents and collected their optical spectra to compare with the aqueous data. The latter results clearly demonstrate that the hexachloro complex only begins to appear at the highest chloride concentrations and is not the majority species in any of our data.

The best-fit values of the formation constants for $PuCl^{3+}$(aq) and $PuCl_2^{2+}$(aq) that we obtained from the constrained fits are significantly smaller than values reported in the literature. We will discuss the reliability and

CP673, *Plutonium Futures — The Science,* edited by G. D. Jarvinen

potential significance of this result. We will also discuss progress and difficulties in determining formation constants for the complexes with higher chloride coordination numbers.

REFERENCES

1. OECD Nuclear Energy Agency, Chemical Thermodynamics of Neptunium and Plutonium, Elsevier, Amsterdam (2001).
2. Giffaut, E., Influence des Ions Chlorure sur la Chimie des Actinides, Ph.D. Thesis, Université de Paris-Sud, Orsay, France (1994).

Stability and Redox Behavior of Plutonium-EDTA and Mixed Pu(IV)-EDTA-L (L = Hydroxide, Carbonate, Citrate) Complexes

Hakim Boukhalfa, Sean D. Reilly, Wayne H. Smith, and Mary P. Neu*

Chemistry and Materials Science and Technology Divisions, Los Alamos National Laboratory, Los Alamos, New Mexico 87545

INTRODUCTION

Amino-carboxylate ligands are nonspecific chelating agents with a strong binding affinity for a variety of metal ions. Upon binding to metal ions, the chelating agents change the composition of the inner coordination shell of the metal ions and thereby alter their chemical speciation and behavior. Because of these properties, ethylenediaminetetra-acetic acid (EDTA) and other amino-carboxylate ligands have been extensively used in the processing of radionuclides. As a consequence, EDTA is codisposed with radionuclides and colocated with radionuclide contamination. If EDTA is released to the environment, it has the potential to significantly affect the solubility and overall behavior of radionuclides.[1] This is particularly true for low-valent actinides, such as plutonium, that would generally hydrolyze and precipitate or sorb to mineral surfaces in water and soils in the absence of strong chelators. The first step in predicting how EDTA would affect the environmental behavior of plutonium and impact the safe disposal of nuclear waste is determining the nature of the complexes formed. Previous work on EDTA complexation of plutonium was focused on acidic conditions and interpretation of bulk solubility studies.[2–4] Our approach was to determine the speciation and thermodynamic stability of species formed, more directly and under near-neutral pH and excess EDTA conditions.

Plutonium can exist in aqueous solution as ions in a range of oxidation states III-VI that can be complexed by EDTA. The specific affinity of aminocarboxylate ligands for Pu(IV) will tend to drive the complexation reaction toward the formation of Pu(IV) complexes through redox processes. Specifically, Pu(IV) EDTA complexes can form through the reduction of the higher oxidation states, as for Pu(VI) and Pu(V), and through the oxidation of Pu(III). Plutonium(IV) is known to form a 1:1 complex with EDTA in aqueous solution; but this hexadentate ligand does not fully encapsulate the metal ion. Two to four waters complete the coordination sphere, providing coordination sites for hydrolysis, polymerization, or the formation of mixed ligand complexes.

RESULTS

We have determined the stoichiometry and the thermodynamic parameters for several Pu(IV)EDTA complexes using *potentiometric* and *spectrophotometric titration* methods (Figure 1). In an acidic solution and at a ratio of 1:1 Pu(IV):EDTA, Pu(IV)-EDTA is formed ($\log\beta_{110} = 27.19$), but at a higher pH Pu(IV)-EDTA(OH) ($\log\beta_{11\text{-}1} = 22.83$) and Pu(IV)-EDTA(OH)$_2$ ($\log\beta_{11\text{-}2} = 16.13$) complexes are formed. Cyclic voltammetric examination of the Pu:EDTA system at a pH below the first hydrolysis (pH <4.3) shows that EDTA forms a very stable complex with Pu(IV), with a reduction potential of $E_{1/2} = 342$ mV (*vs* NHE). The resulting Pu(III)-EDTA complex undergoes rapid protonation and dissociation within the pH range examined. The measured potentials also indicate that in acidic media and in the presence of O_2, EDTA will promote the oxidation of Pu(III) to Pu(IV). This oxidation is due to the difference in the EDTA affinity for Pu(IV) relative to Pu(III) as described by Nernst equation applied to this system, Eq. (1).

CP673, *Plutonium Futures — The Science,* edited by G. D. Jarvinen

$$E^{Pu(IV/III)-EDTA} = E^0_{aq} - 59.15 \log \frac{\beta_{Pu(IV)-EDTA}}{\beta_{Pu(III)-EDTA}} \quad (1)$$

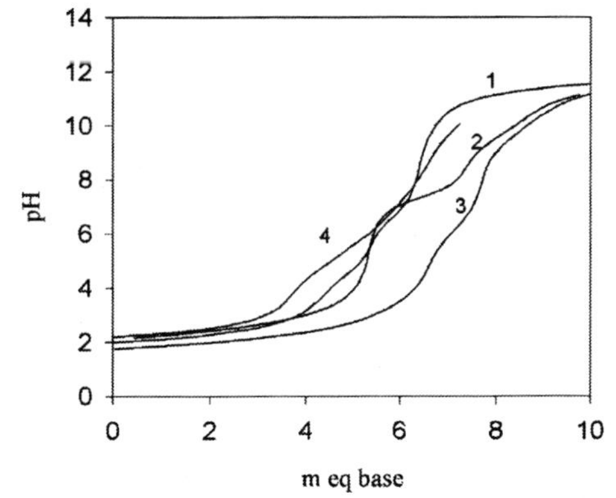

FIGURE 1. Potentiometric titrations of Pu(IV)-EDTA complexes. (1) [Pu(IV)] = 2.5 mM; [EDTA] = 2.53 mM, (2) [Pu(IV)] = 2.49 mM; [EDTA] = 5.0 mM, (3) [Pu(IV)] = 2.49 mM; [EDTA] = 2.5 mM, [citrate] = 2.52 mM and (4) [Pu(IV)] = 2.49 mM; [EDTA] = 2.5 mM, [carbonate] = 2.52 mM, where m represents the number of moles of NaOH added per mol of plutonium(IV).

In the presence of excess EDTA relative to Pu(IV), Pu(IV)-$(EDTA)_2$ is formed at neutral pH ($\log\beta_{120}$ = 35.43). In the presence of additional ligands like citrate or carbonate, mixed complexes Pu(IV)-EDTA-citrate ($\log\beta_{120}$ = 31.54) and Pu(IV)-EDTA-carbonate ($\log\beta_{120}$ = 34.26) are formed. The composition of the inner coordination sphere around Pu is probably very similar in all these complexes, where the remaining aquo ligands in the Pu(IV)-EDTA complex are displaced by the incoming ligand. Examination of these complexes by cyclic voltammetry reveals irreversible behavior (Figure 2). The reduction wave is shifted by about –400 mV as compared to the reduction wave of the complex Pu(IV)-EDTA, but the oxidation wave is observed at the same potential for all complexes. These results agree with the potentiometric and spectrophotometric results, suggesting that the Pu(IV) inner coordination sphere is more electron rich for the mixed-ligand complexes than the initial Pu(IV)-EDTA complex. The nonreversible CV indicates that Pu(IV) can accommodate a higher coordination number from oxygen donor groups than can Pu(III).

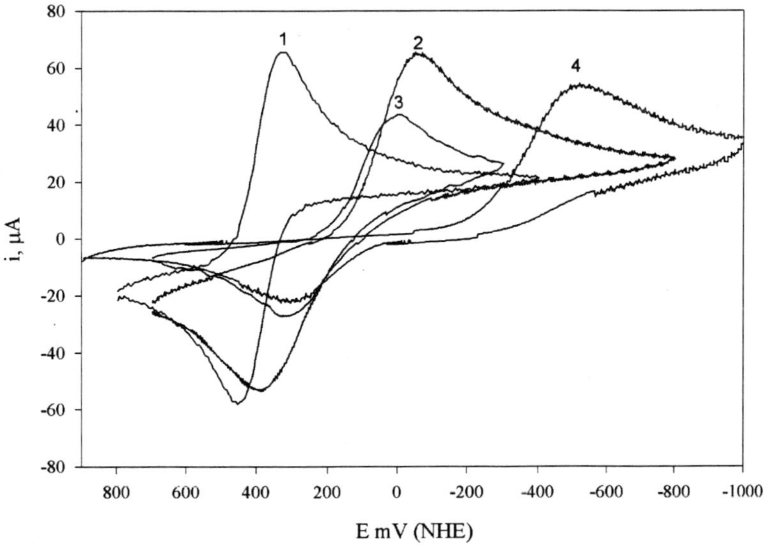

FIGURE 2. Cyclic voltammograms of 2.05 mM Pu(IV):EDTA solution with addition of 1 equivalent of either EDTA, citrate or carbonate. Curve (1) Pu(IV)-EDTA at pH 2.3, Curve (2) Pu(IV)-$(EDTA)_2$ at pH 6.75, Curve (3) Pu(IV)-EDTA-citrate at pH = 7.8, Curve (4) Pu(IV)-EDTA-carbonate at pH 8.2.

DISCUSSION

Our results show that EDTA effectively stabilizes Pu(IV) discriminating Pu(III). The stability of the Pu(IV) EDTA complexes relative to other oxidation states depends on the pH and Pu:EDTA ratio. In excess EDTA over Pu(IV), the complexes formed have a 2:1 EDTA:Pu(IV) stoichiometry. In the presence of competing ligands like carbonate or citrate, the addition of one ligand to Pu(IV)-EDTA has been shown. Ternary Pu(IV)-EDTA-L complexes are formed when EDTA, OH-, citrate, or carbonate are present at a sufficiently high concentration. The formation constants for Pu EDTA species (Pu(IV)-EDTA, Pu(IV)-$EDTA_2$, Pu(IV)-EDTA(L) L = citrate or carbonate, Pu(IV)-EDTA(OH), and Pu(IV)-EDTA$(OH)_2$) have been determined and can now be included in thermodynamic databases and used to help predict the fate and transport of actinides in the environment.

REFERENCES

1. Kakem, N. L., Allen, P. G., and Sylwester, E. R. J., Radioanalytical and Nuclear Chemistry 250(1), 47–53 (2001).
2. Rai, D., Bolton, H. Jr., Moore, D. A., Hess, N. J., and Choppin, G. R., Radiochim. Acta 89, 67–74 (2001).
3. Cauchetier, P., and Guichard, C., Radiochim. Acta 19, 137–146 (1973).
4. Cauchetier, P., and Guichard, C., J. Inorg. Nucl. Chem. 37, 1771–1778 (1975).

Engendering a Reactive Uranium(III) Center with a Single Pocket for Reactivity: A Combined Synthetic, Spectroscopic, and Computational Study

Ingrid Castro-Rodriguez and Karsten Meyer

Department of Chemistry and Biochemistry, University of California, San Diego
9500 Gilman Drive MC 0358, La Jolla, CA 92093-0358.

INTRODUCTION

Although bonding in *f*-elements is traditionally described as mainly electrostatic, the issue of covalency remains an important subject of debate.[1,2] In order to answer fundamental questions regarding trends in bonding and the reactivity of uranium and other actinide metal compounds, the discovery and detailed investigation of new molecules is necessary. In our efforts to identify and isolate uranium complexes with enhanced reactivity relevant to binding, activation, and functionalization of small molecules, we are currently investigating the coordination chemistry of uranium metal centers stabilized by classical Werner-type ligands. The stabilizing ability of "classic" macrocyclic amines has made this class of chelators an indispensable tool for transition-metal coordination chemistry.[3] However, the coordination chemistry of uranium complexes with macrocyclic polyamine ligands remains largely unexplored.[4] This is in contrast to the thoroughly investigated organometallic chemistry of uranium with cyclopentadienyl ligands and their derivatives[5] as well as the recently developed amido chemistry of uranium.[6–8]

On the occasion of the Plutonium Futures 2003 conference, we wish to report on the detailed synthetic, spectroscopic, and computational investigations of analogous uranium III, IV, V, and VI derivatives supported by the tris-aryloxide functionalized triazacyclononane ligand, 1,4,7-tris(3,5-di-*tert*-butyl-2-hydroxybenzyl)-1,4,7-triazacyclononane ($(ArO)_3$tacn).[9] We will demonstrate that the ancillary polyamine macrocycle provides a unique platform for enhanced reactivity at an electron-rich uranium center. Specifically, we will show that the introduction of ($(ArO)_3$tacn) to a uranium(III) metal center results in the formation of stable core complexes leaving a single reactive coordination site available for ligand substitution reactions and redox events associated with small molecule or organic functional group activation.[10,11]

RESULTS

We recently have been successful in isolating the highly reactive mononuclear uranium(III) complexes [($(ArO)_3$tacn)U^{III}] (**1a**)[10] and [($(ArO)_3$tacn)$U^{III}(NCCH_3)$] (**1b**) (Figure 1).[11]

In a combined synthetic, spectroscopic, and computational study, we investigated an entire series of complexes resulting from these reactive starting materials. In this series of analogous complexes, the oxidation state varies from +III to +VI, but the complex core structure remains unchanged (Figure 2). High-valent uranium complexes with formal oxidation states U(V) and U(VI) have not previously been stabilized by aryloxide ligands. The electronic absorption spectra as well as SQUID magnetization measurements (2–300 K at 1T) of the series of complexes with an [$N3O3$-N_{ax}] core structure and oxidation states +III, +IV, and +V were recorded and will be discussed in the context of the electronic structures.

CP673, *Plutonium Futures — The Science,* edited by G. D. Jarvinen

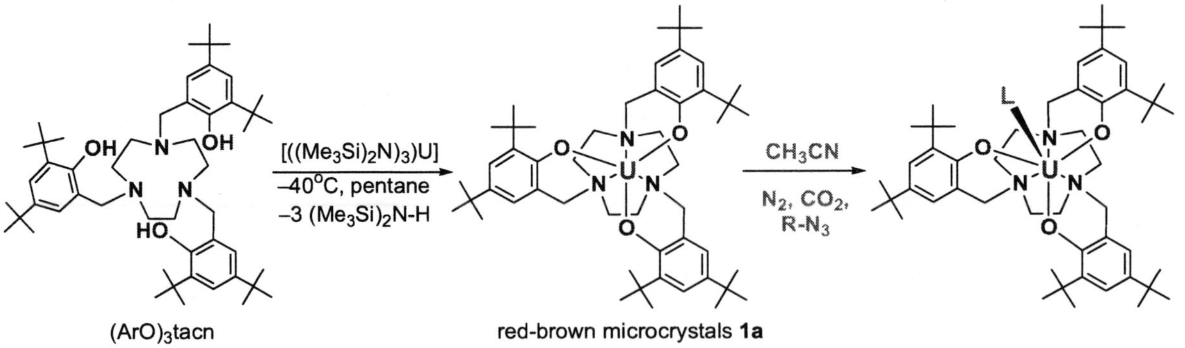

FIGURE 1. Synthesis of uranium complexes with the general formula $[((ArO)_3tacn)U(L)]$.

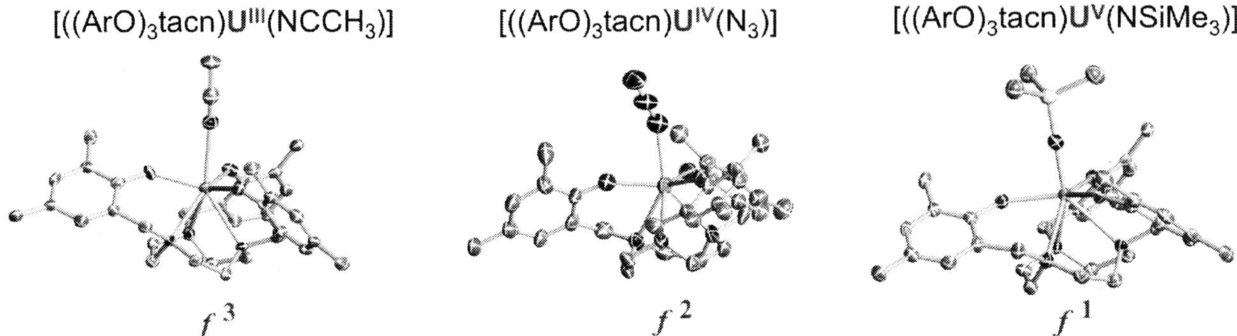

FIGURE 2. Solid-state molecular structures of uranium complexes with formed oxidation states varying from +III to +V (tert.-Butyl groups are omitted for clarity).

We found that uranium(III) complexes **1a** and **1b** react with organic azides to yield uranium(IV) azido and uranium(V) imido complexes, $[((ArO)_3tacn)U^{IV}(N_3)]$ (**2**) and $[((ArO)_3tacn)U^{V}(NR)]$ (**3**) with R = $SiMe_3$ (**3a**), Ad (**3b**), and CPh_3(**3c**), respectively.

Solutions of uranium(V) imido complexes **3a**, **3b**, and **3c** exhibit a one electron oxidation wave at potentials varying from –225 (**3a**) to –355 mV (**3b**, **3c**) versus the ferrocene/ferrocenium couple. The chemical oxidation of samples of **3** was thus conveniently accomplished with silver tetrafluoroborate, a regimen providing diamagnetic uranium(VI) species.

Interestingly, complex **1a** reacts with dinitrogen to yield a μ-dinitrogen-bridged diuranium species $[\{((ArO)_3tacn)U^{III}\}_2(\mu{:}\eta^1,\eta^1\text{-}N_2)]$ (**4**) with an unprecedented N_2 binding mode. Complex **1a** also reacts with carbon dioxide to provide the dinuclear oxo bridged $[\{((ArO)_3tacn)U^{III}\}_2(\mu\text{-}O)]$ (**5**) and carbon monoxide.[12] Single-crystal x-ray diffraction, and the spectroscopic and computational studies of all the above-mentioned uranium tris-aryloxide complexes supported by triazacyclononane were performed and will be presented.

DISCUSSION

The reported uranium tris-aryloxide derivatives stabilized by triazacyclononane represent rare examples of uranium coordinated to a macrocyclic amine chelator. To the best of our knowledge, the uranyl compound $[(18N6)UO_2](SO_2CF_3)_2$ is the only other uranium complex in which a macrocyclic polyamine chelator is coordinated to a uranium center.[4] The hexadentate tris-anionic ligand, $((ArO_3)tacn)^{3-}$, binds to the large uranium ion in unprecedented fashion, engendering a coordinatively unsaturated, highly reactive low-valent uranium center. The macrocyclic chelator occupies six coordination sites, with the three aryloxide pendent arms forming a trigonal plane at the metal center. This is in strong contrast to the coordinatively saturated octahedral geometry found in most *d*-block transition metal complexes of this ligand. DFT quantum mechanical methods[13,14] were applied to rationalize the reactivity and to elucidate the electronic structure of the six-coordinate trivalent **1a**. The three most energetic electrons of system **1a** were found to be uranium based. The energy differences, δ, between SOMO-1/SOMO-2 and SOMO-2/SOMO-3 are small (δ ~3.6 kJ/mol); thus, the orbitals can be considered almost degenerate. SOMO-3 is largely *f* in character. Its electron density is within the plane of the three aryloxide atoms. In contrast, SOMO-2 and -1 possess an unpaired spin density that is perpendicular to this plane, shielded by the triazacyclononane backbone on one side but pointing out into the open reactivity cavity on the other side. Accordingly, complex **1a** behaves like a di-radical and is very reactive towards small molecules, such as carbon dioxide, organic nitriles, and azides (Figure 3).

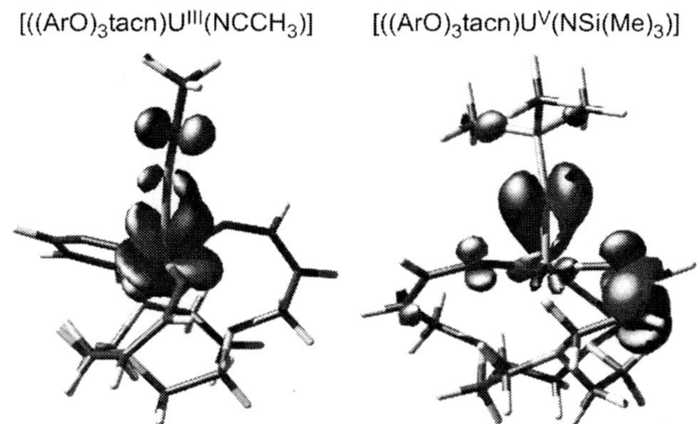

FIGURE 3. Selected molecular orbitals in 1a and 3a.

Furthermore, our calculations show that the deeply colored uranium(III) and uranium(V) species **1b** and **3** are stabilized through significant π-bonding interaction, involving uranium *f*-orbitals and the axial acetonitrile and imido ligand, respectively. In contrast, the bonding in the colorless uranium(IV) azido complex is purely ionic in nature.

REFERENCES

1. Raymond, K. N., and Eigenbrot, C. W., Accounts of Chemical Research 13, 276 (1980).
2. Burns, C. J., and Bursten, B. E., Comments Inorganic Chemistry 9, 61 (1989).
3. Lindoy, L. F., The Chemistry of Macrocyclic Ligand Complexes, Cambridge University Press, Cambridge, New York, Port Chester, Melbourne, Sydney (1989).
4. Nierlich, M., Sabattie, J. M., Keller, N., Lance, M., and Vigner, J. D., Acta Crystallographica Section C-Crystal Structure Communications 50, 52 (1994).
5. Conejo, M. D., Parry, J. S., Carmona, E., Schultz, M., Brennann, J. G., Beshouri, S. M., Andersen, R. A., Rogers, R. D., Coles, S., and Hursthouse, M., Chemistry—A European Journal 5, 3000 (1999).
6. Stewart, J. L., and Andersen, R. A., Polyhedron 17, 953 (1998).
7. Roussel, P., and Scott, P., Journal of the American Chemical Society 120, 1070 (1998).
8. Odom, A. L., Arnold, P. L., and Cummins, C. C., Journal of the American Chemical Society 120, 5836 (1998).
9. Chaudhuri, P., and Wieghardt, K., in Progress in Inorganic Chemistry (2001), Vol. 50, p. 151.
10. Castro-Rodriguez, I., Olsen, K., Gantzel, P., and Meyer, K., Chemical Communications 2764 (2002).
11. Castro-Rodriguez, I., Olsen, K., Gantzel, P., and Meyer, K., Journal of the American Chemical Society (2003), in press.
12. Castro-Rodriguez, I., and Meyer, K., unpublished preliminary results (2003).
13. ADF 2002.01, SCM, Theoretical Chemistry, Vrije Universiteit, Amsterdam, The Netherlands.
14. Integrated molecular orbital/molecular mechanics calculations were performed using the Amsterdam Density Functional theory-based ADF2002.01 program suite. The calculations were performed using the IMOMM scheme proposed by Maseras et al. (1995). In general, the core of the molecule was placed in the QM region; the rest of the molecule, including sterically demanding substituents, was calculated in a Molecular Mechanic scheme. The Sybyl Force Field was used for the organic part of the molecule, but the Universal Force Field was implemented for the uranium metal center. The ADF package was used for geometry optimization of the core structure. The Vosko, Wilk, and Nusair (VWN) local density approximation was used. Becke's (1988) exchange correlation and Perdew (1986) correlation for the gradient correction were also applied. The spin-unrestricted option was used together with ZORA relativistic formalism. No symmetry was specified in the calculation of the species, starting either from the x-ray refined geometry of crystallographically characterized molecules or a geometry-optimized structure: U: triple-ζ basis set with polarization and frozen core 6p. N: triple-ζ basis set with polarization, O: triple-ζ basis set with polarization, C: triple ζ-basis set with polarization, and H: triple-ζ basis set with polarization, including H-polarization.

Early Actinide Organonitrile Compounds and Their Reactivity

Alejandro E. Enriquez, John H. Matonic, Brian L. Scott, and Mary P. Neu*

Actinide, Catalysis, and Separations Chemistry (C-SIC), Chemistry Division Los Alamos National Laboratory, Los Alamos, New Mexico, 87545, USA. Fax: (505) 667-9905; Tel: (505) 667-7717; E-mail: mneu@lanl.gov

INTRODUCTION

The actinide trihalides are an important class of materials. Anhydrous AnX_3 compounds have been prepared through high-temperature synthetic techniques, but they exist as polymeric solids, which are insoluble in organic solvents.[1–3] Transuranic complexes $AnI_3(THF)_x$ have been prepared by the reaction of metals with $C_2H_4I_2$ in THF solutions (An = Pu, x = 5; An = Np, x = 4)[4] and can be viewed as a net oxidation of An metal with 1.5 equivalents of molecular iodine. A convenient preparation of organic solvent soluble trivalent actinide compounds has been described where 1.5 equivalents of I_2 are added to actinide (An) metal turnings (An = Pu, U, Np) suspended in an aprotic organic coordinating solvent.[5,6] Because virtually all nonaqueous An(III) complexes have been prepared through the triodide-tetrahydrofuran complex, we are interested in preparing complexes with solvents other than tetrahydrofuran. And, because uranium is often used as a surrogate for studying the chemistry of transuranic actinides, we are seeking to elucidate the chemical differences between the transuranic elements and uranium. In this presentation, we describe new Pu(III) complexes prepared by plutonium metal oxidation with either iodine or metal hexafluorophosphate salts in acetonitrile. We also revisit the oxidation of uranium metal by iodine in acetonitrile.[7–9] Finally, we also describe a new, versatile, mononuclear U(IV) product, $UI_4(NCPh)_4$, obtained from the oxidation of uranium metal by iodine in benzonitrile.

RESULTS

The treatment of plutonium metal suspended in acetonitrile by iodine generated a blue solution. We attribute the blue color to *in situ* generation of $PuI_3(NCMe)_x$ (**1**). Although we have been unable to characterize **1**, addition of a single equivalent of the f-element extractant *tpza* (*tpza* = tris[2-pyrazinyl)methyl]amine = $N(CH_2(C_4H_3N_2))_3$) to the blue solution resulted in isolation of the Pu(III) derivative, (*tpza*)$PuI_3(NCMe)$ (**2**). Green crystals of **2** were characterized by x-ray crystallography.

Oxidation of excess uranium metal with 1.5 equivalents of iodine in acetonitrile has been reported to provide $UI_3(NCMe)_4$ (**3**).[7] We were able to isolate a single crystal from this reaction after the solution was cooled to –38°C. X-ray crystallography identified this product as the mixed valence U(IV)/U(III) compound $[UI_6][U(NCMe)_9][I]$ (**4**). The cationic portion of the molecule is a homoleptic U(III) ion surrounded by nine acetonitrile molecules. This nine-coordinate cation is flanked by a noncoordinating iodide and the U(IV) dianionion UI_6^{2-}. When the oxidation of uranium metal was carried out in benzonitrile, the mononuclear U(IV) product $UI_4(NCPh)_4$ (**5**) is obtained in ~80% yield.

The addition of three equivalents of either $AgPF_6$ or $TlPF_6$ to an acetonitrile suspension of ^{239}Pu metal gradually provided a blue solution as plutonium metal dissolved and either silver or thallium metal precipitated. Filtration of the solution followed by slow evaporation of acetonitrile formed blue crystalline $[Pu(NCMe)_9][PF_6]_3{\cdot}MeCN$ (**6**) in ~80% yield. Attempts to prepare the nine-coordinate homoleptic acetonitrile uranium analogue of **6** by treatment of uranium turnings suspended in acetonitrile with either $AgPF_6$ or $TlPF_6$ were unsuccessful. No reaction was observed even after days of stirring. However, dissolution of blue $UI_3(thf)_4$ in acetonitrile resulted in a dark green solution

CP673, *Plutonium Futures — The Science,* edited by G. D. Jarvinen

which, upon cooling to –38°C, provided dark-green crystals of the nine-coordinate U(III) acetonitrile salt $[U(NCMe)_9][I]_3$ (**6**). Crystals of **6**, unlike the plutonium analogue **5**, are thermally unstable and rapidly decompose at room temperature to an undetermined black residue (Figure 1).

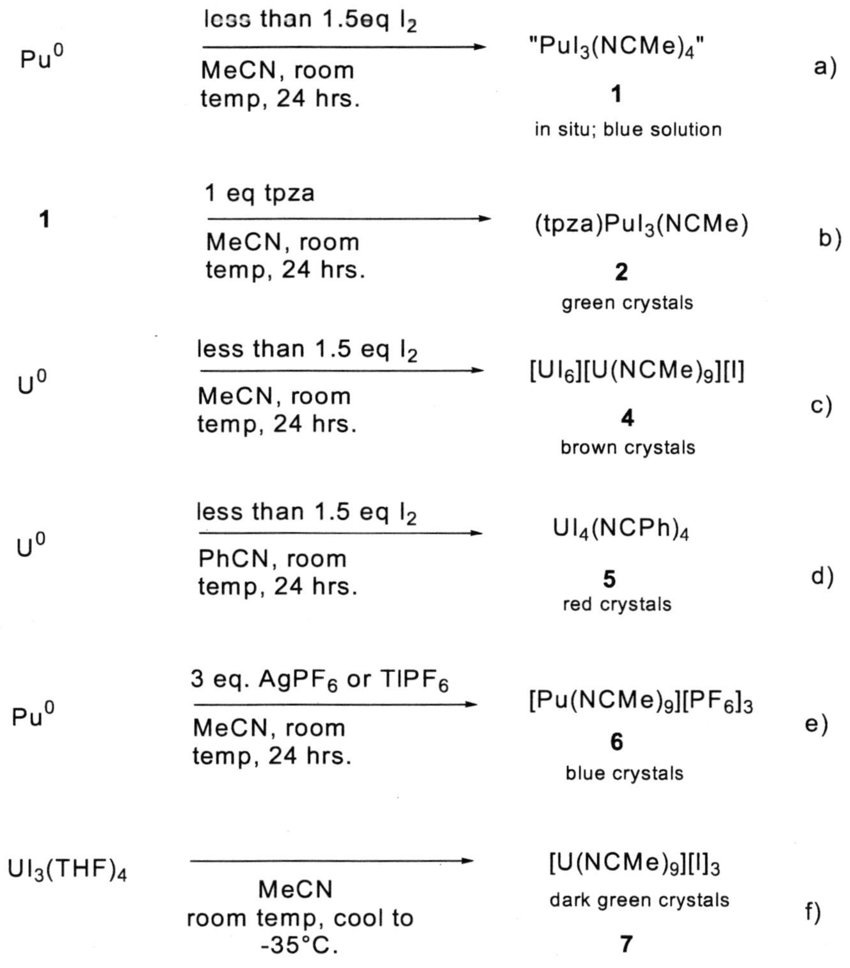

Reactant	Conditions	Product	
Pu^0	less than 1.5eq I_2 → MeCN, room temp, 24 hrs.	"$PuI_3(NCMe)_4$" **1** in situ; blue solution	a)
1	1 eq tpza → MeCN, room temp, 24 hrs.	$(tpza)PuI_3(NCMe)$ **2** green crystals	b)
U^0	less than 1.5 eq I_2 → MeCN, room temp, 24 hrs.	$[UI_6][U(NCMe)_9][I]$ **4** brown crystals	c)
U^0	less than 1.5 eq I_2 → PhCN, room temp, 24 hrs.	$UI_4(NCPh)_4$ **5** red crystals	d)
Pu^0	3 eq. $AgPF_6$ or $TlPF_6$ → MeCN, room temp, 24 hrs.	$[Pu(NCMe)_9][PF_6]_3$ **6** blue crystals	e)
$UI_3(THF)_4$	→ MeCN room temp, cool to -35°C.	$[U(NCMe)_9][I]_3$ dark green crystals **7**	f)

FIGURE 1. Preparation of Pu (III), U (III) and U(IV) nitrile complexes.

DISCUSSION

The uranium and plutonium nitrile compounds isolated from An^0 oxidation reactions in polar aprotic solvents are consistent with known differences between the reactivities of uranium and plutonium metal. For example, plutonium oxidation by iodine in either thf or pyridine solutions gives well-characterized $PuI_3(sol)_4$ products (these Pu(III) have been characterized by TGA, ^1H NMR, IR and UV/vis/near-IR spectroscopies). Our efforts to isolate and structurally characterize **1**, the intermediate prepared by oxidation of Pu^0 with iodine in MeCN, have been unsuccessful. However, **1** is clearly Pu(III) because it can be trapped by tpza to form the Pu(III) compound **2**. Plutonium metal also undergoes a unique reaction in acetonitrile when it is treated with silver and thallium hexafluorophosphate salts to provide the homoleptic Pu(III) acetonitrile salt **6**. We have not yet prepared a Pu(IV) compound by direct oxidation of Pu^0 in polar aprotic solvents.

Unlike plutonium, uranium metal can be oxidized by iodine in polar aprotic solvents to provide both U(III) and U(IV) products. When excess metal is present, U^0 oxidation by 1.5 equivalents of iodine in thf, pyridine, or dme solvents gives the well-characterized U(III) products $UI_3(sol)_x$ (thf and py, x = 4; dme, x =2) (although $UI_3(thf)_4$ is the only one to be structually characterized). Preparation of $UI_3(NCMe)_4$ in an analogous fashion has been reported. Our attempts to prepare this complex by direct oxidation of U^0 with iodine in acetonitrile gave the dark brown crystalline mixed valence U(IV)/U(III) salt **4**. Curiously, oxidation of excess uranium metal in benzonitrile with iodine cleanly provides crystalline **5** in high yield. No evidence for a U(III) product is observed. Complex **5** is soluble and stable in toluene, benzene, and methylene chloride for days. However, when **5** is dissolved in acetonitrile, a reaction occurs. The crystalline material isolated from the reaction is a brown product characterized by x-ray crystallography as $[UI_6][U(NCMe)_7I_2]$. We are currently examining the reactivity of **5** towards neutral and anionic chelating ligands.

REFERENCES

1. Katz, J. J., Seaborg, G. T., and Morss, L. R., Editors, The Chemistry of the Actinide Elements, Vol.1 and 2. 2nd Ed, 1986.
2. Levy, J. H., Taylor, J. C., and Wilson, P. W., Acta Crystallogr., Sect. B: Struct. Crystallogr. Cryst. Chem. B31, 880–882 (1975).
3. Taylor, J. C., and Wilson, P. W., Acta Crystallogr., Sect. B: Struct. Crystallogr. Cryst. Chem. B30, 2803–2805 (1974).
4. Karraker, D. G., Inorg. Chim. Acta 139, 189–191 (1987).
5. Avens, L. R., Bott, S. G., Clark, D. L., Sattelberger, A. P., Watkin, J. G., and Zwick, B. D., Inorg. Chem. 33, 2248–2256 (1994).
6. Clark, D. L., and Sattelberger, A. P., Inorg. Synth. 31, 307–315 (1997).
7. Drozdzynski, J., and du Preez, J. G. H., Inorg. Chim. Acta 218, 203–205 (1994).
8. Du Preez, J. G. H., and Zeelie, B., J. Chem. Soc. Chem. Comm. 743 (1986).
9. Du Preez, J. G. H., and Zeelie, B., Inorg. Chim. Acta 118, L25–L26 (1986).

Diffuse Reflectance Spectroscopy of Plutonium Ions in Zirconolite and Perovskite

K. S. Finnie, Y. Zhang, B. D. Begg, and E. R. Vance

Materials Science and Engineering, ANSTO, Menai, NSW 2234, Australia

The disposition of Pu-bearing radioactive waste is of general interest. The current work first reviews our previous studies on the various valence states that Pu can assume in the titanate phases zirconolite and perovskite, and then details a diffuse reflectance (DR) study of trivalent and quadrivalent Pu in these phases.

In zirconolite, $CaZrTi_2O_7$, and perovskite, $CaTiO_3$, Pu^{3+} and Pu^{4+} can be substituted for Ca, while Pu^{4+} but not Pu^{3+} can be substituted for Zr in zirconolite. The substitution of trivalent and tetravalent Pu in the Ca sites can be made by charge-compensating additions of Al in the Ti sites or by cation vacancies, together with control of the firing atmosphere. For instance, the formulae of such materials containing 0.02 formula units (f.u.) of Pu^{3+} substituted in the Ca site of zirconolite could be written as $Ca_{0.98}Pu_{0.02}ZrTi_{1.98}Al_{0.02}O_7$ or as $Ca_{0.97}Pu_{0.02}\square_{0.01}ZrTi_2O_7$ where $\square$ is a cation vacancy. Samples have been made containing 0.001-0.1 f.u. of Pu in the different available sites in the two phases. For substitutions of < 0.01 f.u., it is very difficult to control the chemistry; therefore, near-edge x-ray absorption spectroscopy is necessary to try to define the Pu valences in the most dilute solid solutions.

From the spectroscopic point of view, we have tried to study the dilute end of the Pu concentration spectrum to make sure any changes of local site symmetry affecting the oscillator strengths of the single-ion electronic absorptions in the visible and near infrared could be documented (in parallel work to be reported shortly, we found new optical absorption bands on U^{4+} in perovskite to occur when the U^{4+} concentration exceeded ~0.01 f.u.).

Solution absorption spectra of Pu^{4+} (aquo) suggest that the lowest electronic transitions of this ion, corresponding to intraconfigurational f-f transitions, occur at ~5,000 cm^{-1}. Zirconolite and perovskite were fired in an air atmosphere at 1,400°C and doped with low concentrations of plutonium, which was shown to be quadrivalent from near-edge x-ray absorption spectroscopy: near-infrared spectra (5,000–12,500 cm^{-1}) were recorded using diffuse reflectance spectroscopy. The solid-state spectra consist of a series of relatively weak absorption bands over the wavelength range. Comparison of the spectra of $Ca_{0.999}Pu_{0.001}TiO_3$ and $Ca_{0.997}Pu_{0.003}TiO_3$ (Figure 1) suggest that the absorption bands approximately scale with Pu content. In contrast, there are absorption bands in the spectrum of $Ca_{0.999}Pu_{0.001}ZrTi_2O_7$ (Figure 2) that do not appear in that of $Ca_{0.997}Pu_{0.003}ZrTi_2O_7$ (Figure 2) and suggest some alteration in local structure even at these low concentrations. Results on samples doped with higher concentrations of Pu will also be presented.

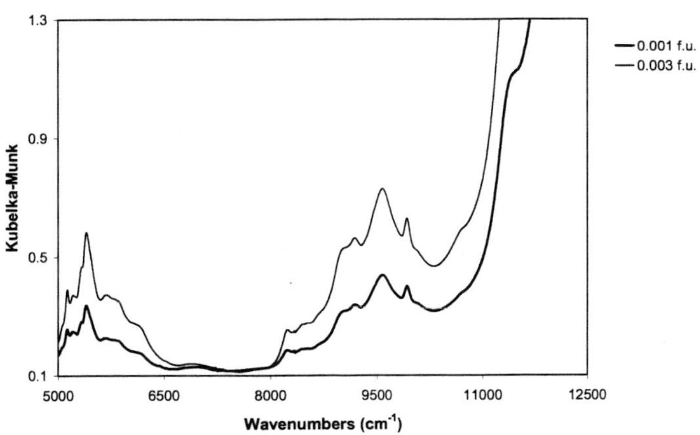

FIGURE 1. DR spectra (5,000–12,500 cm^{-1}) of perovskite with different Pu contents.

CP673, *Plutonium Futures — The Science,* edited by G. D. Jarvinen

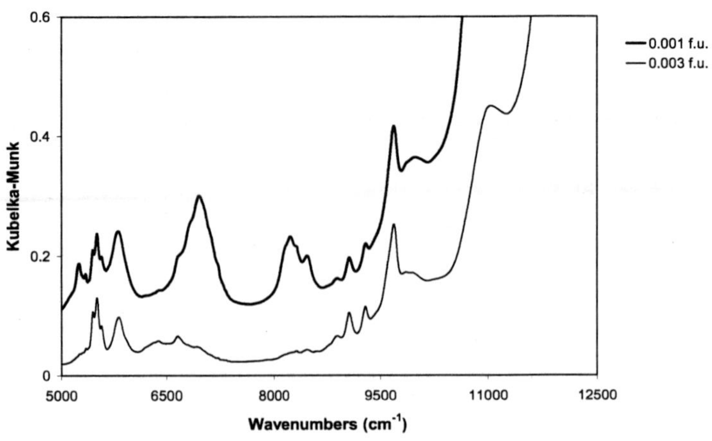

FIGURE 2. DR spectra (5,000–12,500 cm^{-1}) of zirconolites with different Pu contents.

Review and Update of Plutonium (IV) Polymer Chemistry Related to Purex Processing Operations

Mark Klasky[1] and James D. Navratil[2]

[1]*Duke Engineering, P.O. Box 31847, Charlotte, NC 28231*
[2]*Department of Environmental Engineering and Science, Clemson University, 342 Computer Court, Anderson, SC 29625-6510*

INTRODUCTION

A Mixed Oxide (MOX) Fuel Fabrication Facility (MFFF) is being designed by and licensed to Duke, COGEMA, Stone & Webster for siting at the Savannah River Site.[1,2] The MFFF is designed to produce completed MOX fuel assemblies for use in domestic, commercial nuclear power reactors. The feed material for the MFFF is uranium dioxide and surplus plutonium dioxide primarily from the Pit Disassembly and Conversion Facility and other sources supplied by the Department of Energy. The impure plutonium dioxide will be processed using a Purex flowsheet to remove impurities. Then the purified plutonium dioxide will be blended with uranium dioxide to form MOX pellets for loading into fuel rods.

To complete the Integrated Safety Analysis for the MFFF, a number of safety studies are necessary to support the development of the design basis per 10 CFR Part 70. The safety studies required a review and update of plutonium (IV) polymer formation conditions. In particular, experiments were found to be needed to define the regime in terms of plutonium concentration, temperature, and normality of nitric acid in which polymerization is possible.

PLUTONIUM (IV) POLYMER

Previous investigations of plutonium (IV) polymer have provided a considerable amount of information about its formation, kinetics, precipitation, and reactions.[3,4] However, there are still many aspects of the physical and chemical properties of plutonium (IV) polymer that are not known.

Plutonium (IV) polymer can be formed inadvertently under conditions of transient instability, and once formed, is difficult to destroy. For example, dilution of an acidic plutonium solution with water can cause polymerization in areas of low acidity. Plutonium polymerization is sometimes caused by water or steam lines leaking into plutonium solutions or by overheating of plutonium nitrate solutions during evaporation. During processing operations, plutonium polymer formation can cause emulsification in solvent extraction, clog transfer lines, and can deposit and concentrate onto equipment surfaces, causing a potential criticality hazard. Therefore, plutonium process conditions should be carefully controlled to prevent plutonium polymerization.

The molecular weight and other physical properties of plutonium (IV) polymer are dependent on its preparation conditions, and data reported in the literature are not constant for all polymer formation conditions. This poster paper will review the chemical and physical properties of plutonium (IV) polymer and the conditions of its formation in light of criticality safety for the planned MOX processing facility using the Purex process. New experimental work underway on comparing the densities of Pu (IV) to Th (IV) polymer as a function of preparation conditions will also be presented.

CP673, *Plutonium Futures — The Science*, edited by G. D. Jarvinen

REFERENCES

1. Aqueous Polishing Unit KDB—Dissolution Process, U.S. Department of Energy, DCS01 KDB CG NTF F 00266E (February 2001).
2. Basis of Design for Aqueous Polishing Process Criteria, U.S. Department of Energy, DCS01-AAJ-CG-DOB-F-04268-C (September 2002).
3. Cleveland, J. M., Plutonium Chemistry, American Nuclear Society, La Grange Park, IL (1979).
4. Advances in Plutonium Chemistry 1967–2000, edited by D. C. Hoffman, American Nuclear Society, La Grange Park, IL (2002).

Plutonium (IV) Complexes of Mixed Pyridine N-Oxide and Phosphinoxide f-Element Extractants

John H. Matonic,[1] Alejandro E. Enriquez,[1] Brian L. Scott,[1]
Robert T. Paine,[2] and Mary P. Neu[1]

[1]*Chemistry and Nuclear Materials and Technology Divisions, Los Alamos National Laboratory, Los Alamos, New Mexico 87545*
[2]*Department of Chemistry, University of New Mexico, Albuquerque, New Mexico 87131*

INTRODUCTION

Analytical and bulk scale separation and processing of aqueous acidic solutions containing f-element ions are regularly accomplished using liquid-liquid extraction (LLE) methods that employ a neutral organic donor ligand dissolved in an organic phase.[1–5] Several monofunctional ligands have been used as LLE reagents, but all display one or more deficiencies[1–5] due to the chemical similarity of the trivalent lanthanides (Ln) to the trivalent actindes (An). Because the trivalent 4f and 5f ions have identical charges, chemical separation agents for these two groups need to differentiate among these hard cations based on their size or chemical bonding preferences. This task is not easy because, as a consequence of the lanthanide and actinide contractions, the Ln and An fission products that need to be separated have similar ionic radii. In order to develop new ligands for the separation process, we must have a fundamental understanding of how these separation agents interact with both Ln and An ions on a molecular level. We have prepared and are studying derivatives of the mixed pyridine N-oxide and phosphinoxide ligand family shown in Figure 1. These compounds have been examined as LLE reagents in chloroform, arene, and alkane solvents, and they show favorable extraction performance for Ln(III) and Am(III) ions from strong HCl and HNO_3 solutions.[6,7] The coordination ability of 2,6-bis[(diphenylphosphino)-methyl]benzene P,P'-dioxide (*POPO*) to Ln(III) ions has also been tested.[8] Here 2:1 complexes are isolated, with *POPO* acting as a bidentate ligand. The extraction performance of *POPO* has not yet been characterized; however, it is expected to be a more weakly coordinating ligand compared to *NOPO* and *NOPOPO* given the absence of the N-oxide donor group.

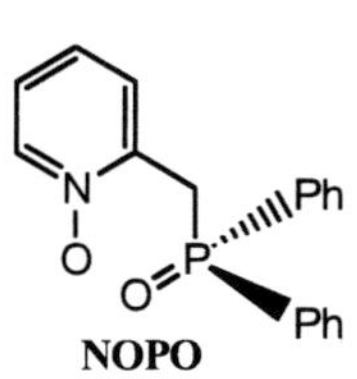

NOPO 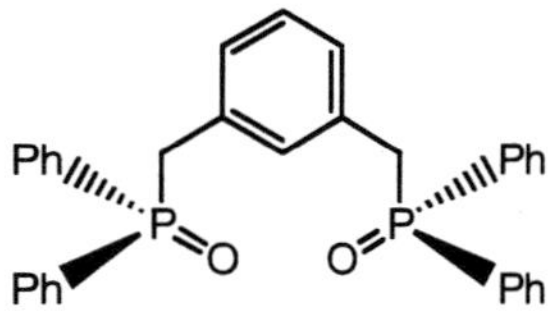 POPO

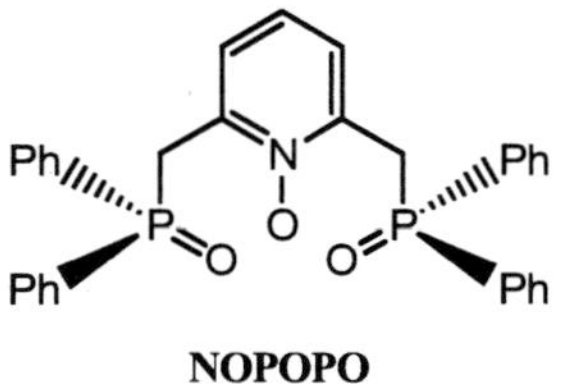

NOPOPO

FIGURE 1. Organic oxide ligands for f-element extractions NOPO = 2-(diphenylphosphino)pyridine N,P dioxide; POPO = 2,6-bis[(diphenylphosphino)-methyl]benzene P,P'-dioxide; NOPOPO = 2,6-bis(diphenylphosphinomethyl) pyridine-N,P,P'-trioxide).

CP673, *Plutonium Futures — The Science,* edited by G. D. Jarvinen

RESULTS AND DISCUSSION

In order to extend this work to tetravalent actinide ions, we are exploring the coordination chemistry of these oxygen donor ligands with Pu (IV). At this point, 2:1 complexes with *NOPOPO*,[9] $[(NOPOPO)_2Pu(NO_3)_2][NO_3]_2$ (**1**), and *NOPO*,[10] $[Pu(NOPO)_2(NO_3)_3][Pu(NO_3)_6]_{0.5}$ (**2**), have been prepared and structurally characterized. In addition, initial Pu(IV) extraction data have been collected with derivatives of *NOPOPO*. These data indicate that Pu(IV) is more strongly extracted than Am(III) and Eu(III),[11] and suggest potentially useful LLE applications for the *NOPOPO* ligand. The coordination chemistry of the bifunctional ligand *POPO* with Pu(IV) has also been investigated. Two Pu(IV) compounds that contain the POPO ligand, $Pu(POPO)(NO_3)_2Cl_2$ (**3**) and $Pu(POPO)(NO_3)_3(OMe)$ (**4**), have been prepared.

The organic oxide ligands shown in Figure 1 were designed to preferentially coordinate actinide ions in order to efficiently separate An(III) and An(IV) from Ln(III) in acid using LLE methods. Strong tridentate coordination by the three oxide donors and the steric bulk of four phenyl groups characterize the *NOPOPO* ligand. The bidentate *NOPO* ligand also displays robust coordination but has much less steric bulk than *NOPOPO*, allowing the formation of 4:1 ligand to metal complexes in some Ln(III) systems.[10] The steric demands of the *NOPOPO* ligand combined with the presence of an additional oxide functionality (which can occupy an additional coordination site on a metal) likely preclude the formation of analogous 4:1 ligand-to-metal *NOPOPO* coordination compounds. In contrast, the *POPO* ligand has approximately the same steric bulk as *NOPOPO*, but occupies one less coordination site because of replacement of the pyridine N-oxide functionality with a noncoordinating arene.

It has been reported that the *POPO* ligand can form 2:1 complexes with Ln(III) ions and structures for $[(POPO)_2(Ln)(NO_3)_2]\cdot 0.5$ MeOH (Ln = Er, Y).[8] In light of these results, we attempted to prepare a 2:1 ligand-to-metal complex of *POPO* with Pu(IV). We discovered that treatment of Pu (IV) in dilute nitric acid with two or more equivalents of *POPO* in methanol produced intractable products after solvent evaporation. However, slight modifications to these reaction conditions allowed us to isolate and characterize two distinct 1:1, POPO:Pu(IV) products.

The first product was obtained when the residue from a 2:1 reaction of *POPO* and Pu(IV) nitrate was dissolved in chloroform. Gradual evaporation of the chloroform produced orange crystals of $(POPO)Pu(NO_3)_2Cl_2$ (**3**). The chloride presumably comes from radiolytic or chemical degradation of chloroform. The structure of **3** was determined by X-ray crystallography and its ORTEP diagram is shown in Figure 2.

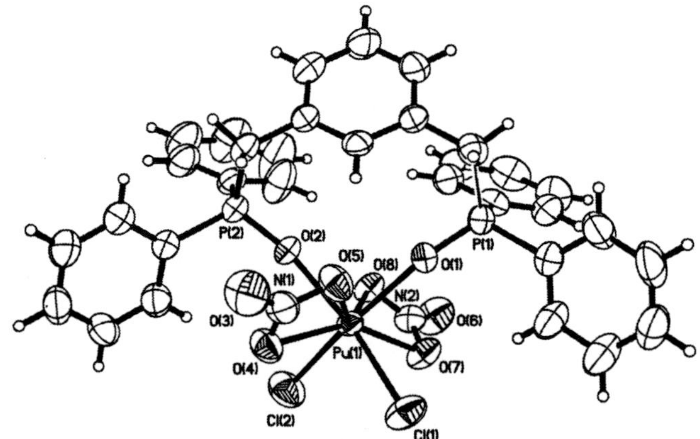

FIGURE 2. Thermal ellipsoid plot of $(POPO)Pu(NO_3)_2Cl_2$ (**3**) at the 40% probability level. Selected bond distances (Å): Pu-O(1) 2.250(3), Pu-O(2) 2.238(3), Pu-O(4) 2.468(4), Pu-O(5) 2.452(4), Pu-O(7) 2.447(4), Pu-O(8) 2.465(4), Pu-Cl(1) 2.6000(3), Pu-Cl(2) 2.572(2).

The second product was prepared by adding excess aqueous sodium hydroxide to a solution of excess *POPO* in methanol spiked with Pu (IV) nitrate stock. The sodium hydroxide was added in an attempt to sequester nitrate as $NaNO_3$ and to encourage coordination of a second *POPO* ligand to the Pu(IV) center, but the sodium hydroxide generates NaOMe instead. The methoxide ion effectively competes with nitrate for one coordination site on the Pu center, and evaporation of the reaction solvent provides crystalline $(POPO)Pu(NO_3)_3(OMe)$ (**4**) as the only product. This result is not unexpected, given the oxophilic nature of Pu(IV) and nucleophilicity of methoxide. However, it is surprising that coordination of two methoxide units is not observed, especially when the synthesis of the dichloride complex **3** is taken into consideration. The structure of **4** was determined by X-ray crystallography and its ORTEP diagram is shown in Figure 3. This presentation will discuss the structural features of all the Pu(IV) complexes derived from the ligands in Scheme 1 and compare them to the Ln and An complexes that have been reported.

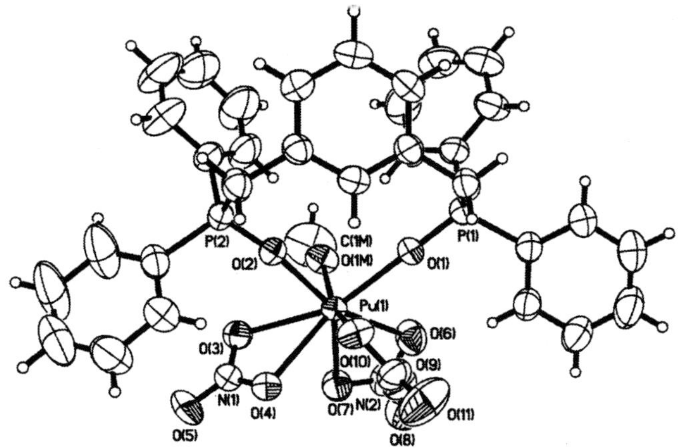

FIGURE 3. Thermal ellipsoid plot of *(POPO)*$Pu(NO_3)_3(OMe)$ (**4**) at the 40% probability level. Selected bond distances (Å): Pu-O(1) 2.301(2), Pu-O(2) 2.276(2), Pu-O(3) 2.497(3), Pu-O(4) 2.503(3), Pu-O(6) 2.530(3), Pu-O(7) 2.456(3), Pu-O(9) 2.487(3), Pu-O(10) 2.544(3), Pu-O(1M) 2.027(3).

REFERENCES

1. De, A. K., Khopkar, S. M., and Chalmers, R. A., Solvent Extraction of Metals, 1970.
2. Navratil, J. D., and Schulz, W. W., Editors, Radioactive Waste Management, Vol. 6; Actinide Recovery from Waste and Low-Grade Sources, 1982.
3. Choppin, G. R., Navratil, J. D., and Schulz, W. W.; Editors Proceedings of the International Symposium on Actinide Lanthanide Separations, Honolulu, Hawaii, 16–22 December 1984 (1985).
4. Nash, K. L., Gschneidner, K. A. Jr., and Eyring, L., Editors, Chapter 21 (1994).
5. Musikas, C. Inorg. Chim. Acta 140, 197–206 (1987).
6. Bond, E. M., Engelhardt, U., Deere, T. P., Rapko, B. M., Paine, R. T., and Fitzpatrick, J. R., Solvent Extr. Ion Exch. 16, 967–983 (1998).
7. Bond, E. M., Engelhardt, U., Deere, T. P., Rapko, B. M., and Paine, R. T., Solvent Extr. Ion Exch. 15, 381–400 (1997).
8. Tan, Y.-C., Gan, X.-M., Stanchfield, J. L., Duesler, E. N., and Paine, R. T., Inorg. Chem. 40, 2910–2913 (2001).
9. Bond, E. M., Duesler, E. N., Paine, R. T., Neu, M. P., Matonic, J. H., and Scott, B. L., Inorg. Chem. 39, 4152–4155 (2000).
10. Matonic, J. H., Neu, M. P., Enriquez, A. E., Paine, R. T., and Scott, B. L., J. Chem. Soc. Dalton Trans. 2328–2332 (2002).
11. Bodrin, G. V., Kabachnik, M. I., Kochetkova, N. E., Medved, T. Y., Myasoedov, B. F., Polikarpov, Y. M., and Chmutova, M. K., Izv. Akad. Nauk SSSR, Ser. Khim. 2572-2575 (1979).

Contribution to the Characterization of Deviations from Ideality of Concentrated Electrolyte Solutions: The Case of Tetravalent Actinide Nitrates

Ph. Moisy and P. Blanc

CEA Valrho DEN/DRCP/SCPS/LCA; BP 17171, 30207 Bagnols sur Ceze – France

INTRODUCTION

As for any model of a chemical system, the behavior of actinides in nitric acid media is based primarily on the law of mass action. Although this law can be used to describe chemical equilibria between species, the activity of the species involved in each chemical equilibrium must be known. In dilute media, a very good approximation of the law of mass action can be obtained by considering that the activity of a species is equivalent to its concentration. Conversely, with concentrated solutions and variable compositions, as in the case of industrial separation processes, the effect of the medium cannot be ignored. A satisfactory model of actinide chemistry in separation processes (PUREX, UREX, etc.) therefore requires allowance for the activity of the species involved. The law of mass action is thus expressed differently than for dilute media because of the inclusion of the activity coefficients.

Numerous models have been proposed in the literature to predict the effects of the medium (Pitzer, SIT, etc.). All are based on the mean stoichiometric activity coefficients for salts that are not fully dissociated and on the mean ionic activity coefficients for fully dissociated salts. The data necessary to calculate the mean stoichiometric and ionic activity coefficients of a constituent in a mixture are generally the physicochemical data for the binary solutions of the electrolytes in the mixture. In most cases, these data are sufficient for a satisfactory description of the media effect. These "binary data" describe, for example, the variation of the activity coefficient of a constituent, of the water activity, or of the density of the binary solution of the constituent according to its concentration.

Because of the limited binary data available in the literature for tetravalent actinides, the concept of chemical analogs is often used instead.

RESULTS

In this study, the mean stoichiometric activity coefficient is calculated from fictive binary data using Mikulin's relation and the concept of "simple solutions." Considering the strong hydrolysis of actinides at oxidation state +IV, concentrated binary solutions cannot be prepared; only ternary $Pu(NO_3)_4/HNO_3/H_2O$ and quaternary $U(NO_3)_4/HNO_3/N_2H_5NO_3/H_2O$ solutions actually exist. Ternary Pu^{IV} and quaternary U^{IV} solutions were prepared experimentally. The compositions of the solutions with the highest actinide concentrations were as follows: (i) $[Pu^{IV}]$ = 3.11 molal, $[HNO_3]$ = 0.85 molal and (ii) $[U^{IV}]$ = 2.22 molal, $[HNO_3]$ = 3.18 molal, $[N_2H_5^+]$ = 0.23 molal.

Water activity and density measurements were performed with these solutions at 25°C over a wide range of concentrations.

DISCUSSION

The measured results were used to calculate the variation of the osmotic coefficient (and the mean stoichiometric activity coefficients) and the density versus the electrolyte concentration. These values were also used to calculate

CP673, *Plutonium Futures — The Science,* edited by G. D. Jarvinen

the interaction parameters for Pitzer's model and for the model based on the specific interaction theory for $Pu(NO_3)_4$ and $U(NO_3)_4$ electrolytes.

The experimental data[1–5] and published data[1,6–9] for $Th^{IV}(NO_3)_4$ provide a basis for discussion of the validity of the concept of chemical analogs for the particular case of tetravalent actinide nitrate salts.

REFERENCES

1. Charrin, N., Moisy, Ph., and Blanc, P., "Thermodynamic Study of the Ternary System $Th(NO_3)_4/HNO_3/H_2O$," Radiochimica Acta 86, 143–149, (1999).
2. Charrin, N., Moisy, Ph., and Blanc, P., "Determination of Fictive Binary Data for Pu(IV) Nitrate," Radiochimica Acta 88, 25–31 (2000).
3. Kappenstein, A. C., Moisy, Ph., Cote, G., and Blanc, P., "Contribution of the Concept of Simple Solutions to Calculation of the Stoichiometric Activity Coefficients and Density of Ternary Mixtures of Hydroxylammonium or Hydrazinium Nitrate with Nitric Acid and Water," Phys. Chem. Chem. Phys. 2, 2725–2730 (2000).
4. Charrin, N., Moisy, Ph., and Blanc, P., "Contribution of the Concept of Simple Solutions to Calculation of the Density of Ternary and Quaternary Solutions of Thorium (IV) or Plutonium (IV) Nitrate: $An(NO_3)_4$ / $UO_2(NO_3)_2$ / HNO_3 / H_2O," Radiochimica Acta 88, 445–451 (2000).
5. Charrin, N., Moisy, Ph., and Blanc, P., "Determination of Fictive Binary Data for U(IV) Nitrate: Preliminary Data," Radiochimica Acta 89, 579–585 (2001).
6. Robinson, R. A., and Levien, B. L., "The Vapour Pressures of Potassium Ferricyanide and Thorium Nitrate Solutions at 25°C," Transactions of the Royal Society of New Zealand 76(3), 295–299 (1946).
7. Lemire, R. J., and Brown, C. P., "Components Activities in the System Thorium Nitrate-Nitric Acid-Water at 25°C," J. Sol. Chem., 11(3), 203–220 (1982).
8. Lemire, R. J., Sagert, N. H., and Lau, W. P., "Osmotic Coefficients of Water for Thorium Nitrate Solutions at 25, 37, and 50°C," J. Chem. Eng. Data 29, 329–331 (1984).
9. Apelblat, A., Azoulay, D., and Sahar, A., "Properties of Aqueous Thorium Nitrate Solutions—Part 2. Water Activities and Osmotic and Activity Coefficients of Thorium Nitrate at 25, 35 and 45°C," J. Chem. Soc. Farad. Trans. 69, 1624–1627 (1973).

Preliminary Data on Np(IV) and Pu(IV) Behavior in Room Temperature Ionic Liquids

Ph. Moisy[1] and S. I. Nikitenko[2]

[1]CEA Valrho DEN/DRCP/SCPS/LCA; BP 17171, 30207 Bagnols sur Ceze – France
[2]Institute of Physical Chemistry, RAS, 31 Leninsky Prosp., Moscow, Russia

Room temperature ionic liquid, RTIL, $[BuMIm]PF_6$, where $BuMIm^+$ is 1-butyl-3-methyl imidazolium, has been synthesized, purified on activated carbon, and characterized by ^1H, ^{13}C NMR. Drying of $[BuMIm]PF_6$ at 80°C in 4 h on a vacuum line (P~5 mbar) allowed the reduction of the water content to 750 ppm.

Complexes $[BuMIm]_2NpCl_6$ and $[BuMIm]_2PuCl_6$, have been obtained by precipitation from Np(IV) and Pu(IV) solutions in concentrated HCl in the presence of [BuMIm]Cl. Both compounds are very soluble in $[BuMIm]PF_6$. UV/VIS/NIR spectra of Np(IV), and Pu(IV) in $[BuMIm]PF_6$ is similar to those in aqueous concentrated HCl or 7M HCl in 75% EtOH and quite different from the spectra of Np(IV) and Pu(IV) in diluted HCl. It was assumed that octahedral hexachlocomplexes $AnCl_6^{2-}$ predominate in this RTIL. Np(IV) and Pu(IV) are slowly hydrolyzed by water in $[BuMIm]PF_6$ solutions, even being stored in a dry atmosphere.

CP673, *Plutonium Futures — The Science,* edited by G. D. Jarvinen

Dissolution of plutonium hydroxide under slightly reducing conditions

H. Nilsson , Y. Albinsson, G. Skarnemark

Nuclear Chemistry, Department of Materials and Surface Chemistry
Chalmers University of Technology, Göteborg, Sweden

INTRODUCTION

In Sweden, the proposed method of disposal of spent nuclear fuel is to deposit it 500 m down in granitic bedrock. Due to lack of oxygen at these depths it is important to study the behaviour of elements present in the spent fuel at reducing conditions.

The dissolution of plutonium (III,IV) hydroxide has been studied in near neutral conditions in 0.1 M NaCl for almost two years. For the study two solids were prepared using precipitation with deoxygenated NaOH followed by filtration and washing of the precipitates with deionised water. Prior to the precipitation of the first solid the plutonium solution was passed through a lead column reductor and then filtered to remove any residues from the column. The resulting solution was verified to be more than 95% Pu(III) using optical spectrometry. The precipitate was initially blue but became khaki after drying. The second precipitate was made directly on the plutonium stock consisting mainly of Pu (IV). This precipitate was bright green. Microgram amounts of the two precipitates were put in 0.1 M NaCl solutions adjusted to pH 5.6, 6.7 and 7.0 (Exp. 1-3) for the Pu(III) hydroxide and pH 5.5 (Exp. 4) for the Pu(IV) precipitate. For the concentration measurements samples were withdrawn and centrifuged at 41000 rpm (288 000 g_{max}). The pH and on occasion Eh was measured directly in the centrifuge tubes after some supernatant had been withdrawn for liquid scintillation counting (LKB Wallac, RackBeta). All handling was performed in an inert (N_2) glove box with active removal of oxygen down to < 10 ppm.

RESULTS

In Figure 1 the total Pu concentrations are plotted against pH. In all four experiments the pH is shifted to lower values with time. However this effect was experienced much earlier for exp.4 than for exp. 1-3. At first there is a difference of about two orders of magnitude in the concentration above the different precipitates but at the end of the experiment they seem to converge as the concentration drops in the Pu(III) experiments. The straight lines represent PuO_2 and $Pu(OH)_4$ solubility data taken from [1] and $Pu(OH)_3$ solubility taken from [2]. Figure 2 shows the concentration versus time for the four experiments and it is fairly clear that these experiments are far from reaching equilibrium.

X-ray powder diffraction (XPD) measurements of both $Pu(OH)_3$ and $Pu(OH)_4$ solids from these experiments have been performed after 136 days and after 665 days. All measurements showed poor crystallinity. In the earlier measurements however, small peaks were observed and could be interpreted as coming from PuO_2. In the latter measurements this indication was substantially less, which could be since these samples were not dried before measurement.

Table 1. Eh versus standard hydrogen electrode (mV).

Exp.	Eh (136 days)	Eh (665 days)
1	821	516
2	689	506
3	614	539
4	512	512

CP673, *Plutonium Futures — The Science,* edited by G. D. Jarvinen

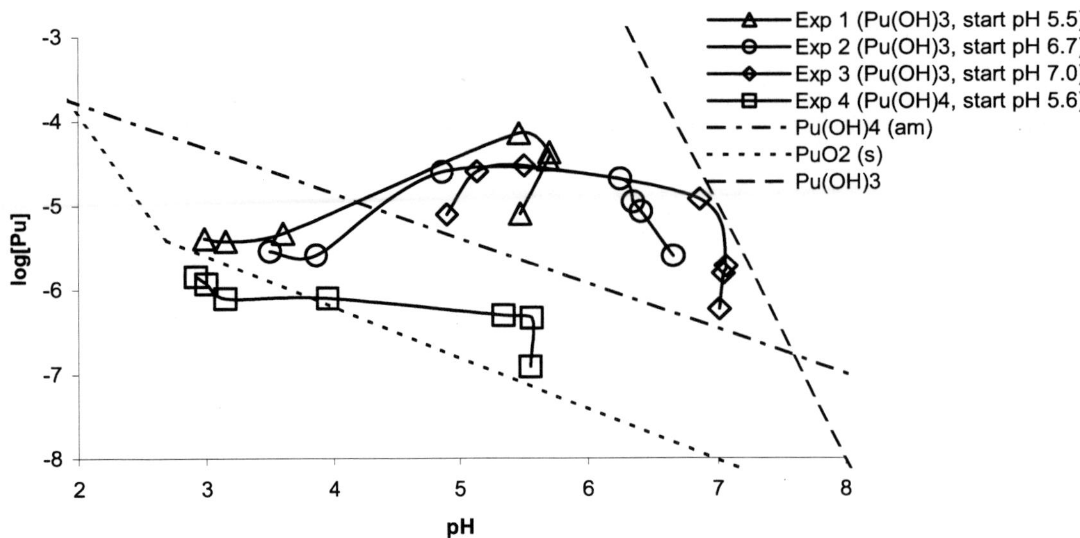

Fig. 1. Total plutonium concentration versus pH

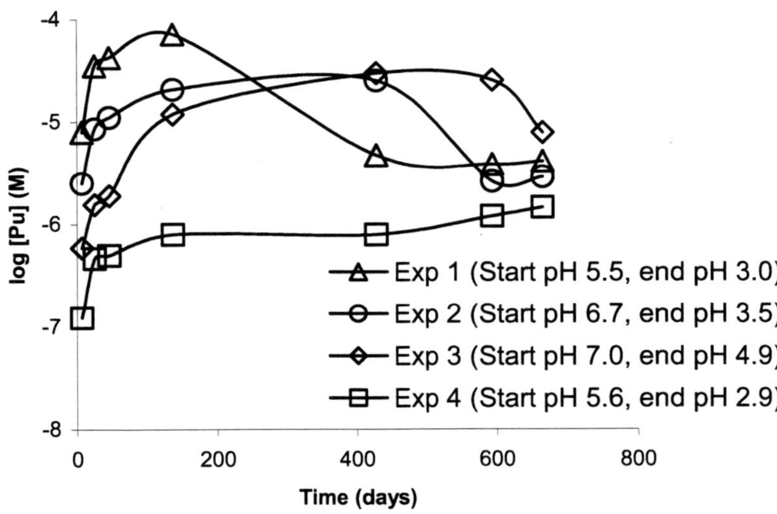

Fig. 2. Total plutonium concentration versus time.

DISCUSSION

The results indicate that the $Pu(OH)_3$ solids are mixtures of Pu(III) and Pu(IV) that are slowly oxidized to become entirely Pu(IV). Thus the concentration initially tries to reach the $Pu(OH)_3$ solubility limit found by Felmy et. al. but then it drops to meet the region where others have measured solubilities for crystalline PuO_2 as gathered by Haschke and Oversby. They propose PuO_{2+x} as the stable phase and that the solid splits water molecules in its pursuit of oxygen and thus $H_2(g)$ or elementary H radicals will be produced at the surface. This could be the explanation for the drift in pH and Eh observed over time.

This work was funded by the Swedish Nuclear Fuel and Waste Management Company, SKB.

REFERENCES

1. Felmy A., Rai D., Schramke J., Ryan J., Radiochimica Acta 48, 29-35 (1989).
2. Haschke J., Oversby V., Journal of Nuclear Materials, 305(2-3), 187-201 (2002).

Equilibrium and Kinetic Studies of Heterogeneous Reactions of Actinides: Hydroxide Compounds in Alkaline Media

I. G. Tananaev, B. F. Myasoedov, and D. L. Clark*

V. I. Vernadsky Institute of Geo and Analytical Chemistry, Russian Academy of Sciences
**G.T. Seaborg Institute for Transactinium Science, Los Alamos National Laboratory*

INTRODUCTION

The problem of radioactive wastes, accumulated as the result of the production of nuclear energy and nuclear weapons in the past and their shortages is nowadays one of the most important ecological problems. In the U.S., highly alkaline radioactive wastes (HARW) exist in aging tanks at Hanford, Savannah River, INEL, West Valley, and Oak Ridge, and many tanks contain actinide elements in quantities of concern for separation, treatment and disposal.[1-3] Knowledge of actinide chemical properties under highly alkaline conditions is essential to understand and predict their solubility, sorption, and phase distribution in tanks to determine whether chemical separation is needed for waste treatment, and the design and optimization of the rational separations process.

The present work is directed to the study of the composition and properties of Am(III–VII) solutions and compounds formed in alkaline media.

EXPERIMENTAL

$Am(OH)_3$ Oxidation by the Ozone

It was found that during a 3.5% vol ozone-oxygen mixture bubbling through the $Am(OH)_3$ pulp in 0.1 M $NaHCO_3$, the pulp is dissolved in just a few minutes while the solution turns red, and in these conditions Am(III) is oxidized consecutively to the formation of Am(IV), Am(V), and Am(VI). It was shown that during ozonization through the $Am(OH)_3$ in 0.1 M $NaHCO_3$, sudden oscillations [Am(V)]↔[Am(VI)] were observed. The $Am(OH)_3$ oxidation rate by the ozone in a 0.1 M Na_2CO_3 + (0.1-2.0)M NaOH mixture, or (0.1–1.0) M NaOH/LiOH does not differ from those described above. On the contrary, in similar conditions for both KOH or CsOH and in concentrated RbF solutions, no noticeable concentration of Am(VI) was recorded.

Interaction of Am(VI) with Water

It was shown that the reduction rate of ^{243}Am(VI) in a basic aqueous solution (~20% per hour) was found to be of the same order as that for ^{241}Am(VI). The values of an apparent constant (k_a) rate of the first-order reaction in 3 M and 1 M NaOH are the same $\sim(5{,}3 \pm 0{,}3)\cdot 10^{-4}\ s^{-1}$ and in 0,1M NaOH k_a are increased to $1{,}7 \cdot 10^{-3}\ s^{-1}$. On the basis of the data obtained, we can conclude that the reduction of Am(VI) by water molecules results in the main contribution in the process above, and a considerable contribution of Am α-radiolysis effects on the americium reduction in these conditions can be excluded.

CP673, *Plutonium Futures — The Science,* edited by G. D. Jarvinen

Am(VII) Species

It was found that an interaction of ~10^{-3} M ^{243}Am(VI) solution in 1M NaOH when mixed at 20°C with a sodium perxenate solution leads to the appearance of the precipitate. The ozonizing of this precipitate leads to the formation of Am green species with specific optical spectrum (Figure 1). The general view is much like the spectra of Np(VII) and Pu(VII) in alkaline media,[7] but shifted to the IR range. It was supposed that in the experiment above the Americium (VII) compound had been synthesized.

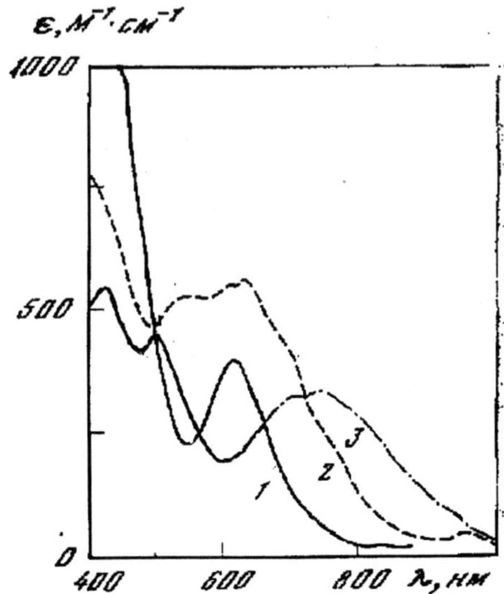

FIGURE 1. Absorption spectra of heptavalent Np (1), Pu (2) in 4M NaOH,[6] and ozonized Am:Xe complex obtained in our work (3).

Interamericium Hydroxide Interactions

It was shown that the mixing of equivalent amounts of ^{243}Am(III) and ^{243}Am(V) hydroxides in 1 M NaOH at 50°C–70°C, or in 7 M NaOH at 25°C during a few minutes caused the appearance of pure Am(IV) species through a reproportionation reaction: Am(III) + Am(V) = 2Am(IV). The absorption spectra of initial mixture and the interaction product are listed in Figures 2 and 3, respectively. The reaction between Am(VI) and Am(III) hydroxide in equivalent amounts brings the same result.

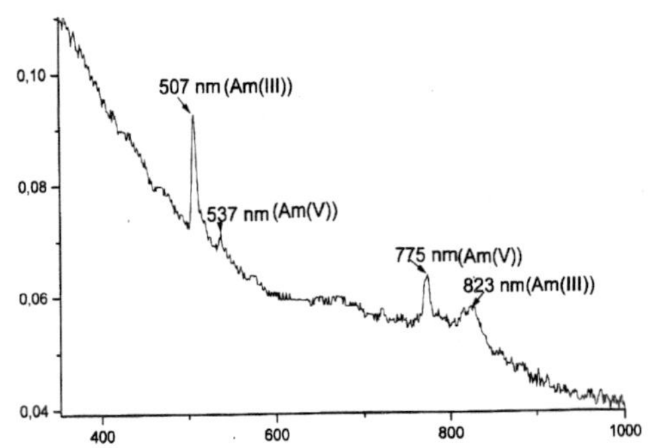

FIGURE 2. The absorption spectrum of the mixture of Am(III) and Am(V) hydroxides separates from a 1 M NaOH solution at 25°C.

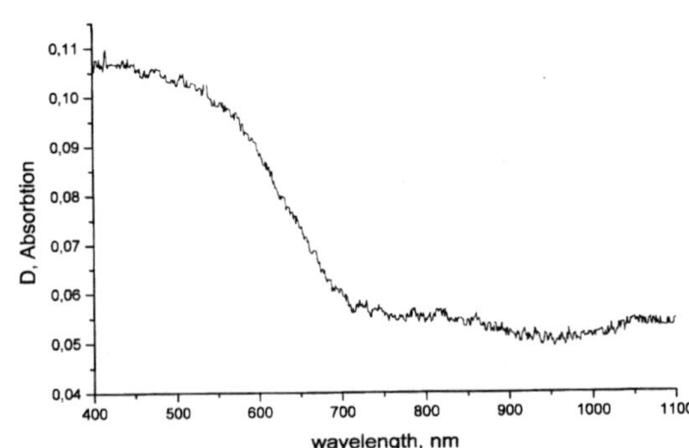

FIGURE 3. The absorption spectrum of the mixture of Am(III) and Am(V) hydroxides separates from a 1 M NaOH at 70°C or in 7 M NaOH at 20°C.

Heterogeneous Reversible Am Redox-Reaction in Alkaline Media

It was found that in an excess of 0.006 M NaOH, $K_3Fe(CN)_6$ does not oxidize Am(III) hydroxide at room temperature. An increase of the [NaOH] to 0.087 M leads to a reaction of 0,067M $K_3Fe(CN)_6$ and 1.5 μmole $^{243}Am(OH)_3$ at 20°C with the formation of $Am(OH)_4$. In a 2 M NaOH solution, the $K_3Fe(CN)_6$ (0.04 M) oxidizes $Am(OH)_3$ to Am(V). It was shown that the redox reaction of Am(VI) and RuO_4^{2-} ions is reversible in basic solutions. Based on the literature, data on the extinction coefficient of ruthenate and perruthenate ions and the redox potential of a RuO_4^-/RuO_4^{2-} pair (0.6 V vis NHE), we obtained an approximate value of a redox potential for an Am(VI)/Am(V) pair. In 1 M NaOH, E^f(Am(VI)/Am(V)) was determined as ≈ 0,7 V (vis NHE).

DISCUSSION

It was found that in high-alkaline solutions, a number of Am(III)-(VII) species in oxidized media were stabilized. Of interest is that the redox potentials have been changed so dramatically by alkaline solutions that α-particle self-irradiation was unimportant. Instead, it was found that Am(VI) is readily reduced by the simple action of water. This is a marked departure from the behavior of Am in other media, where α-particle self-irradiation generates redox-active species that affect the Am oxidation state. Other redox agents, notably $Fe(CN)_6^{3-}$ (present in tank environs) were shown to oxidize Am(III) to Am(IV) with subsequent precipitation as relatively insoluble $Am(OH)_4$. In acidic and near-neutral solutions Am(IV) is unstable, so this observation is quite important, and demonstrates that redox agents in the tanks will have a definite effect on the americium oxidation state, and hence the overall solubility. These results have identified a number of important reactions affecting the oxidation state and stability of Am in alkaline waste environments. Such information will be invaluable to improved separations approaches during remediation. Better methods for separating actinide elements from alkaline solutions could lead to reduced HARW volumes and lower disposal costs. Concentrating the actinides in these wastes could reduce the number of canisters needed by at least tenfold, saving tens of billions of dollars.

ACKNOWLEDGMENTS

The work was supported by the U.S.DOE-OBES, Project RC0-20004-SC14. This work was provided with Drs. M. V. Nikonov and A. V. Gogolev, and we are especially grateful their technical support.

REFERENCES

1. Gephart, R. E., and Lundgren, R. E. "Hanford Tank Cleanup: A Guide to Understanding the Technical Issues," Pacific Northwest Laboratory, PNL-10773 (1995).
2. Agnew, S. F., "Hanford Defined Wastes: Chemical and Radionuclide Compositions," Los Alamos National Laboratory document LA-UR-94-2657, Rev.2 (1995).
3. Congress, U.S., "Long-Lived Legacy: Managing High-Level and Transuranic Waste at the DOE;" Egorov, N. N., "The Condition of the Problem of Treatment of Radiowastes and Exhausted Nuclear Fuel in Russia;" Voprosy Radiatsionnoy Bezopasnosti. (In Russian) 4, 3–8 (1997).
4. Melnikov, N. N., Naumov, V. A., Konjukhov, V. P. et al., "Radioecologic Aspects of the Safety of Underground Burying of Radiowastes and Exhausted Fuel in the European North of Russia," Apatity (2001), 193 pp.
5. Glagolenko, Yu. V., Dzekun, E. G., Drozhko, E. G, Medvedev, G. M., Rovny, S. I., Suslov, A. P., "Strategy of the Treatment of Radiowastes in "Mayak" Industrial Complex," Voprosy Radiatsionnoy Bezopasnosti. (In Russian) 2, 3–11 (1996).
6. Krot, N. N.; Gelman, A. D., Mefod'eva, M. P.; Shilov, V. P.; Peretrukhin, V. F.; and Spitsyn, V. I., Available in English as "Heptavalent State of Neptunium, Plutonium and Americium," UCRL-Trans-11798, Lawrence Livermore National Laboratory, Livermore, California (1977).

Spectroscopic Investigations of the Electronic Structure of Neptunyl Ions

Marianne P. Wilkerson, John M. Berg, and Harry J. Dewey

Los Alamos National Laboratory
Los Alamos, NM 87544

Molecular electronic structures are innately sensitive to the geometric and chemical environments around the metal center of coordination compounds. However, the interrelationships between the electronic structures and molecular geometries of actinide species, which often contain more than one electron in the 5*f* valence shell, are quite complex because of the large numbers of possible electronic states and high densities of vibronically enabled transitions.[1,2] Investigations of the optical signatures of simple, well-defined molecular systems should provide the most straightforward approach for unharnessing these fundamental relationships, and in particular, systems with a single electron in the valence 5*f* shell, such as the neptunyl ion (NpO_2^{2+}), should provide the most viable means for characterizing actinide electronic structure. Furthermore, 5*f* orbital-occupied actinide systems exhibit not only visible and ultraviolet ligand-to-metal charge-transfer spectral bands, but also near-infrared 5*f*-5*f* transitions resulting from promotion of a 5*f* electron to an orbital of primarily 5*f* character.

We will focus here on an analysis of the luminescence spectra of one structurally characterized neptunyl system, $NpO_2Cl_4^{2-}$. We have determined the molecular structure of $Cs_2NpO_2Cl_4$ from single-crystal X-ray diffraction analysis.[3,4] The compound crystallizes in the monoclinic space group *C2/m*, and the neptunium metal is coordinated in a pseudooctahedral fashion, in which two oxo ligands occupy transaxial positions and four equatorially coordinated chloride ligands saturate the metal center (Figure 1). Possible luminescence quenching could be repressed through doping of the analyte into a matrix that has sites of appropriate size for accommodation of $Cs_2NpO_2Cl_4$ and that lacks any modes for radiationless deactivation, and therefore, isostructural $Cs_2UO_2Cl_4$ can be used as the host matrix material.[5] The doped material, $Cs_2U(Np)O_2Cl_2$, is readily prepared from dissolution of the two compounds in mineral acid followed by recrystallization at room temperature.[6] Theoretical studies of the ground and excited state energies of both the NpO_2^{2+} and neptunyl tetrachloride $NpO_2Cl_4^{2-}$ ions have been reported, and polarized single-crystal absorption spectra of the LMCT and 5*f*-5*f* transitions of $Cs_2U(Np)O_2Cl_4$ have been measured.[6-9] However, luminescence offers an inherently sensitive and selective spectroscopic approach for the measurement of optical transitions.

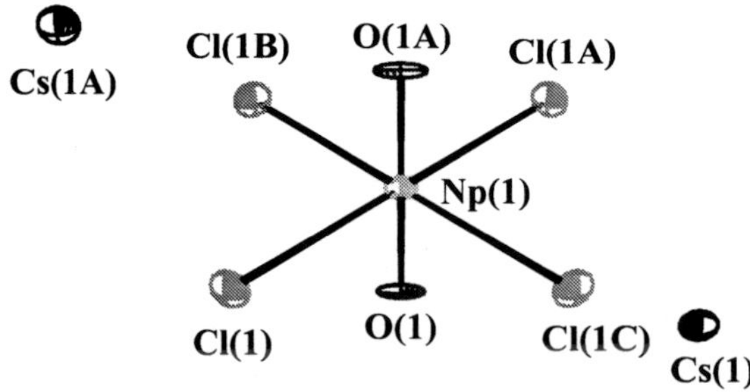

FIGURE 1. Molecular structure for $Cs_2NpO_2Cl_4$ (50% probability ellipsoids).

In an effort to understand the relationship between electronic structures and molecular geometries of actinide molecules, we have discovered that the simple $5f^1$ neptunyl ion luminesces in the near-infrared region following

CP673, *Plutonium Futures — The Science,* edited by G. D. Jarvinen

visible excitation. Although pure *5f-5f* electronic transitions are electric dipole forbidden, they may be allowed by magnetic dipole or electric quadrupole mechanisms. At low temperature the emission peaks exhibit the moderately narrow linewidths typical of *5f-5f* transitions. Less expected is our observation that it also luminesces intensely at room temperature, with some broadening of the bands. The profiles of the two spectra are similar. By comparing with prior theoretical and absorption characterizations, we assign the most intense band at ~6,880 cm^{-1} as the origin of the second excited state. Initial assignments of vibronic bands, provided in Figure 2, are based upon the energies of ground-state vibrational modes of $Cs_2NpO_2Cl_4$.[6] In addition to an analysis of the other potential *5f-5f* transitions, studies of excitation behavior and excited state lifetimes are underway.

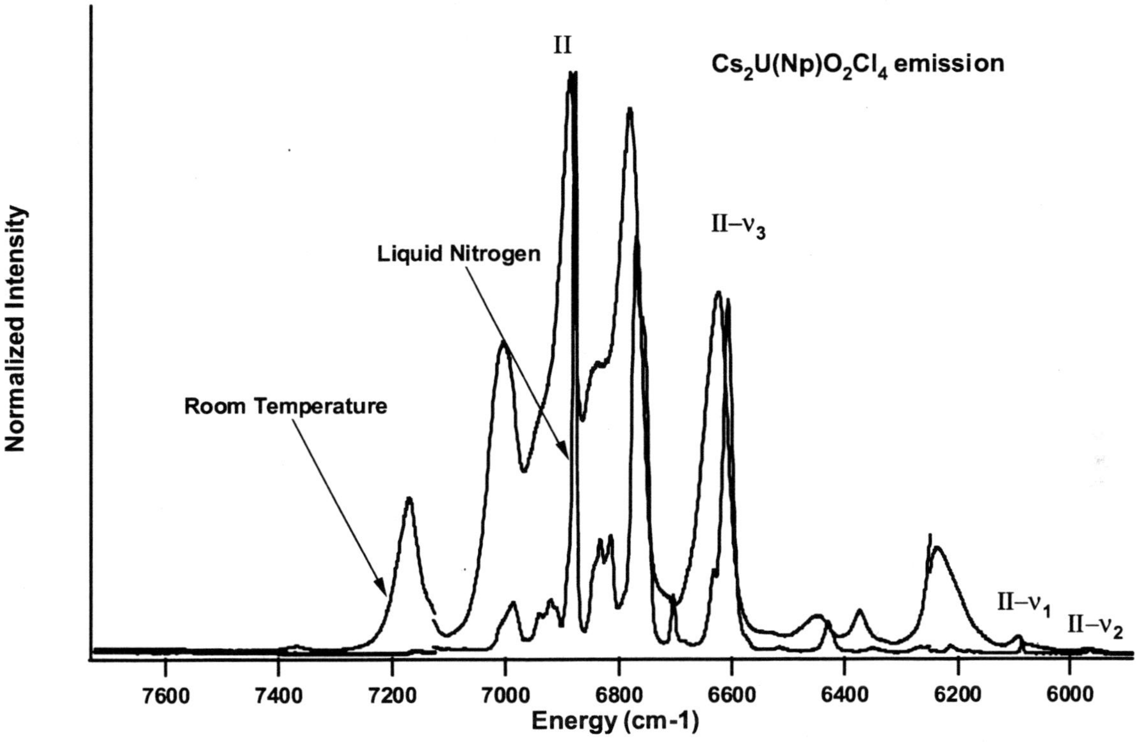

FIGURE 2. Luminescence spectra of $Cs_2U(Np)O_2Cl_4$ following 535 nm excitation.

A picture of the luminescence structure of this neptunyl system will provide a benchmark for comparing with, or perhaps predicting, the emission signatures from other neptunyl structures. The long-term promise of this work is a practical means for beginning to develop a more systematic understanding of actinide excited state properties, which could possibly be extrapolated to the intricate but pervasive $5f^n$ (n>1) actinide species containing coordination geometries and ligand sets accessed in common chemical environments, radiologically contaminated environmental wastes, and nuclear power streams.

REFERENCES

1. Carnall, W. T., and Crosswhite, H. M., Chemistry of the Actinides, edited by J. J. Katz, G. T. Seaborg, and L. R. Morss, Chapman and Hall: London, Vol. 2, Chapter 16 (1986), pp. 1235–1276.
2. Berg, J. M., Sattelberger, A. P., Morris, D. E., Van Der Sluys, W. G., and Fleig, P. Inorg. Chem. 32, 647 (1993).
3. Wilkerson, M. P., Dewey, H. D., and Scott, B. L., manuscript in preparation.
4. Weigel, F., Werner, G. D., and Kalus, C., Physica 102B, 308 (1980).
5. Watkin, D. J., Denning, R. G., and Prout, K. Acta Cryst. C 47, 2517 (1991).
6. Denning, R. G., Norris, J. O. W., and Brown, D., Mol. Phys. 46, 287 (1982).
7. Denning, R. G., Norris, J. O. W., and Brown, D., Mol. Phys. 46, 325 (1982).
8. Matsika, S., and Pitzer, R. M., J. Phys. Chem. A 104, 4064 (2000).
9. Matsika, S., and Pitzer, R. M., J. Phys. Chem. A 105, 637 (2001).

Transuranic Actinide Reactions with Simple Gas-Phase Molecules

Stephen P. Willson,[1] D. Kirk Veirs,[1] and Joseph P. Baiardo[2]

NMT-11, Los Alamos National Laboratory, Los Alamos, NM 87544
NMT-16, Los Alamos National Laboratory, Los Alamos, NM 87544

INTRODUCTION

The intent of this research is to conduct an experimental study of f-element chemistry for the purpose of identifying reaction trends and mechanisms of the early actinide metals with simple-gas phase molecules. Previous research has elucidated some of the fundamental chemistry of the 4f elements;[1-5] however, more complex chemistry is expected for the 5f series due to the inclusion of the 5f electrons in the valence shell. The matrix isolation approach, which is well-suited to the experimental study of transient species, will be used for sample collection, and IR/NIR/VIS spectroscopy will be employed to interrogate deposited matrices. The strength of this method lies in the use of isotopes of reactants, which permits the identification of guest molecules in a noble gas matrix by observation of vibrational frequency shifts and patterns upon isotopic substitution. Using this technique at the University of Virginia, the first noble gas-actinide bond has recently been identified, a weak U-Ar bond on the CUO molecule.[6] Uranium has similarly been observed to bond to krypton and xenon, whereas thorium and the lanthanides have not exhibited this activity. It is expected that plutonium will be even more reactive in this respect. We will extend the body of actinide experimental evidence to include the transuranic elements neptunium, plutonium, and americium reacted with isotopes of oxygen, nitrogen, hydrogen, carbon monoxide, and carbon dioxide.

RESULTS

As of February 2003, the experimental apparatus has been tested, and final authorizations are forthcoming. Installation is expected to be completed in March 2003, and analysis of the plutonium and oxygen reaction system will be completed by May 2003. The presentation will include background on the experimental setup and presentation of data and conclusions from the initial plutonium and oxygen study.

DISCUSSION

Much of the fundamental reaction chemistry of actinides remains unexplored. Greater knowledge of reaction pathways of actinides with common atmospheric constituents and characterization of the products of such reactions will have broad applicability, ranging from elucidation of corrosion mechanisms during a reactor accident to understanding the environmental fate of actinides released to the atmosphere. The spectroscopic data of these fundamental reaction intermediates and products is needed to identify corrosion products of actinide metals by providing an infrared fingerprint of particular actinide-containing chromophores and compounds. The reactivity data are needed for the development of models applicable to the corrosion process, as well as those applicable to the fundamental reactions at the solid-gas interface where radiolysis reactions are important.

The matrix isolation technique can be used to study the formation of actinide-containing molecules, including the direct study of short-lived reaction intermediates in well-defined and reproducible reaction systems. In past studies,

CP673, *Plutonium Futures — The Science,* edited by G. D. Jarvinen

the most commonly used gaseous reactants have been atmospheric constituents such as O_2, N_2, H_2, H_2O and CO_2, and common pollutants NO, and CO. [1-7] Reaction trends of the lanthanides with several of these reactants were identified from results of experiments across the lanthanide series.[5] Attempts to study the actinides at the University of Virginia have been limited to thorium and uranium because of safety concerns. This study seeks to identify similar trends in the early actinides. In this study, the matrix isolation technique will be applied to reactions of the actinides Np, Pu, and Am with O_2, N_2, H_2, CO, and CO_2.

In concord with its relevance to the exploration of actinide reactivity, this direct experimental study of fundamental f-element chemistry will also be of importance to the theoretical community, providing experimental benchmarks for computation and, in conjunction with calculations, identification of the ground states of actinide-containing species, which is often complicated by the abundance of low-lying electronic states. Many novel ionic and neutral actinide compounds will be identified spectroscopically and substantiated by theoretical calculations.

REFERENCES

1. Willson, S. P., and Andrews, L., J. Phys. Chem. A 103, 3171 (1999).
2. Willson, S. P., and Andrews, L., J. Phys. Chem. A 102, 10238 (1998).
3. Willson, S. P., and Andrews, L., J. Phys. Chem. A 104, 1640 (2000).
4. Willson, S. P., Andrews, L., and Neurock, M., J. Phys. Chem. A 104, 3446 (2000).
5. Willson, S. P., Ph.D. Thesis, University of Virginia, Charlottesville, Virginia (1999).
6. Li, J., Bursten, B. E., Liang, B., and Andrews, L., Science 295, 2242 (2002).
7. Andrews, L., Zhou, M., Liang, B., Li, J., and Bursten, B. E., J. Am. Chem. Soc. 122, 11440 (2000).

NUCLEAR FUEL CYCLE

Theoretical and Experimental Research in Neutron Spectra and Nuclear Waste Transmutation on Fast Subcritical Assembly with MOX Fuel

D. A. Arkhipkin,[1] V. S. Buttsev,[1] S. E. Chigrinov,[2] R. Kh. Kutuev,[1] A. Polanski,[1] I. L. Rakhno,[2] A. Sissakian,[1] R. Ya. Zulkarneev,[1] and Yu. R. Zulkarneeva[1]

[1]Joint Institute for Nuclear Research, 141980 Dubna, Moscow region, Russia
[2]Radiation Physics and Chemistry Problems Institute, 220109 Sosny, Minsk, Belarus

Abstract. The paper deals with theoretical and experimental investigation of transmutation rates for a number of long-lived fission products and minor actinides, as well as with neutron spectra formed in a subcritical assembly driven with the following monodirectional beams: 660-MeV protons and 14-MeV neutrons. In this work, the main objective is the comparison of neutron spectra in the MOX assembly for different external driving sources: a 660-MeV proton accelerator and a 14-MeV neutron generator. The SAD project (JINR, Russia) has being discussed. In the context of this project, a subcritical assembly consisting of a cylindrical lead target surrounded by a cylindrical MOX fuel layer will be constructed. Present conceptual design of the subcritical assembly is based on the core with a nominal unit capacity of 15 kW (thermal). This corresponds to a multiplication coefficient, $k_{eff} = 0.945$, and an accelerator beam power of 0.5 kW. The results of theoretical investigations on the possibility of incinerating long-lived fission products and minor actinides in fast neutron spectrum and formation of neutron spectra with different hardness in subcritical systems based on the MOX subcritical assembly are discussed. Calculated neutron spectra emitted from a lead target irradiated by a 660-MeV protons are also presented.

INTRODUCTION

The fast subcritical assembly consists of a central cylindrical lead target surrounded with a cylindrical layer of mixed-oxide (MOX) fuel. As the first step in the studies of the characteristics of ADS, the "Pluton" project was proposed,[1–4] based on metallic weapon-grade plutonium fuel. But the results of the calculations have shown[5] that MOX fuel (25% PuO_2 + 75% UO_2) is better than metallic plutonium for this subcritical assembly. Present conceptual design of the subcritical assembly in Dubna (SAD) is based on a core with a nominal unit capacity of 15 kW (thermal). This corresponds to a multiplication coefficient, $k_{eff} = 0.945$, and an accelerator beam power of 0.5 kW. A blanket based on MOX fuel of the BN-600 type of Russian manufacture will be used, with an average density of fuel, 8.64 g/cm^3. The content of ^{239}Pu in PuO_2 is not less than 93%. Uranium in the oxide is a natural one. The fuel is placed in a stainless-steel vessel. In addition, beryllium and lead reflectors will be used in the radial and longitudinal directions, respectively. A concrete shielding surrounds the blanket. The proton beam will be transported horizontally to the target through a vacuum track provided inside the concrete shielding. The experimental electronuclear installation will include the following:

- 660-MeV proton accelerator,
- beam bending magnets,
- spallation target with different materials (Pb, W, Pb-Bi, Hg),
- subcritical blanket based on BN-600 type fuel elements,
- beryllium reflectors, concrete shielding, and
- protective control and measuring systems.

The installation will be placed in an experimental hall of the accelerator surrounded with a concrete wall with thickness not less than 2 meters. The hall of the accelerator is equipped with a special ventilation system for the

CP673, *Plutonium Futures — The Science,* edited by G. D. Jarvinen

control of radioactive aerosols. For transportation of the extracted beam of protons to blanket a transportation line will be created. To measure the spectra of the neutrons leaving the subcritical assembly, we propose to use a spectrometric method of neutron slowing down time in a block of lead.

The main parameters of the ADS facility are given in Table 1. The CASCADE, LAHET, and MCNP computer codes[6,7] have been used for the calculations.

TABLE 1. Main Parameters of the ADS Facility with the Lead Target.

Parameter	Value
Proton beam energy	660 MeV
Beam power	0.5 kW
k_{eff}	0.945
Energetic gain	30
Fission power	15 kW
Core length of a fuel element	30 cm
Core diameter (with Be reflector)	80 cm

RESULTS OF CALCULATIONS

In this work, our main objective is comparison of neutron spectra in the MOX assembly for different external driving sources: a 660-MeV proton accelerator and a 14-MeV neutron generator. It is well known that in the high-energy region (several GeV/A), nuclear reactions are described in the context of a three-stage mechanism, including cascade, preequilibrium, and equilibrium stages. Energy distributions of secondary particles formed in nuclear reactions in this region depend weakly on the projectile type and energy, the mass number of target nucleus, and the angle of emission. It is possible to extract preequilibrium and equilibrium stages in nuclear reactions induced by 14-MeV neutrons as well. In this case, energy distributions of secondary particles are similar to that observed for high-energy projectiles. When considering thick heavy-metal targets, one should take into account slowing down the particles and fission of nuclei in a wide-energy region. It was shown[8] that energy distributions of escaped neutrons for such targets bombarded with high-energy protons and 14-MeV neutrons are the same. This, in principle, offers the possibility of investigating the different characteristics of accelerator-driven subcritical systems by means of low-energy accelerators and neutron generators.

In Figure 1, calculated neutron spectra averaged over small volumes arranged along the radius of the subcritical assembly are presented. The three volumes are situated inside the core of the assembly in its middle (along the Z-axis) cross section. Several conclusions follow from the presented results: (i) neutron spectra inside the core are the same regardless of whether projectiles are relativistic protons or low-energy neutrons; (ii) different neutron spectra can be formed inside the assembly—from hard to almost thermal; (iii) neutron fluxes about 10^{12} $cm^{-2}s^{-1}$ can be obtained inside the core; (iv) neutron spectra calculated by means of the computer codes LAHET and SONET[7] are similar not only qualitatively but also quantitatively.

To describe quantitatively the scaling between protons and neutrons as projectiles, different calculated reaction rates are presented in Table 2. One can see from the table that the average sealing factor is equal to 6.6 with local deviations being approximately equal to calculated statistical uncertainty (2σ). This value can be easily understood when taking into account the number of neutrons generated in the lead target by a 660 MeV proton ($\cong$12) as well as the reaction (n,2n) for a 14-MeV neutron incident on the target.

Another interesting conclusion from the table is that for the actinides considered, the transmutation rates as a result of (n,γ) reactions significantly exceed those caused by fission reactions even for the hard spectrum observed near the boundary between the lead target and the core. In conclusion, one can state that such an assembly fueled with commercially available MOX fuel enables us to investigate nuclear waste transmutation in different neutron spectra by means of the currently available proton accelerators and neutron generators.

NEUTRON SPECTRA EMITTED FROM THE LEAD TARGET IRRADIATED BY 660-MEV PROTONS

This work is part of the JINR program for systematic research in the angular and energy distributions of the nucleons emitted from massive targets at energy of initial protons in the interval of (0.6 ÷ 2.0) GeV.[1,4] The research of the spectra is needed to design nuclear reactors of a new generation, with a subcritical assembly driven by external proton beams of high energy and intensity.

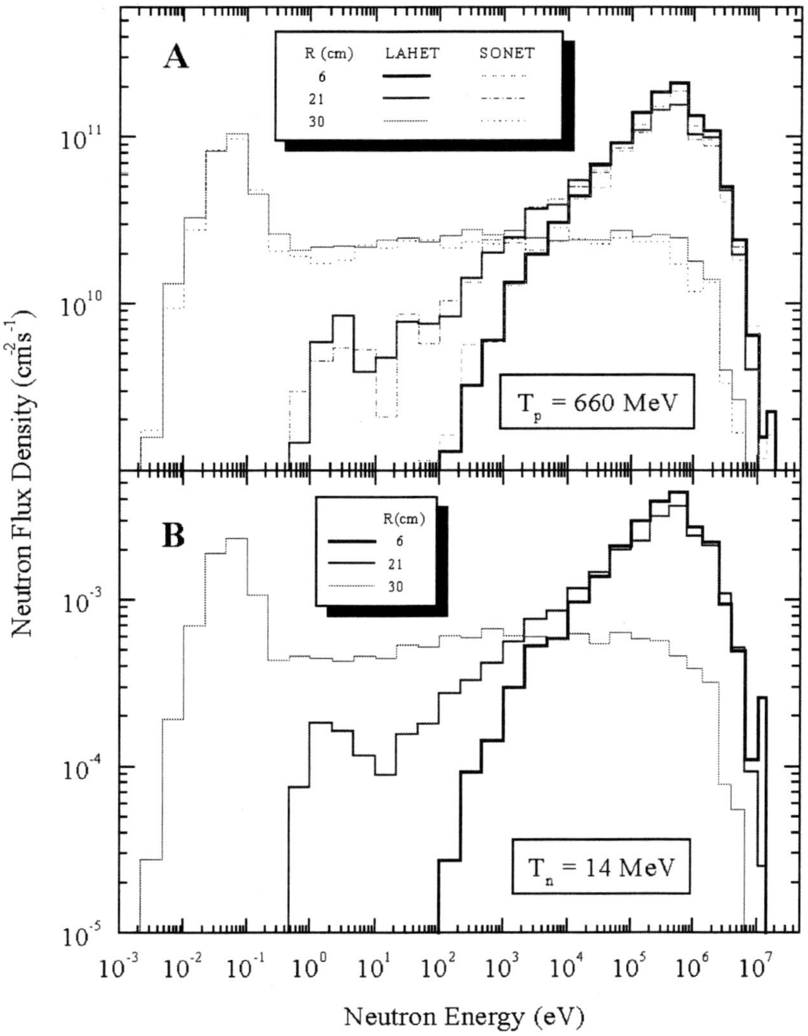

FIGURE 1. Calculated neutron spectra averaged over small volumes (1cm^3) arranged along the radius of the subcritical assembly with k_{eff} = 0.95, irradiated with monodirectional proton (A) and neutron (B) beams. Computer codes LAHET,[6] MCNP,[7] and SONET[9] were used. Normalization was performed per 1 μA proton current (a) and per one incident neutron per second (B)

TABLE 2. Calculated Reaction Rates for Two Positions inside the Subcritical Assembly Irradiated with Proton (A) and Neutron (B) Beams. Computer codes SONET and MCNP were used. Normalization was performed per one target nucleus and per one incident particle per second.

Reaction Rate (10^{-24} s^{-1})	T_p = 660 MeV R = 6 cm R = 30 cm	T_n = 14 MeV R = 6 cm R = 30 cm	(Reaction Rate)$_p$/ (Reaction Rate)$_n$ R = 30 cm
$^{99}Tc(n,\gamma)^{100}Tc$	2.0*	$2.9\cdot10^{-1*}$	6.7
$^{129}I(n,\gamma)^{130}I$	1.2	$1.6\cdot10^{-1}$	7.0
$^{135}Cs(n,\gamma)^{136}Cs$	$6.0\cdot10^{-1}$	$8.5\cdot10^{-2}$	7.1
$^{137}Cs(n,\gamma)^{138}Cs$	$6.0\cdot10^{-3}$	$9.4\cdot10^{-4}$	6.4
$^{237}Np(n,\gamma)^{238}Np$	$2.4\cdot10^{-1*}$ 10.2	$3.2\cdot10^{-2*}$ 1.5	7.0
$^{241}Am(n,\gamma)^{242}Am$	$2.9\cdot10^{-1**}$ 31.6	$6.1\cdot10^{-2**}$ 4.5	7.0
$^{243}Am(n,\gamma)^{244}Am$	$3.8\cdot10^{-1**}$ 9.0*	$6.5\cdot10^{-2**}$ 1.5*	5.9
$^{237}Np(n,f)$	$9.7\cdot10^{-2}$ $1.2\cdot10^{-2}$	$1.4\cdot10^{-2}$ $1.9\cdot10^{-3}$	6.3
$^{241}Am(n,f)$	$8.5\cdot10^{-2}$ $1.8\cdot10^{-1}$	$1.3\cdot10^{-2}$ $2.7\cdot10^{-2}$	6.8
$^{243}Am(n,f)$	$6.7\cdot10^{-2}$ $1.9\cdot10^{-2}$	$1.0\cdot10^{-2}$ $3.0\cdot10^{-3}$	6.2

Note: Statistical uncertainties (1σ) for most of the values in the table do not exceed 5%.
*) Statistical uncertainty is within the range 5%–10%.
**) Statistical uncertainty is within the range 20%–30%.

Here we present preliminary experimental results of research in the distribution of the neutrons emitted at 60° from a cylindrical lead target (∅80 × 300 mm^2) exposed to a 660 MeV proton beam.[9]

The measurements were done for neutron spectra energy ranging from 50–200 MeV. In the experiment, the neutrons emitted at 60° from the target were selected by collimator 1 (see Figure 2), anticoincidence counters A1,2, and then they were scattered by an additional CH_2 target (T).

Secondary products of the interaction were registered with a neutron detector system n, A2, charge particle counters P1, P2, and a total energy absorption detector NaI (Tl). All detectors were arranged in accordance with the elastic np-scattering kinematics for fixed energies of 50, 100, 150, and 200 MeV.

The results of the signal processing of these counters on flight time and amplitude allowed us to identify the registered particles by their mass, to determine the proton recoil energy, and to select those scattering events corresponding to the kinematics of elastic scattering. That allowed us to reconstruct the initial energy of the neutrons scattered by the target.

The results of the neutron spectrum measurement are shown in Figure 3 (by open circles). One can see a qualitative agreement between experimental data and those calculated with the GEANT code (solid line). However, the experimental errors are very large. We expect to continue our measurements and to reduce the experimental errors in the future.

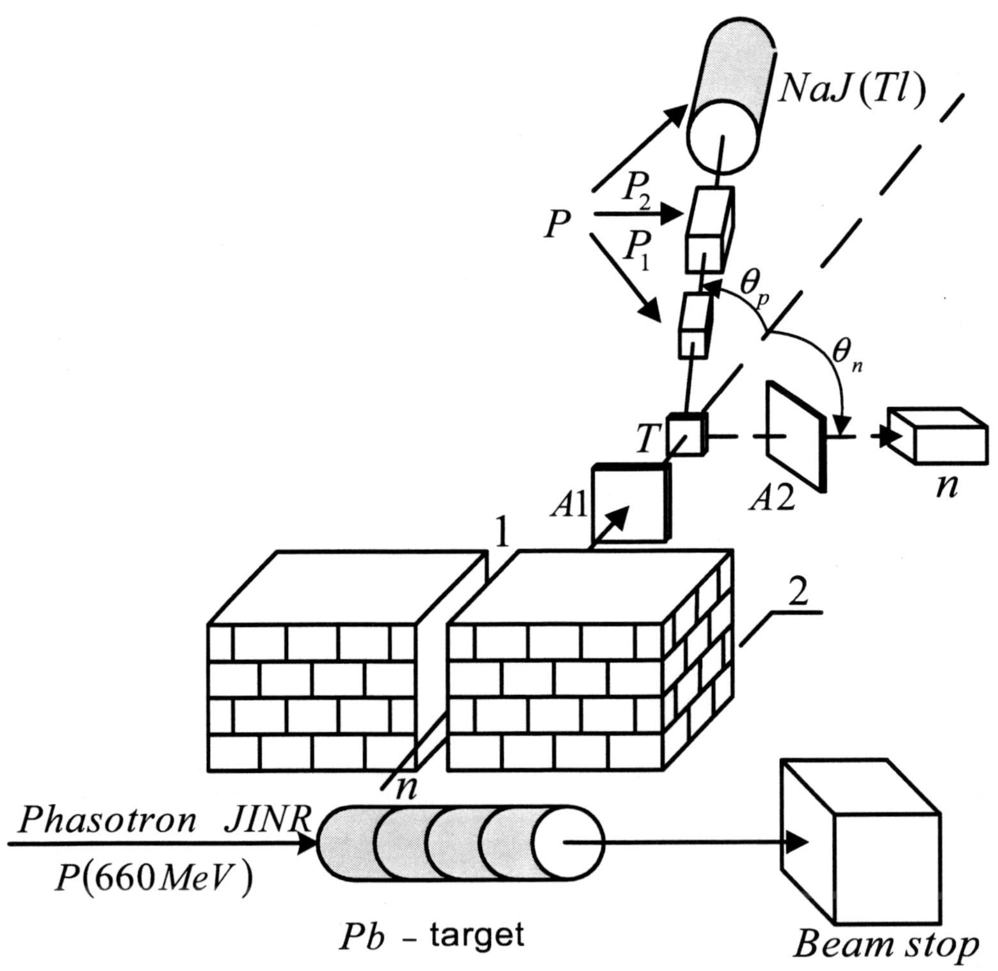

FIGURE 2. The experimental setup.

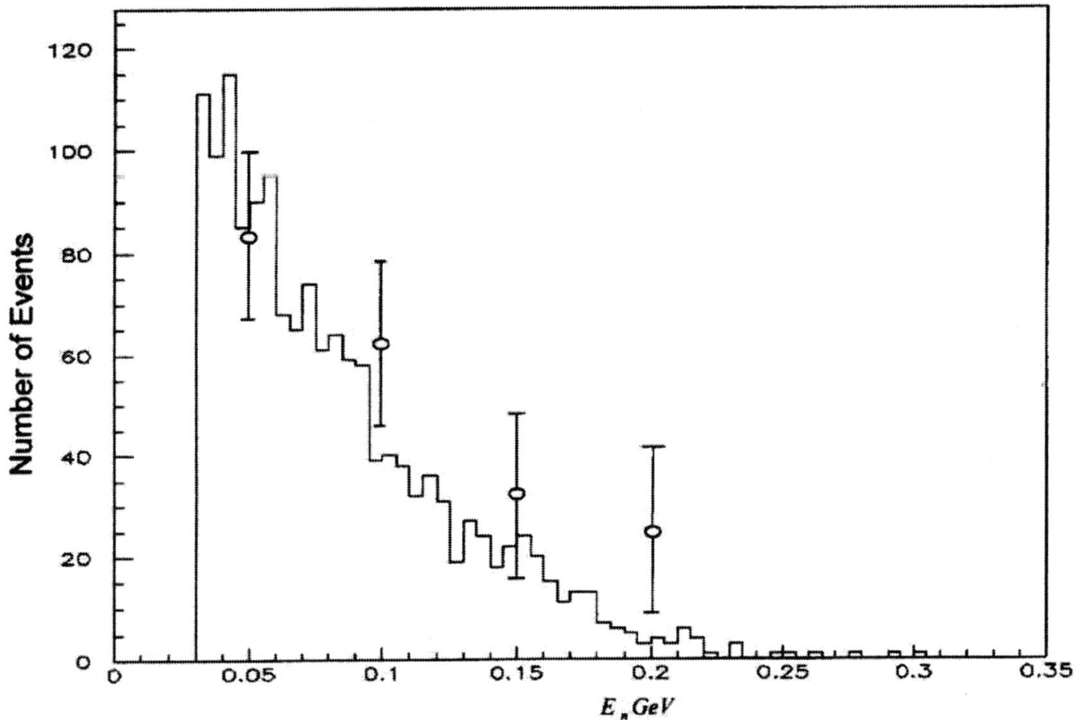

FIGURE 3. Energy spectrum of fast neutrons emitted at 60° from a lead extended target. Open circles: experiment results; histogram: GEANT simulation.

REFERENCES

1. Barashenkov, V. S., Buttsev, V. S., Buttseva, G. L., Dudarev, S. Ju., Polanski, A., Puzynin, I. V., and Sissakian, A. N., "Research Program for the 660-MeV Proton Accelerator Driven MOX-Plutonium Subcritical Assembly," Plutonium Futures—The Science, Santa Fe, New Mexico, USA, July 10–13, 2000, AIP Conference Proceedings 532, 194–197.
2. Arkhipov, V. A., Barashenkov, V. S., Buttsev, V. S., Chultem, D., Dudarev, S. Ju., Furman, V. I., Gudowski, W., Janczyszyn, J., Maltsev, A. A., Onischenko, L. M., Pogodajev, G. N., Polanski, A., Popov, Yu. P., Puzynin, I. V., Sissakian, A. N., and Taczanowski, S., "Accelerator Driven System Based on Plutonium Subcritical Reactor and 660-MeV Phasotron," Experimental Nuclear Physics in Europe, Sevilla, Spain, 1999, AIP Conference Proceedings 495, 478–481.
3. Arkhipov, A., Barashenkov, V. S., Buttsev, V. S. et al., "Research Program for the 660-MeV Proton Accelerator Driven Plutonium Subcritical Assembly, " Experimental Nuclear Physics in Europe (ENPE99), Sevilla, Spain, June 21–26, 1999, edited by AIP, pp. 478–481.
4. Barashenkov, V. S., Buttsev, V. S. et al., "Research Program for the 660-MeV Proton Accelerator Driven M0X-Plutonium Subcritical Assembly," Topical Conference on Plutonium and Actinides, Plutonium Futures—The Science, Santa Fe, New Mexico, USA, July 10–13, 2000, pp. 194–197.
5. Polanski, A., "Monte Carlo Modeling of Electronuclear Processes in Experimental Accelerator Driven Systems," XXVI Mazurian Lakes School of Physics, September 1–11, 1999, Krzyze, Poland, A NATO Advanced Research Workshop, September 2–4, 1999, Acta Phys. Polonica B 31, 95 (2000).
6. Chigrinov, S. E., Kievitskaia, A. I., Rakhno, I. L. et al., "Experimental and Theoretical Research on Transmutation of Long-Lived Fission Products and Minor Actinides in a Subcritical Assembly Driven by a Neutron Generator," 3rd Int. Conf. on Accelerator Driven Transmutation Technologies and Applications, June 7–11, 1999, Prague, Czech Republic (CD-ROM edition).
7. Chigrinov, S. E., Kievitskaia, A. I., Rakhno, I. L., and Rutkovskaia, C. K., "The Code SONET to Calculate Accelerator-Driven System Performance," 3rd Int. Conf. on Accelerator Driven Transmutation Technologies and Applications, June 7–11, 1999, Prague, Czech Republic (CD-ROM edition).
8. Prael, R. E., and Lichtenstein, H., "User Guide to LCS: The LAHET Code System," Los Alamos National Laboratory report LA-UR-89-3014 (1989).
9. "Phasotron at the Laboratory of Nuclear Problems, JINR, and its Beams," E-92-232, Dubna (1992).
10. "MCNP—A General Monte Carlo N-Particle Transport Code," edited by J. F. Briesmeister, Los Alamos National Laboratory report LA-12625-M (1997).

Behavior of Zircon-Based Ceramic Doped with ^{238}Pu under Self-Irradiation

B. E. Burakov, M. A. Yagovkina, and A. S. Pankov

Laboratory of Applied Mineralogy and Radiogeochemistry, the V.G. Khlopin Radium Institute, 28, 2-nd Murinskiy ave., St.Petersburg, 194021, Russia; fax: (7)-(812)-346-1129;e-mail: burakov@riand.spb.su

INTRODUCTION

Zircon, (Zr,Hf,Pu,...)SiO_4, has been proposed as a durable actinide host phase for the immobilization of weapons grade Pu and other actinides.[1–3] Some effects of zircon self-irradiation were studied using a polycrystalline sample doped approximately 8.85 wt % ^{238}Pu.[4] It was demonstrated recently[5] on the basis of ^{239}Pu-doped zircon samples that the loading of the zircon structure by Pu amounts higher than 6.9 wt % el. might cause partial Pu precipitation into the PuO_2 phase during zircon synthesis. It was decided to use the same equipment and conditions as for the synthesis of ^{239}Pu-doped zircon-based ceramic[5] to obtain homogeneous ^{238}Pu-doped polycrystalline zircon for the precise recurrent observation of radiation damage effects as a function of cumulative dose.

RESULTS AND DISCUSSION

Homogeneous methanol-aqueous solutions of $Si(OC_2H_5)_4$, Pu-nitrate and zirconyl-oxynitrate were prepared with Zr in excess of zircon stoichiometry. The experiment was designed to provide a final yield of more that 80 wt.% Pu-doped zircon, (Zr,Pu)SiO_4, with the remainder as Pu-doped zirconia, (Zr,Pu)O_2, thereby avoiding the formation of silica or plutonium silicate. Ammonium hydroxide was added into solution. The precipitated material was then dried, calcined, milled in an agate mortar, and cold pressed into pellets of 5 mm in diameter. Ceramic synthesis was done by sintering in air at 1,500°C for 3 hours. Synthesized samples were studied by scanning electron microscopy (SEM) and electron probe microanalysis (EPMA). In order to study radiation damage effects, XRD analyses of the ceramic were carried out after different cumulative doses. Modified MCC-1 leach tests were performed on the ceramic specimen, which was placed on Pt-foil in Teflon™ vessels with deionized water at 90°C for 28 days. Normalized Pu mass loss, NL(Pu), was calculated for MCC-1 leach tests as follows: $NL(Pu) = A \times (A_0)^{-1} \times W \times S^{-1}$, where: A is the total activity of Pu in solution after leaching, Bq; A_0 is the initial activity of Pu in the specimen, Bq; W is the initial mass of the specimen in grams; and S is the nominal surface area of the specimen without correction for porosity, m^2. The same sample was used for leach tests after different cumulative doses.

The results of the XRD analyses (Figure 1) have shown that the ceramic sample consists of 80 wt % zircon and 15 wt % tetragonal zirconia. No separated Pu phases were identified by methods of optical microscopy, SEM and XRD. The average chemical composition of zircon from EPMA is (in wt % el.): Zr-46.4; Si-13.5; Pu-5.7. Taking into account Pu isotope composition, this corresponds to 4.6 wt % ^{238}Pu content in the zircon phase. The sizes of zirconia grains in the ceramic were too small (less than 1–2 μm) to avoid analytical overlap with the zircon during EPMA analysis, precluding accurate compositional analysis. However, the zirconia phase undoubtedly incorporated some Pu amount that caused stabilization of the tetragonal crystalline structure instead of a monoclinic polymorph. Self-irradiation of zircon causes its amorphization; however, the complete destruction of the zircon structure was not achieved even after a relatively high cumulative dose of 1.13×10^{25} alpha decays/m^3 (Figure 1, spectrum 7). No evidence of matrix swelling or cracking was found by optical microscopy at the same cumulative dose. Tetragonal zirconia has demonstrated significantly higher resistance to radiation damage in comparison with zircon (Figure 1). Although after a cumulative dose of 3.1×10^{24} alpha decay/m^3, self-irradiation caused a significant increase of Pu

CP673, *Plutonium Futures — The Science,* edited by G. D. Jarvinen

release from the ceramic matrix in deionized water, this tendency then slowed down at a dose of 4.3 x 10^{24} alpha decay/m^3 and stopped at a dose of 6.6 x 10^{24} alpha decay/m^3 (Table 1).

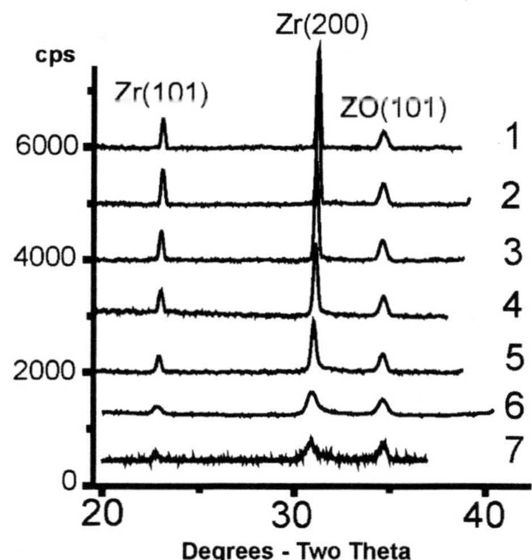

FIGURE 1. XRD patterns of ^{238}Pu-doped ceramic based on zircon (**Zr**) and the minor phase, tetragonal zirconia (**ZO**), after a cumulative dose (in alpha decay/m^3 x 10^{23}): (1) **3**; (2) **13**; (3) **30**; (4) **51**; (5) **65**; (6) **91**; and (7) **113**.

TABLE 1.

Cumulative Dose (alpha decay/m^3 x 10^{23})	NL(Pu) (g/m^2)	Equal Years of Storage Calculated for Zircon Doped with 4.6 wt % ^{239}Pu
7	0.005	30
31	0.036	150
43	0.046	210
66	0.041	330

The investigation of ^{238}Pu-doped zircon based ceramic is not yet completed. However, the results obtained so far allow us to confirm that Pu-doped zircon remains chemically durable after significant damage of its crystalline structure caused by self-irradiation.

The normalized Pu mass loss, NL(Pu) from the zircon-based ceramic after a 28-day test in deionized water at 90°C is dependent on cumulative dose (without correction for ceramic porosity). The ceramic density is 4.4 g/cm^3; the ^{238}Pu content is 4.6 wt % el.; the bulk Pu content in the zircon phase is 5.7 wt % el.

REFERENCES

1. Burakov, B., "A Study of High-Uranium Technogeneous Zircon $(Zr,U)SiO_4$ from Chernobyl "Lavas" in Connection with the Problem of Creating a Crystalline Matrix for High-Level Waste Disposal," Proc. SAFE WASTE'93, 13-18/06/1993, Avignon, France, **2**, 19–28 (1993).
2. Anderson, E. et al., "A Creation of Crystalline Matrix for Actinide Waste in Khlopin Radium Institute," Proc. SAFE WASTE'93, 13-18/06/1993, Avignon, France, **2**, 29–33 (1993).
3. Ewing, R. et al., "Zircon: A Host-Phase for the Disposal of Weapons Plutonium," J. Mat. Res. **10**, 243–246 (1995).
4. Weber, W., "Self-Radiation Damage and Recovery in Pu-Doped Zircon," Radiation Effects and Defects in Solids **115**, 341–349 (1991).
5. Burakov, B. et al., "Synthesis and Study of ^{239}Pu-Doped Ceramics Based on Zircon, $(Zr,Pu)SiO_4$, and Hafnon, $(Hf,Pu)SiO_4$," Mat. Res. Soc. Sym. Proc. **663** (2001).

Influence of Plutonium on the Dissolution Behavior of the Spent-Fuel Matrix

J. Cobos,* T. Wiss, T. Gouder, and V.V. Rondinella

*CIEMAT, A^vda Complutense 22, E-28040 Madrid, Spain. cobos@itu.fzk.de
European Commission, Joint Research Centre, Institute for Transuranium Elements,
Postfach 2340, 76125 Karlsruhe, Germany.

INTRODUCTION

The aim of this work is to characterize the oxidation state and the dissolution behaviour of Pu and U from the matrix of simulated spent nuclear fuel during storage in a geologic repository. The quantification of the effects of Pu on the dissolution of the fuel matrix in contact with groundwater will provide necessary information to establish a comprehensive prediction of the long-term corrosion behaviour of the fuel.

Pellets consisting of alpha-doped UO_2, i.e., UO_2 containing ~10 and ~0.1 $^w/_o$ ^{238}Pu, and unirradiated MOX containing ~10 $^w/_o$ ^{239}Pu, with the Pu homogeneously distributed in the matrix, were used in leaching experiments. Static sequential leaching tests were performed in demineralized water and granite groundwater at room temperature under oxic conditions. The leached surfaces were examined by x-ray photoemission spectroscopy (XPS), and scanning electron microscopy (SEM). The composition of the additive oxide, the fabrication method of the pellets, together with an extensive description of the experimental procedures and of the characterization techniques used[1-6] has been described previously. The evolution with time of relevant properties of the alpha-doped materials under the effect of accumulating alpha-decay damage was also reported elsewhere.[1-5]

RESULTS AND DISCUSSION

Figure 1 shows the uranium concentrations measured in the leachates as a function of leaching time. The values are normalized to the geometric surface area. The experiments in demineralized water show that the amount of uranium in the leachate for the Pu-containing materials was higher than that for undoped UO_2. Saturation and precipitation of U on the walls of the leaching vessels occurred, especially in the case of the material with the highest alpha activity. After longer leaching times, the amount of uranium found in solution from the two alpha-doped materials was converging independently from the type of leachant. In the case of the material containing ^{239}Pu, however, after the initial contact periods, the concentration of ^{238}U in solution remained stable. The pH remained ~7.5 during all tests.

XPS analysis of the oxidation state of the surface for U and Pu was performed for the samples leacheated in ground water (Figure 2). A relatively rapid increase of the O/M ratio (derived from the U(VI)/U(IV) ratio obtained by XPS[5,6]) was observed for UO_2-10%^{238}Pu. After leaching times longer than 10 h mostly U(VI) (and a very small peak for Pu) were detected on the surface of the UO_2-10%^{238}Pu. The trend of the U(VI)/U(IV) ratio for UO_2-0.1%^{238}Pu was characterized by a slower increase of the U(VI) component. The O/M evolution for UO_2-10%^{239}Pu was very similar to that for UO_2-0.1%^{238}Pu. As expected, the ratio for UO_2 did not progress beyond values corresponding to ~$UO_{2.33}$.[6-8]

The presence of carbonate in the groundwater had no major effects on the trends observed. The experiments indicated possible Pu effects on fuel dissolution.

CP673, *Plutonium Futures — The Science,* edited by G. D. Jarvinen

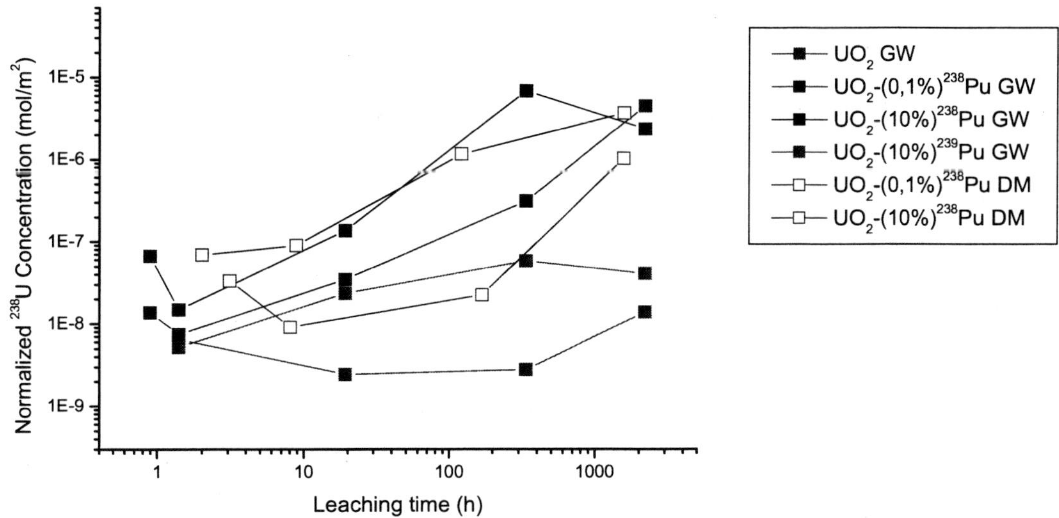

FIGURE 1. ^{238}U concentration in solution released during the leaching experiments in natural granite groundwater (GW) and demineralized water (DM).

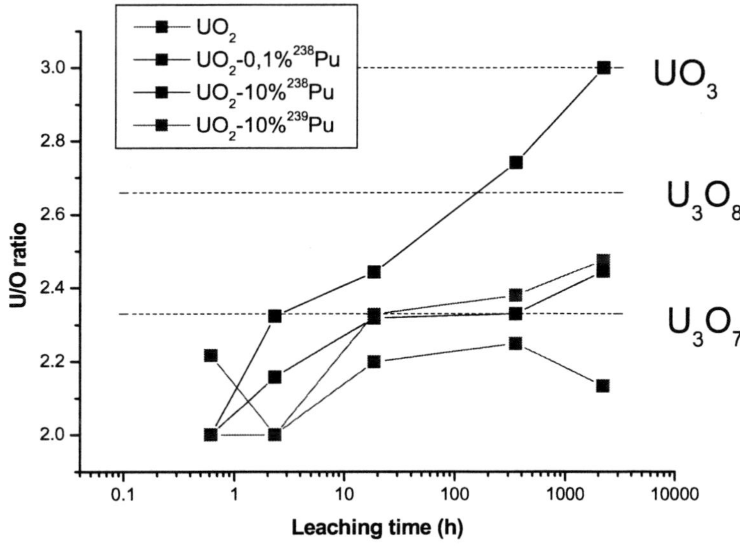

FIGURE 2. O/M ratio measured on the surface of the samples as a function of leaching time.

The alpha radiolysis of water caused enhanced dissolution as previously observed. However, a comparison between UO_2-0.1%^{238}Pu and UO_2-10%^{239}Pu, characterized by alpha-activity levels of the same order of magnitude, showed similar surface oxidation behaviour, but highlighted an apparent dissolution-inhibiting effect that may be related to the presence of a relatively high fraction of Pu in the fuel. More work is necessary to fully characterize such effect.

SEM characterization of surfaces leached in demineralized water was performed. The corrosion process during leaching caused the precipitation of uranium (VI) oxide hydrated crystals on the surface of the pellet, resulting in enhanced formation of studtite under high ^{238}Pu concentration and schoepite in the remaining cases. No precipitation of Pu-rich phases was observed. The surface of the reference UO_2 pellet leached under the same conditions presented only rare crystallites of schoepite, showing for the most part an essentially pristine surface. Dramatic alterations of the surface morphology occurred for the UO_2-10%^{238}Pu, with extensive grain boundary corrosion and relatively rapid covering of the surface by oxidized precipitated phases. The yellow crust formed on this specimen (Figure 3a) consisted of interwoven acicular crystallites sized about 5 μm in length. The oxidation to U(VI) in such conditions is known to produce Studtite $[UO_2|O|(OH)_2] - 3H_2O$.[9]

FIGURE 3. SEM micrographs of the a) UO_2-10%^{238}Pu surface after the longest leaching time showing precipitated radial aggregates constituting a crust and b) UO_2-0.1%^{238}Pu after the longest leaching time showing stellar (1), hedgehog (2) shaped aggregates and acicular crystallites (3) on the surface.

The UO_2-0.1%^{238}Pu did not show such extensive modifications. SEM examination revealed the presence of precipitated phases of various shapes and distribution, with radial aggregates up to ~10 μm of length, some of them being assembled in stellar shapes (Figure 3b(1)). Acicular crystallite (see Figure 3b(3)) could also be observed on the surface of this specimen as well as dense hedgehog-shaped aggregates of ~1 μm grains (see Figure 3b(2)).

The main constituents of these precipitates are dehydrated Schoepite crystallites ((UO_3-n(H_2O), n<2). The formation of the aggregates occurred preferentially at grain boundaries, but acicular crystallites were found distributed on the surface. This behaviour could be a combination of several factors.

REFERENCES

1. Rondinella, V. V., Matzke, Hj., Cobos, J., and Wiss, T., Mat. Res. Soc. Symp. Proc. 556, 447 (1999).
2. Rondinella, V. V., Cobos, J., Matzke, Hj., and Wiss, T., Radiochimica Acta, 88, 527–532 (2000).
3. Cobos, J., Matzke, Hj., Rondinella, V. V., Martinez-Esparza, A., and Wiss, T., Proc. GLOBAL '99 Int. Conf. on Future Nuclear Systems, August 29–September 3, 1999, Jackson Hole, USA, A.N.S.
4. Rondinella, V. V., Cobos, J., Matzke, Hj., Wiss, T., Carbol, P., and Solatie, D., Mat. Res. Soc. Symp. Proc. 663 (2001).
5. Cobos, J., Havela, L., Rondinella, V. V., de Pablo, J., Gouder, T., Glatz, J.-P., Carbol, P., and Matzke, H., Radiochimica Acta, 90, 597–602 (2002).
6. Cobos, J., Wiss, T., Gouder, T., and Rondinella, V. V., Mat. Res. Soc. Symp. Proc., Scientific Basis for Nuclear Waste Management XXVI, in press.
7. de Pablo, J., Casas, I., Giménez, J., Marti, V., and Torrero. M. E., J. Nucl. Mater. 232, 138–145 (1996).
8. Casas, I., Gimenez, J., Marti, V., Torrero, M. E., and de Pablo, J., Radiochimica Acta 66 and 67, 23 (1994).
9. Guilbert, S., Guittet, M. J., Barré, N., Gautier-Soyer, M., Trocellier, P., Gosset, D., and Andriambololona, J., J. Nucl. Mater. 282, 75–82 (2000).

Fabrication of Dispersed CERamic-CERamic and Ceramic-METallic pellets for the Transmutation of Actinides

A. Fernández, D. Haas, R. J. M. Konings, and J. Somers

European Commission, Joint Research Center
Institute for Transuranium Elements
Postfach 2340, 76125 Karlsruhe, Germany

Abstract. This paper describes the development of fabrication technology for target materials to be used in irradiation experiments, in the PHENIX and HFR reactors. Several target concepts will be tested: micro- as well as macrodispersed composites of $(Am,Y,Zr)O_2$ in MgO (cercer) and macrodispersed composites of $(Pu,Y,Zr)O_2$ in Stainless Steel (cermet) material. Results of the completed fabrication campaigns for cermet and cercer will be presented.

INTRODUCTION

Transmutation of long-lived radionuclides is a potential option for reducing the impact of the back-end of the nuclear fuel cycle. In recent years, different concepts have been proposed for fuels and targets for transmutation of (minor) actinides.[1] They can be categorized in terms of fuel material (metal, ceramic, or salt), fuel form (either homogeneous or heterogeneous composites) and fuel packing (pellet, particle, liquid salt). In all cases, however, the transmutation efficiency must be maximized, a condition best achieved if uranium-free fuels are considered.

The choice of the inert matrix for uranium-free fuels is a key aspect of the research, and a significant amount of work has been done in the past decade.[2,3] The actinide phase and the matrix can be combined in a homogeneous fuel form, in which the actinides form a solid solution with the matrix,[4] as is well known for uranium-plutonium mixed oxide fuels. However, most inert-matrix mixed oxides of this type generally have a relatively low thermal conductivity. This can be overcome if dispersions of actinide oxides in a ceramic (cercer) or metallic (cermet) uranium-free matrix are considered. Thus the matrix improves the thermal properties of the fuel, which either permits higher actinide loading, or lower operating temperatures. In addition, the irradiation-induced property changes in the fuel pellets are potentially minimized by localizing the fission damage in a limited geometric domain within the fuel.

The fabrication of composite pellets is considerably more difficult than solid-solution oxide pellets. This is a result of the specific requirements of size and homogeneous distribution of the dispersed actinide phase. At the Institute for Transuranium Elements (ITU), major efforts have been made to develop procedures to fabricate both cercer pellets (e.g. $(An,Y,Zr)O_2$ – MgO) and cermet pellets (e.g. $(An,Y,Zr)O_2$ – Stainless Steel). Particles containing the actinide phase are prepared by a combination of the external gelation, GSP,[5] and infiltration methods,[6] followed by mixing of the particles and the matrix powder by conventional blending methods. The effect of important parameters related to composite pellet fabrication such as the volume fraction of ceramic (actinide phase) and the size of the ceramic particles, have been systematically investigated.[7,8]

In a first step, highly porous spheres of a yttria-stabilized zirconium oxide $Y_{0.17}Zr_{0.83}O_{1.92}$ (YSZ), were produced by the external gelation method, which guarantees a solid solution as the final product. These YSZ spheres have a polydisperse size distribution in the 40- and 150-μm range, and a specific surface area of 67.2 $m^2 \cdot g^{-1}$. After their calcination at 1073 K in air, the YSZ spheres were sieved and specific size fractions (40–60 and 100–125 μm for cercer compounds and 60–80 μm, for cermet composites) were selected. Each size fraction was infiltrated with a plutonium nitrate solution (216 $g \cdot l^{-1}$) to reach the required metal content (i.e., Pu/(Zr+Y+Pu) equal to 0.20 for cercer

CP673, *Plutonium Futures — The Science,* edited by G. D. Jarvinen

and 0.24 for cermet composites). To reach these Pu concentrations, two infiltration steps were required with an intermediate thermal treatment at 1073 K for two hours in air to evaporate excess water and convert the infiltrated nitrate phase to the corresponding oxide. The final Pu content was determined by gravimetric analysis of the spheres before infiltration and after the calcination step.

The $(Pu,Y,Zr)O_2$ phase was mixed with the required amount of either MgO (CERAC M-1017) or stainless steel (ALFA AESAR, SS 316L type with the weight composition: Fe 67.5%, Cr17 %, Ni 13%, Mo 2.5%, and particle size 45μm). Following addition of zinc stearate as lubricant, the mixtures were compacted into pellets. Sintering was performed at 1923 K for 8 hours in argon for MgO cercer pellets and at 1548 K for 3 hours in argon for steel cermet pellets.

MgO-based cercer pellets with a Pu density of 0.7 $g{\cdot}cm^{-3}$ have been fabricated without cracks and with a random distribution of isolated spheres, regardless of the sphere size (Figure 1). A stainless-steel-based cermet fuel with a Pu density of 0.9 $g{\cdot}cm^{-3}$ has also been fabricated and characterized for an irradiation experiment in the HFR-Petten reactor (SMART). The final pellets have a density of 87.2 ± 0.2% TD. As the porosity is principally in the ceramic phase, accommodation of fission gases during irradiation is achieved and therefore swelling mitigated. The optical micrograph shown in Figure 2 indicates that all particles are uniformly distributed in the matrix.

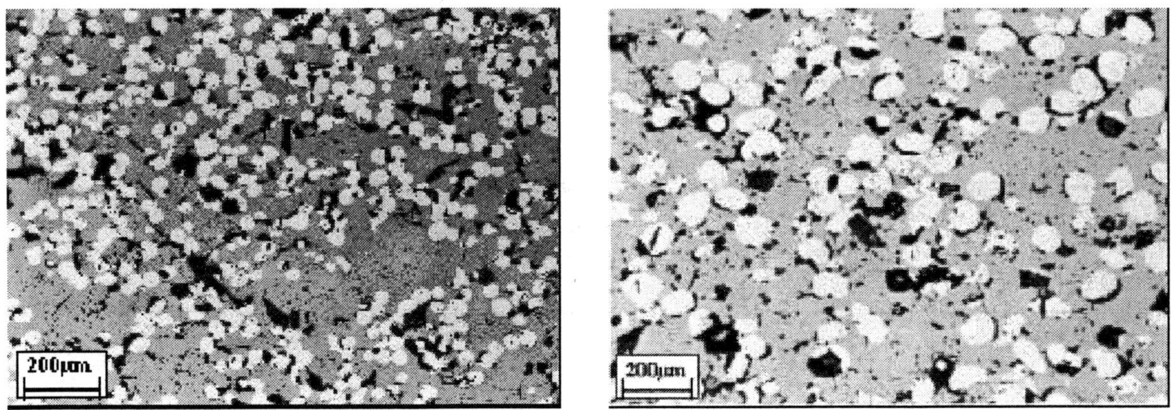

FIGURE 1. Optical microscopy of an axial section of a $Pu_{0.20}Y_{0.13}Zr_{0.67}O_2$-MgO pellet. Spheres size (a) 40–60 μm (b) 100–125 μm.

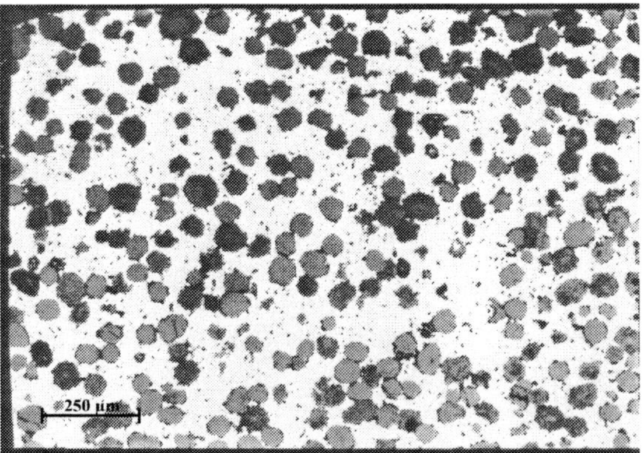

FIGURE 2. Optical microscopy of an axial section of a $Y_{0.13}\ Pu_{0.24}Zr_{0.63}O_2$-SS pellet. Spheres size 60–80 μm.

CONCLUSIONS

The results presented in this paper have demonstrated the feasibility of the fabrication of $(Pu,Y,Zr)O_{2-x}$-based magnesia and stainless steel composite pellets, by a combination of GSP, infiltration, and conventional blending techniques. The main advantage of this fabrication method is its high flexibility for selecting the particle size and volume fraction of the ceramic phase. Materials exhibiting a homogeneous dispersion of the actinide ceramic-bearing phase have been produced. The actual fabrication of Am-containing materials will be performed in a set of specially constructed shielded cells, which are under construction at present.

ACKNOWLEDGMENTS

The authors wish to acknowledge Y. Croixmarie for the fruitful discussion and A. Accarier, C. Boshoven, H. Hein, M. Holzhäuser, F. Naisse, and R. Voet for their supporting activities in this project.

REFERENCES

1. Fernández, A., Haas, D., Konings, R. J. M., and Somers, J., Fuel/Target Concepts for Transmutation of Actinide," in 6th Information Exchange Meeting, 11–13 December 2000, Madrid.
2. Warin, D., Conrad, R., Haas, D., Heusener, G., Martin, P., Konings, R. J. M., Schram, R. P. C., and Vambenepe, G., "10 YEARS EFTTRA: 1992–2001, GLOBAL'2001" (on cd-rom).
3. Chauvin, N. et al., "In Pile Studies of Inert Matrices with Emphasis on Magnesia and Magnesium Aluminate Spinel," J. Nucl. Mater. 274, 91–97 (1999).
4. Fernández, A., Haas, D., Konings, R. J. M., and Somers, J., "Transmutation of Actinides," J. Am. Ceram. Soc. 85(3), 694–696 (2002).
5. Fernández, A., Richter, K., and Somers, J., "Preparation of Spinel ($MgAl_2O_4$) Spheres by a Hybrid Sol-Gel Technique," Advances in Science and Technology 15, 167–174 (1999).
6. Richter, K., Fernández, A., and Somers, J., "Infiltration of Highly Radioactive Materials: A Novel Approach to the Fabrication of Targets for the Transmutation and Incineration of Actinides, J. Nucl. Mater. 249, 121 (1997).
7. Fernández, A., Konings, R. J. M., and Somers, J., "Design and Fabrication of Specific Ceramic-Metallic Fuels and Targets," J. Nucl. Mater. (submitted, 2003).
8. Croixmarie, Y., Abonneau, E., Fernández, A., Konings, R. J. M., Desmouliere, F., and Donnet, L., "Fabrication of Transmutation Fuels and Targets: The ECRIX and CAMIX-COCHIX Experience," J. Nucl. Mater. (submitted, 2003).

Plutonium Distribution in the Presence of Hydroxamic Acids

O. Danny Fox and Robin J. Taylor

[1]*BNFL Technology Centre, Research and Technology, BNFL Sellafield, Seascale, CA20 1PG, UK*

The separation of Pu from U in spent fuel processing by the reductive stripping of Pu(IV) from a TBP solvent phase is a very well established technology. Typical reducing agents include ferrous sulphamate, uranous nitrate/hydrazine, and hydroxylamine. Much fundamental data are available, and plants using these processes have been successfully operated for many years. Current Purex reprocessing plants use several solvent extraction cycles to purify the U and Pu products.[1] Requirements for the development of future Purex processes for LWR and fast reactor fuel processing, or waste conditioning (e.g., the UREX process[2]) include the use of single-cycle flowsheets, intensified (centrifugal) contactors and the flexibility to meet different scenarios (e.g., coprocessing or low DFs on the Pu stream).[3] In such "advanced" flowsheets, one option is to replace the reductive stripping of Pu by complexation—Pu(IV) is selectively complexed by a hydrophilic ligand and stripped in to the aqueous phase. The advantages of complexation include fast kinetics, temperature insensitivity compared to redox reactions and no reoxidation of Pu(III); therefore, there is no need for a stabilizer and likely improved criticality control.[3] Indeed, work on Pu and Np stripping using complexants, namely, hydroxamic acids, has been undertaken by BNFL for over a decade.

Previous papers have reported the applications of hydroxamic acids in Advanced Purex flowsheets for both Np and Pu control and also various aspects of their process chemistry (see Reference 4 and references therein). Hydrophilic hydroxamic acids (XHA) are of particular interest because of their ability to selectively complex tetravalent actinides in HNO_3, but also they are redox active, capable of reducing Np(VI) and Pu(VI, V, IV).

Recently, Laidler and coworkers have reported progress made on the development of the UREX process.[2] This is part of a proposed spent-fuel management strategy and utilizes acetohydroxamic acid to remove Np and Pu from spent fuel, directing them to the highly active raffinate. The objective is to meet criteria for reclassifying the U, which comprises the bulk of the material, as lower-level waste. Also, applications of hydroxamic acids in spent processing are being considered by Russian[5] and Chinese[6] researchers.

Cost-effective flowsheet development and optimization requires a modeling and simulation capability. Process models need to incorporate all relevant chemical and physical parameters, including solvent extraction algorithms to describe Pu distribution between aqueous and solvent phases and chemical kinetics.

This paper reports recently measured Pu(IV)—XHA distribution data at high Pu concentrations, particularly relevant to fast reactor fuel processing and further progress made in understanding and modelling the reduction of Pu ions by hydroxamic acids.

REFERENCES

1. Denniss, I. S., and Jeapes, A. P., "Reprocessing Irradiated Fuel," in The Nuclear Fuel Cycle, edited by P. D. Wilson, Oxford Science Publications, Oxford (1996).
2. Laidler J. J. et al., "Chemical Partitioning Technologies for an ATW System," Progress in Nuclear Energy 38, 65 (2001).
3. Taylor, R. J., Denniss, I. S., and Wallwork, A. L., Nuclear Energy 36, 39 (1997).
4. Taylor, R. J., Proc. Actinides 2001 Conference, J. Nucl. Sci. Tech., to be published.
5. Zilberman, B. Ya. et al., Proc. Int. Symp. NUCEF 2001—Scientific Bases for Criticality Safety, Separation Process, and Waste Disposal," JAERI-Conf 2002-004, edited by T. Banba and Y. Tsubata, Japan Atomic Energy Research Institute (2002).
6. Jiang, H., Chang, Z. Y., Pan, Y. J., and Zhu, J. M., J. Nucl. Radiochem. 22, 1 (2000).

CP673, *Plutonium Futures — The Science,* edited by G. D. Jarvinen

Fundamental Thermodynamics of Actinide-Bearing Mineral Waste Forms

Ubaldo F. Gallegos, Theresa A. Lee, and Luis A. Morales

Los Alamos National Laboratory
MS G730 Los Alamos National Laboratory, Los Alamos, NM 87545

The end of the Cold War raised the need for the technical community to be concerned with the disposition of surplus nuclear weapon material. The United States Department of Energy has determined that of 50 tons of surplus weapons plutonium and clean metal; 17 tons will be incorporated into a ceramic material and then placed in a geologic repository.[1] The form of that ceramic material is a solid solution between four end-member oxide phases: $CaHfTi_2O_7$, $CaUTi_2O_7$, $CaPuTi_2O_7$, and $Gd_2Ti_2O_7$.[2] The stability and behavior of plutonium in the ceramic materials as well as the phase behavior and stability of the ceramic material in the environment is not well established. Recent studies into the fundamental thermodynamics of actinide substitution into surrogates of these phases have begun to provide a basis for technically sound solutions to the issue of a safe, secure, and environmentally acceptable waste material.[3] However, the stability and behavior of plutonium in the proposed ceramic and end-member materials have only been estimated. With the recent installation at LANL (CMR wing 2) of the world's only high-temperature oxide-melt solution calorimeter, work on actual plutonium-containing materials may be conducted.

In order to provide technically sound solutions to these issues, thermodynamic data are essential in developing an understanding of the chemistry and phase equilibria of the actinide-bearing mineral waste forms proposed as immobilization matrices. Mineral materials of interest include zircon, zirconolite, and pyrochlore. High-temperature solution calorimetry is one of the most powerful techniques, sometimes the only technique, for providing the fundamental thermodynamic data needed to establish optimum material fabrication parameters, and more importantly, to understand and predict the behavior of the mineral materials in the environment.[4] The purpose of this project is to experimentally determine the enthalpy of formation of actinide-containing mineral phases and develop an understanding of the bonding characteristics and stability of these materials.

This report summarizes work after completion of a three-year project and our current ongoing efforts. Research efforts at UC Davis have focused on establishing the thermodynamic properties of zirconolite and pyrochlore and the synthesis of other minerals relevant to storage of nuclear material. Heat capacity, entropy, enthalpy of formation, and free energy of formation data were established for zirconolite, $CaZrTi_2O_7$, in the range from 0 to 1,500 K. The heat capacity, entropy, enthalpy of formation, and free energy of formation at 298 K for zirconolite are 211.9 J/K mol, 193.3 J/K mol, –3,713.8 kJ/mol, and –3,514.6 kJ/mol, respectively. Solution calorimetry experiments with cerium pyrochlore, $Ca_{0.8}Ce_{1.2}Ti_2O_7$, are complete. Heat capacity data and confirmation of the pyrochlore composition are required for final data analysis. Synthesis and characterization of $CaHfTi_2O_7$, $CaZr_{0.5}Hf_{0.5}Ti_2O_7$, $Gd_2Ti_2O_7$, and $CeTi_2O_6$ are complete.

Research efforts at Los Alamos have focused on establishing synthesis techniques for actinide-bearing minerals and preparation of the calorimetry laboratory. The preparation of Pu-pyrochlore, nominally $CaPuTi_2O_7$, has been achieved by Ebbinghaus at Lawrence Livermore National Laboratory. A sample of this material has been sent to Putnam at Los Alamos National Laboratory.

FUTURE OPPORTUNITY AND OBJECTIVES

Putnam et al.[3] recently completed a three-year study of the formation energetics of phases related to the ceramic disposition form that has been suggested for surplus weapons plutonium, with the notable exception of Pu-bearing

CP673, *Plutonium Futures — The Science*, edited by G. D. Jarvinen

materials. Because of radiological considerations, surrogates for the plutonium-bearing phases were used to estimate the thermodynamic stability of these key materials, specifically Pu-pyrochlore and Pu-zircon. The estimates are crude and contain large errors.[3] Actual measured values would greatly enhance the models used to forecast the waste form performance in a waste repository, as well as provide additional data for refining the models used to estimate the stabilities of other actinide-containing materials. Ebbinghaus at Lawrence Livermore National Laboratory has recently succeeded in synthesizing a nearly phase pure sample of $CaPuTi_2O_7$, Pu-pyrochlore. This sample is an excellent candidate for the determination of its thermodynamic stability using high-temperature oxide-melt solution calorimetry. LANL is in the unique position to provide an actual measured value of the formation enthalpy of $CaPuTi_2O_7$ because of the installation of a calorimeter capable of studying actinide-bearing materials. This high-temperature oxide-melt solution calorimeter is the world's first such instrument with actinide capabilities. This instrument provides all of the infrastructure needed to safely and accurately determine the fundamental thermodynamic quantities needed to certify proposed ceramic materials as technically sound solutions to the issue of a safe, secure, and environmentally acceptable waste material for surplus weapons plutonium, pursuant to DOE's record of decision.[1]

REFERENCES

1. U.S. DOE ROD 2000, "Record of Decision for the Surplus Plutonium Disposition Final Environmental Impact Statement" (January 2000).
2. Ebbinghaus, B. B., Armantrout, G. A., Gray, L., Herman, C. C., Shaw, H. F., and Van Konynenburg, R. A., "Fissile Materials Disposition Program, Plutonium Immobilization Project; Baseline Formulation" (September 2000).
3. Putnam, R. L., Ebbinghaus, B. B., Navrotsky, A., Helean, K. B., Ushakov, S. V., Woodfield, B. F., Boerio-Goates, J., and Williamson, M. A., Ceramic Transactions 119, edited by D. R. Spearing, G. L. Smith, and R. L. Putnam, American Ceramic Society, Westerville, Ohio (2001).
4. Navrotsky, A., Phys. Chem. Min. 24, 222–241 (1997).

In-situ Chlorination for Molten-Salt Extraction

Devin Gray

Los Alamos National Laboratory, Los Alamos, NM 87545

Los Alamos National Laboratory uses a novel approach to perform molten-salt extraction. Previously, an oxidant such as magnesium chloride was used with a cover salt like calcium chloride. Later developments used plutonium trichloride (generated in a separate process) with less calcium chloride cover salt. Currently, chlorine gas is mixed with argon gas and bubbled into liquid plutonium metal, producing the plutonium trichloride. The in-situ generated plutonium trichloride salt equilibrates in the melt, acts as the cover salt, and it preferentially reacts with any americium metal producing the desired product, americium chloride salt. The americium salt and other metal salts will separate from the plutonium metal upon solidification, achieving the desired purification, which is the basis of molten-salt extraction. The current status and future plans will be discussed.

CP673, *Plutonium Futures — The Science,* edited by G. D. Jarvinen

Enthalpies of Formation of Cerium Zirconate: (Ce,Zr)O_2 Fluorite and $Ce_2Zr_2O_7$ Pyrochlore

K. B. Helean,[1] S. V. Ushakov,[1] C. E. Brown,[1] A. Navrotsky,[1] J. Lian,[2] R. C. Ewing,[2] T. Lee,[3] and R. Haire[4]

[1]*Thermochemistry Facility, Department of Chemical Engineering and Materials Science, The University of California at Davis, Davis CA*
[2]*Department of Nuclear Engineering and Radiological Sciences, The University of Michigan, Ann Arbor MI*
[3]*Los Alamos National Laboratory, Los Alamos, NM*
[4]*Oak Ridge National Laboratory, Oak Ridge TN*

Zirconium-doped ceria, $Zr_xCe_{1-x}O_2$, with the fluorite (F*m3m*) structure has many technologically important applications due to its refractory nature and high oxygen ion conductivity. When doped with transition metals such as Co, Cu, Ni or Mo, Zr-doped CeO_2 acts as a catalyst for the reduction of SO_2 by CO.[1] Zirconium doping has the effect of stabilizing smaller CeO_2 crystallite sizes (<10 nm) and increases the material's resistance to sintering.[1] The result is a higher reactive surface area. Ceria reduces at high temperatures with significant reduction reported at >700°C. When the cerium is reduced, the zirconium and cerium order, forming the fluorite derivative pyrochlore (F*d3m*) structure.[2] Pyrochlore-structured titanate materials comprise three of the four end-members of the ceramic proposed for plutonium immobilization.[3] The substitution of zirconium for titanium has been shown to dramatically increase the radiation-damage resistance of REE-pyrochlore.[4,5] During heavy ion bombardment, REE-zirconate pyrochlore transforms to fluorite and then resists amorphization.[4,5] Despite the relative importance of these materials, virtually no durability data, chemical or thermal, exists for zirconia-ceria solid solutions. Therefore, a thermochemical investigation of (Ce,Zr)O_2 fluorite and $Ce_2Zr_2O_7$ pyrochlore is warranted.

High-temperature oxide melt solution calorimetry was used to measure drop solution enthalpies of the cerium zirconate samples plus their binary oxide components.

The samples were prepared at ORNL and kept under vacuum in glass vials and handled for calorimetry in an Ar-filled glove box to prevent hydration. Two solvents were used in this study, $3Na_2O \bullet 4MoO_3$, at 976 K and $2PbO \bullet B_2O_3$ at 1078 K. Drop solution enthalpies, ΔH_{ds}, were measured by dropping pellets (≈5 mg) of the powdered samples from room temperature into the solvent at calorimeter temperature. Oxygen was bubbled through the melt to aid in the dissolution of the pellets and to provide high oxygen fugacity that ensured the oxidation of Ce^{3+} to Ce^{4+}. The measured values of drop solution enthalpies were used in the appropriate thermodynamic cycles to calculate the enthalpies of formation from the oxides.

The drop solution enthalpy for $Ce_2Zr_2O_7$ in sodium molybdate at 976 K, ΔH_{ds} (kJ/mol) = –33.9 ± 1.5. This value is more endothermic than a trend of other REE-zirconate pyrochlore data would predict. This suggests that despite our best handling efforts, the sample was hydrated. Steps to dehydrate the pyrochlore are underway and new calorimetric data will be presented.

$\Delta H_{f\text{-}ox}$ (kJ/mol), at 298 K for (Ce,Zr)O_2 was derived using data collected in both sodium molybdate and lead borate solvents giving +15.2 ± 1.7 and +18.6 ± 1.7, respectively. The slightly higher $\Delta H_{f\text{-}ox}$ derived from lead borate data may be a result of the greater dissolution difficulties of Ce-containing materials. (Ce,Zr)O_2 is unstable in enthalpy with respect to an assemblage of its binary oxides. This is consistent with observations of zirconia segregation from (Ce,Zr)O_2 after heat treatment at 1,000°C for one hour.[1]

CP673, *Plutonium Futures — The Science*, edited by G. D. Jarvinen

REFERENCES

1. "Advanced Byproduct Recovery: Direct Catalytic Reduction of Sulfur Dioxide to Elemental Sulfur," report to DOE, Arthur D. Little, Inc. reference 54177 (March 1997).
2. Subramanian, M. A., Aravamudan, G., and Rao, G. V. S., Prog. Solid State Chem. 15, 55 (1983).
3. Helean, K. B. et al., J. Nuclear Materials 303, 226 (2002).
4. Lian, J., Wang, L. M., Wang, S. X., Chen, J., Boatner, L. A., and Ewing, R. C., Phys. Rev. Lett. 87, 145901 (2001).
5. Lian, J., Zu, X. T., Kutty, K. V. G., Chen, J., Wang, L. M., and Ewing, R. C., Phys. Rev. B 66, 054108 (2002).

Chop-Leach Dissolution of Commercial Reactor Fuel

G. F. Kessinger and M. C. Thompson

773-A, C-140, Savannah River Technology Center, Aiken, SC 29803

INTRODUCTION

The Advanced Fuel Cycle Initiative (AFCI) program[1] is being developed to address the disposal of commercial spent nuclear fuel and improve the performance of the geologic repository. To decrease the volume of waste sent to the repository, commercial fuel would be processed and separated into multiple components: a transuranic (TRU) stream to be converted to AFCI fuel and transmuted by fissioning to generate electrical power; separate ^{99}Tc and ^{129}I streams to be converted into targets for transmutation to short-lived nuclides; a U stream that meets the criteria for Class C low-level waste; and a high-level waste stream for repository disposal.

To achieve the AFCI separation objectives, new and innovative processes are under development. One such process is the uranium extraction (UREX) process.[2] The UREX process is based on the well-known PUREX process (Pu extraction from irradiated U); however, in UREX, U and Tc are extracted into the organic phase (U and Tc are subsequently stripped into separate streams), but the other actinides and nonvolatile fission products reside in the aqueous raffinate.

The primary goal of the present work was to demonstrate the use of the chop-leach (with nitric acid) process to produce a feed solution for the UREX process (nominally 290 g/L U and 1 M H^+). The secondary goal was to test the efficacy of this treatment for leaching actinides from Zircaloy cladding.

RESULTS

The fuel, approximately 4.5 kg of Zircaloy 2 clad, UO_2 fuel (23,480 MWD/MT burnup) from the Dresden reactor (Morris, IL), consisted of rubble and 3-in. lengths of clad fuel. It was remotely dissolved in a glass dissolver (as three discreet batches). Each fuel batch was charged into a 1 L heel of 4 M HNO_3. After the initial temperature spike (due to the UO_2-HNO_3 reaction) subsided, the temperature was maintained near 363 K for 8–12 hours by the addition of heat and/or 10 M HNO_3 until dissolution was complete. After each of the first two dissolutions, the cladding was leached for 2–6 hours near 363 K in the 1 L of heel acid added for the next dissolution. After the third dissolution, the cladding was heated in the dissolver product solution for about 7 hours in an attempt to further leach fuel components from the cladding; 711 g of cladding were recovered subsequent to the three dissolutions. Analysis of a cladding sample (Table 1) showed that actinides are present in the cladding at a concentration of about 5000 ηCi/g.

The combined dissolver product was filtered and 341 g of finely divided residue were recovered; ICP-MS analysis suggested this material was roughly 25% U and 1% Pu. The residue was heated to near 363 K with 140 mL of 10 M HNO_3 in a stainless-steel beaker for several hours while stirring. There was no visible sign of reaction during heating, and extensive solids were still present, so KF and 0.5 M HNO_3 were added to form 500 mL of solution, and the mixture was again heated and stirred. This treatment resulted in a solution 424 g/L in U and 17.94 g of filtered residue. The leachate was combined with the products of the three batch dissolutions to produce 7.75 L of solution, which was over 450 g/L in U and 1.45 M in H^+. This combined solution was diluted to produce about 13 L of UREX feed 302 g/L in U and 0.84 M in H^+ (Table 2).

CP673, *Plutonium Futures — The Science,* edited by G. D. Jarvinen
2003 American Institute of Physics 0-7354-0140-3

TABLE 1. Chemical Composition of Zircaloy 2 Cladding (excluding Zr).

Element	Result/(µg/g)
Ag	5340
Al	786
B	5050
Ba	35.4
Ca	<20.2
Cd	<3.70
Cm-244	0.00305
Ce	4380
Co -60	0.00248
Cr	815
Cs-137	1.87
Cu	72.5
Eu-154	0.00426
Fe	1500
Gd	174
K	<826
La	76.3
Li	<37.9
Mg	57.8
Mn	21.0
Mo	461
Na	3410
Ni	<11.9
P	<60.7
Pb	<28.1
Pu-239/240	4.08
Pu-238/Am-241	0.292
Sb	<333
Sb-125	0.00130
Si	44000
Sn	13000
Sr	423
Ti	28.4
U	623
Zn	25.6

TABLE 2. Chemical Composition of Diluted Feed Solution.

Component	Result/(g/L)
H^+	0.85
U	302
Pu	2.16
Np	6.8 E-05
Am	0.23
Cm	0.005
^{99}Tc	1.68 E-04
^{137}Cs	1.31
^{90}Sr	0.070
^{154}Eu	3.87 E-04
Ag	0.78
Ba	0.50
Ce	1.47
Cu	0.37
Fe	0.07
Gd	0.64
La	0.49
Mn	0.02
Na	0.33
Ni	0.60
Sn	0.41
Ti	0.11
Zn	0.05

DISCUSSION

Although the primary goal of this study was realized, issues that require further study were also identified. These include the secondary goal of actinide removal from the Zircaloy cladding and the large amount of insoluble residue recovered subsequent to the dissolution experiments. The cladding sample analyzed was roughly 50 times too high in actinide content to qualify as an LLW. It is possible that the low HNO_3 concentration in the dissolver product (it was initially 4–10 M in H^+, but after dissolution it was 1.45 M in H^+) did not effectively leach the actinides from the cladding. Furthermore, after the HNO_3 dissolutions the amount of insoluble residue was much greater than expected. Analysis of this residue showed that it was high in U and Pu. Treatment with the mixed F^-/NO_3^- media resulted in dissolution of more than 90% of the insoluble material (which included roughly 200 g U).

The greater-than-expected mass of insoluble residue and the amount of actinides present in the cladding suggest that more aggressive chemistry may be required to completely dissolve the U in the fuel and to extract actinides from the cladding. Future development work on this method should be aimed at addressing these two issues.

REFERENCES

1. Laidler, J. J., Burris, L., Collins, E. D., Duguid, J., Henry, R. N., Karell, E. J., McDeavit, S. M., Thompson, M., Williamson, M. A., and Willit, J. L., Prog. Nucl. Ener. 38, 65–79 (2001).
2. Thompson, M. C., Norato, M. A., Kessinger, G. F., Pierce, R. A., Rudisill, T. S., and Johnson, J. D., "Demonstration of the UREX Solvent Extraction Process with Dresden Reactor Fuel Solution," WSRC-TR-2002-00444, Revision 0, Westinghouse Savannah River Company (September 30, 2002).

An Optimization Study on a Fast Reactor Core Design for Pu and Minor Actinides Transmutation

Yong Nam Kim,[1] Jong Kyung Kim,[1] and Won Seok Park[2]

[1]Hanyang University, 17 Haengdang, Sungdong, Seoul 133-791, Korea, captain@nural.hanyang.ac.kr
[2]Korea Atomic Energy Research Institute, 150 Dukjin, Yusong, Taejon 305-353, Korea

In many countries and organizations, various concepts are being proposed for TRU (plutonium and minor actinides) transmutation. The fast neutron spectrum is being regarded as superior to the thermal spectrum because of its own attributes of the destruction capability of TRU nuclides through direct fission. In the pervious study, the sensitivities of neutron spectrum and core performances to fuel pin-cell configurations of lead-bismuth cooled core were investigated.[1] The results revealed that core performances are highly dependent on the neutron spectrum shift caused by changes of various core design options. From this result, we can consider a question still open. How much hardened spectrum meets the best performance to incinerate TRU safely and efficiently? To say in other words, even though the liquid metal cooled core has been selected for a fast neutron spectrum as an optimal option, another decision must be given about how much hardened spectrum should be employed. Concerning this question, the subsequent study investigated how the recycling of TRU is influenced by the neutron spectrum shift in a lead-bismuth cooled core.[2] Considering two lead-bismuth cooled cores with the soft and the hard spectrum, respectively, the mechanism of TRU recycling toward the equilibrium was analyzed and compared with each other in terms of burn-up reactivity and the dynamics parameters, such as the Doppler coefficient, the coolant-loss reactivity coefficient, and the effective delayed neutron fraction.

In this study, the incineration characteristics are discussed, aiming at the ultimate goal of the optimal design for the best potential of the lead-bismuth cooled core for TRU incineration from the viewpoint of transmutation capability as well as the safety. A critical core is constructed as the reference model, based on the design parameters of HYPER (HYbrid Power Extraction Reactor), a subcritical lead-bismuth cooled reactor under development for TRU transmutation at KAERI (Korea Atomic Energy Research Institute). The mean neutron energy in this core (Soft Core) was calculated as 97.3 keV. In order to perform the case study for neutron spectrum effect analyses, the core size is decreased gradually. Because the neutron moderation is not saturated in the reference core, the neutron spectrum is hardened considerably with a decrease in the core size. In order to diminish the neutron moderation by the core material and attain the hard spectrum core, the P/D (Pitch-to-Diameter) of the fuel pin cell is reduced from 1.5 to 1.2. In addition, all the geometrical dimensions, such as the core height and the assembly pitch, were scaled down to half the size of the soft core. The TRU concentration and the cycle length were adjusted to the design criteria consistent with the soft core. From the standpoint of the thermal-hydraulic and material limitations, the average linear power of the fuel rods and the average burnup of TRU are kept equal to the value of the soft core. After all, because the volume of the active fuel region was reduced to one-eighth, the reactor power was scaled down to 125 MW (thermal). The neutron spectrum was hardened considerably, and the mean neutron energy was calculated as 271.8 keV. The burn-up calculations are carried out, and the incineration capabilities are compared with each other.

The calculations results give us the following information about the neutron spectrum effects on TRU transmutation characteristics. In the soft core, the neutron spectrum with the average neutron energy around 100 keV cannot destruct out any nonfissile TRU such as nonfissile plutonium and MA (minor actinides), despite the fact that lead-bismuth is employed as a coolant for the fast spectrum. In other words, the reference core burns out only the fissile plutonium, but most of nonfissile TRU nuclides are accumulated considerably as recycling progresses.

CP673, *Plutonium Futures — The Science,* edited by G. D. Jarvinen

In the case of the hard spectrum core, the nonfissile plutonium, ^{242}Pu, and most of MA such as ^{243}Am and ^{244}Cm are still accumulated by recycling. However, ^{240}Pu is not accumulated any longer, but is incinerated out. Because ^{240}Pu has a larger fraction in the fresh TRU in comparison with the other nonfissile TRU nuclides, the total inventory of nonfissile nuclides is decreased.

From the results obtained from the preliminary calculations, it can be inferred that the harder spectrum is desirable from the standpoint of incineration capability of the nonfissile TRU nuclides, as is well known. In a word, the neutronic characteristics of the recycled core are very sensitive to the neutron spectrum shift as a result of changes in the design options of the fast reactor for TRU transmutation. The previous study showed that the response of core safety to the neutron spectrum shift is bidirectional, positive and negative at the same time.[2] Therefore, the more careful evaluations for the neutron spectrum effect on TRU recycling are being conducted in order to find out the optimal point of tradeoff between the core safety and the nonfissile TRU nuclides incineration capability and to obtain the optimal spectrum that has the best potential for the safe and efficient incineration of TRU.

REFERENCES

1. Kim, Y. N. et al., "A Study on the Neutron Spectrum Optimization for TRU Transmutation Reactor," Transactions of the American Nuclear Society 86, 428 (2002).
2. Kim, Y. N. et al., "An Investigation of TRU Recycling with Various Neutron Spectrums," Proceedings of 7th Information Exchange Meeting on Actinide and Fission Product Partitioning and Transmutation, Jeju, South Korea, October (2002).

Analysis of Thermomigration in Irradiated U-Pu-Zr Fuel

Yeon Soo Kim,[1]* G. L. Hofman,[1] S. L. Hayes,[2] and Y. H. Sohn[3]

[1]*Argonne National Laboratory, 9700 S. Cass Ave, Argonne, IL 60439, USA*
Tel: (630) 252-3173; fax: (630) 252-5161; e-mail: yskim@anl.gov
** Corresponding Author*
[2]*Argonne National Laboratory, Idaho Falls, ID 83403-2528, USA*
[3]*University of Central Florida, 4000 Central Florida Blvd., Orlando, FL 32816-2455, USA*

INTRODUCTION

Constituent redistribution in a metallic alloy fuel takes place under various driving forces; namely, temperature gradient, concentration gradient, phase change, and irradiation. As a result, the originally uniform alloy converts to inhomogeneous alloy. The inhomogeneity in a metallic nuclear fuel causes phase transformation, solidus temperature change, and local enrichment in fissile atoms, which lead to changes of the physical and mechanical properties of fuel and to potentially unfavorable fuel performances.

This phenomenon for metallic nuclear fuels was observed in the 1960s. During the course of the IFR (Integral Fast Reactor) fuel development, more irradiation tests were conducted for U-base alloy fuels to study this phenomenon in detail.[1–4] The observations from these tests confirmed some features known before and revealed new facts. The postirradiation test fuels commonly showed three distinctive concentric zones: Zr-enriched center zone, Zr-depleted and U-enriched intermediate zone, and slightly Zr-enriched outer cooler zone. The annular zone structure was a result of not only the difference in porosity but also the constituent redistribution (i.e., composition changes). Although Pu was observed to be relatively immune to migration, the presence of Pu greater than 8 wt % in composition enhanced Zr and U migration.

In order to examine the extent of migration, the concentration of the constituents needs be measured. To analyze, the generalized ternary diffusion equation needs be solved for the diffusion properties.

The objective of the present work is to assess diffusion properties of the irradiated U-Pu-Zr fuel based on the redistribution profiles measured at the end of fuel life. The constituent interdiffusion fluxes at the given time were calculated without the need of knowing diffusion properties. The interdiffusion fluxes were then used to obtain diffusion properties. In order to justify the results, the concentration profiles were constructed by using the newly obtained data and compared with the measured ones. Excellent agreement was found.

The results were used to suggest and predict thermal migration of minor actinides (MA) such as Am and Np in the U-Pu-Zr-MA system. The U-Pu-Zr-MA system is one of the primary candidate fuels for the minor actinide burner system.

RESULTS

As an example, the calculated diffusion properties (i.e., diffusion coefficient and heat of transport) based on the measured concentration redistribution data are given in Figure 1. The main and cross diffusion coefficients $\widetilde{D}^{Pu}_{ZrZr}$, $\widetilde{D}^{Pu}_{ZrU}$, $\widetilde{D}^{Pu}_{UZr}$, and $\widetilde{D}^{Pu}_{UU}$ are also provided in Table 1. In Figure 2, the predicted concentration profiles are compared with measured ones. The two are generally in good agreement for all concentrations except Zr toward the fuel surface. The slight discrepancy toward the fuel surface is attributed to the complexity of the fuel matrix as a result of swelling and the existence of the Zr reaction layer at the fuel surface.

CP673, *Plutonium Futures — The Science,* edited by G. D. Jarvinen

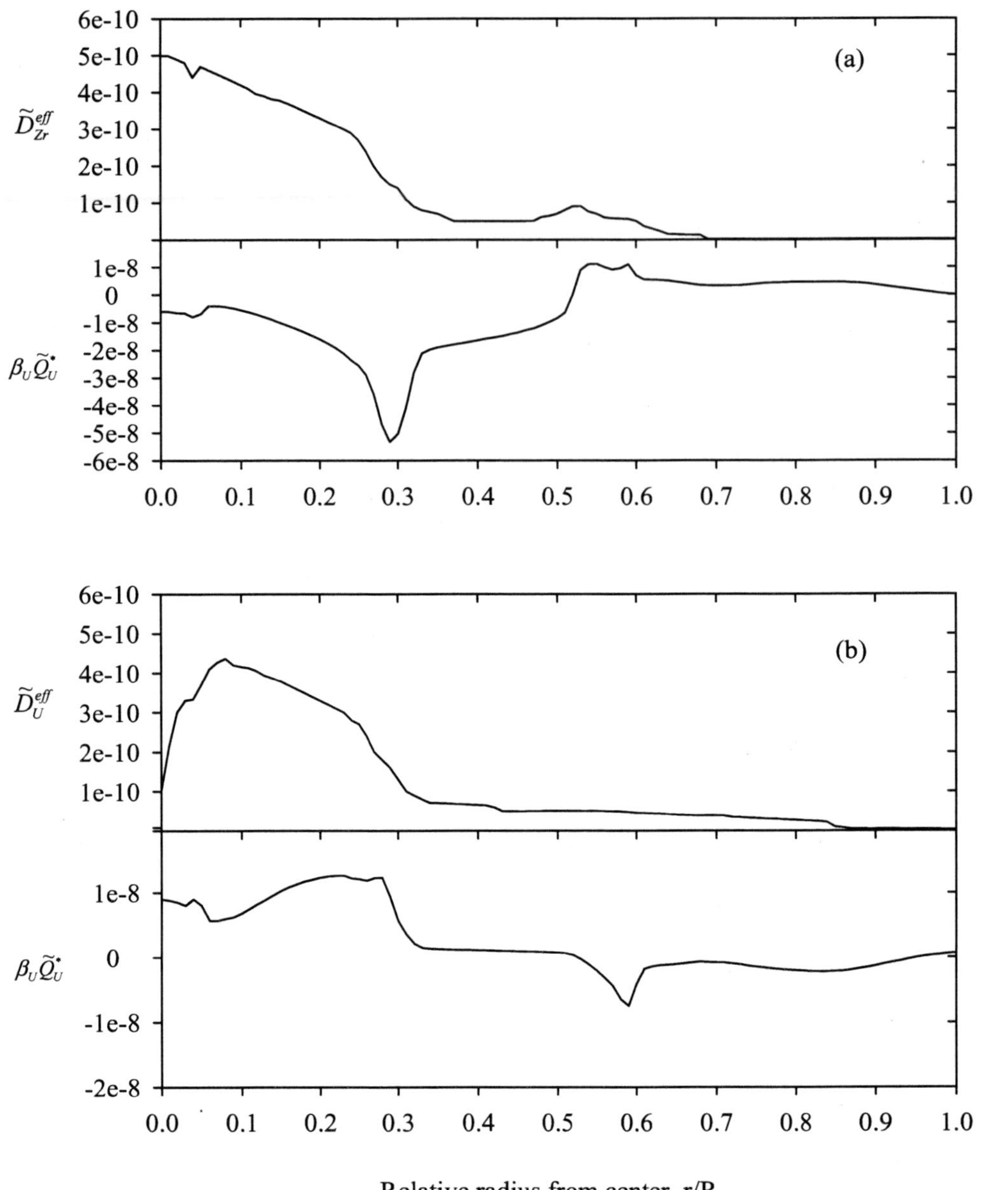

FIGURE 1. Caclulated interdiffusion properties: (a) Zr, (b) U. $\widetilde{D}_i^{eff}$ is the interdiffusion coefficient in cm^2/s and $\beta_i \widetilde{Q}_i^*$ is the product of mobility (β_i) and heat of transport ($\widetilde{Q}_i^*$) in cm^2/s.

TABLE 1. Main and Cross Interdiffusion Coefficients Calculated for the Redistribution of the Constituents and Temperature. These values were calculated at the local concentration maximum and minimum locations.

Radial Location (r/R)	$\tilde{D}_{ZrZr}^{Pu}$ (10^{-14} m^2/s)	$\tilde{D}_{ZrU}^{Pu}$ (10^{-14} m^2/s)	$\tilde{D}_{UZr}^{Pu}$ (10^{-14} m^2/s)	$\tilde{D}_{UU}^{Pu}$ (10^{-14} m^2/s)
0.39	2.9	0.25	na	na
0.44	na	na	0.44	3.7
0.69	0.38	0.068	na	na
0.70	na	na	0.11	0.69
0.86	0.15	0.085	na	na
0.87	na	na	0.10	0.16

na: not available.

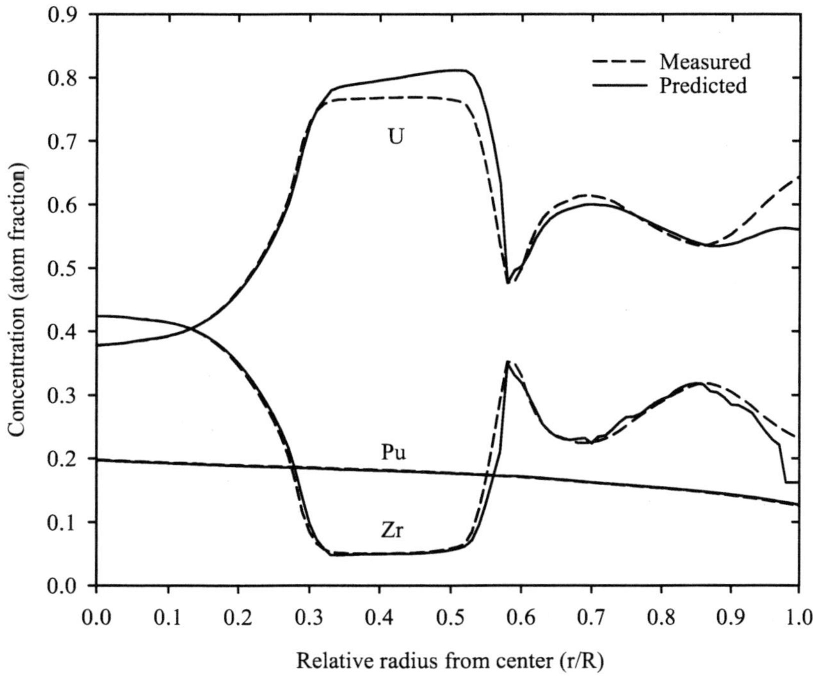

FIGURE 2. Comparison of measured and predicted concentration profiles at 1.9 at. % burnup.

The postirradiation data of the minor actinide (MA) burning experiments in the EBR-II were available where U-Pu-Zr fuels with a small amount (1.2–1.3 wt %) of MA addition were irradiated.[5] Figure 3 shows a postirradiation optical image and constituent redistribution profiles. These tests revealed that the addition of MA did not change the behavior of U, Pu, and Zr from U-Pu-Zr fuels. Am showed vaporization and condensation. Am condensation occurred at the fuel center and periphery regions where the Zr concentration built up. Np, however, showed a virtually uniform and flat signal, indicating negligible migration and no precipitation behavior like Am. This is consistent with the finding in the literature.

DISCUSSION

Based on the postirradiation measured redistribution profiles, the interdiffusion fluxes of Zr, U, and Pu in U-Pu-Zr fuel at a particular time were calculated without the need to know the diffusion properties of the fuel at the given condition. These fluxes were used to assess interdiffusion coefficients and heat of transport as a function of radial positions. The enthalpy of solution for two-phase regions (i.e., $\gamma + \zeta$ and $\delta + \zeta$) was also calculated.

The data obtained from the analyses were examined by constructing the concentration profiles with the measured data. The model predictions for concentration redistributions were very consistent with the measured ones, confirming the predictability of the present model.

Small amounts of minor actinide addition in U-Pu-Zr had a negligible effect on major constituent redistribution. Am was observed to evaporate and condense throughout the fuel cross section. The Am redistribution profile, more specifically, the population of condensation particles, showed that it generally followed the Zr migration; the only difference was its conglomeration showing large peaking signals at the fuel center and surface regions. The Np behavior was approximately similar to that of Pu. From this observation, it is recommended, in analyzing the U-Pu-Zr-MA fuel, that Am should be considered as Zr, and Np considered as Pu for preliminary calculations.

ACKNOWLEDGMENTS

Argonne National Laboratory's work was supported by the U.S. Department of Energy, Office of International Policy and Analysis (NA-241), National Nuclear Security Administration, under contract W-31-109-ENG-38.

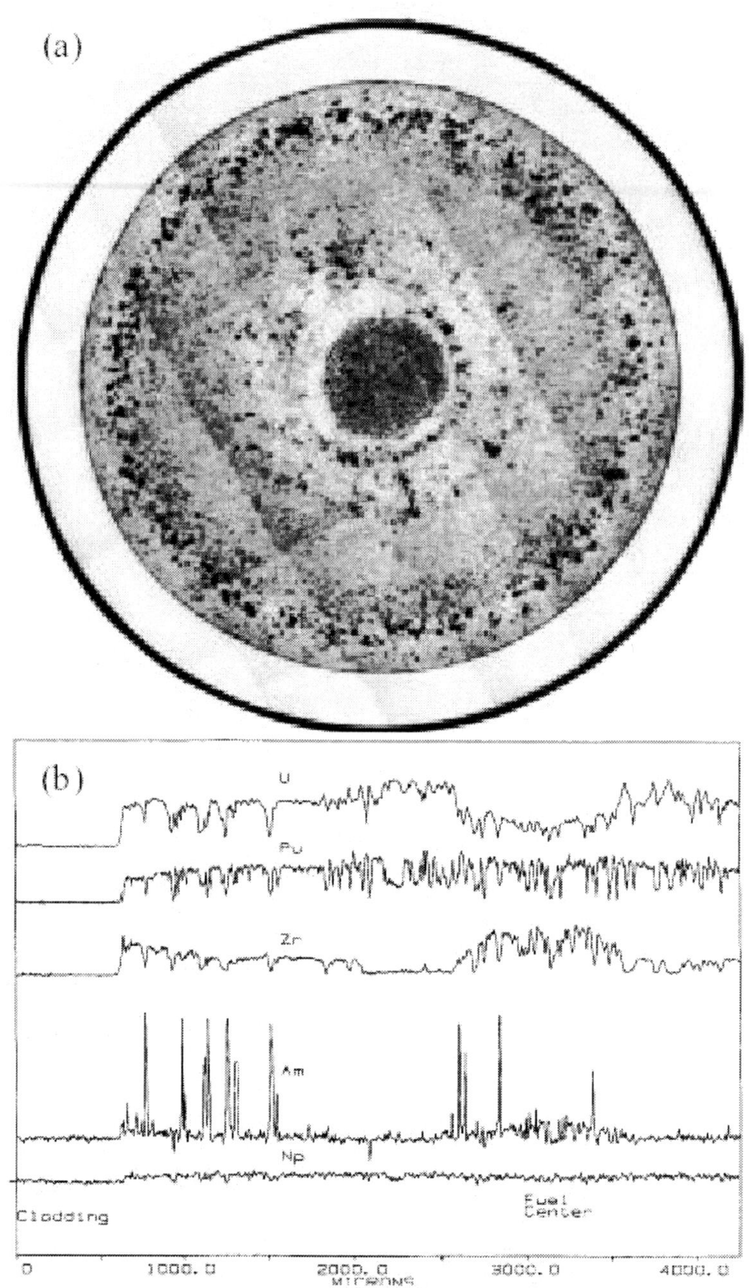

FIGURE 3. Postirradiation results of X501 test: (a) Postirradiation optical image, (b) Wavelength dispersive spectroscopy (WDS) signals. The fuel was composed of U-20.2Pu-9.1Zr-1.2Am-1.3Np (wt %) and irradiated to ~6 at. % burnup.[5]

REFERENCES

1. Pahl, R. G., Lahm, C. E., Villareal, R., Beck, W. N., and Hofman, G. L., in Proc. Int'l Conf. on Reliable Fuels for Liquid Metal Reactors, Tucson, Arizona, USA, September 7–11, 1986, pp. 3–36.
2. Pahl, R. G., Porter, D. L., Lahm, C. E., and Hofman, G. L., Metallurg. Trans. A 21A, 1863 (1990).
3. Porter, D. L., Lahm, C. E., and Pahl, R. G., Metallurg. Trans. A 21A, 1871 (1990).
4. Hofman, G. L., Hayes, S. L., and Petri, M. C., J. Nucl. Mater. 227, 277 (1996).
5. Meyer, M. K., Hayes, S. L., Crawford, D. C., Pahl, R. G., and Tsai, H., Proc. ANS Conf. on Accelerator Applications in the New Millennium, Reno, Nevada, November 11–15, 2001.

Mechanical Activation as an Effective Method for Zirconate Ceramics Preparation

O. I. Kirjanova,[1] S. V. Stefanovsky,[1] N. P. Mikhailenko,[1] S. V. Chizhevskaya,[2] S. V. Yudintsev,[3] and B. S. Nikonov[3]

[1]*SIA Radon*
[2]*D. Mendeleev University of Chemical Technology*
[3]*Institute of Geology of Ore Deposits RAS*

INTRODUCTION

Gadolinium zirconate $Gd_2Zr_2O_7$ has excellent radiation stability[1] and high chemical durability.[2] However, to produce this phase, very high temperatures are required. Even in thoroughly prepared oxide mixtures, reactions are not completed at 1,550°C for 40 hours.[3] Samples prepared using sol-gel-derived precursors sintered in air at 1,200°C for 12 hours, crushed, repeatedly sintered at 1,500°C for 30 hours, and finally hot-isostatically pressed at 1,600°C for 2 hours at 200 MPa were also found to be inhomogeneous.[2] We proposed a mechanical activation in a planetary mill as an alternative route for precursor preparation.[4] Samples from mechanically activated precursors were sintered at lower temperature and/or shorter sintering time than the samples with the same formulations prepared by different methods.[4–6] The goal of this work is to investigate the possibility of rare earth zirconate formations at lowered temperatures for a shorter time using a mechanical activation route for precursor preparation.

RESULTS

The zirconate pyrochlore formulation was chosen to provide immobilization of the Zr-REE-actinide fraction of HLW and was as follows (in wt %): La_2O_3–7.1, Ce_2O_3–14.0, Pr_6O_{11}–6.8, Nd_2O_3–22.2, Sm_2O_3–4.1, EuO–2.0, Gd_2O_3–0.9, UO_2–3.1, ZrO_2–39.8. This composition corresponds to formula

$$(La_{0.26}Ce_{0.51}Pr_{0.24}Nd_{0.79}Sm_{0.14}Eu_{0.04}Gd_{0.03})(U_{0.07}Zr_{1.93})O_7$$

[generally $(REE)_2(Zr,U)_2O_7$ (REE = La...Gd)]. Gd was used as a Am and Cm surrogate. One of the samples (M/24) was synthesized by sintering at 1,550°C of oxide mixture milled in an agate mortar. The product of heat-treatment for 6 hours was remilled and sintered again for 6 hours. This operation was still repeated two times. The mixture was compacted in pellets under pressure of 200 MPa before sintering. Total sintering duration was 24 hours. The other ceramic samples were prepared from oxide powders premechanically activated in a planetary mill—an activator with hydrostatic yokes (AGO-2U) for 5 or 10 min. Activated samples were cold-pressed at 200 MPa, heated to 1,550°C, and kept at this temperature for 3 to 6 hours.

As follows from XRD (Figure 1) and SEM/EDS data (Figure 2), the sample M/24 produced without mechanical activation was predominantly composed of pyrochlore structure phase but pyrochlore composition was varied some and unreacted, and partially reacted zirconia and neodymium-oxide-based grains occurred. The sample A/5/3 prepared from the oxide mixture mechanically activated in the AGO-2U unit for 5 min and sintered at 1,550°C for 3 hours is composed of the pyrochlore structure phase only. But pyrochlore composition is slightly varied. Longer sintering for 6 hours (A/5/6) or longer mechanical activation for 10 min (A/10/3 and A/10/6) resulted in the formation of the pyrochlore with a fixed composition that was close to that specified and the pyrochlore structure phase had approximately the same unit cell parameter, a = 10.607 Å. A similar pyrochlore lattice parameter (a = 10.593 Å) was reached after the sintering of manually milled (nonactivated) oxide mixture at the same temperature

CP673, *Plutonium Futures — The Science,* edited by G. D. Jarvinen

for 24 hours. A comparison of the texture of the samples shows that the ceramics prepared using the mechanical activation are more uniform, the shape of the pyrochlore grains is more regular, and the porosity is lower than in the sample M/24 prepared from a nonactivated mixture with a longer sintering duration (Figure 2).

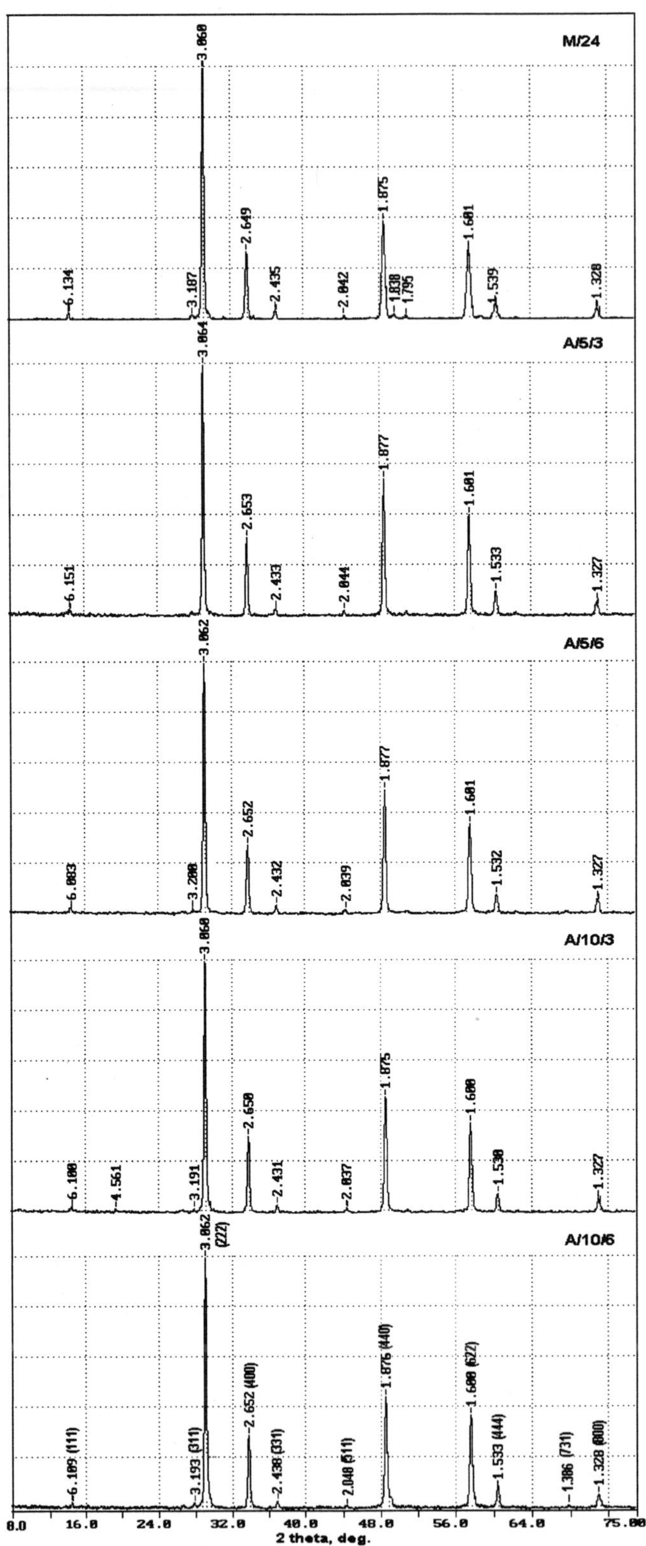

FIGURE 1. XRD patterns of the ceramic samples sintered at 1,550°C. B—baddeleyite, the rest of the peaks are due to the pyrochlore structure phase.

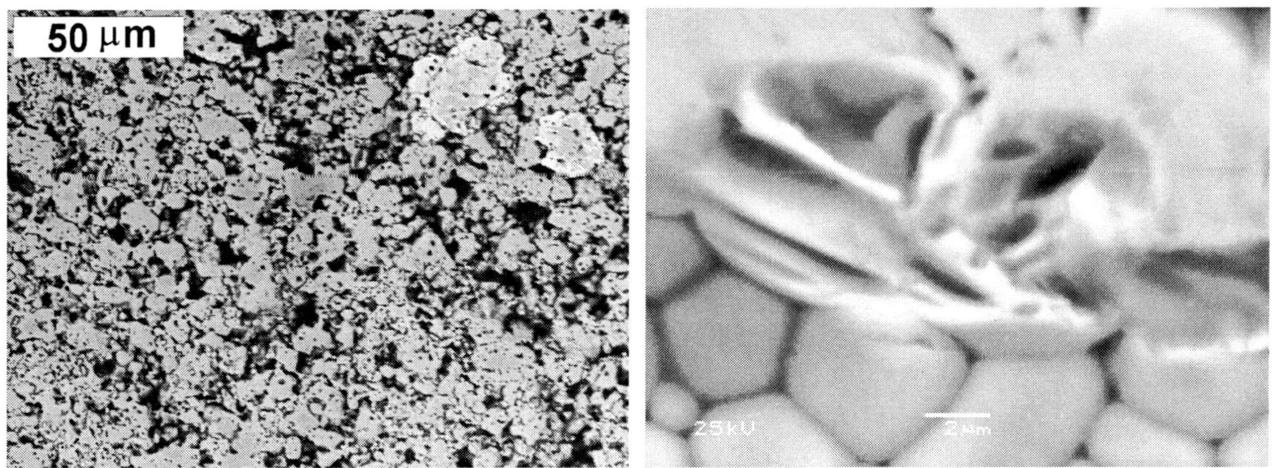

FIGURE 2. SEM images of the samples M/24 (left) and A/10/6 (right)

As has been expected, mechanical activation and a longer sintering duration increase the apparent density (the samples with density up to 90% of theoretical were produced) and reduce open porosity (up to 0.5%–1.0%) and water uptake (<0.1%) of the ceramics. However, the difference in elemental leach rates between the ceramics produced from activated and nonactivated oxide mixtures is not so prominent. Leach rates of major elements measured using the MCC-1 test from the samples are similar and as a rule have the same order of magnitude (10^{-5} for La and Ce, 10^{-5}–10^{-6} for Pr, 10^{-6}–10^{-7} for Nd and Sm, and 10^{-7} for Zr and U). The sample A/10/6 composed of well-formed pyrochlore grains (Figure 2, right) have the lowest porosity and a minimum surface subjected to leaching.

The data obtained demonstrate the suitability of a mechanical activation route to the precursor and some advantages over different methods such as lower sintering temperature and duration. To achieve the densification (~93% of theoretical density for the given formulation) required for high leach resistance, the sintering temperature of zirconate formulations must be at least 1,500°C.[2] In our work, we sintered ceramics at 1,550°C with a simple method of precursor preparation done in one step. We have also demonstrated that dense REE-zirconate ceramics requiring a higher sintering temperature than a Ca-bearing formulation cannot be produced without additional mechanical activation of oxide mixtures. It is clear from these results that mechanical activation is an effective route for sintering dense and chemically durable zirconate ceramics considered as candidate Pu-bearing waste forms under "lightened" technological conditions as compared to different methods.

REFERENCES

1. Wang, S. X., Wang, L. M., Ewing, R. C., Govidan Kutty, K. V., and Weber, W. J., in Plutonium Futures—The Science. Conf. Trans., Santa Fe, New Mexico, July 10–13, AIP Conf. Proc. (2000), pp. 13–14.
2. Icenhower, J. P., Weber, W. J., Hess, N. J., Thevuthasen, S., Begg, B. D., McGrail, B. P., Rodriguez, E. A., Steele, J. L., and Geiszler, K. N., Mat. Res. Soc. Symp. Proc. 753 (2003), in press.
3. Yudintsev, S. V., Stefanovsky, S. V., Jang, J. N., and Chae, S., Glass and Ceramics (Russ.) 7, 18–22 (2002).
4. Chizhevskaya, S. V., and Stefanovsky, S. V., in Plutonium Futures—The Science, Conf. Trans. Santa Fe, New Mexico, July 10–13. AIP Conf. Proc. (2000), pp. 148–150.
5. Stefanovsky, S. V., and Chizhevskaya, S. V., in Waste Management '00, Proc. Int. Symp., February 27–March 2, 2000, Tucson, Arizona, Rep. 55-4 (CD-ROM).
6. Stefanovsky, S. V., Chizhevskaya, S. V., and Yudintsev, S. V., in Waste Management '02, Proc. Int. Conf. Tucson, Arizona, February 24–28, 2002, Rep. 310 (CD-ROM).

Photochemical Oxidation of Oxalate in Pu-238 Process Streams

Kristy M. Long, Doris K. Ford, Leonardo Trujillo, Gordon D. Jarvinen

Nuclear Materials Technology Division, Los Alamos National Laboratory

INTRODUCTION

For over 40 years, NASA has relied on plutonium-238 in Radioisotope Thermoelectric Generator (RTG) units and Radioisotope Heater Units (RHUs) to provide power and heat for many space missions including Transit, Pioneer, Viking, Voyager, Galileo, Ulysses and Cassini. RHUs provide heat to keep key components warm in extremely cold environments found on planets, moons, or in deep space. RTGs convert heat generated from the radioactive decay of plutonium-238 into electricity using a thermocouple. Plutonium-238 has proven to be an excellent heat source for deep space missions because of its high thermal power density, useful lifetime, minimal shielding requirements, and oxide stability.

At Los Alamos, a plutonium-238 aqueous scrap recovery facility is in the final stages of approval at Technical Area 55 (TA-55). This glovebox facility will purify Pu-238 from scrap material and residues from past production operations to provide RTGs and RHUs for additional NASA missions. The scrap material is ground to reduce its mean particle size and then dissolved in a mixture of nitric acid and hydrofluoric acid. The solution undergoes chemical pretreatment for purification using an ion exchange column, if necessary, or it is sent directly to oxalate precipitation. During the oxalate precipitation step, urea is added to scavenge nitrite, hydroxylamine nitrate is added to adjust the valence of plutonium to Pu(III), and oxalic acid is added to precipitate plutonium as $Pu_2(C_2O_4)_3$. The plutonium oxalate is isolated and calcined to produce the PuO_2 used in the RTGs and RHUs. The filtrate after oxalate precipitation still contains a significant amount of Pu-238, most of which is precipitated by addition of sodium hydroxide to a pH of 10-13. A polymer filtration process is used to selectively bind with most of the remaining plutonium in the solution, reducing the alpha radioactivity below the TA-55 Caustic Line discard limit of 4.5 millicuries of total alpha activity / liter. The hydroxide precipitation and the polymer filtration process are both more efficient if the oxalate is not present to compete with hydroxide or polymer binding sites for the plutonium. We are investigating using UV irradiation and hydrogen peroxide to oxidize the oxalate (along with urea and hydroxylamine nitrate). Workers at Lawrence Livermore National Laboratory have reported some initial studies on oxalate destruction using this method in hydrochloric and nitric acid solutions.[1]

RESULTS AND DISCUSSION

A surrogate solution comprised of oxalic acid, urea, and hydroxylamine nitrate in a nitric acid matrix was used for experimentation. Two to three liters of the surrogate solution were placed in a glass reactor containing a water-cooled quartz immersion well. A 1200-watt medium-pressure mercury vapor lamp was used to irradiation the solution and compressed air was used to continuously sparge the solution. Hydrogen peroxide was pumped into the solution during irradiation and samples for analysis were removed using a peristaltic pump. The oxidation rates of oxalic acid and urea were measured throughout each irradiation period by monitoring the total organic carbon content of the solutions. The rate of oxidation has been monitored as a function of a number of variables to establish potential processing conditions including concentrations of oxalic acid, urea, hydroxylamine and hydrogen peroxide, temperature, and type of sparge gas. For example, the following graph plots the total organic carbon content of the solution as a function of irradiation time using different starting concentrations of hydrogen peroxide.

CP673, *Plutonium Futures — The Science,* edited by G. D. Jarvinen
2003 American Institute of Physics 0-7354-0140-3

Photochemical Destruction of 0.19M Oxalic Acid, 0.16M urea, and 0.19M Hydroxylamine Nitrate in a 2L solution of 1M Nitric Acid

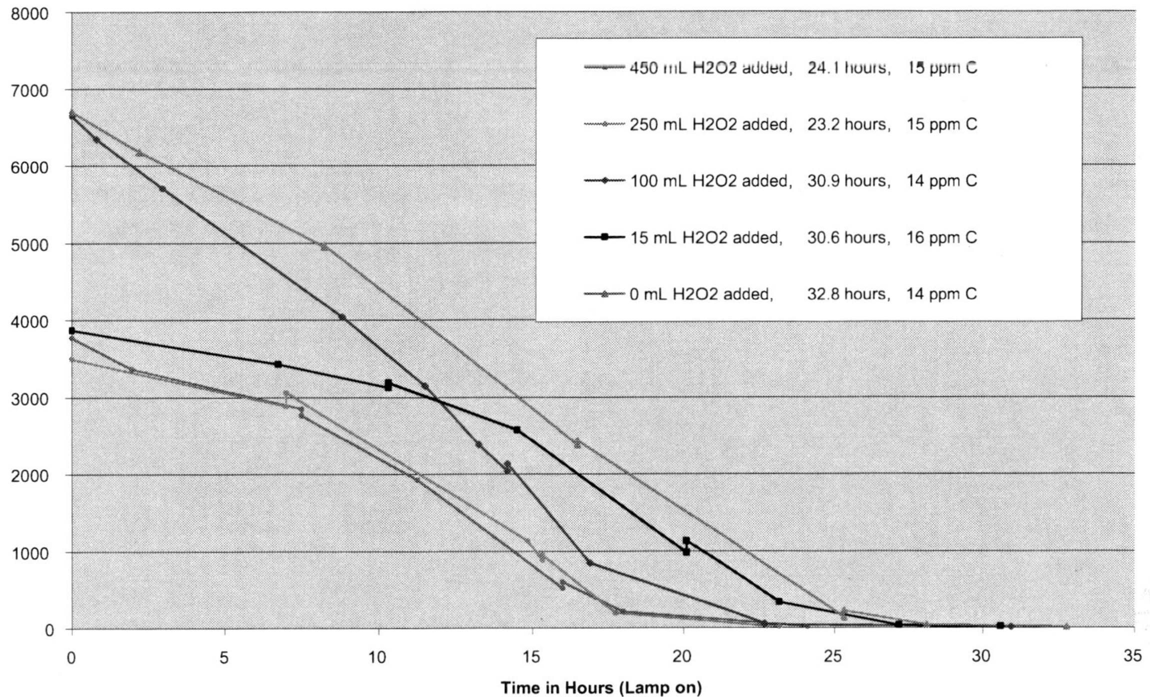

REFERENCES

Based on initial experimentation, hydrogen peroxide does increase the oxidation rate, but the increase may not be enough to justify using hydrogen peroxide. Photochemical oxidation of the nitric acid solution alone appears to explain the bulk of the oxidation reaction. Additional results of the experimental testing of the process for use in the scrap recovery operation will be presented.

1. F.T. Wang and B.Y. Lum, "Photolytic destruction of oxalate in aqueous mixed waste," Proceedings of the ASME Biannual Mixed Waste Symposium, Aug. 7-11, 1995.

Plutonium-239 in Synthetic Brines after Equilibration with Magnesium Oxide Backfill

Ningping Lu and James Conca

EES-12, Earth and Environmental Science Division, LANL-Carlsbad Office, MS A141, Los Alamos National Laboratory, Carlsbad, NM 88220, 505-628-3018, jconca@lanl.gov

INTRODUCTION

Plutonium, with its multiplicity of oxidation states, is one of the primary actinides of concern for long-term disposal and storage of nuclear waste, particularly transuranic waste. The chemical behavior of plutonium is strongly influenced by its oxidation state, which determines the strength of its complexation reactions, solubility, formation of colloids, and sorption processes.

Many radioactive waste disposal programs use, or anticipate using, backfill materials to enhance the containment of the radioactive waste. Backfill can provide well-defined chemical conditions or buffering capacity, favorable hydraulic or permeability properties, and desirable physical characteristics (strength, protection against rockfall, space-filling) to an underground disposal facility. Backfill materials have been classified into two groups according to their primary properties; (1) chemical backfills, e.g., cementitious materials, lime, phosphates, reduced-Fe minerals, and magnesium oxide, and (2) hydrological/physical backfills such as clay, salt, and cementitious materials. Chemical backfills can buffer the chemistry of the repository to conditions that favor low radionuclide solubility or can increase sorption of radionuclides and retard their release from the near field of the repository.

Magnesium oxide (MgO) has been proposed as a backfill material for some geologic repositories because of its capacity to maintain the pH of the repository at basic levels (>8) in order to reduce the solubility of radionuclides, its ability to sequester carbon dioxide that may be produced by microbial degradation of organic materials in the waste, adsorb free water in the repository through hydration reactions, and immobilize Pu.[1-10] The specific advantage depends upon the aqueous chemistry of the repository, e.g., MgO may react with intruding Mg-Cl brines, buffer the pH to 8.5, and form a variety of solids such as brucite [$Mg(OH)_2$], bischofite [$MgCl_2 \bullet 6H_2O$], tachyhydrite [$CaMg_2Cl_6 \bullet 12H_2O$], körshunovskite [$Mg_2(OH)_3Cl \bullet 4H_2O$], or other Mg-hydroxychlorides, or, if sufficient CO_2 exists, nesquehonite [$MgCO_3 \bullet 3H_2O$], hydromagnesite [$Mg_5(CO_3)_4(OH)_2 \bullet 4H_2O$], or magnesite [$Mg(CO_3)$].[8,9] Some of these solids will be thermodynamically stable, and others may only be intermediates. This study investigated the changes in Pu behavior after interactions of Pu-contaminated brines with MgO backfill materials under the low-water-content conditions that represent the intrusion of brine into a salt repository.

Two brines, synthetic brines G and E (Table 1) with Pu at 10^{-7} *M* (about 11 nCi), were infused into MgO backfill at three different water contents and equilibrated for 68 days. After analyses of the aqueous chemistry and Pu activity, the Pu-loaded MgO-brine agglomerates were subjected to desorption batch tests for up to 110 days, in the presence and absence of NaOCl up to 10^{-3} *M*. The Pu solutions (consisting of 96% ^{239}Pu and 4% ^{240}Pu) were ozonated to produce over 99% Pu(VI) immediately before testing (quantified using the 830-nm Pu(VI) peak) to represent a worst case Pu(VI) production in the repository as an effect of radiolysis. The activity of the two working brine solutions were 60,300 Bq L^{-1} for Pu(VI)-Brine G, and 62,571 Bq L^{-1} for Pu(VI)-Brine E. The MgO backfill is produced by calcining magnesite ore (a $MgCO_3$ marine sediment with Si, Fe, and Al impurities that have been calcined in a rotary kiln at high temperatures to drive off CO_2). The resultant material consists of angular grains of materials ≤3/8 inch in diameter, much of the material being powder. X-ray diffraction gave a mineralogical composition of 87.6% periclase (MgO), 8.7% of forsterite (Mg_2SiO_4), 0.74% of lime (CaO), 2.92% monticellite ($CaMgSiO_4$) and trace amounts of ulvöspenel (Fe_2TiO_4) and spinel ($MgAl_2O_4$), with mineralogy varing slightly with

CP673, *Plutonium Futures — The Science,* edited by G. D. Jarvinen

particle size and among different batches. The elemental composition of the MgO backfill material is 92.97% MgO, 4.8% CaO, 0.02% Na_2O, 0.33% K_2O, 0.01% SiO_2, 0.03% Cl, 0.35% SO_4, 0.01% B, and trace amounts of Br.

TABLE 1. Pu-Brine Interactions with MgO Backfill at Various Water Contents.

Element (mg/L)	(Traditional Batch Tests) No Backfill (Initial Brine)	(Traditional Batch Tests) Water to Backfill Ratio of 10:1	(UFA Tests) Water to Backfill Ratio of 0.25:1	(UFA Tests) Water to Backfill Ratio of 0.15:1
Brine G				
Li	23.7	23.5	27.5	23.5
Na	65,000	72,150	83,710	79,300
K	13,940	15,612	30,947	43,632
Mg	20,930	20,449	167.4	125.6
Ca	437.7	442.1	11,476	15,446
B	1,528	1,360	183.3	183.4
Br	1,645	1,694	1,727	1,760
Cl	159,300	164,100	167,270	180,000
SO_4	13,790	13,542	468.9	372.3
Alk (meq/L)	77.4	108.7	36.5	7.4
pH	7.09	7.56	8.39	8.55
Pu (nCi/g)	11.9	7.32	0.357	0.036
Brine E				
Na	93,370	104,570	102,710	101,770
K	2,972	2,972	3,358	27,580
Mg	482.5	338.7	1.93	9.17
Ca	90.7	339.2	256.7	328.3
B	584.5	485.1	379.9	333.1
Br	644.0	663.3	676.2	689.1
Cl	140,000	141,400	144,200	158,200
SO_4	13,720	13,290	13,930	18,770
Alk (meq/L)	24.3	68.9	176.6	143.0
pH	7.83	8.81	11.8	12.0
Pu (nCi/g)	10.8	1.53	< 0.001	0.009

RESULTS

Equilibration experiments were conducted at MgO-to-water ratios of 1:0.15 (a partially indurated repository, wetted down by infusion), 1:0.25 (a fully indurated repository, wetted down by infusion) and 1:10 (traditional batch tests). Release of Pu from Pu-loaded MgO agglomerates was determined in the presence and absence of hypochlorite, under agitated and nonagitated conditions. After Pu(VI) brines equilibrated with MgO backfill for 68 days, the solution pH and alkalinity changed dramatically, while 99% to 100% of the Pu was removed from the brines (Table 1). After 36 days of agitation in batch tests, between 0.1% and 30% of the adsorbed Pu released from the Pu-loaded MgO, more from the MgO-Brine G than from the MgO-Brine E, and more from the low water content conditions, although much more sorbed under low water contents to begin with (Table 1). However, under nonagitated batch conditions that are relevant to a repository, no Pu released from the Pu-loaded MgO-Brine E even after 110 days and even in the presence of OCl^- up to 10^{-3} *M,* and only a small amount of Pu (<0.3% after 100 days) was released from the Pu-loaded MgO-Brine G (Figure 1). XAS results for the Pu-loaded MgO-agglomerates indicate that the Pu was incorporated by precipitation, and the oxidation state of the Pu was primarily Pu(IV). Separate experiments in these brines with pH adjustment without MgO indicated that pH change alone could not account for the removal of Pu from the MgO-Brine G solutions, but pH change alone may account for the removal of Pu from MgO-Brine E solutions. Therefore, in NaCl brines, the buffering capability of the MgO backfill alone may be adequate to prevent Pu migration, while in MgCl brines, the backfill is performing a sorptive function in addition to simple pH buffering that enhances its ability to retard Pu migration.

Percentage of Pu-239 released from Mgo-brine agglomerates under non-agitated conditions

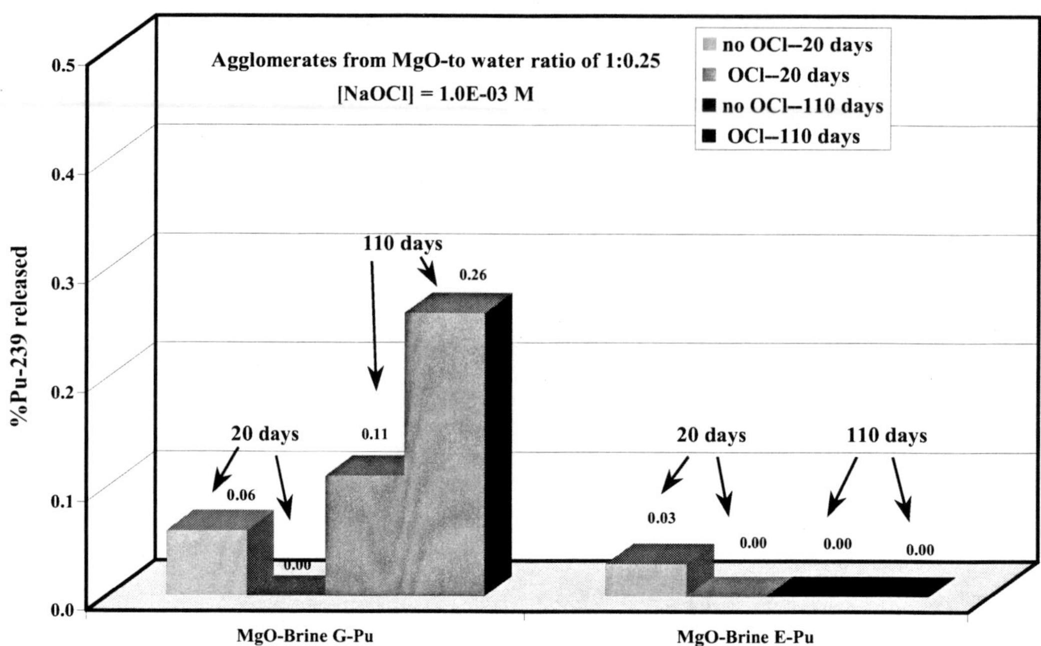

FIGURE 1. Brine desorption tests on Pu-loaded brine-MgO backfill.

ACKNOWLEDGMENTS

We greatly appreciate the efforts of Glenn Bentley, Mark Walthall and CEMRC, Jeff Terri, Thomas Hartmann, Dale Counce, Doug Ware, and Wayne Lemons.

REFERENCES

1. Papenguth, H. W., Krumhansl J. L., Brodsky, N. S., Kelly, J. W., Lucero, D. A., Coffey, D. S., Wang, Y., Anderson, H. L., and Crenshaw, T. B., "Behavior of MgO Backfill in the WIPP Repository," Status Report for Laboratory Research Conducted in 1998, Sandia WIPP Central Files, SWCF-A:1.1.01.2.7:DPRPI:NF:TD:QA:MgO Research Tech. Studies, WPO#49064 (1998).
2. Atkins, M., Glasser, F. P., and Kindness, A., Cement and Concrete Research 22, 241–246 (1991).
3. Berner, U. R., Radiochiica Acta 44/45, 387-393 (1988).
4. Bonen, D., J. Am. Ceram. Soc 75, 2904–2906 (1992).
5. Bynum, R. V., Stockman, C. T., Papenguth, H. W., Wang, Y., Peterson, A. C., Krumhansl, J. L., Nowak, E. J., Cotton, J., Patcher, S. J., and Chu, M. S. "Identification and Evaluation of Appropriate Backfills for the Waste Isolation Pilot Plant (WIPP)," In International Workshop on the Uses of Backfill in Nuclear Waste Repositories, R&D Technical Report P178, Carlsbad, New Mexico, USA, May 1998.
6. Pokrovsky, O. S., Schott, J., and Thomas, F., Geochim. Cosmochim. Acta 63, 863–880 (1999).
7. Zheng, L., Xuehua, C., and Mingshu, T., Cement and Concrete Research 21, 1049–1057 (1991).
8. Grambow, B., Kienzler, B., Bauer, A., Vjemelka, P., and Rudoph, G., "Backfill and Sealing Materials as Geochemical Barriers for Radionuclide Containment in German Salt Formation," In International Workshop on the Uses of Backfill in Nuclear Waste Repositories, R&D Technical Report P178, Carlsbad, New Mexico, US, May 1998.
9. Papenguth, H. W., Krumhansl, J. L., Bynum, R. V., Wang, Y., Kelly, J. W., Anderson, H. A., and Nowak, E. J., "Status of Research on Magnesium Oxide Backfill," In International Workshop on the Uses of Backfill in Nuclear Waste Repositories, R&D Technical Report P178, Carlsbad, New Mexico, US, May 1998.
10. Farr, J. D., Schulze, R. K., and Honeyman, B. D., Radiochim. Acta 88, 675–678 (2000).

Influence of Radiolysis Byproducts on the Actinide Chemistry in Brines from Geological Saline Repository

J-F. Lucchini, A. Rafalski, M. Riggs, and J. Conca

Los Alamos National Laboratory, EES-12 Carlsbad Operations, MS A141, Carlsbad, NM 88220
(505) 628-3257, fax: (505) 628-3238, lucchini@lanl.gov

Because geological salt formations are considered possible or operational sites for radioactive waste disposal, actinide chemistry in brines has been the subject of numerous investigations within the last two decades. Oxidation states, redox behavior, solubility, and speciation of actinides in solution are the main chemical processes involved in the potential migration of actinides in the environment.[1] In the near-field chemistry of a salt repository of nuclear waste, ionizing radiations can strongly affect these processes, and thus can affect the actinide mobility in concentrated saline solution. Therefore, the effects of radiolysis on high-saline brine under simulated repository conditions, and the actinide behavior towards radiolytic byproducts are of particular importance.[2]

The aim of this work is to give a nonexhaustive review of the brine radiolysis effects on the actinide chemistry.

The chemistry of concentrated saline solutions under ionizing radiations is extremely complex. Radiolysis can locally modify the redox conditions, and therefore the speciation and the solubility of the actinides compounds.

Concerning water radiolysis, the same amounts of oxidizing and reducing species are generated.[3] The main oxidizing species are hydrogen peroxide (H_2O_2) and hydroxyl radical (•OH). The main reducing species are hydrogen molecules (H_2), thermalized hydrated electrons (e^-_{aq}), and the hydrogen atom (•H). Because oxidizing species are highly reactive and H_2 is expected to escape from the solution, the local conditions may become oxidizing.

The radiolysis of brines saturated with chloride salts results in the formation of a transitory equilibrium system, made of $Cl_3^-/Cl_2/HClO/ClO^-/Cl^-$ species. These species can also increasingly modify the redox conditions, and thus may have an impact on the chemistry of the actinides compounds in solution.

The higher oxidation states of actinides are known to be more soluble in solution than the lowest ones. Most of the studies report that hypochlorite (ClO^-), generated by alpha irradiation of a concentrated sodium chloride (NaCl) solution, oxidizes Am(III) to Am(V), Pu(IV) to Pu(VI), and Np(IV) to Np(V).[4–8] However, in the case of an underground disposal, we must keep in mind that the chemical conditions of the local environment and the reducing radiolytic byproducts can also play a significant role on the stability of actinide oxidation states in the aqueous phase and on the overall actinide mobility.

Unfortunately, publications concerning the influence of radiolysis byproducts on the speciation of actinides are few, especially in synthetic brine and in conditions as representative as possible of geological salt repository conditions. The complex chemistry of transuranic waste in contact with brines needs to be experimentally investigated by examining the subeffects influencing the overall pcH-Eh system, which includes radiolysis effects.

This work presents the mechanism involved in radiolysis of chloride-ion aqueous solutions, and gives some predictions of the possible reactions and rate constants in the solutions. A summary of the significant reaction processes taking place during the radiolysis of NaCl solution is discussed.

The interaction of actinides with radiolytic species can significantly affect solubility predictions. Because radiolysis is likely the major driving force in oxidizing actinides, it needs to be studied and understood in order to be considered in risk assessments of nuclear repositories in salt formations, and to evaluate more precisely and efficiently the long-term evolution of geological nuclear waste repositories.

CP673, *Plutonium Futures — The Science,* edited by G. D. Jarvinen

REFERENCES

1. Runde, W., "The Chemical Interactions of Actinides in the Environment," Los Alamos Science 26, 330 (2000).
2. Paviet-Hartmann, P., and Hartmann T. "Radiolysis Effects on Actinide Speciation under Adapted Real Waste Scenario," Los Alamos National Laboratory document LA-UR-02-2725 (2002).
3. Spinks, J. W. T., and Woods, R. J., An Introduction to Radiation Chemistry, Third Edition, Wiley-Interscience Publication (1990).
4. Magirius, S., Carnall, W., and Kim, J. I., "Radiolytic Oxidation of Am(III) to Am(V) in NaCl Solutions," Radiochimica Acta 38, 29 (1985).
5. Kim, J. I., Lierse, C., Buppelmann, K., and Magirius, S., "Radiolytically Induced Oxidation Reactions of Actinide Ions in Concentrated Salt Solutions," Mat. Res. Soc. Symp. Proc. 84, 603 (1987).
6. Buppelmann, K., Kim, J. I., and Lierse, Ch. "The Redox Behavior of Pu in Saline Solutions under Radiolysis Effects," Radiochimica Acta 44/45, 65 (1988).
7. Fukasawa, T., Lierse, C., and Kim J. I., "Radiolytic Oxidation of Np in NaCl Solutions," Nuclear Sc. Technol. 33(6), 486 (1996).
8. Kelm, M., and Bohnert E., "Radiation Chemical Effects in the Near Field of a Final Disposal Site I: Radiolytic Products Formed in Concentrated NaCl Solutions," Nuclear Technol. 129, 119 (2000).

Continuous Production Process of Granulated Powders of Mixed Uranium and Oxides, according to "Granat"-Method Results of Investigations, Carried Out at "Granat-M" Continuous Experimental Plant

V. E. Morkovnikov, V. V. Revyakin, V. A. Astafiev, A. P. Pavlinov, V. N. Revnov, V. P. Varikhanov, A. E. Glushenkov, T. F. Voronina, S. K. Zavialov, V. A. Chernov, and N. V. Morozov

SSC RF VNIINM (Russia)

An enormous scientific and technical potential, created by Russian scientists for a period of the previous twenty years, has not been realized in the commercial fabrication of mixed oxide fuel for nuclear power reactors. However, the activities on improving these methods of MOX fuel fabrication are in progress at experimental plants of Russian scientific centers. The production of granulated mixed oxides of uranium and plutonium by means of the hydrogranulation process (coprecipitation of uranium and plutonium hydroxides in the presence of high molecular compounds) is one of such methods.[1,2] In order to check this method experimentally under continuous operation, a pilot plant[3] called "Granat-M," was constructed (Figure 1) in SSC RF VNIINM, named after the academician A. A. Bochvar. Continuous processes are investigated at this plant by means of continuous equipment of nuclear safe geometry,[4] created especially for this method.

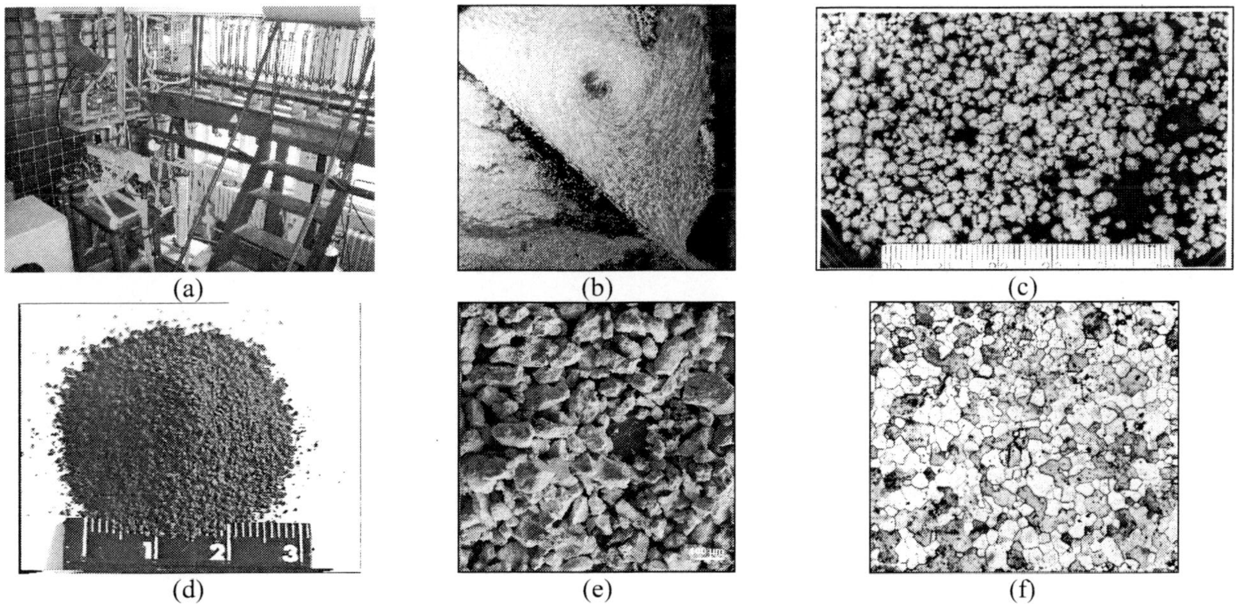

FIGURE 1. Production stages of granulated powders of uranium and plutonium oxides: (a) view of "Granat-M" experimental plant; (b) hydrogranulation process; (c) granules on the basis of uranium and plutonium hydroxides after the rounding process; (d) granulated powder of ceramic grade after the sizing process; (e) microstructure of the powder; and (f) microstructure of a sintered sample.

CP673, *Plutonium Futures — The Science,* edited by G. D. Jarvinen

The report deals with the main findings obtained at the given experimental continuous plant. They include the test results of continuous coprecipitation, strengthening of the produced granules of metal hydroxides, their thermal treatment, and sizing for the preset granulometric composition. The results of morphological investigations of the produced powders as well as pressure and sintering ability tests with the estimation of the quality of the resulting pellets are also given in the report.

REFERENCES

1. Zakharkin, B. S., Revyakin, V. V., Morkovnikov V. E. et al., "Equipment Relevant Solution of Utilizing Weapon's Grade Plutonium into MOX Fuel for Power Reactors," in International Conference "RECOD'91," Njce, France (1991).
2. Revyakin, V. V., Zakharkin, B. S., Borisov, L. M., Varikhanov, V. P. et al., "Development of Production Technology of Uranium-Plutonium Press-Powders by Coprecipitation of Hydroxides in the Presence of Surfactants ("Granat" Technology)," in International Conference GLOBAL 97, Japan (1997).
3. Morkovnikov, V. E., Revyakin, V. V., Pavlinov, A. P., Revnov, V. N., Chernov, V. A., Varikhanov, V. P . et al., "Continuous Process of Powder Production for MOX Fuel Fabrication According to "Granat" Technology. Part 1," in International Conference "Plutonium Future—Science," Santa Fe, New Mexico, USA, July 10–13, 2000.
4. Morkovnikov, V. E., Pavlinov, A. P., Revnov, V. N., and Chernov, V. A., "Implementation of Production Process of Granulated Powders for Mixed Uranium-Plutonium Fuel Fabrication for Power Reactors ("Granat-M" Pilot Plant)," in International Conference "GLOBAL 2001," Paris, France, September 9–11, 2001.
5. Morkovnikov, V. E., Revyakin, V. V., Astafiev, V. A., Pavlinov, A. P. et al., ""Granat-M" Pilot Plant for Ceramic Grade Powder Production, Used for MOX Fuel Fabrication," in the International Conference, devoted to the 60-anniversary of SverdNIIKhimmash, Ekaterinburg, Russia (2002).

Gamma Radiolysis Effects on $(U,Pu)O_2$ Alteration in Water

A. Poulesquen, C. Jegou, V. Broudic, and J. M. Bart

CEA / VALRHO / MARCOULE, Nuclear Energy Division
Confinement Research and Engineering Department
Waste Confinement and Vitrification
arnaud.poulesquen@cea.fr

INTRODUCTION

For a potential performance assessment of the direct disposal of spent fuel in a nuclear waste repository, the chemical reactions between the fuel and possible intruding water must be understood, and the resulting radionuclide release must be quantified. Although beta/gamma irradiation will become much less significant than alpha radiation over the long term, the currently available leaching data for spent fuel several years after removal from the reactor include the effects of α, β, and γ irradiation. Therefore, a study specifically targeting gamma irradiation, together with experiments on alpha-doped samples, should eventually allow these effects to be deconvoluted. Moreover, in the spent nuclear fuel dissolution model, the beta/gamma irradiation is considered to be important in relation to alpha irradiation if, during the first millennium, a failure of containers happens.

An experimental approach designed to quantify the influence of beta/gamma irradiation on UO_2 alteration was therefore developed, and the results will be described. This approach is based on leaching experiments with UO_2 fuel pellets (standard or doped with alpha emitters) and spent fuel submitted to gamma irradiation. In parallel, Chemsimul software[1] is used to model the experimental results obtained.

MATERIALS AND LEACHING PROTOCOLS

Leach tests were conducted in a shielded cell at room temperature (25°C) in an aerated medium, with deionized water in static mode (without solution renewal). Two following types of materials were altered under gamma irradiation at a dose rate of 650 $Gy.h^{-1}$:

- UO_2 pellets lightly doped with ^{238}Pu (1,500-year batch), annealed and preleached in order to observe the release of ^{238}Pu (alteration tracer) by radiochemical analysis.
- Spent fuel fragments (about 1 g of UOX fuel with a burnup of 60 $GWd.t_{HMi}^{-1}$) washed to eliminate the transferable inventory for comparison with UO_2 and to study the effect of the fuel chemistry on the uranium release.

The S/V ratio estimated from the geometric surface area was about 2 m^{-1}. The fuel fragments were selected to ensure the same S/V ratio for both experiments. Solution samples were taken over a 16-day period at the following intervals: 1 hour, 3, 7, 10, and 16 days. On completion of the test, the leachate was acidified (1 N HNO_3, 24 h), as were the components of the leaching device (titanium liner, sample holder, etc.). Uranium was analyzed by laser-induced phosphorescence, ^{238}Pu by alpha spectrometry, and cesium in the fuel leachate by gamma spectrometry.

CP673, *Plutonium Futures — The Science,* edited by G. D. Jarvinen

RESULTS AND DISCUSSION

Figure 1 shows the normalized mass loss versus time for the UO_2 pellets (1,500-year batch). In apparent contradiction with the intermediate sampling results, analysis of the acidified leachate (U acidif + ^{238}Pu acidif) showed that the pellet dissolution was linear over time under these alteration conditions. The alteration rate was 85 $mg.m^{-2}.d^{-1}$.

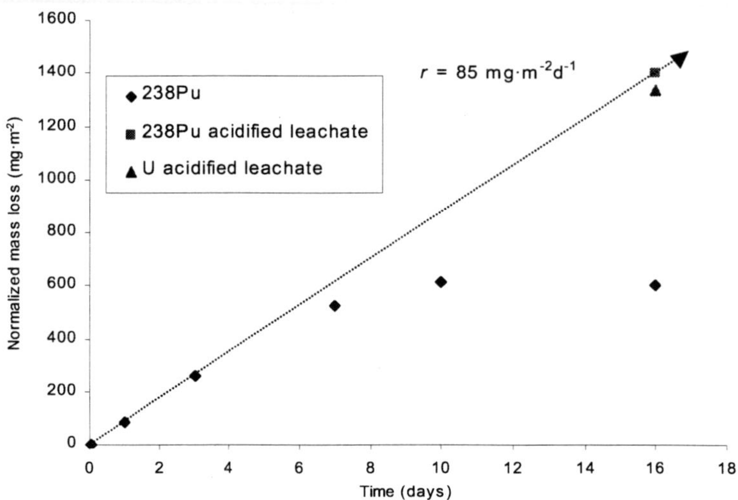

FIGURE 1. Normalized mass loss versus time for UO_2 subjected to gamma irradiation (650 $Gy.h^{-1}$).

The rate determined experimentally under gamma irradiation was of the same order of magnitude as the value proposed by Christensen and Sunder[2] (R = 30 $mg.m^{-2}.d^{-1}$) for UO_2 with a gamma dose rate of 600 $Gy.h^{-1}$.

Table 1 shows experimental results concerning the influence of alpha/gamma irradiation on the uranium release. To have an idea of a reference state, the uranium release for UO_2 without irradiation was also determined. We notice that the dissolution rate strongly increases under irradiation. Furthermore, this rate is the same for the two materials undergoing alpha/gamma irradiation. Otherwise, experimental results obtained show that the alpha irradiation effect is masked by the gamma irradiation in this medium (aerated). Moreover, no effect of fuel chemistry is observed on the uranium release. In this medium, the gamma irradiation seems to be the preponderant parameter in the matrix dissolution in relation to the alpha irradiation.

TABLE 1. Experimental Results of Gamma Irradiation.

Material	D_α ($Gy.h^{-1}$)	D_γ ($Gy.h^{-1}$)	Medium	R ($mg.m^{-2}.d^{-1}$)	$U_{released}$ (µg)
UO_2	0	0	Aerated	0.5	2
UO_2 (1,500-year batch)	110	650	Aerated	85	375
Spent Fuel	1,010	650	Aerated	74	330

Concerning the modeling aspect, the global model of radiolysis developed by Christensen[3] and Chemsimul software were used to account for the influence of gamma irradiation on matrix alteration. This radiolysis model was adapted to our system, and it is possible to describe our experimental values of uranium release by using one adjustable parameter: the water layer affecting UO_2 oxidation.

In Table 2, a comparison between experiments and calculations is made. We noticed that the correlation is quite good for a global water layer of 90 µm. The modeling proves to be an important tool to emphasize experimental results.

TABLE 2. Comparison between Experiments and Calculations.

Material	D_α ($Gy.h^{-1}$)	D_γ ($Gy.h^{-1}$)	$R_{experimental}$ ($mg.m^{-2}.d^{-1}$)	$R_{calculated}$ ($mg.m^{-2}.d^{-1}$)
UO_2	0	0	0.5	0.25
UO_2 (1,500-year batch)	110	650	85	82.4
Spent Fuel	1,010	650	74	83.3

REFERENCES

1. Kirkegaard, P., and Bjergbakke, E., "CHEMSIMUL: A Simulator for Chemical Kinetics," Risø National Laboratory, Roskilde, Denmark (October 2002).
2. Christensen, H., and Sunder, S., "Current State of Knowledge of Water Radiolysis Effects on Spent Nuclear Fuel Corrosion," Nuclear Technology 131, 102–123 (2000).
3. Christensen, H., "Calculations Simulating Spent-Fuel Leaching Experiments," Nuclear Technology 124, 165–174 (1998).

Expansion of Green Is Clean (GIC) Program at Nuclear Materials Technology (NMT) Division Facilities

Susan Ramsey,[1] Robert Dodge,[1] Kathleen Gruetzmacher,[1] Ed Horst,[1] and Steve Myers[2]

[1]*Los Alamos National Laboratory*
[2]*Eberline Services*

INTRODUCTION

The Los Alamos National Laboratory (LANL) Plutonium Facility (TA-55) and the Chemistry and Metallurgy Research (CMR) building are radiological facilities operated by the Nuclear Materials Technology (NMT) Division. The operations within the radiological control areas (RCAs) of NMT Division generated approximately 262 m^3 of low-level wastes (LLW) per year from the routine disposal of laboratory room trash. The process for the generation and disposal of LL and GIC wastes from RCAs is shown in Figure 1.[1] It is estimated that as much as 50% of LLW boxes are free of radioactive contamination and are suitable for inclusion in the GIC waste program.

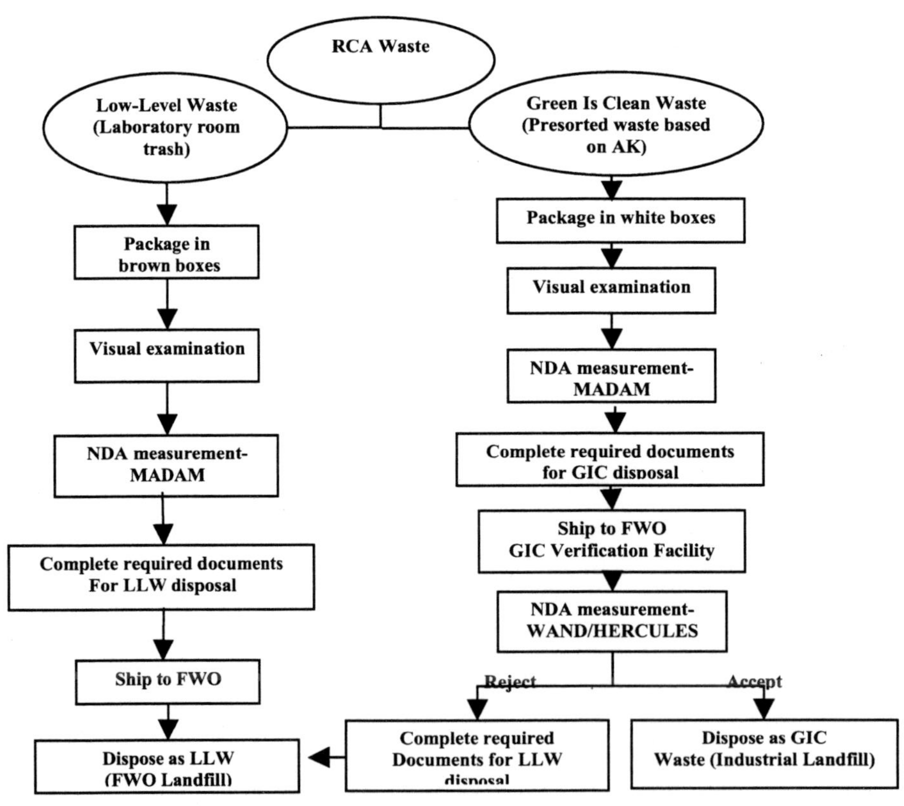

FIGURE 1. Process for the generation and disposal of LL and GIC wastes from RCAs.

CP673, *Plutonium Futures — The Science,* edited by G. D. Jarvinen

The GIC program minimizes the generation of "suspect" LLW by providing a standardized method of waste segregation that uses the concept of acceptable knowledge (AK) to free release waste from an RCA. A nondestructive assay (NDA) technique is used to verify the generator's AK.[3]

RESULTS

A pilot project was conducted in collaboration with Facility Waste Operation (FWO) Division to determine which RCAs generate the greatest volumes of potential GIC wastes. FWO personnel measured the radioactive content of approximately 1000 low-density LLWs generated at NMT's RCAs in a three-month period from August through November 2002 using the High Efficiency Radiation Counter for Ultimate Low Emission Sensitivity (HERCULES) system. HERCULES is a highly sensitive NDA instrument and is optimized to detect minute quantities of plutonium, americium, and uranium for low-density waste.[2] Preliminary results were tabulated and organized by the originating room and the total amount of activity detected. The preliminary results indicate that 40% of the LLW generated are free of radioactive contamination without presegregation by the generator.

CONCLUSION

The expansion of the GIC program to include laboratory room trash that is considered to be "suspect" LLW can significantly reduce the total volume of LLW generated at LANL and thus significantly reduce the waste disposal cost for LLW.

REFERENCES

1. Carlson, Bryan, "Pollution Prevention at Los Alamos National Laboratory Green Is Clean," Los Alamos National Laboratory document LA-UR 02-2603 (May 2002).
2. Gruetzmacher, K. M., and Foxx, C. L., "Operating the WAND and HERCULES Prototype Systems," Los Alamos National Laboratory document LA-UR-01-0445 (2001).
3. Arnone, G. J., Foster, L. A., Foxx, C. L., Hagan, R. C., Martin, E. R., Myers, S. C., and Parker, J. L., "Status of the WAND (Waste Assay for Nonradioactive Disposal) Project as of July 1997," Los Alamos National Laboratory report LA-13432-SR (March 1998).

An Investigation into the Effect of Specific Surface Area on the Reaction Between Silver(II) and Plutonium Dioxide

P. J. W. Rance[1] and G. P. Nikitina[2]

[1] *British Nuclear Fuels, Sellafield, Seascale, United Kingdom*
[2] *V.G. Khlopin Radium Institute, 2nd Murinsky Prospekt, St. Petersburg, Russia*

INTRODUCTION

The ability of the silver(II) ion to dissolve plutonium dioxide has been known for some twenty years; however, there is still little quantitative information available regarding the effect of such parameters as surface area or temperature on the dissolution rate.

The electromediated dissolution process essentially involves the generation of the highly oxidizing Ag(II) ion at the anode of an electrochemical cell; this unstable species then brings about the dissolution of the otherwise insoluble PuO_2. Silver(II) is a very strong oxidizing agent, the standard electrode potential of the Ag(II)/Ag(I) couple (+1.98 V in 4 M nitric acid) is only exceeded by species such as ozone and fluorine. In addition to the greater simplicity of working with silver compared to fluorine or ozone, the process has the significant advantage that much less than stoichiometric quantities of silver are required as the oxidizing form of the silver, Ag(II), is regenerated by electrolysis following its reduction as a result of reaction with PuO_2.

In simple terms, the process can be considered to be

$$Ag^{+} \rightarrow Ag^{2+} + e^{-} \tag{1}$$

$$Ag^{2+} + PuO_2 \rightarrow Ag^{+} + PuO_2^{+} \tag{2}$$

$$2PuO_2^{+} + 4H^{+} \rightarrow Pu^{4+} + PuO_2^{2+} + 2H_2O \tag{3}$$

$$Ag^{2+} + Pu^{4+} + 2H_2O \rightarrow Ag^{+} + PuO_2^{+} + 4H^{+} \quad . \tag{4}$$

Being a heterogeneous reaction between aquated silver(II) ions and solid plutonium dioxide particles, it might be expected that, in the absence of another controlling effect, the specific surface area of the plutonium dioxide might have an influence on its dissolution rate. As the specific surface area of plutonium dioxide is affected by its calcination temperature, powders of different origin might be expected to show significantly different dissolution characteristics. A number of experiments using different plutonium powders have been performed in an attempt to determine the sensitivity of the dissolution kinetics to the specific surface area of the powder.

EXPERIMENTAL

The electrochemical cell used had a working volume of 100 ml; the anode was made from a 40 x 40 mm platinum foil, 0.1 mm thick; and the cathode was a Pt/Ti strip (4 cm^2) housed in a 10-mm-diameter glass tube, the lower end of which was closed with a glass frit above which was laid a bed of silicic acid gel to prevent egress of the catholyte and cathodic reaction products. Experiments were conducted at a temperature of 10°C in 4M nitric acid containing 0.01 M Ag at a current density of 0.006 A/cm^2. A series of PuO_2 powders having a particle size of approximately 1 micron and produced at different sintering temperatures between 450°C and 800°C were used. These powders had specific surface areas, measured by the BET technique, of between 31.0 and 7.6 m^2/g. Reaction

CP673, *Plutonium Futures — The Science,* edited by G. D. Jarvinen

progress was monitored by sampling the anolyte periodically and determining the dissolved plutonium concentration by alpha counting and the Ag(II) concentration by analyzing the amount of Ce(IV) produced by reaction between Ag(II) and Ce(III).

RESULTS AND DISCUSSION

The results from dissolution of a series of PuO_2 powders are shown in Figure 1; these rates can be resolved to give a straight line plot in coordinates of ln{PuO_2} vs time to give pseudo first-order rate constants for the dissolution reaction of PuO_2 as shown in Table 1. Although the straight line correlations give good fits (with R^2 values of 0.98 or better), there is no significant trend due to surface area and although a linear fit to a plot of initial dissolution rate against initial surface area gives a positive gradient, the correlation is only 0.65 (R^2 = 0.43) and is even worse when forced through the origin.

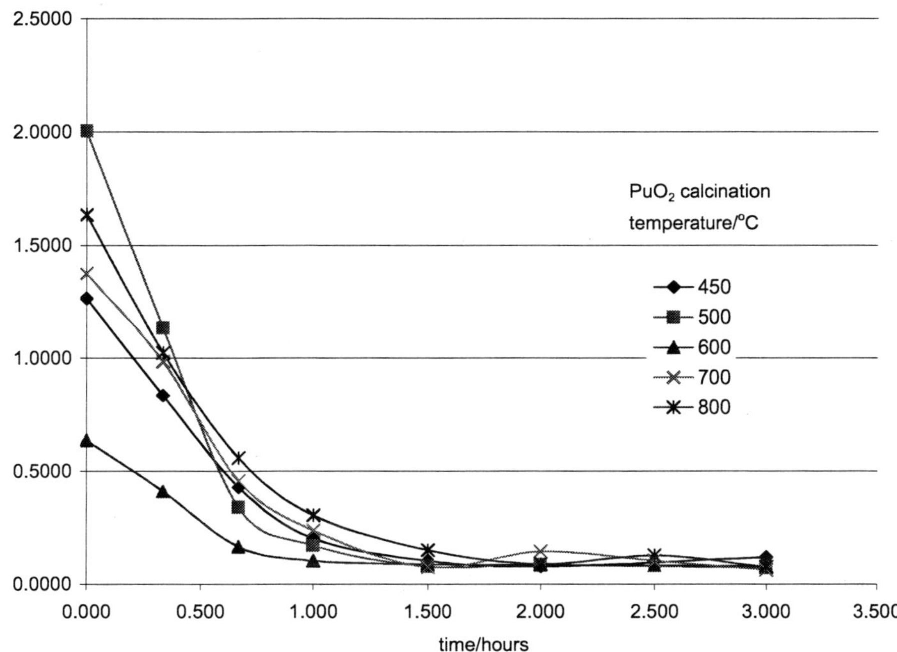

FIGURE 1. Effect of PuO_2 calcination temperature on its dissolution rate.

TABLE 1. Experimental Summary.

Annealing Temperature (°C)	Specific Surface Area (m^2/g)	PuO_2 (mg)	$k_{ef} \times$ [Ag(II)] (h^{-1})	R^2	$[Ag(II)]_{plateau} \times 10^3$ (M)
450	31.0 ± 0.1	34.2	1.75 ± 0.10	0.990	6.83
500	22.7 ± 0.6	54.4	2.26 ± 0.22	0.977	7.69
600	13.8 ± 0.8	17.2	1.91 ± 0.19	0.980	6.33
700	9.0 ± 0.2	37.2	1.99 ± 0.13	0.987	7.42
800	7.6±0.2	44.3	1.63±0.05	0.997	7.33

In addition, the pseudo first-order rate constants assume a constant silver(II) concentration throughout the dissolution process that, as Figure 2 shows, was clearly not the case in these experiments. The silver(II) concentration is shown to increase as the remaining PuO_2 decreases, as might be expected for conditions in which the reaction between Ag(II) and PuO_2 is significant. However, the rates of increase in silver(II) concentration with time for different experiments are similar in each case despite there being a more than five-fold difference in initial PuO_2 surface areas between the different experiments. If the silver(II) concentration were indeed being controlled by its reaction with PuO_2, it might be expected to increase at a greater rate in the presence of a smaller PuO_2 surface area.

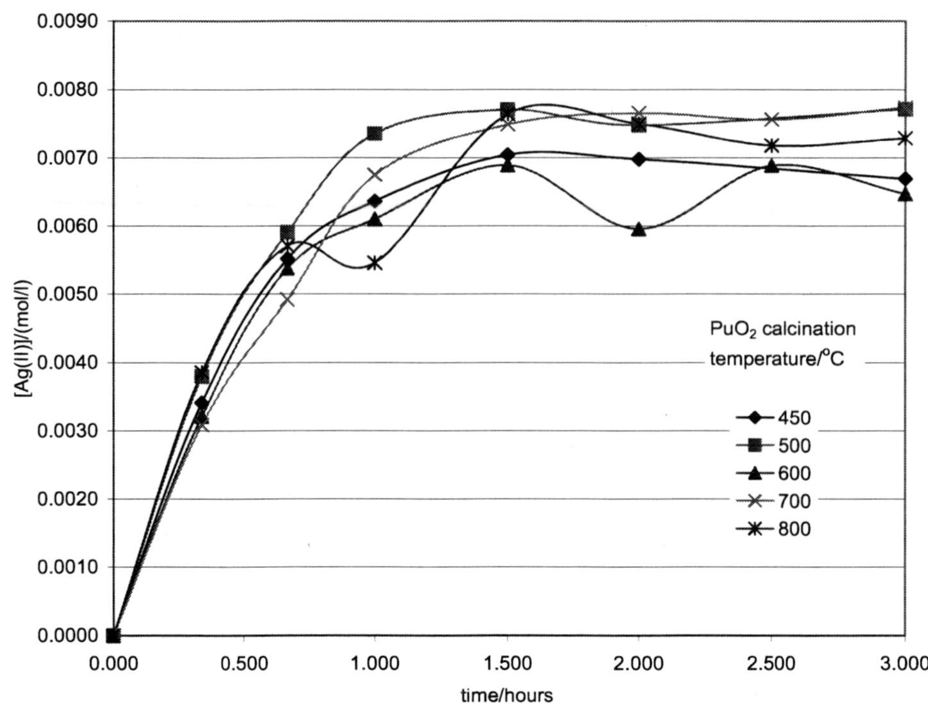

FIGURE 2. Effect of PuO_2 calcination temperature build up of Ag(II).

One possible explanation is that the Ag(II) as analyzed by this technique is not the active oxidation species and that a more short-lived species is actually responsible for the dissolution. In addition, this short-lived species comes into equilibrium much faster and presumably at much lower concentrations than the silver(II) measured in this experiment. Alternatively, the effective particle surface area for dissolution might not be equivalent, or even proportional, to the surface area as determined by the BET technique. We have previously reported results[1] that show that under conditions of steady-state silver(II) concentration the dissolution kinetics for high specific surface area material are best described in a surface area context (assuming spherical particles) and that low specific surface area material is better described in a mass context. This is somewhat counterintuitive as decreasing specific surface area more closely approximates to a model sphere and perhaps supports the view that the high surface areas available for gas adsorption are not all available for reaction with aquatic species. However, work on reductive dissolution of PuO_2 has shown a very strong correlation between reaction yield and PuO_2 surface area as measured by the BET technique.[2]

SUMMARY

The rate of silver(II) electromediated dissolution of PuO_2 powders calcined at temperatures between 450°C and 800°C shows no clear evidence of being controlled by the specific surface area of the powder. Similarly, the rate of silver(II) buildup in such systems shows no correlation with either PuO_2 surface area or mass. No entirely satisfactory explanation for these findings has been found, but other controlling parameters are possible. Future work will attempt to determine the controlling factors and extend this work to lower specific surface area materials that are known to suffer from slower dissolution kinetics.

REFERENCES

1. Rance, P. J. W., and Nikitina, G. P., Proceedings of the International Symposium, NUCEF 2001, Japan Atomic Energy Research Institute (2002), pp. 519–525.
2. Madic, C. et al., in Transuranium Element—A Half Century, edited by L. R. Morss and J. Fuger, American Chemical Society (1992), pp. 457–468.

Glovebox Decontamination

Jay S. Samuels, Deborah J. Dale, Mary Ann B. Abeyta, Pamela K. Trujillo, Donivan R. Porterfield, J. Kevin Barbour

Los Alamos National Laboratory

The U.S. Department of Energy (DOE) continually seeks effective and safer decontamination technologies for use in decontamination and decommissioning (D&D) of nuclear facilities. To this end, the Deactivation and Decommissioning Focus Area (DDFA) of the DOE's Office of Science and Technology sponsors large scale Demonstration and Development Projects (LSDDP's) in which developers and vendors of improved and innovative technologies show-case products that are potentially beneficial to DOE projects and to others in the D&D community.

Benefits sought include reducing health and safety risks to personnel and the environment, increasing productivity, and decreasing the cost of operation. Los Alamos National Laboratory (LANL) waste inventory includes approximately 200 "legacy" Transuranic (TRU) waste gloveboxes in temporary storage at TA-54. These gloveboxes will be processed through the LANL Decontamination and Volume Reduction System to separate the Low Level Waste (LLW) and TRU waste components. The TRU components will be packaged and certified for disposal at the Waste Isolation Pilot Plant in Carlsbad, New Mexico. At this time, it is anticipated that the gloveboxes will have sufficient surface activity to be classified as TRU unless they can be decontaminated to LLW.

The cost benefits are exceptionally good. Waste items / components classified as TRU waste are costly to dispose of, with an estimated cost of approximately $140,000 for an average sized glovebox. When the LANL gloveboxes are decontaminated to LLW (i.e., < 100 nanocuries per gram [nCi/g]), the disposal cost will be reduced to approximately $6,500, or about by 95 percent. In addition, to cost savings decontamination may enable the reuse of gloveboxes that are not considered obsolete by design. It also may allow for other gloveboxes at LANL to be stored temporarily within the Plutonium Facility at TA-55 pending reuse. Finally, LLW categorized gloveboxes of no future utility have an immediate path forward to disposition – they may be disposed of at approved LLW sites.

The Actinide Analytical Chemistry group of the Chemistry Division (C-AAC) at LANL volunteered the use of two plutonium-contaminated gloveboxes to test three decontamination techniques. The demonstration at LANL included comparison of the Environment Alternative Inc. (EAI) technology, cerium nitrate solution, and the baseline technique (nitric acid solution) in use at LANL. All of the decontamination techniques involved the application of solutions onto the metallic, interior, surfaces of the glovebox, scrubbing it with an abrasive 'Scotch-Brite'®™ pad, and wiping it down with polypropylene rags. The surface activity was measured prior to decontamination and after each successive application of the solutions.

Cerium nitrate (in nitric acid) is a strong oxidizer capable of oxidizing and dissolving plutonium and stainless steel components such as nickel, chrome, and iron. The solution is sprayed onto the surfaces, scrubbed in, allowed to react, and then rinsed with water. While it is reacting, it strips a few microns from the metal surface of the glovebox resulting in the removal of fixed radiological surface contamination.

The EAI decontamination technology requires applying and removing (by rinsing) three separate chemical formulations to the contaminated surfaces in a specified sequence. Each formulation is customized based on the metal to be decontaminated and the isotopes present. Each formulation solution is applied in low volumes, usually as a spray. The solution then remains on the surfaces for a defined period of time, rinsed clean with an EAI customized rinsate solution, and then removed. The technology is not dependent on adequately scrubbing the surface to be effective. The application and removal of all three formulations (and associated rinsing) to the contaminated

CP673, *Plutonium Futures — The Science,* edited by G. D. Jarvinen
2003 American Institute of Physics 0-7354-0140-3

surfaces consists of one cycle of the process, and typically requires one day (24 hours) to complete. This cycle is repeated as needed until the desired residual decontamination levels are achieved. The EAI solutions are reportedly compatible with all glovebox surfaces including windows and plastic seals and gaskets.

Prior to conducting the demonstration, the inner surfaces of the glovebox were cleaned and wiped down with Fantastik®™. All points measured on the glovebox surface showed an activity above one million counts per minute (2857 kilodisintegrations per minute [kdpm]/100 square centimeters [cm^2]). To be successful, the demonstration had to show a reduction of free and fixed contamination on all contacted surfaces to below 50-kdpm/100 cm^2. The operation times from start to finish of each task, alpha survey measurements for surface activity, and waste volumes generated during the demonstration were recorded. Results from each technique will be presented.

Uranium-Plutonium Separations by Ion Exchange

Alice Slemmons, John FitzPatrick, and Jeff Treasure

C-AAC, Los Alamos National Laboratory, Los Alamos, NM 87545

INTRODUCTION

Many programs in the nuclear complex require uranium processing to be performed in plutonium-contaminated enclosures. Understandably, this work produces trace plutonium contamination in the resulting uranium residues. Before reprocessing the uranium residues, the plutonium concentration must be reduced from approximately 10 g Pu / kg U to 100 μg Pu / kg U.

The feasibility of batch or semibatch ion exchange systems to perform this separation has been investigated. As part of this study, contact experiments were performed to examine how well uranium and plutonium ions [U(VI) and Pu(IV)] were removed from nitric acid systems by various ion-exchange materials.

A number of commercially available ion exchange materials were included in this study. These resins include the following: Reillex HPQ Polymer (70% and 100%), Dowex Marathon 11, Lewatit MP500, U/TEVA resin, TRU-spec resin, Actinide Resin, Diphonix, Diphosil, CMPO-pan, and Duolite C-467.

RESULTS

In order to study the uptake kinetics, the uranium and plutonium distribution coefficients and decontamination factors were determined as the resin-solution contact time was varied. The results of the contact-time studies were used to select appropriate contact times both for the subsequent experiments and potentially for the residue treatment itself.

For most of the ion exchangers studied, equilibrium had been reached by one hour. However, a number of the resins did exhibit an increase in distribution coefficients again when very long (approximately 24-hour) contact times were used.

Because the kinetics at one hour of resin contact were fairly stable for most of the resins, this length of time was selected for mixing the subsequent experiments (described below). However, longer mixing times were sprinkled throughout these experiments in order to verify that the actinide uptake followed the same general trends at both contact times.

The ensuing set of batch experiments examined the effect of varying nitric acid strengths on U(VI) and Pu(IV) extraction. The contact time was held constant as discussed above, while the nitric acid concentration was varied from 0.1 to 10 molar. Assorted data (distribution coefficients, separation factors, etc.) depicting the nitric acid molarity dependencies for each resin are presented. Figure 1 illustrates the nitric acid dependencies of the strong base resins in the study. The nitrate dependency of the extractions was also explored by adding NO_3^- to solutions of constant nitric acid molarity.

For some of the preferred ion exchangers, select experiments were repeated at different uranium and plutonium concentration ratios in order to determine whether, and at what point, competition between the two elements could become a problem when treating the residues. Additionally, the comparisons of actual loading capacities and literature loading capacities were performed.

CP673, *Plutonium Futures — The Science,* edited by G. D. Jarvinen

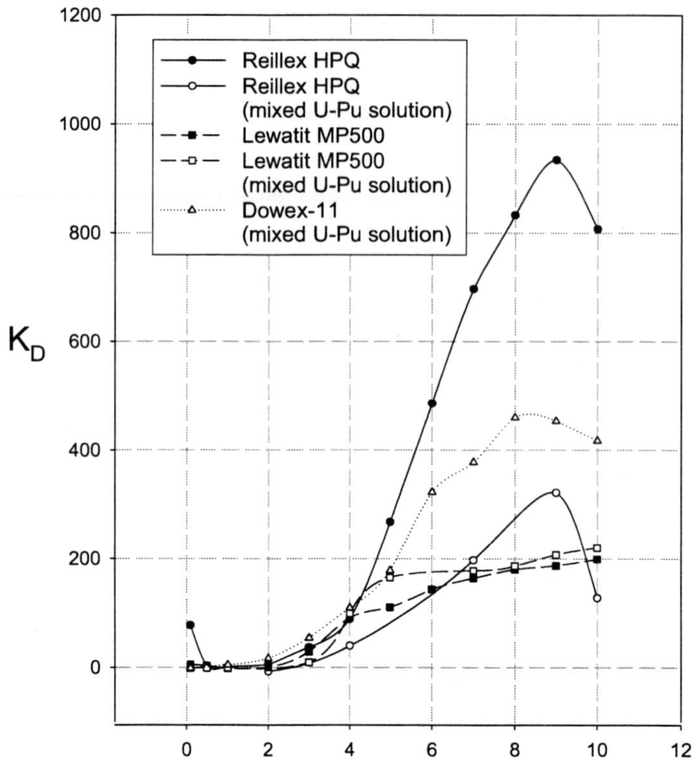

FIGURE 1. Comparison of plutonium uptake by the strong base resins, both with and without uranium present.

DISCUSSION

The data resulting from this work answer a number of questions such as:

- How does the performance of one resin compare to another under identical conditions?
- How does U uptake compare to Pu uptake for each resin?
- What separation schemes become apparent through the data (i.e., optimal Pu loading molarity, optimal Pu elution molarity, potential U elution schemes based on acid molarity, etc.)?
- How much resin will be necessary to reach desired decontamination factors?
- Will there be uranium vs. plutonium competition issues?

This information helps to enable the selection and design of appropriate tools for cleaning the residues in question.

Gallium Behavior in a Low-Temperature Molten Salt System $[C_6H_{11}N_2][N(SO_2CF_3)_2]$

V. V. Smolenski,[1] A. A. Khokhryakov,[2] A. L. Bove,[1] and A. G. Osipenko[3]

[1]*Institute of High-Temperature Electrochemistry, the Urals Branch of Russian Academy of Sciences, Ekaterinburg, Russia, tel. 7-3432-507183; e-mail: smolenski@etel.ru*
[2]*Institute of Metallurgy, the Urals Branch of Russian Academy of Sciences, Ekaterinburg, Russia*
[3]*FSUE SSC Russian Federation Reactor Institute of Atomic Reactors, Dimitrovgrad, Russia*

Currently, there are no industrial-scale processes capable of converting surplus weapons plutonium to usable MOX fuel. The current baseline processes (aqueous and pyroprocessing) suffer from several distinct disadvantages. Low-temperature alternatives based on newly developed room temperature ionic liquids may overcome many of these problems, while maintaining many of the advantages.

Room temperature ionic liquid $[C_6H_{11}N_2][N(SO_2CF_3)_2]$ or (EtMeIm-Tf_2N) was used as a solvent. It is thermally resistant up to 670 K, melts at 270 K, possesses hydrophobic properties, an enhanced conductivity for this group ($\sim 1.3 \times 10^{-2}$ S/cm), and has a wide electrochemical window (~5.0 V).[1–6]

The chemical and electrochemical behavior of gallium was studied using the procedures of linear and cyclic voltammetry, IR spectroscopy, and measurement of the working electrode potential change after its short-term polarization.

As an example, Figure 1 shows the electrochemical windows obtained by cyclic voltammetry for pure solvent and with addition $GaCl_3$. It indicates the availability of a wide electrochemical window and the possibility of using the given electrolyte as a prospective solvent. When gallium trichloride is introduced into the melt, new current peaks are produced on the cathode branch of the curve at a potential of –1.7 V and at 0.2 V on the anode branch. Recording the potential-time dependencies, obtained after a short-term polarization of a working electrode revealed that when a cathode more negative than –1.7 V is polarized, a plateau appears at a potential of –0.75 V. This potential is equal to the equilibrium potential of gallium metal relative to a reference tungsten electrode.

Oscillation spectra EtMeIm ($C_6H_{11}N_2$) and Tf_2N ($N(SO_2CF_3)_2$) and their structural models are described in detail in References 7–10. Comparison of the experimental results of these works with our data for solvent shows good conformity of the obtained oscillation frequencies.

Gallium trichloride introduction into ion liquid EtMeIm-Tf_2N results in changes in oscillation frequencies below 500 cm^{-1}. The changes concern the intensity and position of the absorption bands 408 and 359 cm^{-1} from group SO_2 and 290 cm^{-1} from group CF_3. Moreover, a new absorption band at 375 cm^{-1} appears in the same range; its intensity depends upon gallium trichloride concentration in the solution. This absorption band lies in the region of triply degenerate oscillation frequency ν_3 of the complex $GaCl_4^-$, symmetry T_d; its oscillation spectra are well known.[11,12]

Considering the fact that changes in the oscillation spectrum mainly occur only with libration oscillation frequencies of group SO_2, we can assume that gallium trichloride, being a Lewis acceptor and interacting with anion Tf_2N^-, produces a heteroligandic gallium complex. One of the oxygen ions of group SO_2 of anion Tf_2N^- is a part of its coordination sphere along with three chloride ions.

Therefore, analysis of IR-spectra of gallium trichloride solutions absorption in ion liquid EtMeIm-Tf_2N allows the assumption that ion Tf_2N^- coordination is performed through one of the oxygens of group SO_2 and that the heterocomplex $GaOCl_3^{2-}$ is generated with local symmetry C_{3v}.

CP673, *Plutonium Futures — The Science,* edited by G. D. Jarvinen

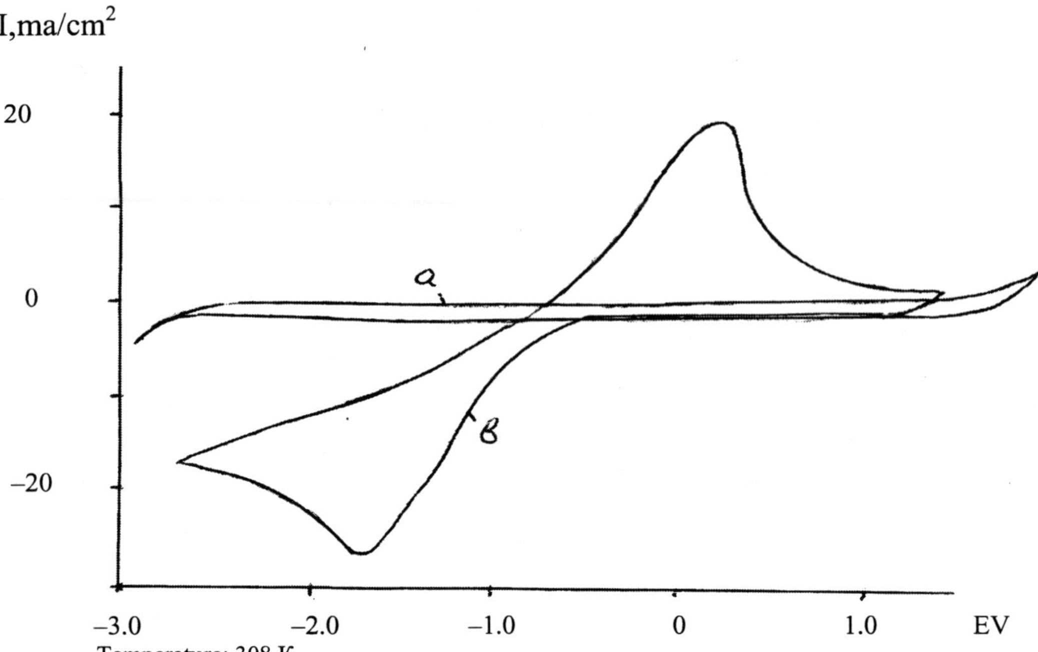

Temperature: 308 K
Working electrode: metal tungsten
Reference electrode: metal tungsten
Counter electrode: glass carbon
Scan rate: 0.1 V/s
(a) without $GaCl_3$; (b) with $GaCl_3$

FIGURE 1. Cyclic voltammetric diagram of ionic liquid $C_6H_{11}N_2$-$N(SO_2CF_3)_2$

Temperature – 298 K
1. $GaCl_3$ concentration – 15,7 mol%.
2. $GaCl_3$ concentration – 20,8 mol%.
3. $GaCl_3$ concentration – 50,2 mol%.

FIGURE 2. IR spectrum of gallium chloride solution in $[C_6H_{11}N_2][N(SO_2CF_3)_2]$.

ACKNOWLEDGMENTS

This work was supported by the U.S. Department of Energy, Office of Basic Energy Sciences, under a Memorandum of Understanding with the Russian Academy of Sciences. The authors wish to thank Los Alamos National Laboratory for financial support and its interest in the work.

REFERENCES

1. Hussey, C. L., Pure and Appl. Chem. 60(12), 1763–1772 (1988).
2. Sun, W., and Hussey, C. L., Inorg. Chem. 28, 2737 (1989).
3. Bonhote, P., Inorg. Chem. 35, 1168–1178 (1996).
4. Heerman, L., De Waele, R., and D'Olieslager W., J. Electrochem. Chem. 193, 289–294 (1985).
5. Olivier-Bourbigou, H., and Magna, L., J. Molecular Catalysis, 1–19 (2002).
6. Schoebrechts, J. P., Gilbert, B. P., and Duyckaerts, G., J. Electrochem. Chem. 145, 139–146 (1983).
7. Dieter, K. M., Dymen, Ch. J., Heimer, N. E. et al., J. Am. Chem. Soc. 110, 2722–2726 (1988).
8. Campbell, T. L. E., Johnson, K. E., and Torkelson, T. R , Inorg. Chem. 33, 3340–3345 (1994).
9. Takahashi, S., Curtiss, L. A., Gosztola, D., Konra, N., and Saboungi, M. L., Inorg. Chem. 34, 2990–2993 (1995).
10. Rey, I., Johanson, P., Lindgren, T., Lasseques, T. C., Croudin, T., and Servant, L., J. Phys. Chem. 102, 3249–3258 (1998).
11. Øye, H. A., and Buss, W., Acta Chemica Scandinavica A 29, 489–498 (1975).
12. Hvistendahl, J., Infrared Emission Spectra of Chloraluminates and Related Melts. Adhandling (1982) 80 pp.

Purification of Aqueous Plutonium Chloride Solutions Via Precipitation and Washing

Mary Ann Stroud, Richard R. Salazar, Kent D. Abney, Elizabeth A. Bluhm, and Janet A. Danis

Actinide Process Chemistry Group (NMT-2)
Nuclear Materials Technology Division
Los Alamos National Laboratory

INTRODUCTION

Pyrochemical operations at Los Alamos Plutonium Facility (TA-55) use high-temperature melts of calcium chloride for the reduction of plutonium oxide to plutonium metal and high-temperature combined melts of sodium chloride and potassium chloride mixtures for the electrorefining purification of plutonium metal. The residual plutonium and americium are recovered from these salts by dissolution in concentrated hydrochloric acid followed by either solvent extraction or ion exchange for isolation and ultimately converted to oxide after precipitation with oxalic acid. Figure 1 illustrates the current aqueous chloride flow sheet used for plutonium processing at TA-55.

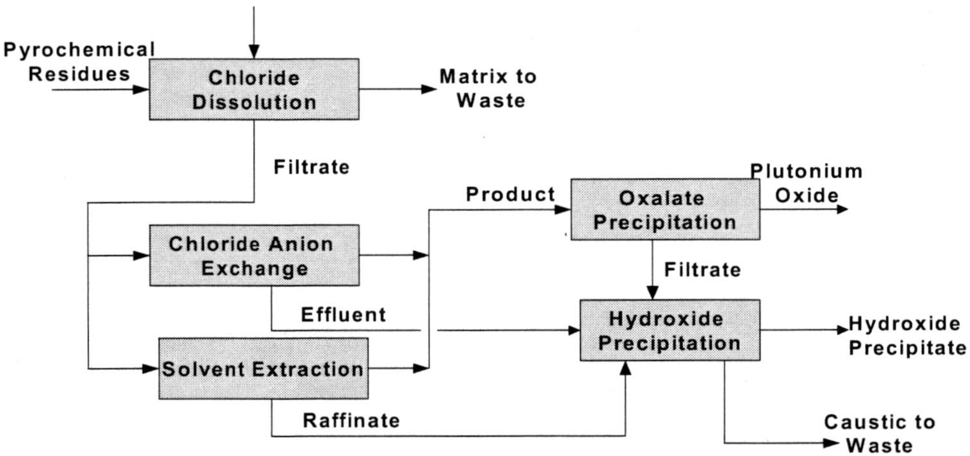

FIGURE 1. Principle unit operations in the current aqueous chloride flow sheet.

Aqueous chloride operations generate copious quantities of secondary wastes and are difficult to manage from a materials corrosion perspective. Two goals of this research are to minimize the waste generated and the glovebox space needed to process chloride residues. These goals may be achieved by modifying the existing flow sheet so that residues can be sent to the mainstream nitrate operations for processing. Before the residues can be sent to nitrate operations, most of the chloride must be removed. The proposed modification to the existing flow sheet, illustrated in Figure 2, would eliminate the chloride anion exchange and solvent extraction steps. After dissolution, plutonium precipitation is conducted, followed by a series of precipitate washings to remove accumulated chloride. The washed plutonium precipitate would be sent for redissolution and purification in the nitric acid flow sheet, where nitric acid recycle is an option that minimizes overall aqueous waste generation.

CP673, *Plutonium Futures — The Science,* edited by G. D. Jarvinen

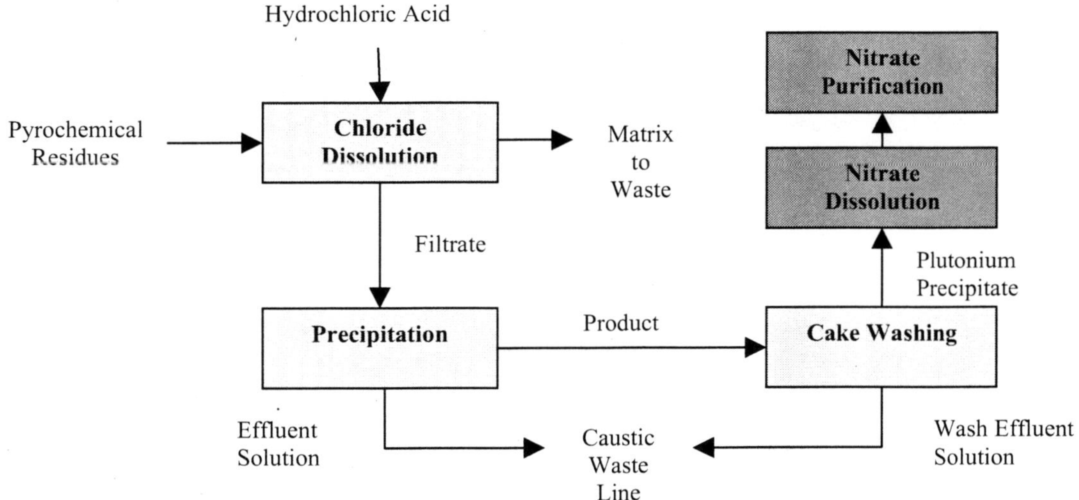

FIGURE 2. Illustration of the proposed modifications to the aqueous chloride flow sheet.

This paper discusses results obtained by processing two direct oxide reduction (DOR) residues, containing plutonium, calcium chloride and calcium oxide, using the flow sheet illustrated in Figure 2. Results are discussed from the perspective of chloride removal efficiency, plutonium recovery and filterability.

EXPERIMENTAL PROCEDURE

Two DOR salt residues were pulverized, dissolved in 6 M HCl, and filtered to form stock solutions for precipitation experiments. Hydroxylamine hydrochloride was added to the stock solution to reduce the plutonium to the +3 oxidation state. The quantity of oxalic acid added was varied to investigate the efficiency of plutonium oxalate formation and washing properties. Potassium hydroxide or sodium hydroxide was added to precipitate any remaining plutonium as a hydroxide. The combined oxalate/hydroxide precipitate was washed with deionized water of varying pHs to remove interstitial chloride. The oxalate/hydroxide precipitate was dried, dissolved in nitric acid, and chloride concentrations of the resulting solutions were measured. Two liters of stock solution containing 63 g of Pu and 480 g of chloride were used for each experiment in Run 1. Two liters of stock solution containing 11 g of Pu and 370 g of chloride were used for Run 2 experiments. Seven experiments with the following varying experimental parameters were conducted.

For Run 1A, a 15% molar excess of oxalic acid was added and the hydroxide precipitation pH was 8.3.

For Run 1B, a 250% molar excess of oxalic acid was added and the hydroxide precipitation pH was 10.1.

For Run 1C, a 125% molar excess of oxalic acid was added and the hydroxide precipitation pH was greater than 11. A considerable amount of calcium hydroxide co-precipitated with the plutonium hydroxide.

Run 2A, a 28% molar excess of oxalic acid was added and the hydroxide precipitation pH was 10.0.

Run 2B, a 275% molar excess of oxalic acid was added and the hydroxide precipitation pH was 8.6.

Run 2C, no oxalate precipitation was conducted. The hydroxide precipitation pH was 9.0.

Run 2D, no oxalate precipitation was conducted. The hydroxide precipitation was conducted at 60°C and pH 9.0.

RESULTS AND DISCUSSION

Results of plutonium recovery and chloride washing efficiency are summarized in Table 1.

TABLE 1. Plutonium Recovery and Chloride Washing Efficiency Results *Values in red were below the target levels and met acceptance criteria.

Sample ID	Total Grams of Precipitate	Initial Grams of Pu	g of Pu in Washed Precipitate	g of Pu in Filtrate and Wash Solutions	Alpha Activity of Combined Solutions (mCi/L)	mg of Cl^- in Washed Precipitate	mg Cl^- per Grams of Pu in Washed Precipitate
Target					<4.5		< 3.0
Run 1A	72	63	41	0.42	4.6	39 (4, 65)	0.95*
Run 1B	148	63	53	2.3	14	*13.4*	0.26
Run 1C	200	63	44	1.2	12	500	11
Run 2A	30	11	11	1.3	15	6.3	0.50
Run 2B	46	11	11	1.7	20	8.1	0.73
Run 2C	23	11	13	.057	1.8	84	7.6
Run 2D	40	11	13	.049	0.88	340	31

For each run, the chloride concentrations in all solutions were measured to determine the efficiency of the chloride removal. The target maximum level of chloride in the washed precipitate was 3 mg Cl^- per gram of plutonium. Chloride concentrations in the plutonium oxalate/hydroxide precipitates were reduced to desired levels by washing with caustic water of varying pH. When washed with identical amounts of solution, chloride concentrations were reduced by at least an order of magnitude more in precipitates containing both oxalate and hydroxide as compared with hydroxide-only precipitates. During the hydroxide precipitation, it was possible to minimize calcium hydroxide precipitate formation by maintaining the pH of the solution to less than 10. When a significant amount of calcium hydroxide was precipitated with the plutonium hydroxide, chloride concentrations in the washed precipitate increased by greater than an order of magnitude.

A goal of the research is to minimize plutonium losses in the chloride processing, maximizing the amount of plutonium sent to nitrate operations for recovery. To ensure minimal plutonium losses, the alpha activity in the filtrate and wash solution should be below the caustic waste line discard limit of 4.5 mCi/L. Table 1 summarizes the plutonium recovery and alpha activity of the combined filtrate and wash solutions for all runs. Results indicate that plutonium oxalate precipitation resulted in increased plutonium losses in the filtrate and washes. This may be due to the small particle size of the oxalate precipitate passing through the 20-micron filter. For oxalate/hydroxide precipitations, all combined solution activities exceeded the discard activity level. For hydroxide-only precipitations, filtrate/wash solution activities were below discard activity level. For Run 1, a large amount of plutonium was not recovered in the solutions or the precipitate. The plutonium loss may have been due to hold-up in the filter paper and filter boat, spillage, or a miscalculation of the initial amount of plutonium present in the stock solution.

For Run 1, filtration rates varied form 190 ml/min for Run 1B, the run with considerable excess oxalate, to 20 ml/min for run 1C, where a significant amount of calcium hydroxide was precipitated with the plutonium hydroxide. Filtration rates were maximized when the amount of calcium precipitated as a hydroxide was minimized and when the amount of plutonium precipitated as an oxalate rather than a hydroxide was maximized.

A goal of the research is to minimize the amount of aqueous waste generated in chloride processing. The chloride washing flow sheet resulted in approximately a 50% reduction in aqueous waste generation, as compared with processing by the standard flow sheet. These results support the feasibility of using the chloride washing processing flow sheet, Figure 2, as an alternative to the existing flow sheet to process chloride residues resulting from plutonium metal production operations.

The Optimum Plutonium Fuel Form in Light Water Reactors

Prof. James S. Tulenko, Michael Savela, and Dr. Gueorgui Gueorguiev

University of Florida, Nuclear & Radiological Engineering Department
Gainesville, FL 32611-8430
352-392-1401 ext. 314; e-mail to tulenko@ufl.edu

Abstract. The University of Florida has underway a research program to validate the benefits of developing a Pu/ZrH/U matrix fuel for the irradiation of the U.S. weapons plutonium and European reprocessed plutonium from an economic, operational, and performance basis. Thermal reactors using plutonium as a fuel are inherently undermoderated because of the large absorption cross sections of plutonium and the presence of large absorption resonances for plutonium in the thermal and near-thermal energy ranges. The use of the proven TRIGA ZrHx-based fuel with plutonium has shown an extremely large (>20%) increase in reactivity over the conventional UO_2/PuO_2 fuel form currently being considered, with an additional major increase in the destruction of plutonium, rendering it an extremely attractive fuel form for plutonium disposition.

The University of Florida is investigating the use of zirconium hydride as a fuel matrix material with plutonium and uranium in order to better thermalize the core flux. Our initial analysis shows a beginning life reactivity of ~1.55 for the fully hydrided fuel versus ~1.35 for the conventional mixed oxide (MOX) fuel currently being pursued. When fully hydrided, the hydrogen atom density in ZrH_x approaches that of water. Because the neutron capture cross section of zirconium is small, the hydride is an excellent moderator. In TRIGA reactors, the hydride is incorporated with uranium in the fuel elements to serve as a moderator. The thermal conductivity of the compacted hydride material approaches that of solid zirconium metal. The TRIGA U-ZrHx fuels contain uranium of various enrichments as a fine metallic dispersion (about 1 um in diameter) in a zirconium hydride matrix. The uranium is incorporated into the zirconium matrix during the melting process of zirconium sponge and enriched uranium metal. After solidification, a homogeneous alloy of uranium-zirconium is ready for hydriding. The final hydrogen-to-zirconium ratio is controlled with temperature and hydrogen pressure during hydriding. The single delta phase of zirconium hydride is normally selected, which eliminates the problems of density changes associated with phase changes. A similar process can be used for plutonium. Most of the irradiation experience to date has been with the uranium-zirconium hydride fuels used in the SNAP and TRIGA reactors. The early TRIGA fuels contained either a nominal 8.5 wt % or 12 wt % uranium. These values are consistent with the plutonium content being used in the University of Florida fuel study.

The TRIGA fuel elements used a stainless-steel (SS) clad because the presence of the hydrogen precludes the use of zircaloy. The UZrHx or PuZrHx fuel is chemically stable. It can be safely quenched at 1,200°C in water. High-temperature strength and ductility of the SS fuel cladding to be used provides total clad integrity at temperatures as high as 950°C.[1] The TRIGA fuel has been taken to a burnup in excess of 80,000 MWD/MTU. The Connecticut Yankee Haddam Neck Reactor operated with SS cladding because it could not meet the LOCA requirements with zircaloy clad. Because of the large absorption cross section associated with SS with its resultant 1% higher uranium enrichment required, it is not an economic cladding for uranium fuel. However, in a previous study,[2] Tulenko et al. showed that SS cladding can be economically used with a plutonium fuel. The high absorption and fission cross section of a plutonium fuel greatly reduces the negative economic effects of the absorption cross section of the SS clad, making stainless steel an economically acceptable clad for a plutonium mixed-oxide fuel (MOX) assembly.

SS cladding can be operationally attractive because the corrosion resistance of SS cladding has been well proven and will accommodate ultrahigh burnup better than zirconium clad. Additionally, the loss of coolant accident (LOCA) becomes far less restrictive for stainless steel because of a minimal SS clad/water reaction. This factor

CP673, *Plutonium Futures — The Science,* edited by G. D. Jarvinen

opens the door for commercial MOX fuel to use the proven TRIGA composite fuel using zirconium-hydride. The MOX lattice is greatly undermoderated, and the hydrogen in the zirconium hydride provides additional moderation, which increases the reactive performance of the MOX fuel. Our studies have shown that the reactivity of the Pu/ZrHx/U fuel is increased by over 20% Δk when the U is replaced fully by ZrH_x (see Figure 1).

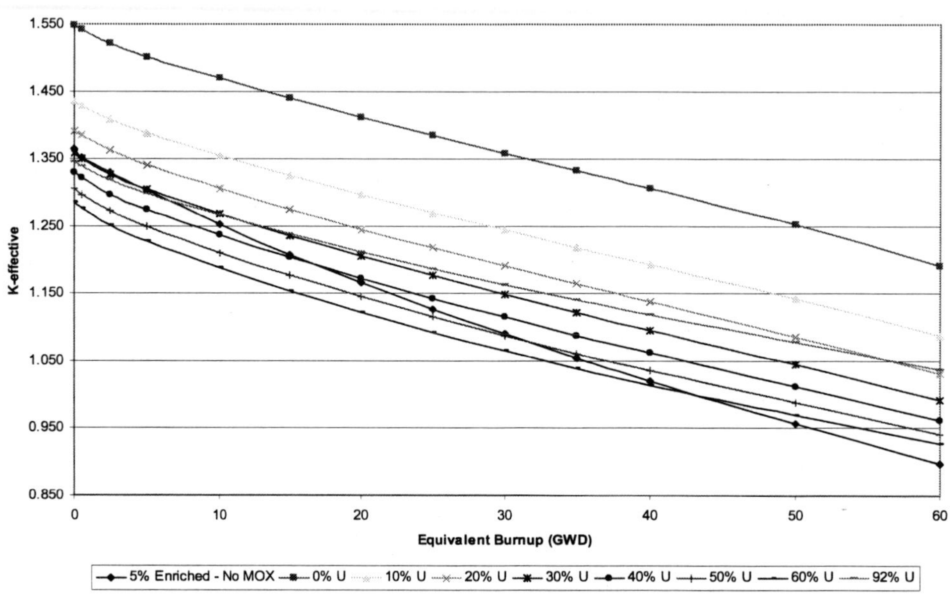

FIGURE 1. K effective vs burnup for plutonium fuel with varying % ZrHx.

Additionally, this composite fuel is attractive to the nuclear weapons nonproliferation program in its efforts to dispose of the weapons plutonium because the matrix material does not breed more plutonium and along with the higher reactivity, more plutonium is destroyed.

REFERENCES

1. Simnad, M. T., "The U-ZrHx Alloy: Its Properties and Use in TRIGA Fuel," Nuclear Engineering and Design 64, 403–422 (1981).
2. Smith, L. A., and Tulenko, J. S., "Fuel Cycle for the 21st Century," Trans. Am. Nucl. Soc., 69, 118–120 (1993).

Development of an Electrochemical Reduction Process of Oxide Fuels

Tsuyoshi Usami and Tadashi Inoue

Central Research Institute of Electric Power Industry
Iwadokita 2-11-1, Komae-shi, Tokyo, Japan

INTRODUCTION

A pyrochemical conversion process of spent oxide fuel to metal has been developed. The process, named the "lithium reduction process," employs lithium (Li) metal as a reductant and lithium chloride (LiCl) as a solvent at 650°C.[1–7] Lithium oxide (Li_2O) is generated from the Li metal and oxygen in the spent fuel, dissolved in the LiCl and finally decomposed by electrolysis when the oxide reduction is finished. The reduction product is sent to an electrorefiner. In order to simplify this reduction process, the development of an electrochemical reduction technology was started. In this process, oxides to be reduced are loaded in a cathode in molten LiCl or LiCl-KCl eutectic salt. Oxygen is electrochemically transported from the oxide to the anode made of platinum. An oxygen source such as Li_2O is previously loaded into the solvent salt as a carrier of oxygen from the spent fuel. In this process, the solvent salt works only as a path of the oxygen while it works as reservoir of oxygen in the lithium-reduction process. Therefore, the electrochemical reduction process has a possibility of decreasing the solvent salt. The LiCl-KCl eutectic is not suitable for the lithium-reduction process because small portions of KCl are reduced by the Li metal. Because no Li metal exists in the electrochemical reduction process, the use of eutectic salt can be an advantage in the sense of being consistent with the electrorefiner, which also employs eutectic salt.

Purposes of this study are as follows:

- To indicate that UO_2 can be reduced directly without generating Li metal, and
- To show suitable operating conditions, such as the composition of the salt, temperature, and the concentration of oxygen in the solvent salt.

RESULTS

Figure 1 shows polarization curves measured in LiCl at 650°C. Anode and reference electrodes are a platinum and Bi-30 mol. % Li alloy, respectively. The cathode is tantalum wire in blank tests, but it was a UO_2 pellet with a 10-mm diameter and a 1.5-mm thickness hung by tantalum wire in real UO_2 tests.

Blank Tests

It can be seen from the graph that the current increased at a cathode potential of –0.78 V. At the anode, the current increased at a potential of 2.40 V in the salt without Li_2O. In tests with Li_2O, the anode current increased at 1.8 V and saturated, increased again at 2.0 V and saturated, and finally increased at 2.40 V. In x-ray diffraction analysis, complex oxide Li_2PtO_3 was observed on the surface of the anode at times. Even after forming a complex oxide, a decrease of the current was not observed.

CP673, *Plutonium Futures — The Science,* edited by G. D. Jarvinen

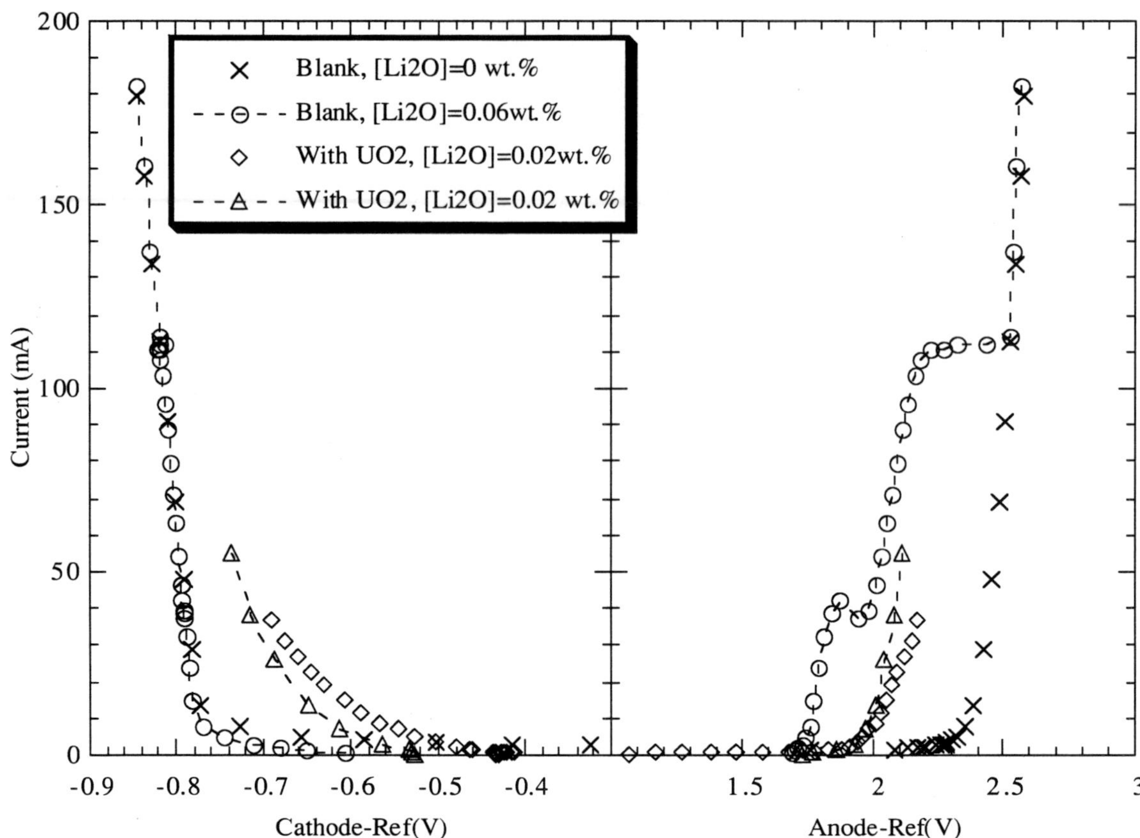

FIGURE 1. Polarization curves in LiCl-Li_2O at 650°C with and without UO_2.uranium tests.

When UO_2 was loaded on the cathode, the cathode current gently increased around –0.5 V. The anode current steeply increased at 2.0 V.

Figure 2 shows polarization curves measured in LiCl-KCl eutectic salt, including various concentrations of Li_2O at various temperatures. It can be seen from the graph that the potential of cathode at which Li deposits tend to become ignoble with an increase of temperature and a decrease of the Li_2O concentration.

DISCUSSION

Because it is observed in pure LiCl, the anode current at –0.78 V is thought to be a deposition of Li. The cathode current at 2.40 V is caused by the generation of Cl_2 gas and/or the dissolution of Pt. The cathode current increasing at 1.80 V is thought to be oxidation of Pt by the following reaction. $Pt + 2Li^+ + 3O^{2-} \rightarrow Li_2PtO_3 + 6e^-$.

Because the electrolysis is continued after forming, the complex oxide is thought to have electron conductivity at 650°C. The cathode current increasing at 2.00 V should generate O_2 gas.

Because the cathode current increased at a more ignoble potential when UO_2 is loaded, the current is thought to be the evidence of the direct reduction of UO_2 without forming Li metal.

The anode potential at which the current increases is independent of the temperature and the Li_2O concentration. On the other hand, the cathode current increases to a more ignoble potential when the temperature is much higher than the melting point. The change of the potential means that Li or K metal can easily deposit on the cathode at a high temperature, and therefore, the electrochemical reduction of oxide becomes difficult. Generally, the reduction rate of the oxide increases with temperature because the rate-determining step is diffusion in the solid being reduced if other parameters are optimized. For these reasons, LiCl-KCl eutectic salt has a disadvantage when compared with LiCl. The reason for the change of the cathode potential was supposed to be caused by decomposition of the KCl because it was not observed in LiCl at the same temperature and because the cathode current at ignoble potential decreased with the addition of Li_2O.

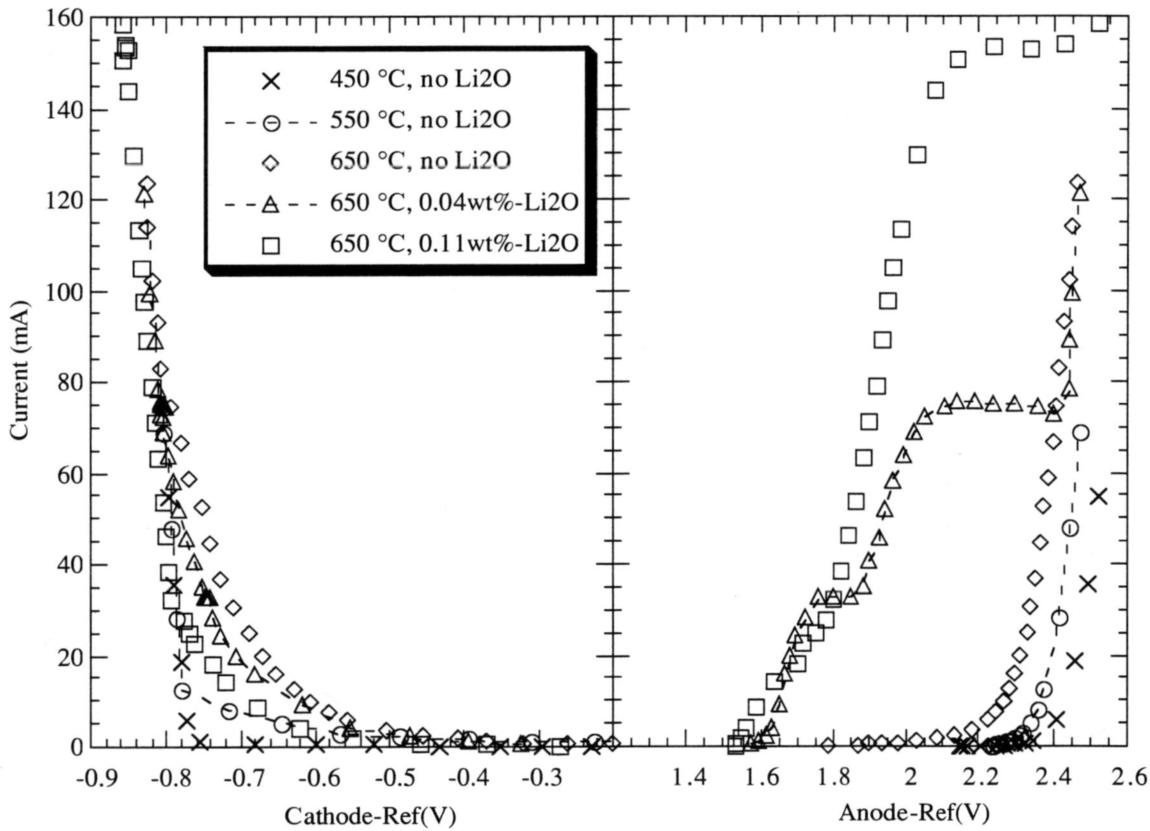

FIGURE 2. Polarization curves in LiCl-KCl without UO_2.

REFERENCES

1. Battles, J. E. et al., in Actinide Processing: Methods and Materials, edited by B. Mishra, The Minerals, Metals and Mater. Soc. (1994), pp. 135–151.
2. Karell, E. J., Pierce, R. D., and Mulcahey, T. P., DOE Spent Nucl. Fuel and Fissile Material Management, presented at ANS, 1996.6.16-20, Reno, Nevada 352-358.
3. Johnson G. K. et al., in Actinide Processing: Methods and Materials, edited by B. Mishra, The Minerals, Metals and Mater. Soc. (1994), pp. 199–214.
4. Usami, T. et al., J. Nucl. Mater. 300, 15–26 (2002).
5. Usami, T. et al., Proc. of GLOBAL 2001, FP164.pdf.
6. Usami, T. et al., J. Nucl. Mater. 304, 50–55 (2002).
7. Usami, T. et al., J. Nucl. Sci. and Technol., Supplement 3, 858–861

Hanford Sr/TRU Decontamination Program: Research from Beaker to Pilot Scale

W. R. Wilmarth, W. D. King, C. A. Nash, F. F. Fondeur, S. W. Rosencrance, D. P. DiPrete, C. C. DiPrete, J. R. Zamecnik, M. A. Baich, M. R. Williams, and T. J. Steeper

Savannah River Technology Center, Westinghouse Savannah River Company

INTRODUCTION

Plutonium and americium are present in the Hanford High-Level Liquid Waste complexant concentrate (CC) waste as a result of the presence of complexing agents, including di-(2-ethylhexyl) phosphoric acid (D_2EHPA), tributylphosphate (TBP), hydroxyethylene diamine triacetic acid (HEDTA), ethylene diamine tetraacetic acid (EDTA), citric acid, glycolic acid, and sodium gluconate.[1] The transuranic (TRU) concentrations approach 600 nCi/g and require processing before encapsulation into low-activity glass. The Savannah River Technology Center (SRTC) has been actively participating in the development of a strontium/TRU decontamination process for the CC waste stored at Hanford in Tanks AN-107 and AN-102.[2–5] The current baseline flowsheet involves the addition of strontium nitrate to effectively remove radio-strontium through an isotopic dilution, followed by a sodium permanganate strike to coprecipitate the actinides in the in-situ-produced manganese solid phases. Demonstration efforts to validate the Sr/TRU flowsheet have included beaker-scale reagent optimization and reaction kinetics testing and engineering scale-up experiments at the multiliter and 1/100th-plant scale (200 gallon) using both simulated and actual waste samples. Several key process engineering and process chemistry needs were identified during a recent Hanford Waste Treatment Plant project risk review.[6]

DESCRIPTION

Testing on the laboratory scale (beaker scale) was performed by adding the precipitating agents to 50 mL of actinide-spiked simulant that had been adjusted to 6 M sodium ion concentration while being heated for 4 h at 50°C. The samples were allowed to cool and aliquots were filtered through a 0.45-micron filter. Samples were submitted for analysis following an additional filtration using a 0.1-micron filter. Simulant multiliter scale tests (~50 gallons) were performed in 55-gallon drums and filtered using a cross-flow filter unit (the Cells Unit Filter or CUF). The CUF uses a 0.01-micron sintered metal filter with a 24-inch length and 3/8-inch diameter. The 1/100th scale facility included a 250-gallon reaction tank and a 150-gallon slurry hold tank. A mechanical agitator and four vertical baffles were used to ensure vigorous agitation for the suspension of entrained solids in the sample and to provide thorough slurry mixing. Heating of the waste simulants was performed using a 208-V electric heater in a recirculating loop. Either a 40-inch by 0.5-inch or 90-inch by 0.5-inch seven-element, cross-flow filter bundle with a nominal porous rating of 0.1 microns was used to process the precipitated simulant batches. Particle size analysis was conducted using a Lasentec M400L chord length analyzer. The method is based on the detection of backscattered laser intensity (Focused Beam Reflectance Measurements—FBRM) from individual particles suspended in the process solution.

CP673, *Plutonium Futures — The Science,* edited by G. D. Jarvinen
2003 American Institute of Physics 0-7354-0140-3

RESULTS

Small-scale testing at SRTC focused on defining the minimum reagent additions required to meet process requirements for strontium, plutonium, and americium using actual tank wastes and waste simulants. Americium-241 is the transuranic limiting isotope. Table 1 lists two flowsheet cases that have received extensive testing. The baseline flowsheet was demonstrated originally using actual radioactive waste from Tank AN-102, and the required Sr/TRU decontamination factors were met (see Table 1).

TABLE 1. Sr/TRU Decontamination Flowsheet Conditions and Decontamination Factors.

Flowsheet Treatment Conditions	**Temperature**	**Hydroxide**	**Strontium**	**Permanganate**
Baseline	50°C	1 M	0.075 M	0.05 M
Optimized	25°C	No Added	0.03 M	0.03 M

Decontamination Factor	**Strontium AN-107 (102)**	**Americium AN-107 (102)**	**Plutonium AN-107 (102)**
Baseline	100 (65)	45 (5.9)	5 (2.1)
Optimized	7 (30)	3.1 (4.3)	3.4 (3.5)
Target	30	5	2

Results with samples from Tank AN-107 revealed that americium and plutonium decontamination factors were far exceeded. Attempts to reduce reagent additions for Tank AN-107 waste simulant provided the optimal conditions (OC) for actinide removal with this sample. However, the optimized flowsheet failed when applied to an actual waste sample from Tank AN-102.

One interesting aspect of the work has been examining the kinetics of particle growth and strontium and actinide decontamination. Figure 1 shows the initial results for kinetics studies of the removal of radio-strontium and americium-241 from a Hanford CC waste simulant. The decontamination was accomplished well within the facility-imposed 4-hour duration. However, the particle growth kinetics (as indicated by total counts measured by FBRM), also shown in Figure 1, reveals that the formation of strontium particles upon the addition of strontium nitrate solution and manganese oxide particles upon the addition of sodium permanganate solution are complete in minutes. Unfortunately, the time scale of the decontamination kinetics experiment was much longer. Most likely, the decontamination occurs during the particle growth stages.

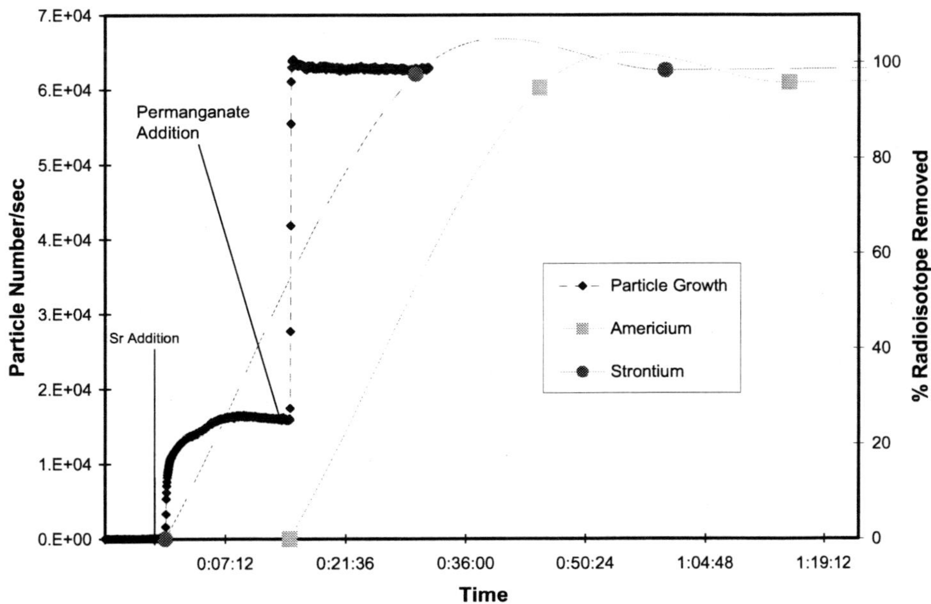

FIGURE 1. Particle growth and decontamination kinetics.

Following a disciplined approach, research into the decontamination of the Hanford CC waste focused next on the multiliter (15 L to 200 L) scale. Program objectives were to establish the behavior of the surrogate slurry while measuring actinide decontamination through the use of lanthanide surrogates. Using actual waste-generated correlation,[7] the decontamination factors obtained for the lanthanides indicated that, had americium and plutonium been present, both the baseline and optimized flowsheet would have produced decontaminated filtrate meeting the design requirements.

The slurry behavior with regards to cross-flow filtration is of utmost importance because of anticipated filter fluxes in the plant operation. Shown in Figure 2 are the cross-flow filtration fluxes produced during various simulant and actual waste flowsheet tests. Three multiliter lab-scale test results are shown where filtration was accomplished using the CUF filter. The results show very good agreement between the filtration run with waste simulant and actual CC tank waste. Filter fluxes rapidly decrease as the filter cake builds and the insoluble solids loading increases in the slurry. However, the time-averaged filter flux exceeded the design target of 0.01 gpm/ft^2. The third multiliter filtration experiment used the reagent-optimized conditions (OC). In general, the filter fluxes agreed with the baseline results, although at higher solids loading, the performance was lower compared to the baseline. These chemical and physical property results strongly suggested that either flowsheet would successfully remove the radio-strontium and actinide constituents of the Hanford CC waste.

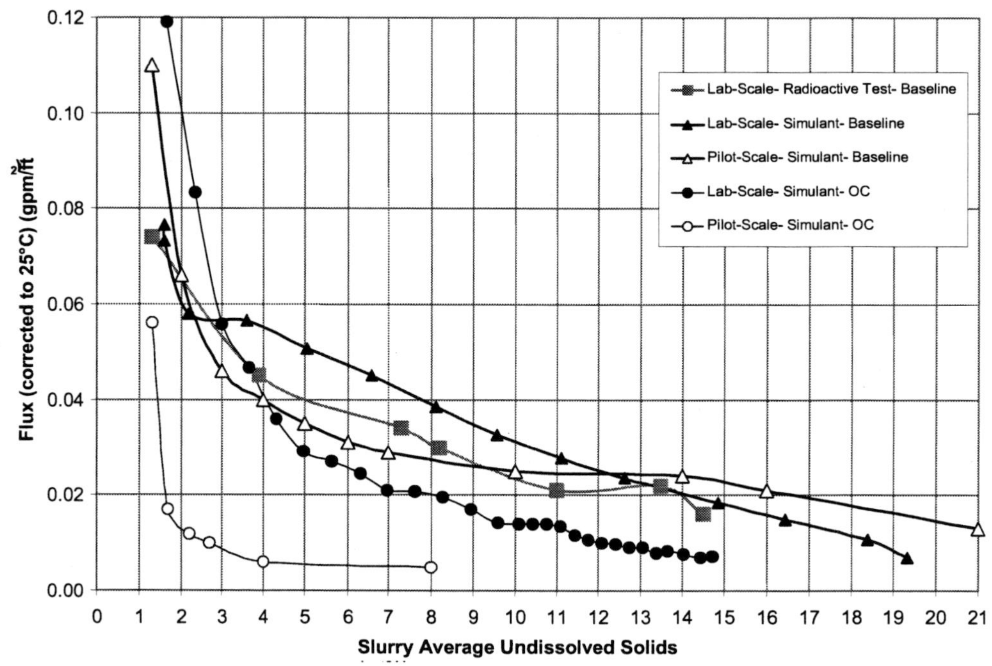

FIGURE 2. Cross-flow filtration results from various Sr/TRU flowsheet experiments.

Also included in Figure 2 are two sets of results from filtering the Sr/TRU precipitate slurry in the 1/100th scale pilot facility. The conditions were the baseline and optimized conditions. A disappointing result was obtained at the pilot scale when the experiment was conducted under the reagent OC. In this experiment, the filtration flux started at a very low value (0.055 gpm/ft^2) and rapidly dropped. Several studies examined the cause of the poor filtration performance. In hindsight, the multiliter experiment under the same reaction conditions showed filtration results about 20%–30% below the baseline flowsheet conditions. Previous beaker scale testing[2] with actual tank waste from this particular tank had shown filtration rate dependence on the waste hydroxide level. This connection between beaker chemistry and pilot scale demonstrations is very beneficial when designing full-scale facilities.

In the experiment conducted under the baseline flowsheet conditions, the slurry filtration behavior mirrored the actual and simulated waste testing at the multiliter scale. Additionally, actinide surrogates were successfully decontaminated at all three experimental scales, indicating a robust Sr/TRU decontamination flowsheet. Although further engineering is required to increase the scale to the plant environment, the strontium and actinide chemistry along with the slurry precipitate behavior strongly indicate that a successful treatment process has been developed.

REFERENCES

1. Klein, M. J., and Wilson, W. G., "Strontium Recovery from Purex Acidified Sludge," ARH-CD-691 (May 1976).
2. Wilmarth, W. R., Rosencrance, S. W., Nash, C. A., and Edwards, T. B., "Sr/TRU Removal from Hanford High Level Waste," J. Sep Sci. 36, 1283 (2001).
3. Wilmarth, R. W., Rosencrance, S. W., Nash, C. A., DiPrete, D. P., and DiPrete, C. C., "Transuranium Removal from Hanford High Level Waste Simulants Using Sodium Permanganate and Calcium," submitted to J. Radioanal. Nucl. Chem.
4. Wilmarth, W. R., Rosencrance, S. W., Nash, C. A., Fonduer, F. F., DiPrete, D. P., and DiPrete, C. C., "Transuranium Removal from Hanford High Level Waste Simulants Using Sodium Permanganate and Calcium," "Plutonium Futures—The Science," American Institute of Physics (2000), p. 53.
5. Nash, Charles A., Rosencrance, Scott W., Wilmarth, William R., and Saito, Hiroshi H., "Transuranic/Strontium Precipitation and Filtration of Hanford Complexant Waste," Proceedings of Waste Management Conference, WM'01 Tuscon, Arizona.
6. Research and Technology Plan, 24590-WTP-PL-RT-01-002, Rev. 1.
7. Wilmarth, W. R., Dukes, V. H., Mills, J. T., Fondeur, F. F., DiPrete, C. C., and DiPrete, D. P., "Optimization Study for Strontium and Actinide Removal from 241-AN-107 Supernate," WSRC-TR-2002-00258 (September 18, 2002).

Radiolytic Effects of Plutonium

Z. P. Zagorski,* J. Dziewinski,** and J. Conca***

*Institute of Nuclear Chemistry and Technology, Warsaw, Poland
**Los Alamos National Laboratory, Los Alamos, New Mexico, USA
***Los Alamos National Laboratory, Carlsbad Operations, New Mexico, USA

INTRODUCTION

Plutonium isotopes, most of them α emitters, cause radiolytic changes in the matrix in which they are embedded. The internal irradiation of Pu^0 metal or its alloys results in physical changes, largely as a result of the formation of helium bubbles, well known to material scientists and weapons specialists. In all other media where plutonium occurs, usually as Pu^{n+} in an ionic form, the results of irradiation are chemical in nature. Homogenous media containing Pu are often aqueous or nonaqueous solutions of plutonium compounds, mostly originating during processing of spent nuclear fuel or from Pu processing. Heterogenous matrices containing plutonium are more complex from the point of view of radiolysis; they usually contain a variety of combinations of common materials contaminated with radionuclides. This class of radioactive materials represents a challenge for the management of plutonium waste. One has to consider a range of time scales for radiolytic effects (and consequently a several-orders-of-magnitude range of the cumulative dose) beginning with waste generation, through packaging and transportation, to the period of final storage. Final storage could be for thousands of years in deep geologic repositories. At every stage of that time scale, radiolysis proceeds continuously and cumulative effects can complicate operating procedures and final disposition. The results presented here have been obtained from experiments that have been irradiated with model materials, which are typically the objects of contamination with plutonium. They were irradiated with linearly accelerated electrons up to very high dose rates, adjusted to simulate any contamination at any point on the time scale.

RESULTS

The wide field of all possible interactions of plutonium radiation with very different chemical compounds has been divided into groups of problems ordered according to the urgency of occurring in practice. The most-needed information was about the radiation yield of hydrogen from the fastest decaying plutonium isotopes. That information is influencing the strategy of preparing waste for transportation. The problem involves investigating various materials, such as organic compounds, including polymers, concretes containing plutonium, and plutonium-contaminated organic debris. Another set of problems includes salt rock in geologic salt repositories, which will eventually come into contact with plutonium after the containment corrodes over the course of thousands of years. Salt exposed to radiation can form electron traps in the crystalline lattice defects. The worst-case scenario involved in modeling the repositories assumes that water will seep into the deposit with irradiated salt. The resulting dissolution of F-centers will free the trapped electrons and cause radiolysis of the brine, separate from radiolysis that will be occurring in the waste. The spectrum of products of this radiolysis is wide and complicated. Besides many chlorine-oxygen compounds formed, also hydrogen peroxide appears in multiionization spurs of comparatively high LET radiation. The seeping water may, but does not have to be, deoxygenated by iron and organic materials present in the waste. The presence or absence of oxygen has a tremendous effect on the outcome of radiolysis and both conditions need to be investigated.

CP673, *Plutonium Futures — The Science*, edited by G. D. Jarvinen

A 10-MeV electron beam (EB) has been applied as the source of radiation in these investigations. It simulates plutonium α radiation, by taking into account easy recalculable effects of different LET radiations. An integrated hardware system of EB irradiation combined with gas chromatography has been developed to handle specific difficulties occurring with the delivery of wide dose ranges, stretching over several orders of magnitude.

It is very important from a waste-management perspective to investigate H_2 production by irradiation, which occurs in all chemical systems containing hydrogen. Hydrogen is energetically easiest to detach, whether bound inorganically or present in organic compounds, especially in polymers. Hydrogenated butadiene rubber has been chosen as a model polymer for elastomers often occurring in the waste as gaskets or tubings. As theoretically expected, a linear relationship of H_2 production versus dose was observed beyond the initial dose of 20 kGy (Figure 1). The reduced yield of hydrogen generation in the initial stage is caused by aromatic additives used routinely in the polymer industry. The phenomenon is welcomed from the point of view of waste management, because it means that the hydrogen generation in the initial stage of waste disposition, i.e., transportation, is less than in the later stages of radiolysis. Other polymers, both synthetic and natural (e.g., cellulose) are under investigation, and very high doses have been administered.

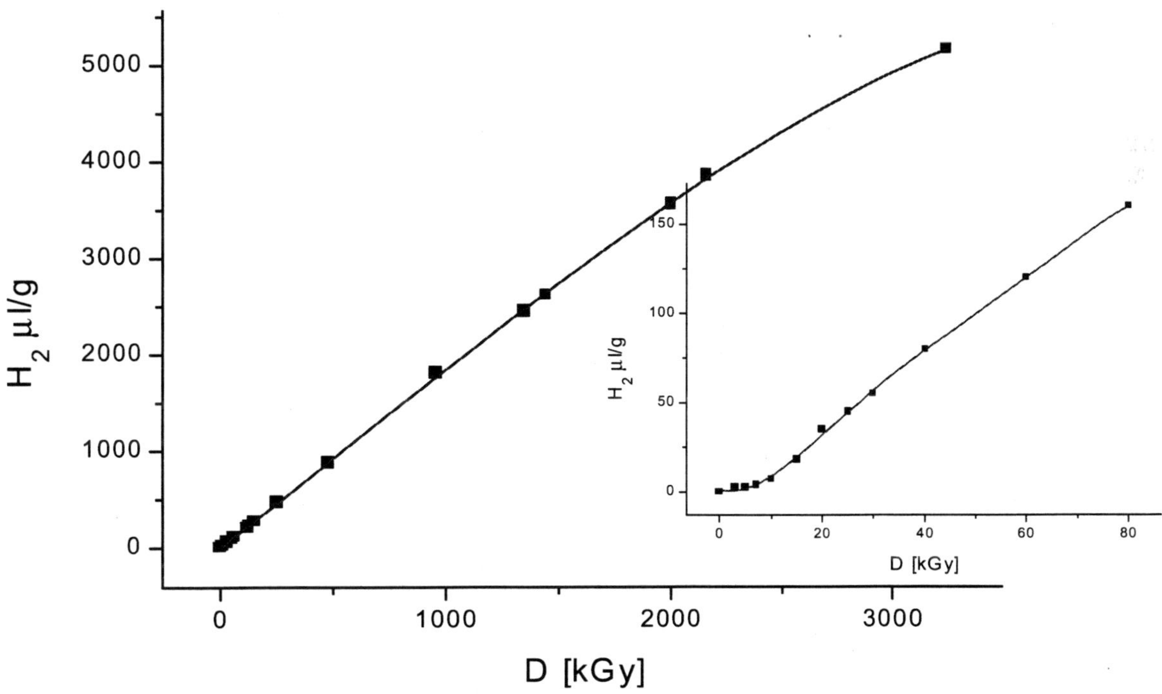

FIGURE 1. Production of hydrogen (µL normal) per 1 g of elastomer in the function of dose. Inset: Production of hydrogen at low doses, showing diminished yield as a result of the presence of stabilizing additives.

Slight changes in the technique are applied to the determination of hydrogen yield from concrete containing plutonium and various organics, involved in the processing of nuclear fuel and the recovery of plutonium from weapons. Although no chemical chain reaction of hydrogen generation has been observed, that question is still open and will be eventually answered as research continues. There is a possibility of a chemical chain reaction in the presence of some materials. Such materials should be avoided during plutonium processing. Different procedures are applied for the experimental investigation of the radiolysis of brines. The analyses of irradiated brines have shown the sensitivity of the radiolysis to the presence of organics (possible humic acids in seeping water) and to the presence of oxygen. Preliminary results were presented and discussed on two international conferences in 2002.[1,2]

DISCUSSION AND CONCLUSIONS

Investigating the chemical aspects of radiation of plutonium is important for waste management for several reasons. It singles out the risks related to the radiolytic products and indicates ways to mitigate and avoid those risks. It also lowers the expense of waste management by lowering the maximum limits of safe packaging, transportation, and storage of waste. The reported investigations, continued in 2003, fit well into the increasing role of radiation chemistry in radioactive waste management, as announced by Zimbrick.[3] A plutonium future is hard to imagine without considerations of radiolytic phenomena occurring when nuclides are in contact with, and embedded in, waste materials. It is part of the responsibility of our generations towards future generations to come.

ACKNOWLEDGMENTS

This work is done and continued under subcontract LANL-INCT 45302-001-02-AA.

Keywords: brines radiolysis, hydrogen generation, plutonium radiation, polymer radiolysis, radiolysis, spurs

REFERENCES

1. Zagórski, Z. P., (a) "Radiation Chemistry in the Management of Radioactive Waste (Selected Topics)," and (b) "Spectroscopic Investigation of the Formation of Radiolysis Products by 10-MeV Linear Accelerator Electron in Salt Solutions," two oral presentations at the 14th Radiochemical Conference, April 14–19, 2002, Marianske Lazne, Czech Republic.
2. Zagórski, Z. P., and Dziewinski, J., "Radiation Chemistry of Polymeric Components of Radioactive Waste," paper presented at the Fifth International Symposium on Ionizing Radiation and Polymers (IraP 2002), September 21–26, 2002, Saint-Adele (Quebec) Canada.
3. Zimbrick, J. D., Rad. Res. 158, 127–140 (2002).

ACTINIDES IN THE ENVIRONMENT AND LIFE SCIENCES

Solubility of Pu, Np, and U from Spent UO_2-Fuel Under Inert/Reducing Conditions

Yngve Albinsson,[1] Virginia Oversby,[2] Arvid Ödegaard-Jensen,[1] and Lars Werme[3]

[1]*Nuclear Chemistry, Chalmers University of Technology, SE-412 96 Gothenburg, Sweden*
[2]*VMO Konsult, Karlavägen 70, SE-114 59 Stockholm, Sweden*
[3]*Swedish Nuclear Fuel and Waste Management Co, Box 5864, SE-10240 Stockholm, Sweden*

Abstract. The overall objective of this program is to improve the scientific understanding of processes that control the release of radioactive species, especially actinides, from spent fuel inside a disposal canister. The Swedish concept has focused on deep burial in the rock, with an iron-lined Cu-canister. Corrosion of the canister iron insert will consume any residual oxygen and provide actively reducing conditions in any fluid phase. Therefore, an investigation of the solubility of different radionuclides under actively reducing conditions) (Fe^{2+}/H_2) has been performed. The solubility of U, Np, and Pu is measured as a function of time for three different conditions: Ar atmosphere, H_2 atmosphere, and H_2 atmosphere with Fe(II) in solution.

EXPERIMENTAL

The current series of tests are conducted in high-pressure vessels lined with PEEK (polyetheretherketone). The liquid phase used for the leaching is a modification of "Allard" groundwater.[1] The modification was done to ensure stability of the composition under anaerobic and reducing conditions. The first cycle of testing used an argon atmosphere of 10 bar in the vessels. Samples were taken after about 1, 4, 8, and 21 days of leaching. The atmosphere in the vessels was then changed to H_2 (10 bar) and an additional 20 ml synthetic groundwater was added to return the volume to approximately 30 ml. Samples were again taken after about 1, 3, 8, and 21 days. At the end of this cycle, the water was removed and replaced with groundwater that had been equilibrated with an Fe strip and an H_2 (10 bar) atmosphere. This gives a concentration of Fe(II) in the solution of approximately 1-0.1 μM. The change of solution was done in a hot cell with an air atmosphere. Another cycle of samples at 1, 4, 8, and 24 days was taken, and the test was terminated. The spent fuel samples were transferred to new PEEK vessels for further testing. The test cycle with Fe-equilibrated water and H_2 atmosphere was repeated, including vessel changes, three more times. The change of vessel after sample number 20 (110 days total testing) was done in an argon atmosphere box.

SUMMARY AND CONCLUSION

After 21 days of leaching, the concentration of U increased from about 10^{-7} to 10^{-6} M and the concentration of Pu increased from 10^{-11} to almost 10^{-9} M. After changing to the H_2 atmosphere, the U began to slowly decrease while the Pu quickly went down to the detection limit (around 10^{-11} M). Changing the solution at day 42 produced a sharp increase in U to 1×10^{-5} M and the Pu to 10^{-7} M , probably due to some oxygen contacting the sample during the change of solution. This peak dropped fast (within 3 days) and at the end of the cycle, the U was below 10^{-8} M and the Pu below 10^{-10} M. For Np no large changes where observed (Np 3×10^{-9} to 10^{-11} M), except for a sharp peak at 42 days as for U and Pu (Np about 10^{-7} M).

CP673, *Plutonium Futures — The Science,* edited by G. D. Jarvinen

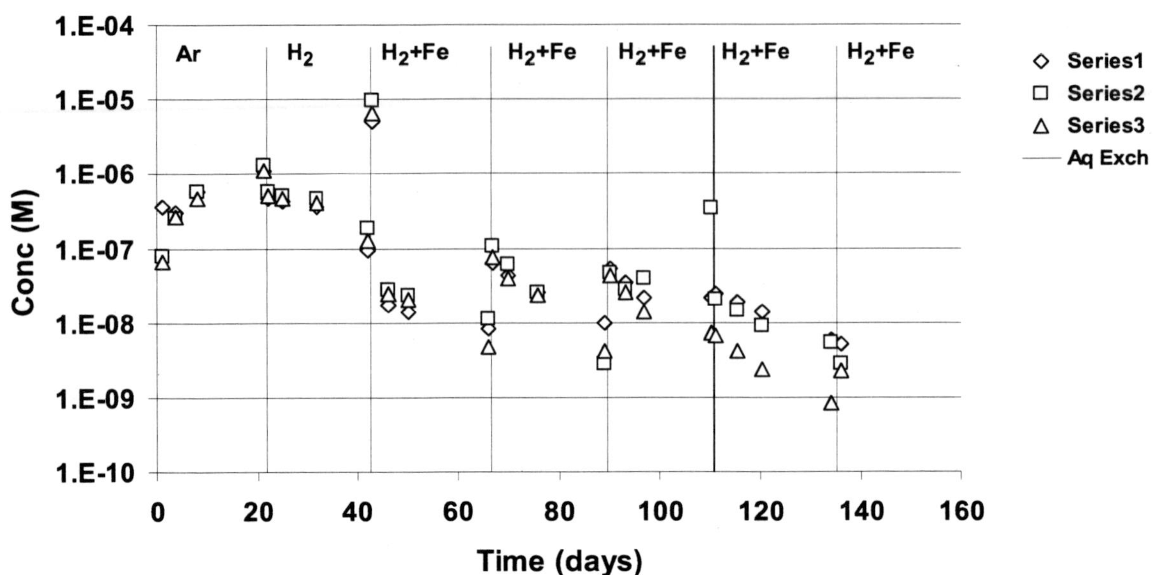

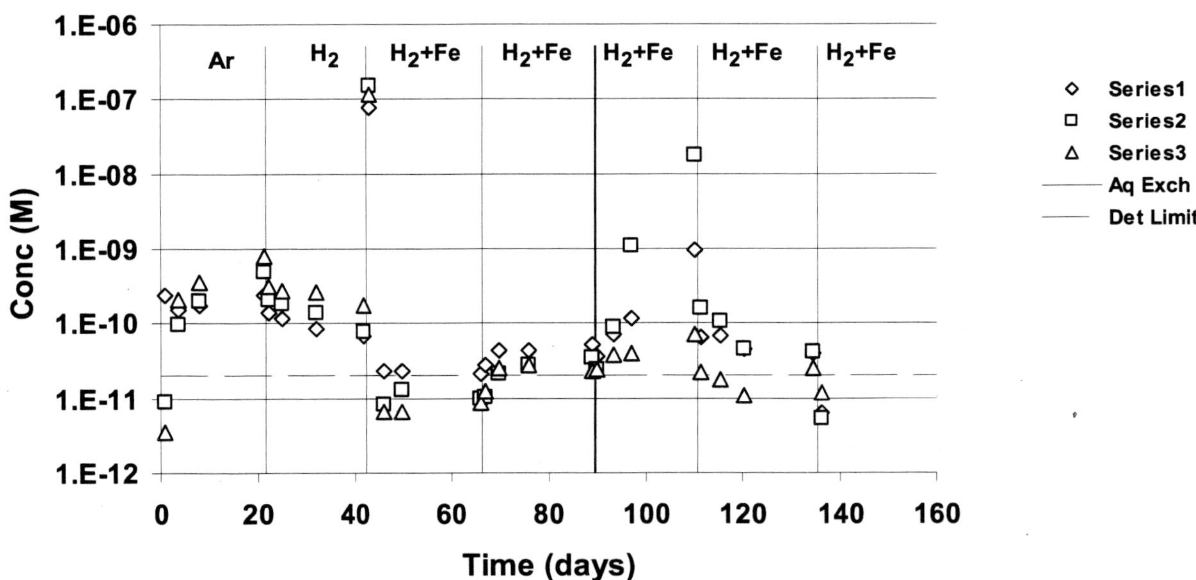

FIGURE 1. Leaching data for U. ICP-MS measurement.

REFERENCE

1. Beall, G.W., and Allard, B., in Adsortion from Aqueous Solutions, edited by P. H. Tewari, Plenum Press, New York (1981), p. 193.

Siderophore Production and Facilitated Uptake of Iron and Plutonium in *P. Putida*

Hakim Boukhalfa, Joe Lack, Sean D. Reilly, Larry Hersman, and Mary P. Neu*

Chemistry and Materials Science and Technology Divisions,
Los Alamos National Laboratory Los Alamos, New Mexico 87545

INTRODUCTION

Bioremediation is a very attractive alternative for the restoration of contaminated soil and ground water. This is particularly true for radionuclide contamination, which tends to be low in concentration and distributed over large surface areas. Microorganisms, through their natural metabolism, produce a large variety of organic molecules of different size and functionality. These molecules interact with contaminants present in the microbe's environment. Through these interactions biomolecules can solubilize, oxidize, reduce, or precipitate major metal contaminants in soils and ground water. We are studying these interactions for actinides and common soil subsurface bacteria. One focus has been on siderophores, small molecules that have a great affinity for hard metal ions, and their potential to affect the distribution and mobility of actinide contaminants. The metal-siderophores assembly can be recognized and taken up by microorganisms through their interference with their iron uptake system. The first step in the active iron transport consists of Fe(III)-siderophore recognition by membrane receptors, which requires specific stereo orientation of the Fe(III)-siderophore complex. Recent investigations have shown that siderophores can form strong complexes with a large variety of toxic metals and may mediate their introduction inside the cell. We have previously shown that a Pu-hydroxamate siderophore assembly is recognized and taken up by the *Microbacterium flavescens* (JG-9). However, it is not clear if Pu-siderophore assemblies of other siderophores are also recognized.

RESULTS

We have examined siderophore production by *Pseudomonas putida* strain ATCC 33015 grown under iron-deficient conditions in minimal succinic media. This microorganism produces a green fluorescent siderophore with high and specific affinity for iron(III). This siderophore belongs to the pyoverdin family, whose structures composition include a peptide chain and a fluorescent chromophore. Most pyoverdins have the same functional groups responsible for metal binding; however, the structures of the peptide chain and the chromophore vary, depending on the strain and the culture media used.

We have followed the production of pyoverdin under iron-deficient conditions over a period of 15 days. Although *pseudomonas putida* reaches the growth stationary phase after 24 h, the siderophore production is continued for over the 2 weeks, and most siderophore production (60%) occurs in the first 4 days. Analysis of the purified culture media by HPLC shows the production of three different siderophores, in which the main siderophore produced (MW: 1091) represents about 70% of the total siderophore production. The two remaining siderophores have molecular weights of 1,126 and 1,144 and represent 25% and 5% of the total siderophore produced, respectively. The separation and purification of the three fractions was achieved by using a C18 reverse-phase column chromatography. The three fractions were eluted with a 1:1 mixture of acetonitrile/0.05 M pyridine acetic acid buffer at pH 5. The changes in the UV-visible spectrum of the main siderophore produced with pH suggest that the presence of two buffer regions are between pH 2.8 and 4.8 and one between pH 6 and 10. This result is in agreement with literature results suggesting the presence of catecholate and hydroxamate functionalities. Spectral changes upon the addition of Fe, Pu, and U are very similar to the changes observed in the first buffer region.

CP673, *Plutonium Futures — The Science,* edited by G. D. Jarvinen

However, the changes observed between pH 6 and 10 for the free siderophore are no longer observed in the presence of Fe, Pu, and U (Figure 1). This observation indicates that the hydoxamate group binds to metal ions at a much lower pH, in agreement with our previous examination of hydroxamate siderohphore metal binding.

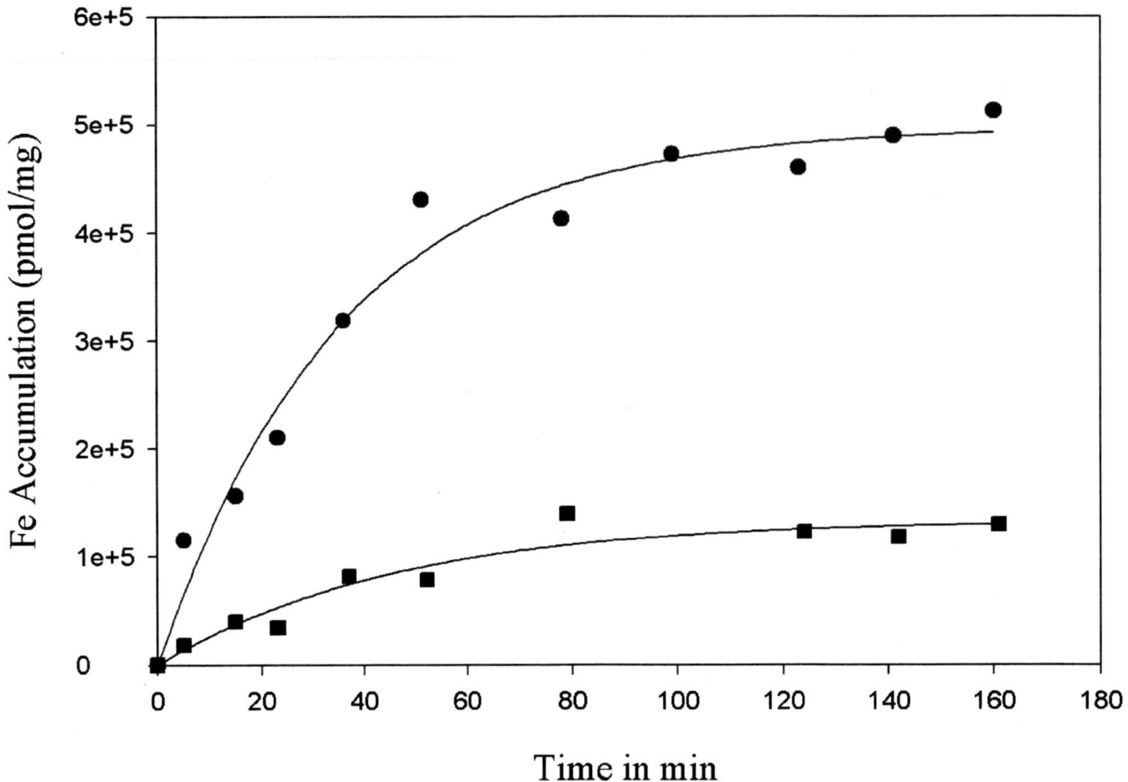

FIGURE 1. Iron accumulation inside the cell as a function of time. Curve 1, 0.3 μM total iron added; Curve 2, 6 μM total iron added.

Pyoverdin-mediated uptake of iron(III) experiments show similar results for the three siderophores produced. The experiments were followed by measuring the amount of ^{55}Fe(III) accumulated inside the cell as a function of time and the form of added iron. The accumulation of iron inside the cell shows specific recognition of Fe-pyoverdin discriminating other forms, such as Fe-EDTA and Fe-hydroxamates. When Fe-pyoverdin is added to the culture media, the concentration of Fe accumulated inside the cell exponentially reaches saturation after about 30 min. The amount of Fe inside the cell depends on the concentration of Fe added. However, the rate constant of the uptake process was ($k = 6.3 \times 10^{-2}$ min^{-1}) independent of the initial Fe-siderophore concentration over a 10^3 M range. When Fe-EDTA or Fe-DFO were used, the rate of Fe uptake was slow and the amount of intercellular Fe was less than 10% the amount observed with Fe-pyoverdin. We also observed that over time (>60 min) the amount of Fe incorporated inside the cell slowly increased, suggesting ligand exchange between Fe-EDTA or Fe-DFO and receptor-bound pyoverdin. Although pyoverdin forms a stable complex with Pu(IV), our uptake experiments showed an insignificant accumulation of plutonium inside the cell, suggesting that the complex formed is not recognized by the Fe-pyoverdin uptake system.

CONCLUSIONS

The siderophore produced by *pseudomonas putida* binds plutonium and uranium and will certainly stabilize their IV and VI oxidation states, respectively. The complexes formed are not taken up by the microorganism, which shows that Pu-siderophore complexes are not always recognized and transported by microorganisms.

Aerosol Mobility of High Specific Activity Alpha-Emitting Materials

Ralph H. Condit

Lawrence Livermore National Laboratory

It is part of the folklore among those who work with highly radioactive materials that they can self-levitate and migrate as some sort of aerosol. This is an inquiry into the possible mechanisms of such migration of alpha-emitting species. The purpose is to estimate the magnitude of such migration and, perhaps, to achieve enough understanding to improve the safe methods for working with materials such as plutonium-238.

It has often been suggested that the emission of an alpha particle in a downward direction could drive a recoil particle of material up from a surface, and this ejection could explain the reported migration of alpha-emitting isotopes. However, ejection of individual daughter nuclei by this mechanism would not yield the quantity of material necessary to explain the effects reported. Each recoil atom would have to ensnare many neighboring atoms in its escape from a surface. However, there do not appear to be plausible mechanisms whereby the original recoil nucleus could link its momentum in the ways that would be required.

We must look to the mechanisms of radiation damage, which have been much studied in the interiors of solids. When such radiation damage occurs near to a surface, it is to be expected that material can be ejected by one of several mechanisms. Evaporation of material may occur from thermal spikes. Material may spall from a surface as shock waves reach that surface from near-surface events. In the case of electrically insulating materials such as oxides, a "coulomb explosion" resulting from a possible charge displacement following thermal spike formation could be another force for ejecting material.

The initial radiation damage is the result of knock-on collisions following the recoil of a decay daughter nuclide. Alpha-emitting nuclides experience recoil energies close to 100 keV, but beta-emitter recoil energies are rarely more than a few tens of eV, particularly among the longer-lived isotopes. The ratio of these recoil energies (as much as 10,000 to 1) means that the magnitudes of material damage in the neighborhoods of a decay event are very different in the two cases. It is for this reason that alpha-emitter radiation damage is greater than for beta emitters, and near-surface damage with ejection of material needs to be studied for the case of alpha emitters.

The volumes affected by thermal spikes are in the order of 30 nm in diameter and involve something like 10,000 atoms. If material is ejected from near-surface spike regions, the ejected particles could not have diameters much larger than this. Therefore, it may be assumed that the elected particles will have a size range below a few tens of nanometers.

Calculations can be made that take into account the specific activity of Pu-238 (6.3×10^{12} dps/cm^3), that assume a depth from which ejecta may originate (10 nm), and that assume the ejection of particles having a diameter of 10 nm following each near-surface decay. It can be concluded that the apparent "evaporation" rate could be 1.6×10^1 dps of Pu-238 activity per cm^2 per second or 5.1×10^8 dps/cm^2/year.

After material is ejected from a surface, its transport through the air, adhesion to surfaces, and possible resuspension needs to be examined. In the case of Pu-238 oxide particles, the specific activity is such that a 500 nm diameter particle would experience about 1 atomic disintegration per second. Thus, the frequency of atomic decays with any ejected particle, which is likely to be around 10 nm will seem to have no effect on the migration of the particle, its adhesion to other airborne particles, its adhesion to surfaces, or resuspension from those surfaces.

From a practical point of view we now need to ask questions about the hazards in handling high-activity materials and in responding to questions that have come up over the span of many years:

- Can migration allow high-activity material to escape from small leaks in a box used for ^{238}Pu, which might be of negligible importance when handling weapons-grade Pu?

CP673, *Plutonium Futures — The Science,* edited by G. D. Jarvinen

- Can active material pass through filters by becoming resuspended from surfaces on which they were first captured?
- Can recoiled material become imbedded in box surfaces and escape at a later date, after the box has been converted to other uses or decommissioned?
- Will recoil fragments adhere to dust already present? Could general dust-suppression inhibit alpha-emitting material migration?

Our further analysis will be directed toward trying to identify possible mechanisms of migration:

- There will be electrostatic consequences of decay—the particle will become charged.
- Atmospheric ionization will play a role in charging or discharging a particle and atmospheric electrical conductivity may be enhanced as a result of the radioactivity.
- The particles will likely not be spherical. The effective density may be much less than theoretical with the consequence that they can be carried farther in moving air than simplistic calculations for quiet air would indicate.

ACKNOWLEDGMENTS

This work was performed under auspices of the U.S. Department of Energy by University of California Lawrence Livermore National Laboratory under contract No. W-7405-ENG-48.

Radioecological Sensitivity of the Rhône Aquatic System (France) Submitted to Forty Years of Plutonium Liquid Releases

F. Eyrolle,[1] B. Rolland,[1] and M. Morello[2]

[1]*Institut de Radioprotection et de Sûreté Nucléaire, DPRE/SERNAT/LERCM, CEN-Cadarache BP1, 13108 Saint Paul lez Durance cedex, France, tel 33 (0)4 42 25 39 27, Fax 33 (0)4 42 25 63 73, frederique.eyrolle@irsn.fr, benoit.rolland@irsn.fr*
[2]*Institut de Radioprotection et de Sûreté Nucléaire, DPRE/SERLAB/LRE, CEN-Cadarache BP1, 13108 Saint Paul lez Durance cedex, France, tel 33 (0)4 42 25, Fax 33 (0)4 42 25, marcel.morello@irsn.fr*

INTRODUCTION

The Rhône watershed extends over one hundred thousand square kilometers, i.e., one fifth of the metropolitan French territory. The Rhône River exports towards the Mediterranean Sea about fifty billion cubic meters of water and more than several million tons of solid matter per year. The Rhône is the major river entering the Western Mediterranean Sea as it generates the main source of sediments and fresh water (50%) to the sea, with an average water discharge of 1,700 m^3s^{-1}. The river input affects primary production significantly in the northwestern Mediterranean area and plays a leading role on the marine ecosystem functioning in the whole Gulf of Lion.

Twenty nuclear reactors are situated along the Rhône valley, representing Europe's biggest concentration of nuclear power plants. All these installations and the spent fuel reprocessing plant of Marcoule have released most of the liquid radioactive wastes downstream, including plutonium isotopes, into the Rhône River.[1] Almost 40 years of radioactive releases into the river have led to a permanent contamination of both the river and marine compartments.[2,3] For the last ten years, the industrial radioactive releases have significantly decreased. This especially relates to plutonium isotope releases because the Marcoule reprocessing plant is being dismantled. Furthermore, plutonium isotopes observed in the Rhône River also originate from the weathering of the catchment basin labelled by the global atmospheric fallout.[4–6] The contribution of this plutonium source term to the Rhône waters was estimated in Reference 5 over the 1960–2000 period.

The long-term radioecological sensitivity of the Rhône River natural aquatic system is rather unknown. Chronological series on $^{239+240}$Pu and ^{238}Pu activities at the lower course of the Rhône River are studied to underline the radioecological sensitivity of a river system over time, including continuous industrial plutonium release periods and a postrelease one.

RESULTS AND DISCUSSION

Referring to the Marcoule plant reports and References 1 and 5, plutonium activity levels introduced in the Rhône River either by soil leaching or the reprocessing plant were able to be estimated since the 1960s (Figures 1 and 2). From 1961 to 1965, $^{239+240}$Pu and ^{238}Pu activities released by the Marcoule reprocessing plant in the Rhône River ranged from about 10 to 70 GBq y^{-1} and from 0.7 to 4 GBq y^{-1}, respectively, because of the reprocessing for military applications. Following a period of lower plutonium releases from 1966 to1978, around 20 GBq of $^{239+240}$Pu and 6 GBq of ^{238}Pu were annually introduced into the waters until 1991 because of an increase in spent-fuel reprocessing. In 1991, a new treatment process led to a two-order-of-magnitude decrease of plutonium discharges in the Rhône River. Today, ^{238}Pu industrial inputs in the Rhône River are still about ten times higher than those

CP673, *Plutonium Futures — The Science*, edited by G. D. Jarvinen

originating from the weathering of the watershed soils, but $^{239+240}Pu$ inputs from each of the source terms are of similar importance, i.e., 1 GBq y^{-1} (Figure 2). Because the Marcoule reprocessing plant has been in the process of being dismantled since 1997, a new consistent decrease of industrial releases is soon expected.

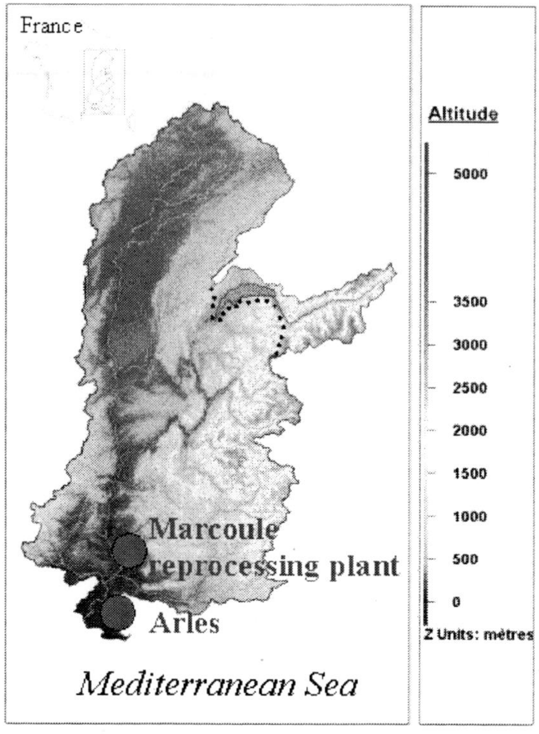

FIGURE 1. The Rhône River catchment basin.

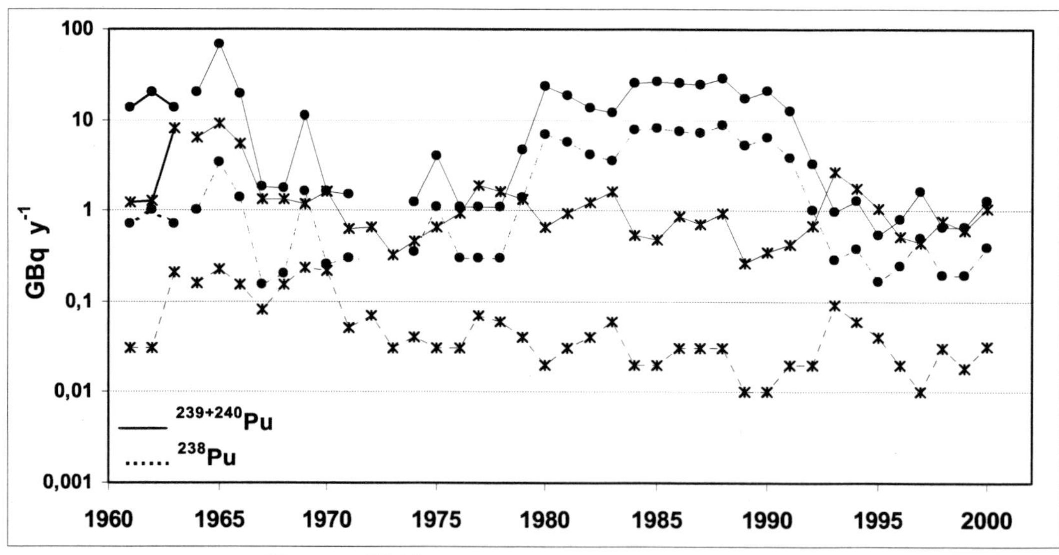

FIGURE 2. Historical account of ^{238}Pu and $^{239+240}Pu$ inputs from industrial[12] and terrigenous[13,14] source terms in the Rhône River, given in GBq y^{-1}. •: Marcoule liquid releases, ✳: Watershed soil leaching.

From 1987 to 2002, plutonium activities were measured in the suspended matter of the Rhône River and/or the dissolved phase at a sampling station in Arles located downstream from both the Marcoule reprocessing plant and the Rhône tributaries. About 40 data are available characterizing $^{239+240}Pu$ and ^{238}Pu activities in water. This chronological series allows us to quantify the radioecological sensitivity of the Rhône River waters subject to varying release levels and origins.

The results show that plutonium isotopes are mainly exported by the suspended load (up to 80%) as it might be expected with respect to the particle-reactive property of this element.[6,7] Over the studied period, plutonium activities in waters decreased proportionally to the Marcoule releases, demonstrating thus a fast response of the River system. Nevertheless, an almost two-order-of-magnitude increase of plutonium activities was observed during flood events. We demonstrated that the increase in plutonium activities for high flow rates was significantly due to the remobilization of sedimentary storages. River sediments act as a delayed-source term of plutonium for the Rhône fresh waters depending on the hydraulic regime and flood events.[8–11] Although the remobilization of radioactive sedimentary stock is being widely studied within the marine environment,[12] little is known regarding continental aquatic systems. We estimate that over the 1987–2002 period of time, about 40% of plutonium isotopes originated from the remobilization of sedimentary storages. As plutonium industrial releases are shut down, the long-term contribution of the sedimentary delayed source of plutonium for the Rhône River waters is attempted.

CONCLUSIONS

The long-term radioecological sensitivity of natural aquatic systems submitted to continuous or accidental releases is attempted. Our preliminary results obtained on the Rhône River demonstrate that the sedimentary compartment may act as an important delayed-source term for fresh waters and may play a fundamental role on the long-term radioecological sensitivity of natural aquatic systems. Because of both the important decrease in the primary industrial source term nowadays and during the next few years, the Rhône River system represents an interesting area for quantifying the role of radioactive sedimentary storages on the long-term radioecological sensitivity of natural aquatic systems. Our results aim at quantifying the role of sedimentary storages on the radioecological sensitivity of the aquatic systems.

REFERENCES

1. Charmasson, S., "Cycle du combustible nucléaire et milieu marin—Devenir des effluents rhodaniens en Méditerranée et des déchets immergés en Atlantique Nord-Est," Thèse d'Etat, Université Aix-Marseille II (1998), 359 pp.
2. Charmasson, S., Arnaud, M., Piermattei, S., and Romero, L., "Sources of Radioactivity in the Mediterranean Sea," Report of Working Group I, in The Radiological Exposure of the Population of the European Community to Radioactivity in the Mediterranean Sea, Marina-Med Project, Proceedings of a Seminar, May 17–19, 1994, Rome, Italy, edited by Cigna, Delfanti, Serro. EUR 155564 EN, 11-70 (1994).
3. Thomas, A. J., "Input of Artificial Radionuclides to the Gulf of Lions and Tracing the Rhône Influence in Marine Surface Sediments," Deep-Sea Research 44(3–4), 577–595 (1997).
4. Noël, M. H., "Le plutonium traceur du transfert et de l'accumulation des apports particulaires du Rhône en Méditerranée Nord Occidentale," Thèse de l'Université Pari XII–Val de Marne (1996), 256 pp.
5. Duffa, C., "Répartition du plutonium et de l'américium dans l'environnement de la basse vallée du Rhône," Thèse de l'Université d'Aix Marseille III, (2001), 179 pp.
6. Vives i Battle, J., "Speciation and Bioavailability of Plutonium and Americium in the Irish Sea and Other Marine Ecosystems," Ph.D. Thesis, National University of Ireland (1993), 347 pp.
7. Eyrolle, F., Goutelard, F., and Calmet D., "Pu-239+240 and Pu-238 Distribution among Dissolved, Colloidal, and Particulate Phases in the Rhone River Waters (France)," Proceedings of an International Symposium on Marine Pollution, Monaco, October 5–9, 1998, IAEA (July 1999), pp. 466–467.
8. Eyrolle, F., Arnaud, M., and Duffa C., "Plutonium Fluxes from the Rhône River to the Mediterranean Sea," in Proceedings of an International Congress on the Radioecology-Ecotoxicology of Continental and Estuarine Environments (ECORAD), September 3–7, 2001, Aix en Provence, France, O4/2 (90) (2001).
9. Eyrolle, F., Arnaud, M., Duffa, C., and Renaud, Ph., "Plutonium Fluxes from the Rhône River to the Mediterranean Sea," Radioprotection-Colloques 37, C1, 87–92 (2002).
10. Eyrolle, F., and Duffa C., 2002, "Sedimentary Beds as a Delayed Source of ^{137}Cs, ^{238}Pu, and $^{239+240}Pu$ for the Rhône River Freshwaters," in Ninth International Symposium on the Interaction between Sediments and Waters, IASWS, Banff Springs Hotel, Canada, May 5–10, 2002.
11. Eyrolle, F., Duffa, C., and Charmasson, S., "Inputs of Plutonium Isotopes from the Rhône River to the Gulf of Lion (North Western Mediterranean Sea) over the 1945–1998 period—Mass Balances, Fluxes and Predictive Trend," in Proceedings of the International Conference on Radioactivity in the Environment, September 1–5, 2002, Monaco (2002), pp. 577–581.
12. Cook, G. T., MacKenzie, A. B., McDonald, P., and Jones, S. R., "Remobilization of Sellafield-Derived Radionuclides and Transport from the North-East Irish Sea," J. Environ. Radioactivity 5, 227–241 (1997).

Microbial Transformations of Plutonium and Other Actinides in Transuranic and Mixed Wastes

A. J. Francis

Environmental Sciences Department
Brookhaven National Laboratory, Upton, New York 11973, USA

The presence of the actinides Th, U, Np, Pu, and Am in transuranic (TRU) and mixed wastes is a major concern because of their potential for migration from the waste repositories and long-term contamination of the environment. The toxicity of the actinide elements and the long half-lives of their isotopes are the primary causes for concern. In addition to the radionuclides, the TRU waste consists of a variety of organic materials (cellulose, plastic, rubber, and chelating agents) and inorganic compounds (nitrate and sulfate). Significant microbial activity is expected in the waste because of the presence of organic compounds and nitrate, which serve as carbon and nitrogen sources and in the absence of oxygen the microbes can use nitrate and sulfate as alternate electron acceptors. Biodegradation of the TRU waste can result in gas generation and pressurization of containment areas, and waste volume reduction and subsidence in the repository. Although the physical, chemical, and geochemical processes affecting dissolution, precipitation, and mobilization of actinides have been investigated, we have only limited information on the effects of microbial processes.

Microbial activity could affect the chemical nature of the actinides by altering the speciation, solubility and sorption properties and thus could increase or decrease the concentrations of actinides in solution. Under appropriate conditions, dissolution or immobilization of actinides is brought about by direct enzymatic or indirect nonenzymatic actions of microorganisms. Dissolution of actinides by microorganisms is brought about by changes in the Eh and pH of the medium, by their production of organic acids, such as citric acid, siderophores, and extracellular metabolites. Immobilization or precipitation of actinides is due to changes in the Eh of the environment, enzymatic reductive precipitation (reduction from higher to lower oxidation state), biosorption, bioaccumulation, biotransformation of actinides complexed with organic and inorganic ligands and bioprecipitation reactions. Free-living bacteria suspended in the groundwater fall within the colloidal size range and may have strong radionuclide sorbing capacity, giving them the potential to transport radionuclides in the subsurface.

The actinides in TRU and mixed wastes may be present in various forms, such as elemental, oxide, coprecipitates, inorganic, and organic complexes, and as naturally occurring minerals depending on the process and waste stream. They exist in various oxidation states, and the ones of concern are III (Am, Pu, U), IV (Th, Pu, U), V (Np), and VI (Pu, U). Microorganisms have been detected in TRU wastes, Pu-contaminated soils, low-level radioactive wastes, backfill materials, natural analog sites, and waste-repository sites slated for high-level wastes.[1] Seventy percent of the TRU waste consists of cellulose and other biodegradable organic compounds. Biodegradation of cellulose under the hypersaline conditions such as in the WIPP repository can produce CO_2 and methane gas, as well as affect the solubility of actinides. Microbially produced gases could have significant ramifications for the long-term stability of the repository (up to 10,000 years). Little is known of microbial degradation of organic constituents in TRU and mixed wastes and its implications on gas generation and the fate of actinides, i.e., mechanisms for the dissolution and precipitation of actinides.

Chelating agents are present in TRU and mixed wastes because they are widely used for decontaminating nuclear reactors and equipment, in cleanup operations, and in separating radionuclides. Plutonium forms very strong complexes with a variety of organic ligands. Naturally occurring organic complexing agents, such as humic and fulvic acids, and microbially produced complexing agents, such as citrate, and siderophores, as well as synthetic chelating agents, can affect the mobility of Pu in the environment. Biotransformation of actinide-organic complexes should result in the degradation of the organic ligand and precipitation of the actinide.

CP673, *Plutonium Futures — The Science,* edited by G. D. Jarvinen

Key microbial processes involved in the mobilization or immobilization of selected actinides of interest are summarized in Table 1. Among the actinides, biotransformation of uranium has been extensively studied, but we have only a limited understanding of the microbial transformations of other actinides such as Th, Np, Pu, and Am present in TRU and mixed wastes.[1,2]

TABLE 1. Biotransformation of Selected Actinides in TRU and Mixed Wastes.

Process	Th	U	Np	Pu	Am
Oxidation[1]	NA	++	ND	ND	NA
Reduction[2]	NA	++++	+?	+?	NA
Dissolution[3]	+	++	?	+	?
Biosorption	?	++++	+	++	+
Biocolloid[4]	+	++	+	+	+

NA—not applicable; ND—not determined.
[1]Dissolution due to oxidation from lower to higher valence state.
[2]Reductive precipitation due to enzymatic reduction from higher to lower valence state.
[3]Dissolution due to oxidation from lower to higher valence state, changes in pH, production of organic acids, and sequestering agents such as siderophores.
[4]Association of radionuclides with suspended bacteria, which can be potentially transported as biocolloids.

The effects of microbial activities on TRU and mixed waste and their potential for treatment of certain waste forms to stabilize the actinides and reduce the volume of the waste have not been fully exploited. Fundamental understanding of the mechanisms of microbial transformations of several chemical forms of actinides under various microbial process conditions such as aerobic, anaerobic (denitrifying, fermentative, and sulfate reducing), and repository relevant conditions would pave the way to assess the microbial impact on TRU wastes and the long-term behavior of actinides in the environment. In this paper, we present the biotransformation of actinides, in particular Pu, under various microbial process conditions.

REFERENCES

1. Francis, A.J., "Microbial Transformations of Plutonium and Implications for its Mobility," in Plutonium in the Environment, edited by A. Kudo, Elsevier Science Ltd., Co., UK. (2001) pp. 201–219.
2. Neu, M.P., Ruggiero, C. E., and Francis, A. J., "Bioinorganic Chemistry of Plutonium and Interactions of Plutonium with Microorganisms and Plants," in Advances in Plutonium Chemistry 1967–2000, edited by D. Hoffman, American Nuclear Society, La Grange Park, Illinois, and University Research Alliance, Amarillo, Texas (2002), pp. 169–211.

A New Measurement Watershed for Environmental Studies of Plutonium: Inductively Coupled Plasma Mass Spectrometry

Wendy J. Hartsock, Kevin M. Hafer, and Michael E. Ketterer

Department of Chemistry, Northern Arizona University, Flagstaff, AZ 86011-5698 USA.

INTRODUCTION

Recent advances have made mass spectrometry (MS) a satisfactory alternative to alpha spectrometry for determining $^{239+240}Pu$ activities in environmental samples. Mass spectrometry has proven useful for measurements of ^{239}Pu, ^{240}Pu, ^{241}Pu, and ^{242}Pu; determination of $^{239+240}Pu$ activity and the $^{240}Pu/^{239}Pu$ atom ratio are relatively straightforward for most fallout-containing samples; ^{241}Pu and ^{242}Pu (and ratios thereof) can also be determined if sufficient numbers of atoms are available. Of the various mass spectrometric techniques, inductively coupled plasma mass spectrometry (ICPMS) is arguably the best-suited technique for routine, high-throughput measurements. ICPMS involves aqueous solution introduction and short acquisition times of 2–10 minutes per sample following relatively straightforward preparations. We have recently described the use of quadrupole ICPMS in studies of fallout Pu in the environment.[1] Using sector ICPMS, we obtain detection limits of 0.1, 0.02, and 0.002 Bq/kg $^{239+240}Pu$ for sample sizes of 0.5 grams, 3 grams, and 50 grams of soil, respectively.

The use of ICPMS, particularly sector ICPMS, as a routine tool for determining Pu activities and isotopic compositions greatly facilitates several areas of study. These consist of: (a) using fallout Pu as a chronological marker in the dating of recent aquatic sediments; (b) using fallout Pu as an indicator analogous to ^{137}Cs in soil erosion studies; (c) routine environmental monitoring of Pu activities for compliance and cleanup objectives; and (d) distinguishing between "global fallout" and local/regional Pu sources through isotopic fingerprinting. The use of ICPMS as a means to "fingerprint" specific Pu sources is especially appealing; these studies necessarily involve comparisons of inventories (Bq/m^2) and atom ratios to known global fallout parameters[2] when probing the presence of "elevated" or "nonglobal fallout" sources.

RESULTS

Sediment Chronology

Sector ICPMS is used to investigate the chronology of a sediment core from Lake Erie's Central Basin. To determine $^{239+240}Pu$, samples of 0.4–0.6 g dry sediment are sintered with potassium pyrosulfate in the presence of ^{242}Pu tracer, and Pu is concentrated with TEVA resin (EIChrom). The $^{239+240}Pu$ activity profile is nearly identical to the gamma-spectrometric ^{137}Cs profile (Figure 1). Both data sets are in agreement for the horizons where bomb nuclides are first detected (ca. 1952) and the horizons of maximum fallout (1963/1964). Although gamma spectrometric ^{137}Cs measurements are nondestructive, about three weeks counting time is required, but the Pu data can be acquired in three hours.

CP673, *Plutonium Futures — The Science,* edited by G. D. Jarvinen

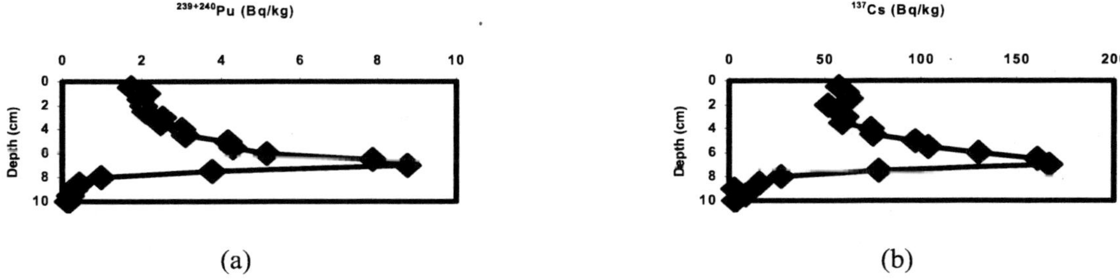

FIGURE 1. (a) $^{239+240}$Pu and (b) ^{137}Cs activity profiles in a sediment core from Lake Erie's Central Basin.

Soil Erosion Studies

To investigate the possible use of fallout $^{239+240}$Pu as a tracer of soil redistribution, the inventories and depth distributions of Pu and ^{137}Cs were investigated in undisturbed soils from Konza Prairie, Kansas. The $^{239+240}$Pu activity profile is in accordance with what is expected for an "undisturbed" soil—namely, the Pu activities are highest at the top and decrease in an exponential fashion with depth. The vast majority of the Pu activity is present in the top 15 cm of soil. The quantities of Pu below 27 cm are negligible, although detectable. Inventories of $^{239+240}$Pu are 82 ± 2 Bq/m^2 for Core KP01-12-7C and 73 ± 2 Bq/m^2 for Core KP01-08-05C. The depth distributions of ^{137}Cs and $^{239+240}$Pu are virtually identical.

"Fingerprinting:" Chernobyl Pu in Poland

Sector ICPMS was used to determine ^{240}Pu /^{239}Pu and ^{241}Pu /^{239}Pu in archived NdF_3 sources prepared from Polish soils. Compared to global fallout, Chernobyl Pu exhibits higher abundances of ^{240}Pu and ^{241}Pu. The ratios ^{240}Pu/^{239}Pu and ^{241}Pu/^{239}Pu covary and range from 0.186–0.348 and 0.0029–0.0412, respectively, in forest soils. A plot of ^{241}Pu/^{239}Pu vs ^{240}Pu/^{239}Pu demonstrates two-component mixing between global fallout and Chernobyl Pu (Figure 2). Chernobyl Pu is most evident in northeastern Poland. The ^{241}Pu activities and/or the ^{241}Pu/^{239}Pu atom ratios are more sensitive than ^{240}Pu/^{239}Pu or ^{238}Pu/$^{239+240}$Pu activity ratios at detecting small Chernobyl $^{239+240}$Pu inputs, found in southern Poland. The mass spectrometric data show that the ^{241}Pu inventory is 40%–62% Chernobyl-derived in southern Poland, and 58%–96% Chernobyl in northeastern Poland.

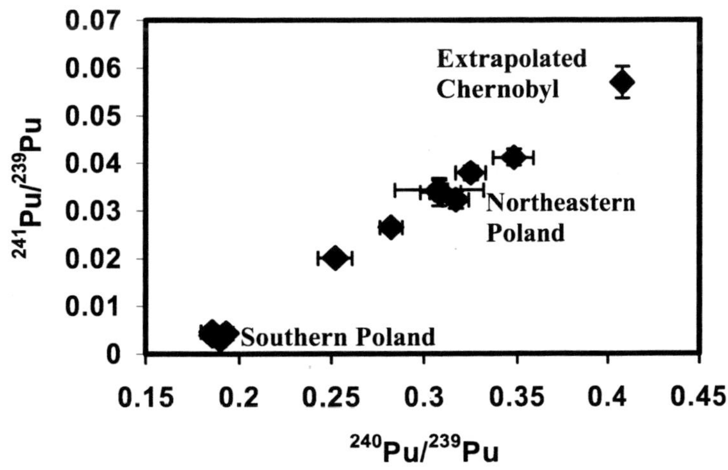

FIGURE 2. Mixing plot of ^{241}Pu/^{239}Pu vs ^{240}Pu/^{239}Pu for Pu in Polish soils.

DISCUSSION

An extensive scientific base already exists for using ^{137}Cs as a tracer of particle erosion, redistribution, and sedimentation. Unfortunately, the use of ^{137}Cs as a tracer is constrained analytically because of counting time requirements and the availability of multiple counters. Since $^{239+240}$Pu activities can be easily measured at fallout levels on large numbers of samples, we contend that $^{239+240}$Pu can essentially replace ^{137}Cs as an indicator of these processes. Based upon the Lake Erie sediments and numerous other examples of aquatic sediments from throughout the world, our results indicate that $^{239+240}$Pu and ^{137}Cs convey the same chronological information about sediment deposition.

There is substantial interest in tracing the deposition of Chernobyl Pu in northern and eastern Europe. In addition to resolving Chernobyl Pu from global fallout Pu by analysis of the isotopic composition, the analysis of large numbers of samples will be required to better understand the distribution and behavior of the first-release Chernobyl actinides plume in the environment. ICPMS truly represents a new "watershed" in the ability to rapidly and effectively address these environmental monitoring and source apportionment needs.

REFERENCES

1. Ketterer, M. E., Watson, B. R., Matisoff, G., and Wilson, C. G., "Rapid Dating of Recent Aquatic Sediments Using Pu Activities and ^{240}Pu/^{239}Pu as Determined by Quadrupole Inductively Coupled Plasma Mass Spectrometry," Environmental Science and Technology 36, 1307–1311 (2002).
2. Kelley, J. M., Bond, L. A., and Beasley, T. M., "Global Distribution of Pu Isotopes and ^{237}Np," The Science of the Total Environment 237/238, 483–500 (1999).

Synthesis of New Plutonium Main-Group Metal Materials

Ryan F. Hess,[a] Kent D. Abney,[a] and Peter K. Dorhout[b]

[a]*NMT-2, Los Alamos National Laboratory, Los Alamos, NM 87545*
[b]*Department of Chemistry, Colorado State University, Fort Collins, CO 80523*

The number of known ternary and even binary plutonium chalcogenides, excluding oxides, is small compared to transition metal and rare earth compounds. There are only eight plutonium chalcogenide crystal structures known, and all are binaries. A series of two- and three-dimensional thorium compounds was synthesized by Ibers and coworkers with the formulas KTh_2Q_6 (Q = Se, Te), $CsTh_2Se_6$, $CuTh_2Te_6$, and $SrTh_2Se_5$. The ternary compounds, KTh_2Q_6 and $CsTh_2Se_6$, are interesting because conventional electron counting does not yield a reasonable oxidation state for thorium. This difficulty is due to the presence of intermediate length Q-Q bonds. It is almost certain, however, that thorium is in the 4+ oxidation state. Experiments were undertaken to determine if it were possible to synthesize a similar compound containing plutonium in place of thorium. A plutonium compound would be particularly interesting because both 3+ and 4+ would be reasonable plutonium oxidation states. This presentation discusses the synthesis and characterization of a new ternary plutonium selenide compound with the formula KPu_3Se_{8-x}. This new ternary compound is compared to both its parent structure, $PuSe_2$, and the related ternary compound, KTh_2Se_6.

XANES experiments were undertaken to gain information on the electron density around plutonium in KPu_3Se_{8-x} and a number of binary plutonium chalcogenides. The energies of the L_{III} absorption edges for the compound examined in this study are listed in Table 1. Figure 1 contains a plot of the second derivative of absorption versus energy, the absorption edge value is recorded where the second derivative first crosses the energy axis. In many systems, XANES is able to determine oxidation-state information about the absorbing, or target, atom. By analyzing the post-edge region, it is possible to learn structural information about the environment around the target atom. One key difference between x-ray diffraction and x-ray absorption experiments is that the former examines the long-range order of a system, but the latter is capable of examining the local region around a specific target atom. Because of the complexity of the theory needed to predict XANES spectra, the determination of oxidation states is largely empirical. That is, to successfully determine an atom's oxidation state, it is important to have available a number of standards in which the target atom is in a similar chemical environment compared to the sample being studied. Unfortunately, the number of plutonium compounds previously studied using XANES is small. This adds some uncertainty in determining an absolute oxidation state number for the compounds examined in this study. What can be learned, however, is something about the relative amount of electron density around the plutonium atoms in these compounds. It is important to remember that in metallic systems the use of the oxidation state formalism becomes less valuable than in valence precise systems. The key result from this study is that the absorption edges of these heavy chalcogenide systems appear close to the value found for plutonium metal (within 1–2 eV). This implies that the electronic environment around plutonium is very similar to that found in the metal. This is significantly different than what is found for PuO_2. The value for the oxide is much closer to what is found for Pu^{4+} in aqueous solutions. Though the overall trend is not unexpected because of the presence of the much softer S, Se, and Te atoms compared to O, the magnitude of the effect and the similarity to plutonium metal is interesting.

TABLE 1. The L_{iii} Absorption Edges of the Compounds Examined in this Study.

Sample	Edge Position (eV)
Pu Metal	18056
$PuSe_2$	18057
KPu_3Se_8	18057
PuO_2	18061

CP673, *Plutonium Futures — The Science,* edited by G. D. Jarvinen

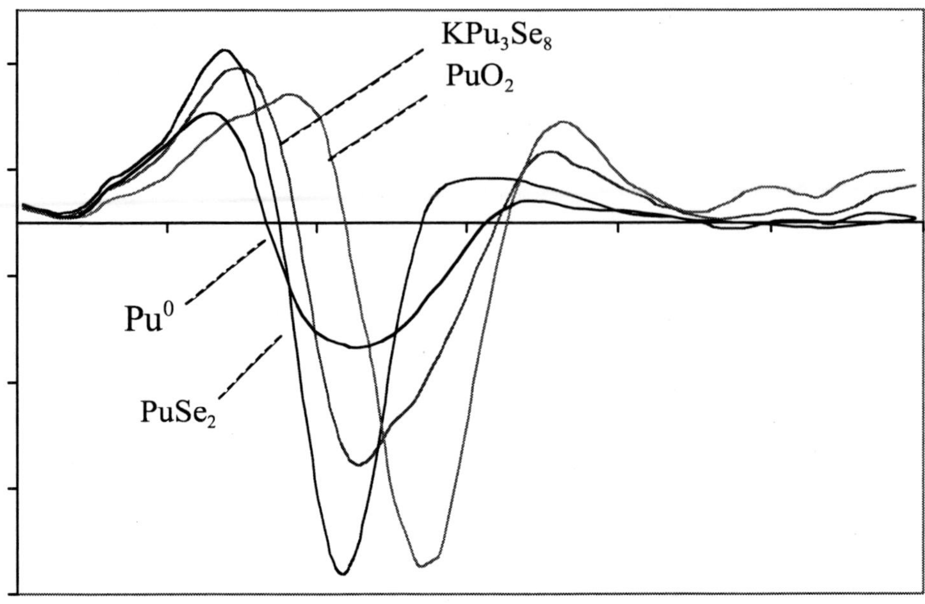

FIGURE 1. The plot of the second derivative of absorbance vs energy for the plutonium compounds examined in this study. The absorption edge is recorded where the second derivative first crosses the energy axis.

Neptunium(V) Sorption by α-FeOOH and γ-Fe_2O_3

A. B. Khasanova,[1] N. S. Shcherbina,[1] St. N. Kalmykov,[1,2]
A. P. Novikov,[2] and S. B. Clark[3]

[1]*Radiochemistry Division, Chemistry Department, Lomonosov Moscow State University, Russia*
[2]*Vernadsky Institute of Geochemistry and Analytical Chemistry RAS, Moscow, Russia*
[3]*Washington State University, WA, USA*

INTRODUCTION

The mechanisms and kinetics of Np(V) sorption on well-characterized goethite (α-FeOOH) and magemite (γ-Fe_2O_3) were studied in batch experiments. Neptunium redox speciation on the mineral surface was studied using a solvent extraction technique and X-ray Photoelectron Spectroscopy (XPS). The distribution of Np(V) was analyzed using FITEQL software, and surface complexation reactions equilibrium constants were determined.

RESULTS AND DISCUSSION

The behavior of actinides in the environment is governed mostly by the reactions at the mineral-water interface. Therefore, the molecular-level description of sorption is required for the assessment of migration abilities of radionuclides. These include actinide redox speciation, determination of mechanisms of sorption (ion exchange, surface complexation or surface precipitation) and calculation of corresponding equilibrium constants. Here we studied the sorption of Np(V) on two common Fe-containing minerals.

The α-FeOOH and γ-Fe_2O_3 were synthesized according to the methods described earlier[1,2] and characterized using powder XRD, SEM imaging, BET surface analysis and potentiometric titration. The sorption of Np(V) was studied at 25 ± 1°C as a function of pH, ionic strength, and total metal concentration under N_2 atmosphere to avoid carbonate complexation.

The kinetics of sorption were studied at different pH values in 0.1 M $NaClO_4$ solutions that are presented in Figure 1 for γ-Fe_2O_3. For constant pH values, the kinetic curves were found to be linear in $-\ln(C_t/C_0) - K{\cdot}t$, where C_t and C_0 are the Np current and initial concentrations, K is the reaction-rate constant, and t is time. The possible explanation of rather slow kinetics of sorption is due to Np(V) diffusion to micropores of solid phases.

Earlier it was found that α-FeOOH could reduce Pu(V) to Pu(IV)[3] and Cr(VI) to Cr(III).[4] For Np redox speciation we used the method described in the paper by A. Morgenstern and G. Choppin[5] using solvent extraction with TTA and HDEHP and X-ray Photoelectron Spectroscopy (XPS). Using both solvent extraction and XPS, it was found that no reduction of Np(V) to Np(IV) occurred.

The sorption was found to be not dependant on ionic strength that indicates the formation of inner sphere complexes in both cases. The equilibrium distributions of Np(V) on solid phases upon pH (see Figure 2) and at different total metal concentrations are used to distinguish the mechanism of sorption and calculate the surface complexation equilibrium constants using FITEQL.[6] The following possibilities were considered: *monodentate complexes* according to the reactions (charges are omitted for convenience):

$$\equiv \text{Fe-OH} + \text{An} \leftrightarrows \equiv \text{Fe-O-An} + \text{H}, \quad (1)$$

$$\equiv \text{Fe-OH} + \text{An(OH)}_n \leftrightarrows \equiv \text{Fe-O-An(OH)}_n + \text{H}, \quad (2)$$

and *bidentate complexes* according to the reactions:

$$(\equiv \text{Fe-OH})_2 + \text{An} \leftrightarrows (\equiv \text{Fe-O})_2\text{-An} + 2\text{H}, \text{ and} \quad (3)$$

$$(\equiv \text{Fe-OH})_2 + \text{An(OH)}_n \leftrightarrows (\equiv \text{Fe-O})_2\text{-An(OH)}_n + 2\text{H}. \quad (4)$$

CP673, *Plutonium Futures — The Science,* edited by G. D. Jarvinen

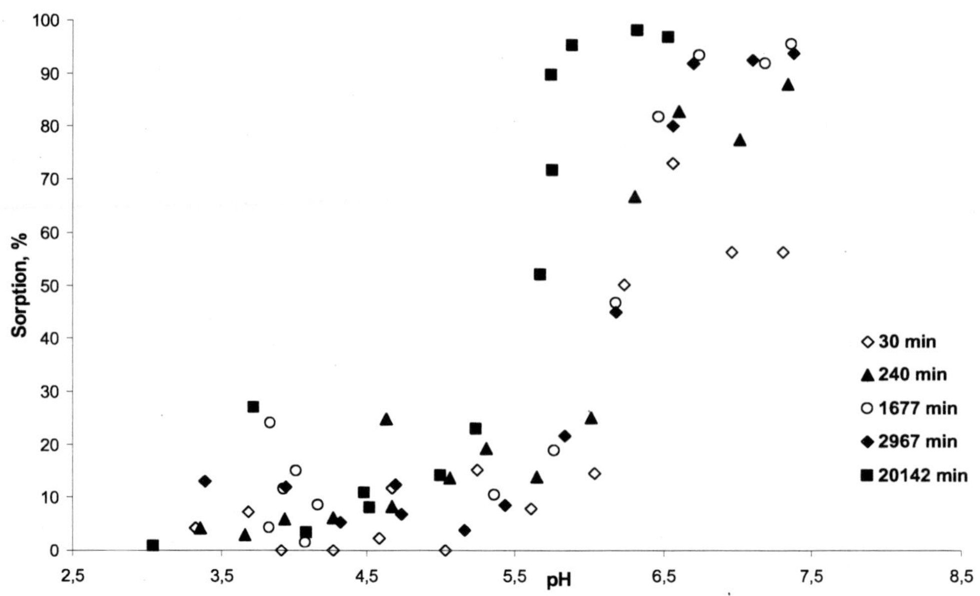

FIGURE 1. The kinetics of Np(V) ($1.48 \cdot 10^{-7}$ M) sorption by γ-Fe_2O_3 ($2.45\ m^2L^{-1}$).

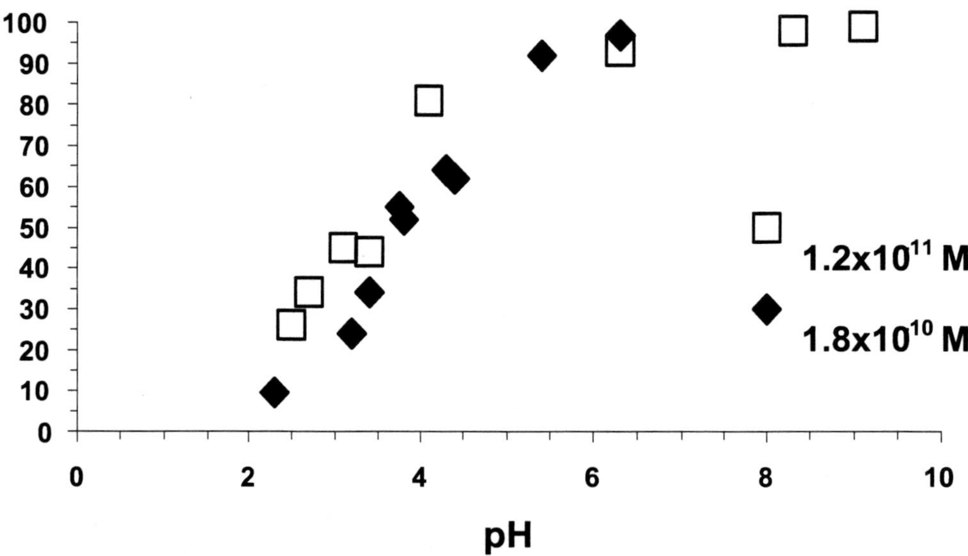

FIGURE 2. The pH dependence of Np(V) sorption on α-FeOOH ($14.97\ m^2L^{-1}$) at two different concentrations.

It was found that in both cases the only reaction that fits experimental data is (1), and equilibrium is described by the diffuse layer model. Table 1 demonstrates the equilibrium constants for Np(V) complexation by α-FeOOH and γ-Fe_2O_3.

TABLE 1. Equilibrium Constants for Np(V) Surface Complexation on α-FeOOH and γ-Fe_2O_3.

Solid Phase	Surface Area (m^2g^{-1})	log K
α-FeOOH	43.3	–1.41
γ-Fe_2O_3	20.4	–2.19

REFERENCES

1. Atkinson, R. J., Posner, A. M., and Quirk J. P., J. Phys. Chem. 71(3), 245 (1967).
2. Park, G. A., and De Bruyn P. L., J. Phys. Chem. 66(2), 967 (1962).
3. Abdel-Samad, H., and Watson P. R., Applied Surface Science 108, 371–377 (1997).
4. Keeney-Kennicutt, W. L., Morse, J. W., Geochim Cosmochim. Acta 49, 2577–2588 (1985).
5. Morgenstern, A., and Choppin G. R., Radiochim. Acta 90, 69–74 (2002)
6. Herbelin, A. L., and Westall, J. C., "FITEQL, A Computer Program for Determination of Equilibrium Constants from Experimental Data," Department of Chemistry, Oregon State University, Corvallis, Ver. 3.1, Report 94-01 (1994).

Plutonium Isotopes in Seas Around the Korean Peninsula

C. K. Kim,[a*] C. S. Kim,[a] B. U. Chang,[a] G. H. Hong,[b] K. Hirose,[c] M. Aoyama,[c] and Y. Igarashi[c]

[a]Korea Institute of Nuclear Safety, P.O. Box 114, Yusong, Taejon 305-338, Korea.
**Correspondence: E-mail: k216kck@kins.re.kr, Tel: (+)82-42-868-0257, Fax;(+)82-42-868-0556*
[b]Korea Ocean Research and Development Institute, Ansan, P.O. Box 29, Kyonggi 425-600, Korea
[c]Meteorological Research Institute, Nagamine 1-1, Tsukuba, Ibaraki, 305-0052, Japan

INTRODUCTION

Plutonium has been released into the environment by atmospheric nuclear weapons testing, various accidents involving nuclear material, and nuclear fuel reprocessing plants.[1] Among those sources, the major source of fallout of plutonium to the environment is atmospheric nuclear weapons testing, which was conducted beginning in 1945 and peaking in 1962 before a partial test-ban treaty in 1963.[2] It is known that the $^{240}Pu/^{239}Pu$ atom ratio of fallout depends upon the specific weapons device design and test yields. The $^{240}Pu/^{239}Pu$ atom ratio is one of the most important pieces of information for analyzing Pu origin in the environment. Generally, the global fallout average of $^{240}Pu/^{239}Pu$ atom ratio is typically 0.18 ± 0.01 in atmospheric aerosols, soil, ice core samples, and marine sediments.[3,4] This value is nearly identical with an average value of 0.176 ± 0.014 obtained from the worldwide survey of terrestrial soils in 1970–1971.[5] $^{239+240}Pu$ concentrations in the seas adjacent to the Korean Peninsula have been reported as reflecting wide regional concern regarding the potential artificial radionuclide contamination from the past dumping activities of the Former Soviet Union and Russian Federation, as well as the potential application of water mass tracers.[6,7] But there are few data on the $^{240}Pu/^{239}Pu$ atom ratio in sea areas around the Korean Peninsula. In this work, we intend to find not only the $^{239+240}Pu$ concentrations but also the sources; and therefore, its isotopic signature in seawater around the Korean Peninsula.

MATERIALS AND METHODS

As shown in Figure 1, surface seawaters were collected at the Yellow Sea, Korea Strait, and East Sea/Sea of Japan around the Korean Peninsula during cruises on board the R/V 'Onnuri' (Korea Ocean Research and Development Institute; KORDI) over the period from 1999 to 2002. Deep waters were collected in the southwestern part of the East Sea (37°25.0′N, 131°30.3′E) using a 100 L volume sampler in October 1999. All seawater samples were filtered with the cellulose nitrate membrane filter (0.45 μm) and subsequently acidified (pH < 2) with 6N HCl for storage. After being purified by an on-line purification system, Pu isotopes were determined by sector field inductively coupled plasma mass spectrometry (SF-ICP-MS).[8]

RESULTS AND DISCUSSION

In summary, the range of the $^{239+240}Pu$ concentrations in surface waters adjacent to the Korean Peninsula are of the same order of magnitude as those observed in the period from 1977 to 1985, showing a little higher concentration than those found in the western North Pacific. The $^{239+240}Pu$ concentrations in surface waters were higher in winter than in summer. The higher surface $^{239+240}Pu$ concentrations in the winter may be explained by the hypothesis that subsurface water containing higher Pu is transported to the surface by winter convection.

CP673, *Plutonium Futures — The Science,* edited by G. D. Jarvinen

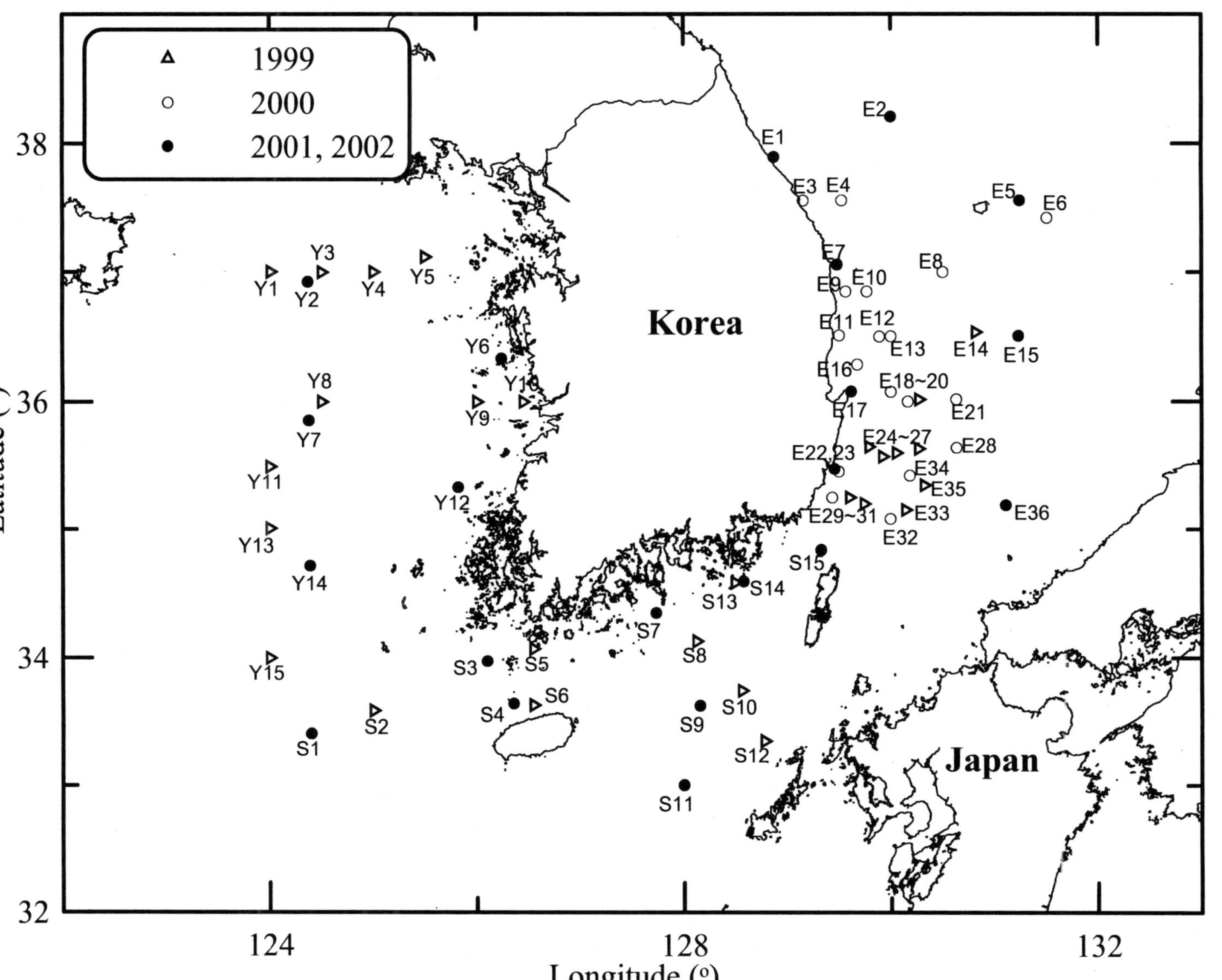

FIGURE 1. Sampling locations of surface waters in sea areas adjacent to the Korean Peninsula.

The $^{240}Pu/^{239}Pu$ atom ratios in seawaters of the East Sea/Sea of Japan, Yellow Sea and Korean Strait ranged from 0.18 to 0.33 with an average value of 0.25 ± 0.03, and were found to be relatively higher than that of stratospheric global fallout (0.18). The alternative source of the elevated $^{240}Pu/^{239}Pu$ in seawater adjacent to the Korean Peninsula would be due to transport of plutonium derived from the Pacific Proving Ground. It is hypothesized that the tropospheric fallout of high $^{240}Pu/^{239}Pu$ atom ratio was first introduced in the 1950s and that the signal was diluted by introduction of stratospheric global fallout plutonium of low $^{240}Pu/^{239}Pu$ atom ratio in the 1960s; after cessation of stratospheric fallout since the 1980s, the contribution of remobilization of Pu from the equatorial Pacific Proving Ground[9-10] became increasingly important. For a more complete understanding of the origin of Pu, we will need to gather a number of data through a wide area of ocean around the Korean Peninsula including the Pacific Ocean, and to furthermore improve the accuracy of the measured $^{240}Pu/^{239}Pu$ atom ratio in seawater with extremely low-level Pu concentrations. The physical/chemical form of fallout Pu is still not identical from all of its sources, despite a major role in determining the geochemical behavior of Pu in the oceans. Therefore, in addition to the Pu sources, we will also need to identify the physical/chemical form.

REFERENCES

1. Nelson, I. C., Thomas, V. W., and Kathren, R. L. "Plutonium in South-Central Washington State Autopsy Tissue Samples—1970–1975," Health Phys. 65, 422–428 (1993).
2. Perkin, R. W., and Thomas, C. W., "Worldwide Fallout," in Transuranic Elements in the Environment, edited by W. C. Hanson (1980).
3. Buesseler, K. O., and Halverson, J. E., "The Mass Spectrometric Determination of Fallout ^{239}Pu and ^{240}Pu in Marine Samples," J. Environ. Radioactivity 5, 425–444 (1987).
4. Kelly, J. M., Bond, L. A., and Beasley, T. M., "Global Distribution of Pu Isotopes and ^{237}Np," Sci. Total Environ. 237/238, 483–500 (1999); Buesseler, K. O., "The Isotope Signature of Fallout Plutonium in the North Pacific," J. Environ. Radioactivity 36, 69–83 (1997).
5. Krey, P. W, Hardy, E. P., Pachucky, C., Rourke, F., Coluzza, J., and Benson, W. K., "Mass Isotopic Composition of Global Fallout Pu in Soils," in Transuranium Nuclides in the Environment. IAEA-SM-199/39, Vienna, IAEA (1976), pp. 671–678.
6. Kim, C. K., Kim, C. S., Yun, J. Y. and Kim, K. H., "Distribution of ^{3}H, ^{137}Cs and $^{239+240}Pu$ in the Surface Seawater around Korea," J. Radioanal. Nucl. Chem. 218, 33–40 (1997).
7. Hirose, K., Miyao, T., Aoyama, M., and Igarashi, Y., "Plutonium Isotopes in the Sea of Japan," J. Radioanl. Nucl. Chem. 252, 293–299 (2002).
8. Kim, C. S., Kim, C. K., and Lee, K. J., "Determination of Pu Isotopes in Seawater by an On-Line Sequential Injection Technique with Sector Field Inductively Coupled Plasma Mass Spectrometry," Anal. Chem. 74, 3824–3832 (2002).
9. Buesseler, K. O., "The Isotope Signature of Fallout Plutonium in the North Pacific," J. Environ. Radioactivity 36, 69–83 (1997).
10. Mitchell, P. I., León Vintró, L., Dahlgaard, H., Gascó, C., and Sanchéz-Cabeza, J. A., "Perturbation in the $^{240}Pu/^{239}Pu$ Global Fallout Ratio in Local Sediments Following the Nuclear Accidents at Thule (Greenland) and Palomares (Spain)," Sci. Total Environ. 202, 147–153 (1997).

The Solubility of Pu(IV) Hydroxide and Pu Dioxide in Simulated Groundwater Solutions (SGW) under Various Conditions

Boris Myasoedov,[1] Yury Kulyako,[1] Dmitry Malikov,[1] Trofim Trofimov,[1] Ai Fujiwara,[2] Satoru Tsushima,[3] and Atsuyuki Suzuki[3]

[1] Vernadsky Institute of Geochemistry and Analytical Chemistry RAS, 19, Kosygin St., 119991, Moscow, Russia
[2] Radioactive Waste Management Funding and Research Centre, 15, Mori Bldg., 2-8-10, Toranomon, Minatoku, Tokyo, 105-0001, Japan
[3] Department of Quantum Engineering and Systems Science, School of Engineering, University of Tokyo 7-3-1, Hongo, Bunkyo-ku, Tokyo, Japan

INTRODUCTION

As a result of weapons plutonium production and functioning of spent nuclear fuel reprocessing plants, there were accumulated great volumes of radioactive waste in some countries. The most dangerous long-lived radionuclides—transuranium elements (TRU)—were released into geologic formations, ground- and seawater. The prediction of an aquatic behavior of TRU required a fundamental database consisting of solubility values of actinide hydroxides and oxides, especially of Pu ones. However, the reported values are often scattered, especially for tetravalent actinide hydroxides and oxides; there are very large differences of several orders of the magnitude between the lowest and highest values.[1–7] Therefore, it is of great interest to obtain additional information, particularly regarding the solubility of $Pu(OH)_4$ and PuO_2 in near-neutral groundwater solutions simulating those expected in the sites of HLW deep geological disposals.

RESULTS

The solubility of fresh-precipitated $Pu(OH)_4$ in simulated groundwater solutions (pH = 6 ÷ 11) under ambient conditions was determined. The maximum solubility value of $Pu(OH)_4$ (6×10^{-6} M) is observed at pH ~6. On increasing the pH value of the heterogeneous system from 6 to ~11, the concentration of the dissolved Pu decreases more than by two orders of magnitude (1.5×10^{-8} M). Solubility values of Pu(IV) hydroxide in substitute seawater are close to 10^{-7} M (pH ~8). At that oxidation state, Pu(IV) does not change in the solid phases of its hydroxide being in contact with both simulated groundwater and substitute seawater. "Polymeric" and "monomeric" samples of Pu(IV) hydroxide were prepared, and their solubilities were determined under ambient conditions in simulated groundwater solutions with various pH values. The solubility of both types of Pu(IV) hydroxide depends on the pH value of groundwater. The solubility of "polymeric" Pu(IV) hydroxide is smaller than that of a "monomeric" one. This difference may be explained by the specificity of structures of the hydroxides under study. The "monomeric" Pu(IV) hydroxide was used in the study on the determination of Pu(IV) hydroxide solubility in simulated groundwater solutions with various pH values under reducing conditions. The reducing conditions and low oxygen contents are more typical for respective geologic stratums of deep underground radioactive waste burials. The solubility of $Pu(OH)_4$ in simulated groundwater solutions under reducing conditions depends on the pH. For the pH values of 6.4 ± 0.3, 8.5 ± 0.2, and 10.6 ± 0.3, solubility values of Pu(IV) hydroxide are equal to $(2.2 \pm 1.6) \times 10^{-6}$ M, $(4.0 \pm 3.0) \times 10^{-7}$ M, and $(4.9 \pm 2.6) \times 10^{-9}$ M, respectively. It should be noted that the process of aging of

CP673, *Plutonium Futures — The Science,* edited by G. D. Jarvinen

"monomeric" Pu(IV) hydroxide results in its transformation into "polymeric" Pu(IV) hydroxide. This is accompanied by both a decrease of the hydroxide solubility and the disappearance of the dependence of hydroxide solubility on the pH value within a pH range of 6–11. Thus, among the factors that influence the solubility of $Pu(OH)_4$, polymerization processes going in their solid phase with time, along with a nature of the solutions (pH, ionic strength), play an important role.

The solubilities of PuO_2 samples prepared by calcinations at 850°C for 8 hours in SGW solutions were studied. It was shown that the solubility of a PuO_2 sample kept under ambient conditions in the course of a year is anomalously high, and for the solid/liquid phase ratios in the suspensions (m/V) 0.4, 4, and 40 mg/mL, it is equal to 3.0×10^{-6} M, 2.1×10^{-5} M, and 1.5×10^{-4} M, respectively. The high solubility of the sample under study is connected with an accumulation of Pu(VI) in its surface layers during storing the sample. On contact with SGW, readily soluble Pu(VI) hydroxide is formed. Therefore, the dependence of hydroxide solubility on solid/liquid phase ratio is observed. For a freshly prepared sample, the dependence of solubility on solid/liquid phase ratio is absent under analogous conditions, and its solubility achieves the level of about 2×10^{-10}M. The results presented below relate to the heterogeneous systems containing 4 mg/mL freshly prepared PuO_2. The solubility of PuO_2 in SGW is equal to about $(2.6 \pm 0.4) \times 10^{-10}$ M during 6 months of keeping the suspensions, and it does not depend on the pH within the pH range of 6–10 in the heterogeneous systems involved. It turned out that the thermodynamic solubility value of the plutonium dioxide in SGW at 25°C (1.4×10^{-10} M) was calculated on the assumption that dioxide powder forms amorphous phases with a corresponding Gibbs free energy of formation and is in good agreement with the solubility value determined experimentally.

The solubility of Pu dioxide in SGW with an initial pH of ~8.5 at 90°C and 150°C under the conditions of contact with ambient air was determined as well. The solubility of Pu dioxide in SGW increases by three orders of magnitude on transition from ambient conditions to t = 90°C. However, on further increasing the temperature from t = 90°C to t = 150°C, the solubility of Pu dioxide in SGW practically does not change. Average solubility values of PuO_2 in SGW at 90°C and at 150°C are equal to $(4.4 \pm 1.7) \times 10^{-7}$ M. Calculations of solubilities of the crystalline PuO_2 in SGW at elevated temperatures were made as well. The calculated solubilities of crystalline dioxide at 90°C and 150°C are much less than those measured experimentally. At that, the trends in the experimental and calculated solubilities are the same for Pu dioxide. Under inert conditions, the solubility of plutonium dioxide in the system "PuO_2 – SGW" is equal to $(2.2 \pm 0.4) \times 10^{-10}$ M. On transition from the "inert" conditions to the ambient ones, PuO_2 solubility in SGW practically does not change and remains equal to $(2.6 \pm 0.4) \times 10^{-10}$ M.

DISCUSSION

The performed study shows that solubilities of plutonium hydroxide and dioxide are controlled by the conditions of their preparation as well as by the conditions of their occurrence in real systems (pH, Eh, ionic strength).

ACKNOWLEDGMENTS

This work was supported by Radioactive Waste Management Funding and Research Centre, Japan, as well as by the Russian Foundation for Basic Research, grants # 99-03-32819 and 00-15-97391.

REFERENCES

1. Nitsche, H., Lee, S. C., and Gatti, R. C., J. Radioanal. Nucl. Chem. Articles 124,171–185 (1988).
2. Nitsche, H., Muller, A., Standifer, E. M., Deinhammer, R. S., Becraft, K., Prussin, T., and Gatti, R. C., Radiochim. Acta 58/59, 27–32 (1992).
3. Novak, C. F., Nitsche, H., Silber, H. B., Roberts, K., Torretto, Ph. C., Prussin, T., Becraft, K., Carpenter, S. A., Hobart, D. E., and AlMahamid, I., Radiochim. Acta 74, 31–36 (1996).
4. Neu, M.P., Hoffman, D.C., Roberts, Kevin, Nitsche, H., and Silva, Robert J., Radiochim. Acta 66/67, 251–258 (1994).
5. Nitsche, H., and Edelstein, N. M., Radiochim. Acta 39, 23–33 (1985).
6. Milligan, W. O., Beasley, M. R., and Lloyd M., Acta Crystall. B 24(7), 979–980 (1968).
7. Pazukhin, V. M., Radiochimiya (in Russian) 31(4), 430–436 (1989).

Biotransformations of Plutonium and Uranium by Naturally Occurring Microorganisms

Mary P. Neu, Christy E. Ruggiero, Hakim Boukhalfa, Sean D. Reilly, Larry E. Hersman, and Joe G. Lack

Chemistry and Biosciences Divisions, Los Alamos National Laboratory, Los Alamos, NM 87545, email: mneu@lanl.gov

INTRODUCTION

Microorganisms and microbial activity affect the biogeochemical cycling and overall environmental behavior of major elements such as C, N, P, S, and transition metals such as Mn and Fe. By extension, microorganisms are expected to affect the environmental behavior of minor and contaminant metals. Recent research has demonstrated that this is indeed true for less common elements, including the actinides.[1–5]

Naturally occurring bacteria can influence the speciation and solubility of actinides through several mechanisms, including the following: (1) intra- or extracellular accumulation, (2) direct or indirect redox, (3) direct or indirect pH changes, (4) production of mineral acids, organic acids, and chelators, (5) biomineralization, and (6) biocolloid formation. In turn, actinides can induce physiological changes to bacteria, such as the excretion of specific agents and inhibiting growth. We are studying these processes both to increase our knowledge of actinide chemistry and to improve our ability to predict the fate of actinides in the environment. A primary goal is to provide the underlying science needed to develop biostabilization and bioremediation technologies to restore actinide-contaminated sites. In this paper we present recent research results on several specific actinide-bacteria systems and conclusions on how those interactions can stabilize or destabilize actinides in subsurface and soil environments.

RESULTS AND DISCUSSION

Siderophore Chelation and Uptake

Siderophores, small molecules that bacteria use to sequester and incorporate Fe(III), have the potential to form highly stable actinide complexes. We have learned that the trihydroxamate siderophores (DFO), desferrioxamine B (DFB) and desferrioxamine E (DFE), are surprisingly slow at solubilizing Pu(IV) oxide/hydroxide solids, but eventually form Pu(IV) siderophore complexes in solution.[6] The Pu(IV)DFO complex forms rapidly and irreversibly at near-neutral pH when DFO is combined with Pu in its main oxidation states (III, IV, V, VI). Once the formation, prevalence, and structure of the Pu(IV)DFO complexes were established, we studied its microbial uptake.[7] Fe(III)-DFB and Pu(IV)-DFB complexes, but not U(VI)-DFB, are taken up by *Microbacterium flavescens* (JG-9).[8] We know the process is siderophore mediated, and not surface biosorption because the siderophore complexes were taken up, but corresponding metal-nitrilotriacetic (NTA) acid complexes were not. (In addition, cells were washed with concentrated EDTA solutions to remove any loosely bound or otherwise accessible metal or metal complex). We discovered that only living, metabolically active bacteria were capable of taking up the Pu siderophore complexes, just as with Fe-siderophore uptake (represented in Figure 1). Our studies also show that the two complexes (Pu(IV)-DFB and Fe(III)-DFB) mutually inhibit the uptake of the other, indicating that they compete for the same binding or transport sites in the microbe. Having demonstrated and quantified the microbial uptake of

CP673, *Plutonium Futures — The Science*, edited by G. D. Jarvinen

Pu-hydroxamate siderophore complexes, we seek to determine if this phenomenon is general. That is, are other types of actinide-siderophore complexes translocated across the cell membrane? Our initial results suggest that, at conditions under which Fe(III)-pyoverdin is accumulated by metabolically active cells of *Pseudomonas putida*, the Pu(IV)-pyoverdin complex does not appear to be taken up. These on-going studies are important to determine if siderophore-mediated uptake and transport are important factors in environmental mobility and Pu entry into the food chain.

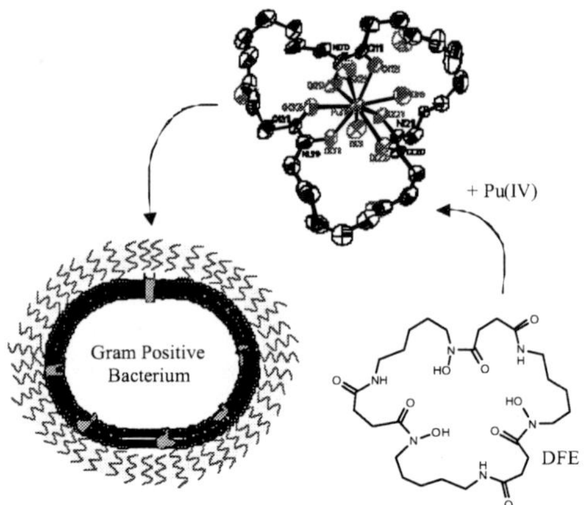

FIGURE 1. Fe-siderophore uptake.

Extracellular Accumulation

Bacteria commonly have polymers on the cell surface that have a variety of roles, including preventing dessication, facilitating mineral attachment, and binding metal ions.[9] These extracellular polymers vary broadly in their molecular weight, functionality, and potential metal-binding groups. We have examined the metal-binding properties of two extracellular polymers to assess their ability to accumulate actinides and potentially form biocolloids. *Bacillus licheniformis* produces a γ-polyglutamic acid (PGA) polymer that binds significant quantities of Fe(III) and other metals.[10] We isolated and purified PGA and determined that it binds 0.12, 0.18, and 0.24 mmol metal/mg PGA for U(VI), Fe(III), Pu(IV), respectively.[11] For comparison, we isolated, purified, and characterized the exopolysaccharide produced by *Rhodococcus erythropolis* (EPS). This type of bacterial extracellular polymer is thought to be very common. EPS binds approximately 60% less than Fe and U than PGA. Notably, the EPS is not a sufficiently strong chelator to outcompete hydrolysis and bind Pu(IV).

In order to ascertain the relative importance of different cell substructures in actinide binding, we studied metal binding by metabolizing whole cells of *B. licheniformis* under conditions used to isolate PGA. Surprisingly, the whole cells bound more of each metal ion than did the PGA on a per mass basis.[12] (Note, the whole cell mass includes a relatively large mass of water in the interior.) These results suggest that general extracellular binding may be more important for U(VI) than for Pu(IV) and that the cell wall and other substructures may be more accessible to metal ions than was previously thought.

Reduction by Dissimilatory Iron-Reducing Bacteria

Two species of iron-reducing bacteria, *Shewanella* sp. and *Geobacter* sp. have been widely studied for their ability to reduce U(VI) and other metals.[13,14] We now demonstrate that whole-cell suspensions of metabolically active *Shewanella putrefaciens*, CN32, rapidly reduce concentrations (up to 10 mM) of Pu(VI). We have also determined that cell growth is inhibited at Pu(VI) concentrations greater than the 6 mM Pu(VI). Coupled cell growth/Pu reduction is currently being investigated and will be compared with previously published results for U(VI) and Tc(VII).

Actinide Toxicity

When considering bioremediation of sites contaminated with radioactive materials, microbial tolerance to the metal contaminant must be considered. We have determined the toxicity of a variety of metals, radionuclides, and organic chelators to several bacteria. We found that actinides are less toxic than most other metals (e.g., Ni(II), Cd(II), Pb(II), Al(III)), inhibiting growth only at concentrations in the micromolar to millimolar concentration range–far above contamination levels reported in the U.S. Furthermore, by testing the toxicity of isotopes with differing specific activity ($^{238, 239}Pu$), we found that actinide toxicity is primarily chemical, not radiological, in nature. For example, the DFB complex of $^{239}Pu(IV)$ inhibited growth of the astonishingly radiation-resistant bacteria *Deinococus radiodurans* at 1,100–1,200 ppm (5.0 mM), the DFB complex of U(VI) inhibited growth at 600 ppm (2.5 mM), and the DFB complex of Np(V) inhibited growth at 260–380 ppm (1.6 mM). However, the order of the specific activity of these actinides is U (0.34 μCi/g) < Np (0.71 mCi/g) << Pu (80 mCi/g). For all these actinides, the radiation dose is orders of magnitude lower than what has been reported to be toxic to *D.rad* based on γ–irradiation studies. Also, *D. radiodurans* showed no growth inhibition at ^{238}Pu (17 Ci/g) concentrations of 8.2 ppm, a radiation dose equivalent to approximately 2,000 ppm ^{239}Pu, but ^{239}Pu was toxic at 1,000–1,200 ppm. The difference in oxidation state and chemical form of the actinides presumably contributed more to toxicity than did the alpha activitity.

REFERENCES

1. Neu, M. P., R. C. E., and Francis, A. J., "Bioinorganic Chemistry of Plutonium and Interactions of Plutonium with Microorganisms and Plants," in Advances in Plutonium Chemistry 1967–2000, edited by D. C. Hoffman, University Research Alliance and American Nuclear Society: La Grange Park, Illinois (2002), pp. 169–211.
2. Gadd, G. M., "Microbial Interactions with Metals/Radionuclides: The Basis of Bioremediation," Radioactivity in the Environment 2 (Interactions of Microorganisms with Radionuclides), 179–203, (2002).
3. Fredrickson, J. K. et al., "Influence of Mn Oxides on the Reduction of Uranium(VI) by the Metal-Reducing Bacterium Shewanella Putrefaciens," Geochimica et Cosmochimica Acta 66(18), 3247–3262 (2002).
4. Anderson, R. T. and Lovley, D. R., "Microbial Redox Interactions with Uranium: An Environmental Perspective," Radioactivity in the Environment 2 (Interactions of Microorganisms with Radionuclides), 205–223 (2002).
5. Gillow, J. B. et al., "Actinide Biocolloid Formation in Brine by Halophilic Bacteria," Materials Research Society Symposium Proceedings 556 (Scientific Basis for Nuclear Waste Management XXII), 1133–1140 (1999).
6. Ruggiero, C. E. et al., "Dissolution of Plutonium(IV) Hydroxide by Desferrioxamine Siderophores and Simple Organic Chelators," Inorganic Chemistry 41(14), 3593–3595 (2002).
7. Neu, M. P. et al., "Structural Characterization of a Plutonium(IV) Siderophore Complex: Single-Crystal Structure of Pu-Desferrioxamine E," Angewandte Chemie, International Edition 39(8), 1442–1444 (2002).
8. John, S. G. et al., "Siderophore Mediated Plutonium Accumulation by Microbacterium Flavescens (JG-9)," Environmental Science and Technology 35(14), 2942–2948 (2001).
9. Beveridge, T. J., and Graham, L. L., "Surface Layers of Bacteria," Microbiological Reviews 55(4), 684–705 (1991).
10. McLean, R. J. C. et al., "Metal-Binding Characteristics of the Gamma-Glutamyl Capsular Polymer of Bacillus Licheniformis ATCC 9945," Applied and Environmental Microbiology 56(12), 3671–3677 (1990).
11. He, L. M., Neu, M. P., and Vanderberg, L. A., "Bacillus Lichenformis Gamma-Glutamyl Exopolymer: Physicochemical Characterization and U(VI) Interaction," Environmental Science and Technology 34(9), 1694–1701 (2000).
12. Johnson, M. T. et al., "Interactions of Microbial Exopolymers and Whole Cells with Actinides," Preprints of Extended Abstracts presented at the ACS National Meeting, American Chemical Society, Division of Environmental Chemistry 40(2), 452–454 (2000).
13. Finneran, K. T. et al., "Potential for Bioremediation of Uranium-Contaminated Aquifers with Microbial U(VI) Reduction," Soil and Sediment Contamination 11(3), 339–357 (2002).
14. Wildung, R. E. et al., "Microbial Methods of Reducing Technetium," in U.S. 2001 (Battelle Memorial Institute, USA), 8 pp., Cont.-in-part of U.S. Ser. No. 99,680.

The Effect on the Pu(IV) Oxidation State in Aqueous Suspensions of UO_2, ThO_2, TIO_2 and MNo_2

Mattias Olsson, Anna-Maria Jakobsson, Hans Nilsson and Yngve Albinsson

Department of Materials and Surface Chemistry (Nuclear Chemistry)
Chalmers University of Technology, Göteborg, Sweden

INTRODUCTION

The behavior of plutonium with regard to redox and chemical properties is of interest because of (amongst other issues) the presence of plutonium in spent nuclear fuel. If placed in a deep geological repository, the consequenses of leakage need to be properly assessed. This includes investigations of the sorption behavior of dissolved waste constituents, such as plutonium.

Metal oxide powders are widely used as the solid phase in sorption studies. It only seems reasonable that such a surface can affect the oxidation state of the ions in the surrounding solution. For plutonium this has been shown in the case of the solid phase MnO_2, which has been reported to have an oxidizing effect on Pu(IV) in solution[1]. Also, it seems probable that UO_2, which is rather easily oxidized itself[2, 3] will reduce Pu(IV) in solution. Such behavior has been suggested for UO_2 in the presence of Np(V)[4].

In this work, the effect on the oxidation state of Pu(IV) in solution has been monitored for the two abovementioned oxides, and also for ThO_2 and TiO_2. The two latter oxides are generally considered non-reactive and only likely to work as catalysts if at all being involved in a redox process.

RESULTS

The solid phases were kept in suspension at pH 0.5 and 10 g/L. The solution used was 2 mM Pu^{4+} (the oxidation state was produced electrochemically using a potentiostat) and the ionic strength was about 0.5 M with regard to $NaClO_4/HClO_4$. Traces of Cl^- were also present. After three days, samples were removed, centrifuged at about 288,000 g and analyzed using an UV-Vis spectrophotometer (observing changes in oxidation state) and liquid scintillation counting (observing the amount Pu left in solution). A reference sample was also kept and analyzed at the start of the experiment and then after three days along with the rest of the samples. At the time of submission, this experiment was still in progress and expected to yield data from two more sample withdrawals.

The reference solution was, not surprisingly, found to change its oxidation state over time. Pu(IV) is partly converted into Pu(III) and Pu(VI), which is shown in Figure 1. In this spectrum, the black curve shows Pu(IV) with a slight presence of Pu(III) at the beginning of the experiment. After three days, a peak at 600 nm has grown, which comes from Pu(III). A new peak has also emerged at 830 nm. This peak is due to Pu(VI) and has a extinction coefficient about one order of magnitude higher than any other peaks mentioned in this work. Reference spectra for the different oxidation states of Pu have been compiled by Cleveland[5]. Figure 1 also shows that the characteristic Pu(IV) peaks at about 470 and 650 nm decrease in intensity, suggesting (as an expected consequence) that the amount of Pu(IV) has decreased with time.

UO_2 and MnO_2 actively affect the oxidation state of Pu in opposite directions. This can be seen in Figure 2. The spectrum for the Pu solution with UO_2 shows a complete Pu(III) dominance: no other oxidation states can be discerned. In the case of MnO_2, the spectrum corresponds to complete Pu(VI) dominance.

CP673, *Plutonium Futures — The Science,* edited by G. D. Jarvinen

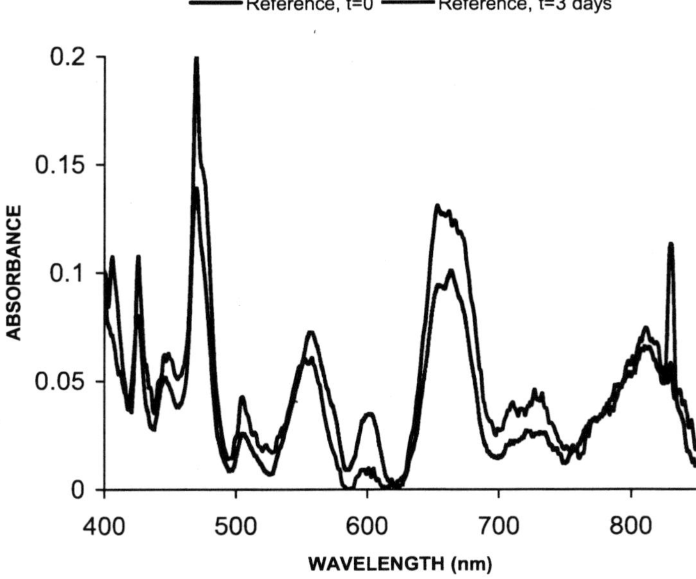

Fig. 1. UV-Vis spectra of the reference Pu solution (5 mM) at the start of the experiment and after three days. The solution was initially (t=0) almost pure Pu(IV).

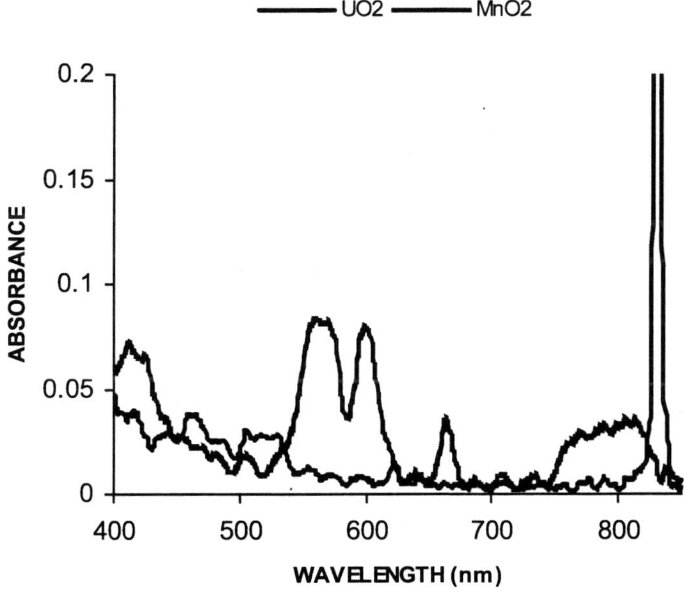

Fig. 2. UV-Vis spectra of two Pu solutions (2 mM) after three days in contact with suspended UO_2 and MnO_2.

The TiO_2 and ThO_2 samples yield overlapping spectra (Figure 3) and the presence of Pu(III, IV, VI) is shown in both cases. These two spectra are similar to the reference spectrum after three days. The difference in absorption values between the reference sample and the other samples is due to a higher concentration of the reference solution measured spectrophotometrically.

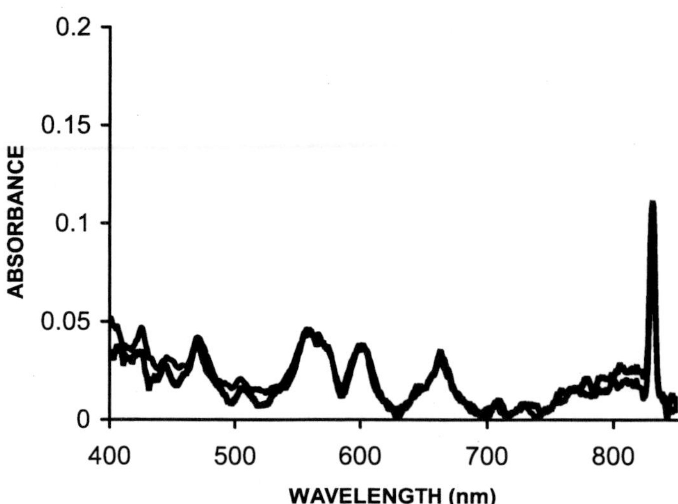

Fig. 3. UV-Vis spectra of two Pu solutions (2 mM) after three days in contact with suspended TiO_2 and ThO_2.

Samples withdrawn for liquid scintillation show that, in the cases with MnO_2 and TiO_2, some of the Pu leaves the solution. After three days, 86% of the original Pu amount remains in solution in the case of MnO_2, compared to 95% with TiO_2.

DISCUSSION

The formation of both Pu(III) and Pu(VI) in the same sample suggests disproportionation. This behavior seems dominant for the reference solution and the ThO_2 and TiO_2 surfaces. Comparing the amount of Pu that has disproportionated in the reference and the other two samples is not recommended since the concentrations of the spectrophotometry samples differ. The reference sample used is about 5 mM Pu. Disproportionation of Pu(IV) can be expected at low pH[6].

The oxidative effect of MnO_2 and the reducing effect of UO_2 are interesting from a repository point of view because of the massive amount of UO_2 in spent nuclear fuel, and also because of the predominance of Mn in natural aquatic redox processes[7].

The Pu missing from the solution for the MnO_2 and TiO_2 samples is believed to have sorbed onto the oxides. This thought may be elaborated further after the coming sample withdrawals.

The reason for this investigation has been dual: to see if sorption experiments with Pu(IV) are possible to do on any of these surfaces, and, through this, also learn whether Pu(IV) is an oxidation state that needs to be regarded in the presence of these oxides.

ACKNOWLEDGEMENT

This work was funded by the Swedish Nuclear Fuel and Waste Management Company, SKB.

REFERENCES

1. A. Morgenstern and G. R. Choppin, Radiochim. Acta, **90** (2002).
2. I. Casas, J. Giménez, V. Martí, M. E. Torrero and J. de Pablo, Radiochim. Acta, **66/67** (1994).
3. M. Olsson, A-M. Jakobsson and Y. Albinsson, J. Colloid Interface Sci., **256** (2002).
4. A-M. Jakobsson, *Measurement and Modelling using Surface Complexation of Cation (II to VI) Sorption onto Mineral Oxides*, thesis (Dept. of Nuclear Chemistry, Chalmers University of Technology, Göteborg, 1999).

5. J. M. Cleveland, *The Chemistry of Plutonium*, p. 13-19 (Gordon and Breach, Science Publishers, New York, 1970).
6. L. M. Toth, J. T. Bell and H. A. Friedman, Radiochim. Acta, **49** (1990).
7. W. Stumm and J. J. Morgan, *Aquatic Chemistry*, p. 448 (2nd ed., Wiley-Interscience, New York, 1981).

Reduction of Plutonium (VI) Species in Brine

D. T. Reed,[1] S. B. Aase,[2] A. J. Kropf,[2] and J. Conca[3]

[1]*Valparaiso University, Valparaiso Indiana 46383*
[2]*Chemical Technology Division, Argonne National Laboratory, Argonne IL 60439*
[3]*Los Alamos National Laboratory, Carlsbad NM*

SUMMARY

The reduction and oxidation state stability of Pu(VI) in brine was investigated in the presence of MgO, organics, aqueous Fe^{2+}, and iron coupons. This was an extension of long-term studies to evaluate the effects of these constituents on Pu(VI) in brine that had been stable for over two years in the absence of reducing agents under anoxic conditions. The addition of these constituents, in all cases, resulted in a net reduction of Pu(VI). That reduction, not sorption, occurred was confirmed by edge-position analysis of the XANES spectra of the Pu precipitates. Fe coupons, i.e., zero valent iron, was the most effective in reducing Pu(VI) and led to essentially complete reduction in < 200 hours. These data show that reduction, even in the presence of significant radiolysis, is the likely path for Pu(VI) when iron is present under anoxic conditions.

INTRODUCTION

The long-term speciation and migration of plutonium when it exists as a subsurface contaminant continues to be an important area of study. In sodium-chloride-magnesium brine, radiolytic pathways exist that can facilitate the oxidation of actinides, including plutonium, to the An(V) and An (VI) oxidation states. In our work, we have shown that Pu(VI) can be stable in brine over a wide range of pH as a hydrolytic/chloride or carbonato species when no reducing agents are present. This was true over a period of two years with ^{239}Pu as the plutonium isotope.

This research, however, did not reflect the potential effects of MgO, organics, and iron species on the speciation of Pu. These constituents are present in the subsurface of brine repositories. Herein we report the results of interactions between these constituents and "stable" Pu(V/VI) in brine. The goal of this research was to establish their potential to reduce the effective solubility of the plutonium.

RESULTS

Stability of Pu(VI) in the Absence of Reducing Agents

Plutonium, which was initially present as Pu(VI), remained predominantly in solution when no reducing agent was present—therefore, no precipitation was observed. The concentration of plutonium, for 0.2 μ-filtered and unfiltered solutions, is shown in Figure 1 for a low magnesium brine at p_cH 8 and 10.

The aqueous speciation of plutonium in the brine experiments was not the same. At pH 8 and 10, in the presence of carbonate (Experiments Pu-E10 and Pu-E8), the absorption spectra are almost identical and correspond to a Pu(VI) carbonate species. At pH 10, in the absence of carbonate (experiment Pu-E10-NC), an undefined Pu(VI)-inorganic complex is present. This is presumably a hydrolytic species, although we cannot discount the presence of a chloride complex. There is no spectroscopic evidence for the reduction of Pu (VI) in these higher p_cH systems.

CP673, *Plutonium Futures — The Science,* edited by G. D. Jarvinen

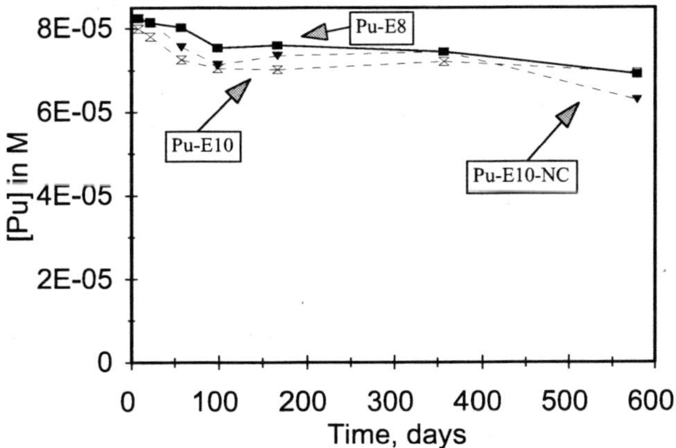

FIGURE 1. Concentration of plutonium as a function of time in ERDA-6 Brine (experiments, Pu-E10, Pu-E8, and Pu-E10-NC). No significant differences in concentration were noted between the unfiltered, 0.2-μ filtered, and 20-nm filtered solutions.

An aquo/chloride Pu(VI) complex is present in the experiments in G-Seep brine at p_cH 7. A small decrease in total plutonium concentration was evident at long times. There was also spectral evidence for the formation of Pu(V), presumably because of autoradiolysis effects. The formation of Pu(V) accounts for the small decrease in total concentration because this would likely lead to the formation of some Pu(IV). At p_cH 5, total concentrations have decreased by ~20%. The speciation, based on the absorption spectra is similar to that observed at p_cH 7 except that a lower concentration of Pu(V) is present, suggesting that Pu(IV) may be forming more rapidly.

Reduction/Precipitation of Pu(VI) in the Presence of Fe, Organics, and MgO

The presence of EDTA, citric acid, and oxalic acid led to a slow reduction of Pu(VI). Typically, this occurred over a period of 1–2 months. Reduction, which was confirmed by absorption spectrometry, led to the formation of Pu(V) and Pu(IV) organic complexes. MgO also led to a slow decrease in the Pu(VI) concentrations, but the mechanism was not clear, and it is possible that a reduction of plutonium did not occur (i.e., this occurred solely as a result of sorption).

The presence of an iron coupon, even under oxygen-free conditions, significantly lowered the plutonium concentration in the brine. These data are shown in Figure 2. The rate of reduction was correlated with the aqueous speciation and was most rapid at the lower pH high-magnesium brines. Carbonate complexation at both p_cH 8 and 10 resulted in a slower decrease in plutonium concentration. The plutonium species in the carbonate-free p_cH 10 ERDA-6 experiment was removed from solution at the slowest rate.

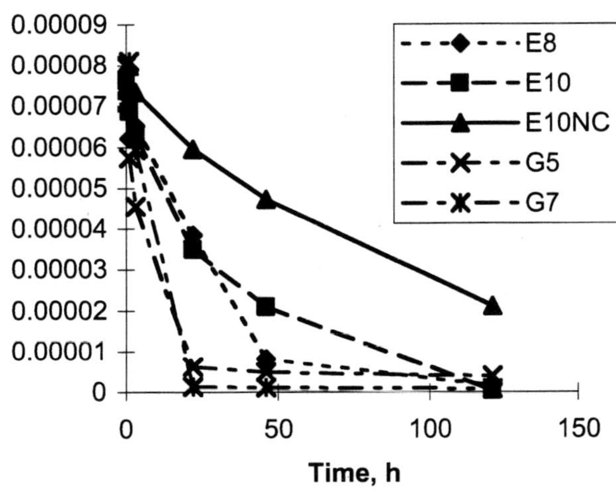

FIGURE 2. Effect of iron coupons on the concentration of plutonium in brine as a function of pH.

Although a number of questions regarding the mechanism of reaction remain, there is no question that the Pu(VI) reduction was significantly increased when iron solids were present. The interaction with Fe coupons led to an almost four order-of-magnitude reduction in the plutonium solution concentrations and caused the precipitation of plutonium phases. XANES analysis of both the plutonium precipitate and the plutonium sorbed onto the iron surface was shown to be Pu(IV) by the edge position analysis that confirmed that reduction had taken place. The EXAFS analysis, in radial space, of the precipitate obtained at pH 7 is shown in Figure 3. Although there is a good match in Pu-O bond lengths with the PuO_2 standard, there is much greater disorder in the brine precipitate, which suggests an amorphous and highly disordered structure in the Pu precipitate.

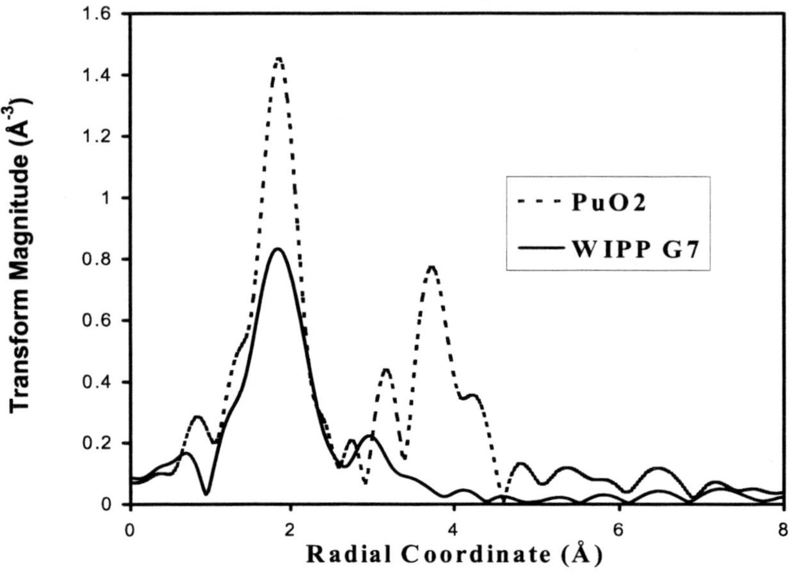

FIGURE 3. XANES of plutonium precipitates collected from the pH 7 WIPP experiments containing iron. The XANES of these unknowns match the spectrum of $Pu(IV)O_2$ and indicate that reduction leading to precipitation was occurring.

DISCUSSION

In the presence of Fe and organic complexants, the reduction of Pu(VI) was observed. This led to significantly lower apparent solubility even under radiolysis-affected conditions. The $\sim 10^{-8}$ M final concentrations noted were consistent with the solubility of a Pu(IV) species in the presence of carbonate (as we had in our experiment). That reduction occurred was confirmed by the XANES analysis of the Pu-sorbed and precipitated species. These data reflect positively on the performance of a subsurface disposal facility in brine where excess iron and iron phases are present.

Plutonium(VI) Sorption to Manganese Dioxide

Sean D. Reilly, William K. Myers, Stephen A. Stout, Donna M. Smith, Matthew A. Ginder-Vogel, and Mary P. Neu*

Chemistry Division
Los Alamos National Laboratory
Los Alamos, NM 87545

INTRODUCTION

Redox-active metal oxides may strongly affect the environmental behavior and mobility of actinides. Manganese oxides are relatively common redox-active soil components, which have a high surface area and which some studies show sorb plutonium selectively over other mineral phases.[1] For plutonium, oxidation states that could exist in the environment include +III to +VI, with Pu(IV) being predominant in the insoluble phase. Plutonium(V), and to a lesser extent Pu(VI), are the stable Pu oxidation states in solution under environmental conditions.[2] We are using synthetic δ-MnO_2 because it is most similar to the common natural manganese oxide mineral birnessite. Previously, we have shown that Pu(V) is oxidized to Pu(VI) in solution by δ-MnO_2, then very effectively sorbed to the mineral. We are now studying Pu(VI) sorption to synthetic δ-MnO_2 in detail to determine its sorption mechanisms and sorption capacity.

RESULTS

A series of sorption experiments were conducted at pH 3 to avoid Pu(VI) hydrolysis and still be at conditions near or above the point of zero charge of δ-MnO_2 so that it has a net negative surface charge. Titration experiments were performed where aliquots of Pu(VI) were added to aqueous solutions containing 1 mg δ-MnO_2/mL. The solutions were allowed to equilibrate a set time after each Pu aliquot addition, and then the concentration of Pu(VI) remaining in the separated solution phase was determined by liquid scintillation counting and visible/near-IR spectroscopy. Batch sorption experiments were also performed of 3×10^{-5} M to 3×10^{-3} M Pu(VI) with mineral suspensions of 1 mg δ-MnO_2/mL. Figure 1 shows the Pu(VI) sorption capacity of δ-MnO_2 as a function of Pu(VI) solution concentration. Experiments suggest the sorption capacity of manganese dioxide at pH 3 is approximately 0.1 mmol Pu(VI)/g δ-MnO_2, with small changes in pH influencing the capacity. The rate of Pu(VI) sorption at pH 3 was also studied and found to be rapid, with approximately 80% of the equilibrium Pu(VI) concentration sorbed onto MnO_2 within the first hour of contact.

The pH dependence of Pu(VI) sorption onto δ-MnO_2 is being studied at variable Pu(VI) concentrations. At 1×10^{-5} M Pu(VI) and pH 5–7, Pu(VI) is essentially completely adsorbed to MnO_2, while some Pu remains in solution at lower and higher pH (Figure 2a). The low percentage Pu(VI) bound below pH 4 is interpreted as a result of a decreasing negative charge on MnO_2, which results in less sorption of PuO_2^{2+}. Pu(VI) binding decreases as the pH is raised above 6.5, where Pu(VI) increasingly exists as neutral and negatively charged hydroxo and carbonato species.

In order to interpret the sorption data, we also need to know the solution speciation of Pu(VI) in the absence of the mineral. We are investigating and characterizing Pu(VI) hydrolysis species using potentiometric and spectrophotometric methods. Our studies on 10^{-2} M to 10^{-3} M Pu(VI) solutions indicate the formation of dimeric hydrolysis species $[PuO_2OH]^{2+}$ and $[PuO_2(OH)_2]_2$. At lower Pu(VI) concentrations, 10^{-4} M, data suggest a change in speciation and the presence of monomeric hydrolysis species. The species diagram Figure 2b, calculated for Pu(VI) at the same concentration as that in sorption experiments shown in Figure 2a, shows that the decrease in Pu(VI) sorption coincides with the onset of the formation of Pu(VI) hydroxo species.

CP673, *Plutonium Futures — The Science,* edited by G. D. Jarvinen

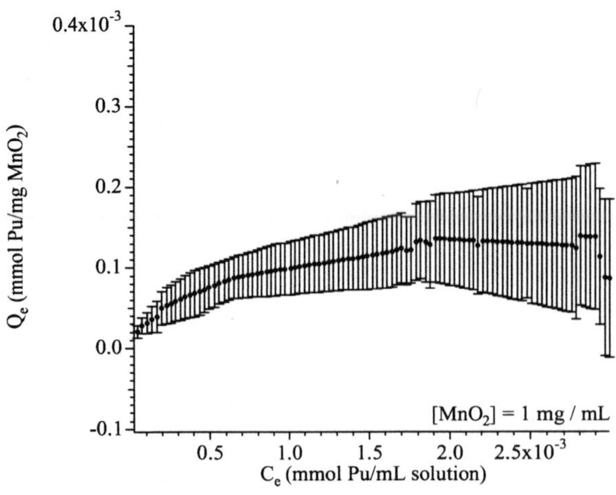

FIGURE 1. Pu(VI) sorption capacity of δ-MnO_2 in a 0.1 M ionic-strength aqueous solution at pH 3. C_e and Q_e are the equilibrium Pu(VI) concentrations in solution and on the MnO_2 adsorbent, respectively.

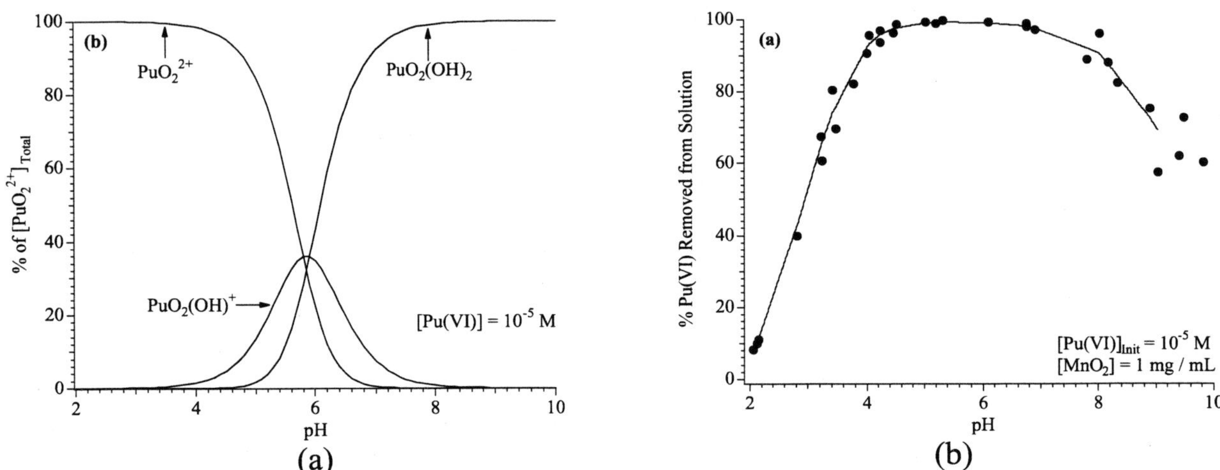

FIGURE 2. (a) pH-dependent sorption of Pu(VI) on δ-MnO_2 in a 0.1 M ionic-strength aqueous solution. Circles represent data from several independent experiments and the line is their average. (b) Pu(VI) hydrolysis species distribution plot as a function of pH.

DISCUSSION

Manganese dioxide has a high surface area and oxidizes Pu(V) to Pu(VI) and rapidly sorbs it. The observed sorption behavior of Pu(VI) onto δ-MnO_2 correlates well with Pu(VI) solution speciation. Plutonium(VI) sorption onto δ-MnO_2 begins at a lower pH than, for comparison, U(VI) sorption onto Fe_2O_3, consistent with the lower point of zero charge of δ-MnO_2.[3] Modeling of this sorption data using site sorption models is in progress. Additional Pu(VI) sorption data will be presented and compared to data collected for U(VI). Comparisons to actinide sorption onto other mineral phases will also be presented.

REFERENCES

1. Duff, M. C., Hunter, D. B., Triay, I. R., Bertsch, P. M., Reed, D. T., Sutton, S. R., Shea Mccarthy, G., Kitten, J., Eng, P., Chipera, S. J., and Vaniman, D. T., Environmental Science & Technology 33, 2163–2169 (1999).
2. Choppin, G. R., Bond, A. H., and Hromadka, P. M., Journal of Radioanalytical and Nuclear Chemistry 219, 203–210 (1997).
3. Davis, J. A., "Surface Complexation Modeling of Uranium(VI) Adsorption on Natural Mineral Assemblages," U.S. Geological Survey, 2001.

Sorption Constants for Pu(III) – Pu(VI) onto Mineral Oxide Surfaces

Jon M. Schwantes [σ] **& W. Batchelor** [β]

[σ] *Lawrence Berkeley National Laboratory, Heavy Elements Group, Nuclear Science Division*
[β] *Environmental Engineering, Texas A&M University*

INTRODUCTION

Researchers have suggested that sorption to mineral surfaces may control the overall mobility of Pu at some contaminated sites[1]. For the development of safe waste management practices, then, modeling this behavior would be advantageous. However, accurate surface complexation constants for Pu(III) – Pu(VI) onto mineral surfaces are needed before adequate predictive tools may be developed.

It may not be possible to derive sorption constants for all of the oxidation states of Pu directly from experimental data. Derivations of these constants require the measurement of redox distributions within solution and on the surface, since Pu may exist simultaneously in four different oxidation states within aqueous solutions. Unfortunately, traditional analytical techniques used to measure these distributions are plagued with high detection limits and may alter the redox distribution of Pu during the procedure[2]. Lacking sufficient measurements to derive constants, some researchers have assumed the dominant oxidation state of Pu on the surface was identical to that within solution[3,4,5]. Alternatively, sorption constants for Pu might be derived from measured data on its redox-stable analogs without making such an assumption.

This work developed binding constants for Pu(III) – Pu(VI) onto mineral surfaces by (1) summarizing published sorption data on chemical analogs of Pu, (2) deriving consistent sorption constants for these analogs, and (3) making inferences regarding sorption behavior of Pu based upon these derived constants. Model simulations were conducted using the PHREEQC[6] geochemical software. The WATEQ4F[7] thermodynamic database was augmented with best available constants for Pu and its analogs as suggested by several extensive reviews[8,9,2]. A standard double-layer surface complexation modeling approach similar to Wang et al.[5], was adopted for consistency and for purposes of comparing binding strengths of different cations. INVRS K, a computer program that couples the geochemical modeling power of PHREEQC with a Gauss-Newton non-linear regression routine[10], was developed and employed for regressing on best available chemical constants from experimental data found in the literature.

RESULTS

A summary of regression results from this work is shown in Table 1. A total of 18 regressions were conducted. The relative standard errors of all model fits to experimental data were less than 10% and typically below 5%. In all but one case, the introduction of a single new surface species would allow for adequate simulations of the observed behavior over a range of pH values.

CP673, *Plutonium Futures — The Science,* edited by G. D. Jarvinen

Table 1. Sorption constants for M(III) – M(VI) on selected mineral surfaces.

Solid[1]	$[M_T]$ (M)	Element	I (M)	Log Pco_2 (atm)	Log K	# of Points	RSE (%)[8]
Quartz	2e-9	Am(III)	0.005	-3.5	-0.18[2]	11	2.4
Quartz	1e-9	Eu(III)	0.05	-3.5	-1.66[2]	12	11
Alumina	5e-10	Am(III)	0.1	-3.5	3.2[2]	10	4.9
Montmorillonite (Si/Al = 1.2)	2e-9	Am(III)	0.005	-3.5	>8[2]	11	0.26
Montmorillonite (Si/Al = 1.2)	1e-9	Eu(III)	0.05	-3.5	12[2]	10	1.2
Biotite (Si/Al = 3)	2e-9	Am(III)	0.05	-3.5	1.36[2]	3	-
TiO_2	2e-8	Pm(III)	0.01	None	2.28[2]	8	1.7
Goethite	2e-8	Pm(III)	0.01	None	3[2]	6	1.7
Hematite	8e-14	Th(IV)	0.1	None	13.5[2]	15	9.7
Alumina	8e-14	Th(IV)	0.1	None	16[2]	9	9.6
TiO_2	3e-8	Th(IV)	0.04-0.005	None	10.2[2]	9	9.4
Goethite	1e-11	Pu(IV)	0.1	None	11.6[3], 23.5[4]	13	4.2
Hematite	1e-6	Np(V)	0.1	None	4.38[5]	15	5.1
Magnetite	1e-6	Np(V)	0.1	None	-2.05[6]	21	3.7
TiO_2	3e-5	Np(V)	0.01	None	-0.17[6]	5	7.3
Goethite	1e-10	Pu(VI)	0.1	None	1.06[6]	11	1.8
γ-Alumina	1e-10	Pu(VI)	0.1	None	-2.48[7]	6	1.7
Alumina	3e-10 - 3e-8	U(VI)	0.1	None	7.5[6]	22	2.4

[1]Silico-aluminates were modeled assuming a stoichiometric ratio of Si and Al surface sites, employing silica and alumina acidity constants to represent protonation and deprotonation at these surfaces, and assuming only Al surfaces participated in the sorption of the metal.

[2-7]The formation constant is defined for the following reaction:

$$^{2}K \qquad \equiv SOH + M^{+z} \Leftrightarrow \equiv SOM^{+z-1} + H^{+}$$

$$^{3}K \qquad \equiv SO^{-} + M^{+z} \Leftrightarrow \equiv SOM^{+z-1}$$

$$^{4}K \qquad \equiv SO^{-} + M(OH)_3^{+(z-3)} \Leftrightarrow \equiv SOM(OH)_3^{+(z-3)}$$

$$^{5}K \qquad \equiv SOH + M^{+z} \Leftrightarrow \equiv SOHM^{+z}$$

$$^{6}K \qquad \equiv SOH + M(OH)^{+(z-1)} \Leftrightarrow \equiv SOM(OH)^{+(z-2)} + H^{+}$$

$$^{7}K \qquad \equiv SOH + M(OH)_2^{+(z-2)} \Leftrightarrow \equiv SOM(OH)_2^{+(z-3)} + H^{+}$$

[8]Relative Standard Error (%).

DISCUSSION

The theory of hard and soft acids and bases provides a tool for only qualitative estimates of binding strength, as binding is a complex function of both ion size and charge. For quantitative estimates, however, the first hydrolysis constant may be used, since implicit within this value is information on both size and charge of a metal ion. Sorption constants derived here for the reaction:

$$\equiv SOH + M^{+z} \Leftrightarrow \equiv SOM^{+(z-1)} + H^{+} \qquad (1)$$

were plotted against the first hydrolysis constants of their respective metal (Figure 1). Binding constants from Wang et al.[5], who first observed this trend for actinides, were also included in the figure for comparison. The log of the sorption constants ($K_{sorption}$) for the metals was linearly related to the log of their first hydrolysis constants (K_{MOH}). The linear function derived here is:

$$LogK_{Sorption} = 2.32x(LogK_{MOH}) + 17.5 \quad (R^2 = 0.91) \tag{2}$$

Equation 2 provides a tool for predicting the adsorption for cations for which aqueous speciation has been characterized but only limited or no sorption data exist. Figure 1 indicates that binding strength for each of the cations does not vary greatly for different types of minerals surfaces. Using this function, the log of the sorption constants for Pu^{+3}, Pu^{+4}, $Pu^{V}O_2^{+}$, and $Pu^{VI}O_2^{+2}$ onto a generic mineral oxide surface were calculated to be 1.5, 15.7, -5.0, and 4.7, respectively.

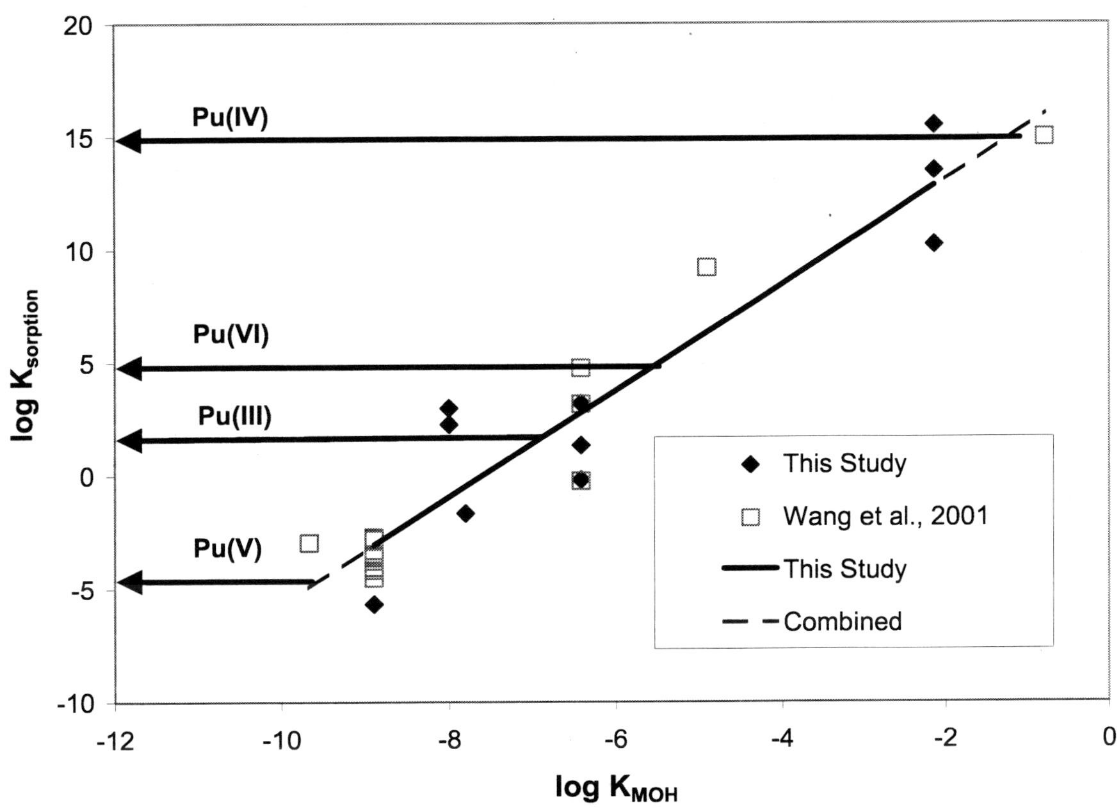

Fig. 1. Order of binding strength onto mineral surfaces for the four oxidation states of Pu.

Using an equilibrium model, redox distributions of Pu onto an aluminum hydroxide surface over a range of pH (4-9) and pe (5-10) values indicative of model environmental systems were predicted. Model systems contained 10^{-10} M Pu and 0.2 g of mineral having a surface area of 130 m^2g^{-1} and a surface site density of 2.31×10^{18} sites/m^2 in the presence of an atmosphere maintaining a $pCO_{2(g)}$ of -3.5. The redox distribution of Pu was controlled by the O^{-2}/O^0 couple, which was in turn controlled by the partial pressure of oxygen in equilibrium with the solution. Results of these simulations indicated that: (1) a complex relationship existed between the redox distribution within solution and that on the mineral surface; (2) Pu(IV) or Pu(VI) dictated (generally >95%) sorption (Figure 2), while Pu(V) dominated (typically >80%) the solution species; (3) sorption of Pu(V) accounted for less than 5% of the total sorbed Pu except within a narrow range of pe and pH (pe < 7 and pH > 7.0) values. This means, traditional solvent extraction techniques would not likely be able to resolve Pu(V) surface species at equilibrium and past studies reporting greater than a few percent of Pu(V) on the mineral surface under oxic environmental conditions are likely observing a kinetic feature due to slow redox processes rather than a thermodynamically stable surface species.

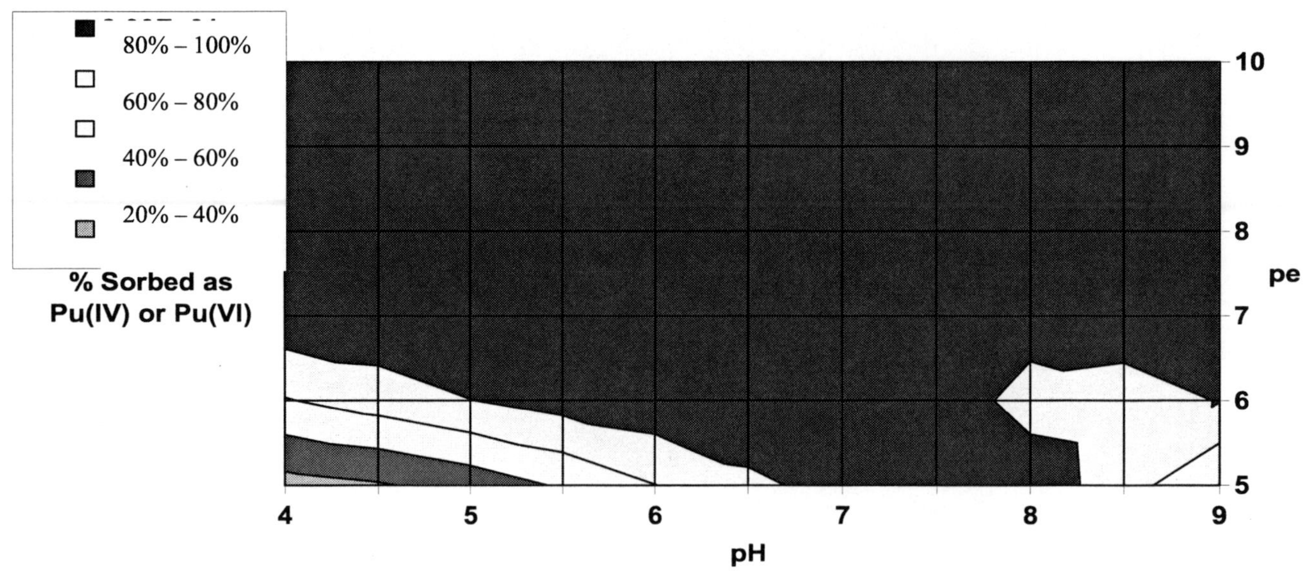

Fig. 2. Percent of the total Pu sorbed as either Pu(IV) or Pu(VI) to an aluminum hydroxide surface as a function of pe and pH; $[Pu]=1x10^{-10}$ M; $pCO_2 = -3.5$; $Al(OH)_{3(s)}$ - 0.2 g/L, 130 m^2/g, $2.31x10^{18}$ sites/m^2.

REFERENCES:

1. Kersting, A.B., Efurd, D.W., Finnegan, D.L., Rokop, D.J., Smith, D.K., Thompson, J.L., 1999. Migration of plutonium in ground water at the Nevada Test Site. Nature, 397, 7, 56-59
2. Schwantes, J., 2002. Equilibrium, kinetic and reactive transport models for plutonium. Ph.D.. Dissertation Civil Engineering, Texas A&M University, 325p.
3. Sanchez, A.L., 1983. Chemical speciation and adsorption behavior of plutonium in natural waters. PhD Dissertation, University of Washinton, University Microfilms International, Ann Arbor, MI, 191p.
4. Sanchez, A.L., Murray, J.W., Sibley, T.H., 1985. The adsorption of plutonium IV and V on goethite. Geochmica et Cosmochmica Acta, 49, 2297-2307.
5. Wang, L.L., Chin, Y.-P., Traina, S.J., 1997. Adsoprtion of (poly)maleic acid and an aquatic fulvic acid by goethite. Geochmica et Cosmochmica Acta, 61(24), 5313-5324.
6. Parkhurst, D.L., 1999. User's guide to PHREEQC, a computer model for speciation, reaction-path, advective-transport and inverse geochemical calculations. U.S. Geological Survey Water-Resources Investigations Report 99-4259, 143 p.
7. Ball, J.W. and Nordstrom, D.K., 1991, WATEQ4F--User's manual with revised thermodynamic data base and test cases for calculating speciation of major, trace and redox elements in natural waters: U.S. Geological Survey Open-File Report 90-129, 185 p.
8. Lemire, R.J., Fuger, J., Nitsche, H., Potter, P., Rand, M.H., Rydberg, J., Spahiu, K., Sullivan, J.C., Ullman, W.J., Vitorge, P., Wanner, H., 2001. Chemical thermodynamics of neptunium and plutonium, Volume 4. In: Thermodynamics, OECD Nuclear Energy Agency, Data Bank, Nuclear Energy Agency Organicsation for economic co-operation and development, Elsivier, New York, NY, p. 870.
9. Ewart, F.T., Cross, J.E., 1991. HATCHES – a thermodynamic database and management system. Radiochica Acta, 52/53, pp. 421-422.
10. Chapra, S.C., Canale, R.P., 1998. Numerical methods for engineers, Third Edition. Mc-Graw Hill, Boston, MA. pp. 924.

Interactions of Plutonium (V) and Plutonium (VI) with Manganese Dioxide, Iron Oxide, and Sediments from the Hanford Site

S. A. Stout, S. D. Reilly, D. M. Smith, W. K. Myers, M. A. Ginder-Vogel, S. Skanthakumar,* L. Soderholm,* and M. P. Neu

**Chemistry Division, Argonne National Laboratory*
Actinide Environmental and Structural Chemistry, Chemistry Division, Los Alamos National Laboratory, Los Alamos, NM 87545

INTRODUCTION

In order to make accurate predictions about the fate of actinides released into the environment and to implement effective remediation strategies, fundamental studies are needed to understand the complex geochemical interactions occurring between actinides and mineral surfaces. Minerals having strong redox potentials, such as Fe and Mn oxides and oxyhydroxides, are of special interest because they are likely to influence the complex redox chemistry of Pu under groundwater conditions. Thus, the interactions of Pu(V) and Pu(VI) with the following redox active synthetic mineral phases were studied: 100% δ-MnO_2, 50% δ-MnO_2:50% α-FeOOH, and 10% δ-MnO_2:90% α-FeOOH (w/w). Electron microscopy data show that the δ-MnO_2 phases effectively coat the α-FeOOH. The binding of Pu(V) and P(VI) to sediments from the Upper Ringold Formation near the DOE Hanford Site were also investigated. Solution concentrations in the range of 1×10^{-5} M to 4×10^{-4} M Pu were used to minimize hydrolysis, and the pH of the final mineral suspensions was adjusted between 0.5 and 10.

RESULTS

Plutonium associated rapidly with the synthetic mineral phases, regardless of its initial oxidation state. The absorbance spectra for Pu(V) solutions in the presence of δ-MnO_2 are shown in Figure 1. The characteristic absorbance of Pu(V) at 569 nm was not observed; however, a strong characteristic absorbance for Pu(VI) (830 nm) is present. This shows that Pu(V) in solution is oxidized to Pu(VI). For all experiments, a decrease in intensity of the 830 nm absorbance peak with time indicated the removal of Pu(VI) from solution. These spectra also suggest that sorption behavior is pH-dependent, as expected. At a lower pH, the Pu presumably competes with protons for available binding sites. A more detailed study of the effect of pH on Pu sorption by δ-MnO_2 is in progress. Previously, Pu has been assumed to associate with MnO_2 phases in a reduced form (III or IV); however, here we report that Pu is bound in an oxidized form (V or VI). X-ray absorption spectroscopy was performed on the δ-MnO_2 following a reaction with Pu(V) or Pu(VI) (Figure 2). The XAS data indicate that the Pu is bound to MnO_2 as a hexavalent inner-sphere complex regardless of the initial Pu oxidation state. To date, we have no evidence to suggest that Pu(V) associates with the mineral surfaces in that oxidation state.

Based on distribution coefficients calculated from sorption data, the concentration of Mn in the mixed mineral phases appeared to have little effect on the amount of Pu bound to the mineral surfaces. The UV/Vis spectra for Pu(V) solutions equilibrated with mixed Mn/Fe phases showed only a slight amount of Pu(V) remaining in solution soon after the addition of the Pu solution (Figure 3A). The mixed-mineral phase containing 50% MnO_2:50% FeOOH appeared to cause a slightly more rapid transformation of Pu(V) to Pu(VI). The increase in baseline at 29h and 7d

CP673, *Plutonium Futures — The Science,* edited by G. D. Jarvinen

indicates the formation of colloidal Pu(IV) hydroxide in solution. When Pu(VI) was reacted with the mixed mineral phases under similar conditions, no Pu(V) peaks were present, and the presence of colloidal Pu(IV) was not indicated (Figure 3B). The XAS for Pu(V) and Pu(VI) reacted with the mixed mineral phases and were very similar regardless of the initial Pu oxidation state or percentage of δ-MnO_2 in the samples. Figure 2 shows the XANES spectrum for a Pu(V) solution reacted with 10% MnO_2:90% FeOOH. The spectrum shows that Pu was associated with the mineral surface as in a mixture of oxidation states.

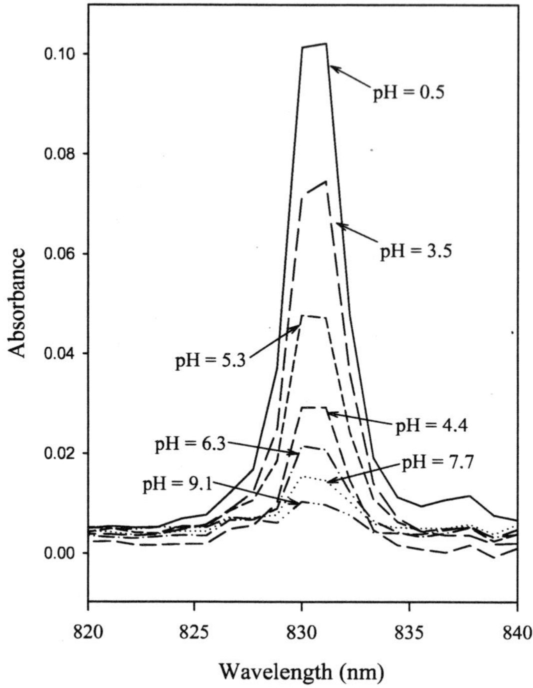

FIGURE 1. UV/Vis spectra of Pu(V) solution in contact with MnO_2. $[Pu]_i$ = 0.160 mmol/g MnO_2 and time = 3d.

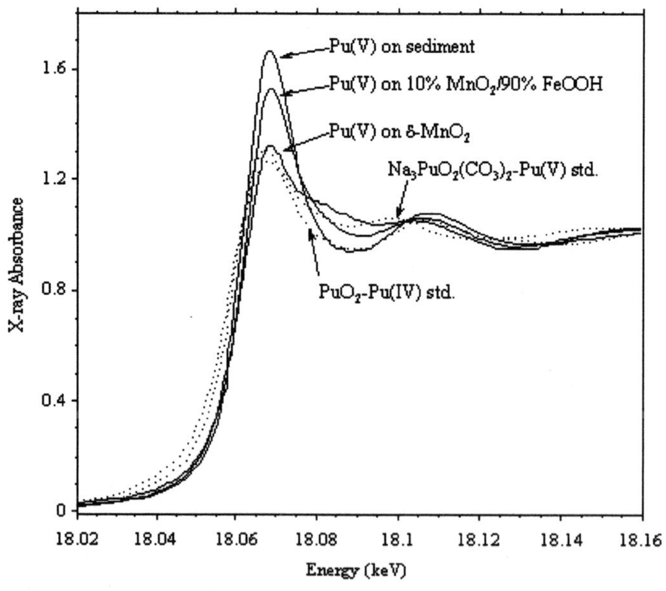

FIGURE 2. X-ray absorbance spectra of Pu (V) reacted with mineral phases and Ringold sediments.

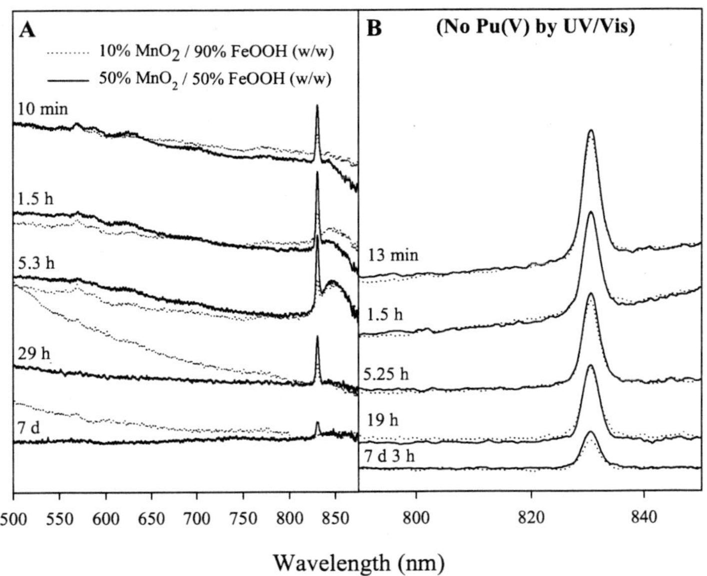

FIGURE 3. UV/Vis spectra of Pu absorption onto mixed MnO_2/FeOOH at pH = 7. (A) Initially 10^{-4} M Pu(V); (B) Initially 10^{-4} M Pu(VI).

When Pu(V) and Pu(VI) were reacted with the Ringold sediments at pH = 7, Pu(VI) hydroxide formed (845 nm) initially. After 7–14 days of equilibration, Pu(V) was present in solution for all samples equilibrated with the sediments. The XAS data for the sediment samples that reacted with Pu(V) and Pu(VI) showed a mixture of Pu oxidation states, but suggested that a large fraction was bound as Pu(IV) (Figure 2).

DISCUSSION

The implication of this research is that current data used for modeling Pu mobility and transfer may be underestimating the actual mobility of the Pu under oxidizing conditions. It was previously believed that the Pu precipitated immediately as PuO_2. Our research indicates that in the presence of oxidizing minerals (such as MnO_2), Pu exists in the more mobile oxidation states of (V) and (VI) or sorbed as Pu(IV). The mobility of the Pu may be enhanced to an even greater extent when Pu(VI) is bound to colloidal Mn and Fe oxides and oxyhydroxides.

A Study of Americium Speciation in the Calcium Carbonate of Mollusk Shells

M. A. Zuykov, M. V. Zamoryanskaya, and B. E. Burakov

Laboratory of Applied Mineralogy and Radiogeochemistry, the V.G. Khlopin Radium Institute, 28, 2-nd Murinskiy ave. , St. Petersburg, 194021, Russia; fax: (7)-(812)-346-1129; e-mail: zuykov@riand.spb.su

INTRODUCTION

The relevance of mollusks in the detection of water pollution has been explored for a variety of environmental contaminants, including Pu and Am.[1,2] Notwithstanding what shell may prove to be a more useful than soft parts of whole mussels as pollutant indicators,[3,4] no special detailed study of radionuclide distribution in the shells has been carried out. The lack of those important data precludes the successful using of mollusk shells for long-term monitoring and understanding the processes of radionuclide migration.

The mollusks build their shells with calcium carbonate, adopting a crystalline structure such as calcite and aragonite. The majority of modern mollusks have shells that consist of aragonite, although in some species the calcium carbonate of both types of crystalline structures is presented in one shell.[5]

The objective of the current investigation is to clarify the distribution of Am in aragonite shells of zebra mussel (*Dreissena polymorpha*) obtained after laboratory experiments in the V. G. Khlopin Radium Institute.[4] Because the relevant method of synthesis of Am-doped aragonite was not found until recently, samples of artificial Am-doped calcite were analyzed in order to compare the incorporation of Am in a calcium carbonate.

EXPERIMENT

The zebra mussel individuals were collected in the eastern part of the Gulf of Finland. The results of XRD analyses have shown that their shells consist of calcium carbonate (with crystalline structure of aragonite) with admixtures (in wt %) of SiO_2 (0.41), Al_2O_3 (0.39), Sr (0.15), S (0.07), P_2O_5 (0.06), Fe_2O_3 (0.04), and K_2O (0.03). The bulk α- and β- radioactivity in pure shells is less than 0.25 Bq/g and 1 Bq/g, respectively. The laboratory facilities include glass aquariums with brackish water (salinity, 4%, temperature 20 ± 2°C), which were placed in a glovebox. The concentration of ^{241}Am in experimental solutions varied from 8 x 10^2kBq/l to 2 x 10^5 kBq/l. The details of radionuclide uptake experiments and obtained data on radionuclide concentration in mollusks (for shell and five organs) are given in our previous work.[4]

Samples of artificial calcite were synthesized at room temperature using a well-known method of calcite single crystal growth.[6] The concentration of ^{241}Am in experimental solutions varied from 81 kBq/l to 8 x 10^5 kBq/l. Calcite crystals pure and doped by ^{241}Am up to 3 mm in size were obtained after 3 weeks of synthesis.[7]

The samples of shells and single crystals of calcite were characterized by x-ray diffraction, optical microscopy, and gamma-spectrometry. The selected samples were mounted in epoxy, polished, coated with carbon (and gold) and examined using scanning electron microscopy and cathodolumenescence (CL). The modification of cathodoluminescence study were made using SEM. The electron beam can have energy of 15 keV, the beam current is 10 nA, and the electron beam diameter is 5 μk and 200 μk.

CP673, *Plutonium Futures — The Science,* edited by G. D. Jarvinen

RESULTS AND DISCUSSION

The cathodoluminescent images on a full section of studied Am-doped shells along a length of valve was characterized by light bands of blue-green color that are oriented in parallel to the shell surface that correspond to different structural layers. The maximum intensity is exhibited in the boundaries between these layers. However, this result should be treated with care because of the large uncertainty in the determination of the americium in organic and mineral components of mollusk shells separately.

Samples of shells and single crystals of calcite obtained in pure (nonradioactive) media were characterized by a weak cathodoluminescent band in the blue region. However, cathodoluminescence of Am-doped samples increases three times. Figures and 1 and 2 show that these bands have a structure, they consist of three or four different bands at 2.0, 2.3, 2.5, and 3.0 eV. These bands can be related to the structure defects of calcite or Am-luminescent centers in calcite. The interpretation of luminescent nature of these bands requires further investigation.

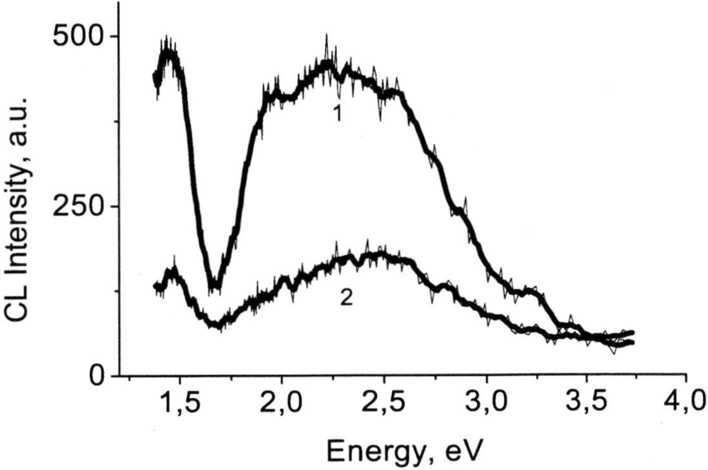

FIGURE 1. The cathodoluminescence spectra of pure (1) and Am-doped (2) mollusk shells of *D.polymorpha*. The content of Am in the studied shell is about 0.00005 wt %.

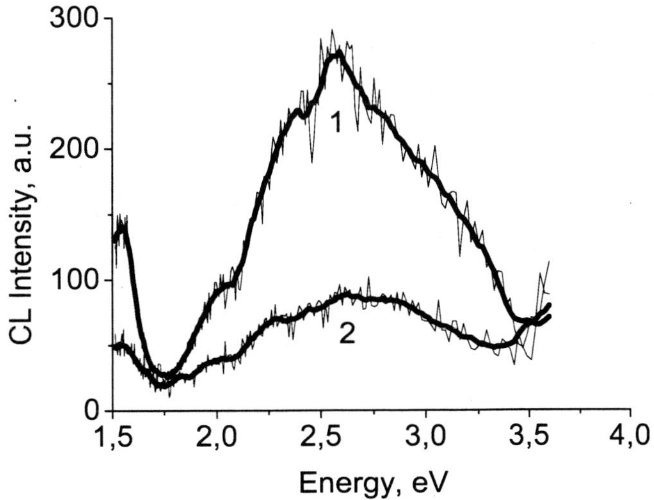

FIGURE 2. The cathodoluminescence spectra of pure (1) and Am-doped (2) artificial calcite.[7] The content of Am in the studied sample of calcite is about 0.01 wt %.

REFERENCES

1. Goldberg, E. D., Bowen, V. T., Farrington, J. W., Harvey, G. R., Martin, J. H., Parker, P. L., Risebrough, R. W., Robertson, W., Schneider, E., and Gamble, E., "The Mussel Watch," Environ. Conserv. 5, 101–125 (1978).
2. Goldberg, E. D., Koide, M., Hodge, V., Flegal, A. R., and Martin, J. H., "U.S. Mussel Watch: 1977–1978 Results on Trace Metals and Radionuclides," Estuar. Coast. Shelf Sci. 16, 69–93 (1983).
3. Koide, M., Lee, D. S., and Goldberg, E. D., "Metal and Transuranic Records in Mussel Shells, Byssal Threads, and Tissues," Estuar. Coast. Shelf. Sci. 15, 679–695 (1982).
4. Zuykov, M., Orlova, M., and Burakov, B. "Experimental Study of Simultaneous and Separate Accumulation of ^{137}Cs, ^{90}Sr, and ^{241}Am by the Freshwater Molluscs," in International Conference on Radioactivity in the Environment, Monaco (2002), CD-ROM.
5. Borisenko, A. Yu., "Evolution of Carbonate Biomineralization and System of Bivalvia," edited by V. N. Dubatolov and A. T. Moskalenko, Principe of Evolution and Historicism in Geology and Paleobiology 236–245 [in Russian] (1990).
6. Gruzensky, P. M., "Growth of Calcite Crystals," in Crystal Growth, Proceedings of an International Conference on Crystal Growth, Boston, June 20–24, 1966, edited by H. S. Peiser, Pergamon Press (1966).
7. Burakov, B., Zamoryanskaya, M., and Zuykov, M., "Investigation of Americium Incorporation by Carbonate Minerals under Conditions of Crystal Growth," in International Conference on Radioactivity in the Environment, Monaco (2002), CD-ROM.

DETECTION AND ANALYSIS

Analyzing Samples Using an IRIS Inductively Coupled Plasma-Atomic Emission Spectrometer (ICP-AES)

Mary Ann Abeyta

Los Alamos National Laboratory, Los Alamos, NM 87545

The determination of trace metallic elements is important to material processing, recovery, environmental sampling, and remediation programs here at Los Alamos National Laboratory. Samples are received in the liquid or solid state. The samples are prepared and analyzed according to approved EPA methods or procedures. This analysis procedure involves the standardization and calibration of the Inductively Coupled Plasma-Atomic Emission (ICP-AES) system with known concentrations of analytes of interest. The calibration and standardization data obtained is then used to determine the concentration of trace metallic elements (analytes) in samples of interest. During the treatment, preparation, segregation, and analysis of samples using the ICP-AES system, significant quantities of residue and chemical waste are generated. Thus the proper disposal of waste and residue according to waste-disposal procedures here at Los Alamos is also an important part of this effort.

CP673, *Plutonium Futures — The Science*, edited by G. D. Jarvinen

Alpha Liquid Scintillation Applied for Actinide Environmental Analyses: What Improvement?

J. Aupiais,[1] C. Aubert,[1] A Reboli,[1] and J. C. Mialocq[2]

[1]*CEA, DASE/RCE, Centre de Bruyères-le-Châtel, BP 12, 91680 Bruyères-le-Châtel, France*
[2]*CEA, DRECAM/SCM/URA 331 CNRS, Centre de Saclay, 91191 Gif-sur-Yvette, France*

Many authors have demonstrated the interest of α-liquid scintillation with pulse shape discrimination for the measurement of α emitters like actinides in environmental samples. Nevertheless, this technique suffers from a lack of resolution. Indeed, the energy resolution is about 300 keV at 5-MeV alpha energy, leading, for instance, to the impossibility of separating americium isotopes. The enhancement of the resolution is a great challenge for the coming years to promote this technique as a reliable and robust analytical method in the determination of α emitters in environmental samples. Two complementary issues are possible: the first is chemical by preparing new more-efficient cocktails in terms of light emission and pulse shape discrimination,[1] the second is technological by testing a new generation of photomultiplier tubes like hybrid P.M.[2] or avalanche diodes.

The light pulse emitted by the cocktail has two components: a prompt signal resulting mainly from a "singlet pathway" and a delayed signal that also involves triplet states. The "singlet pathway" is well described by the Förster theory of long-range Coulombic energy transfer,[3] which states that the rate constant of energy transfer is proportional to the square of the dipole-dipole interaction energy. The latter is proportional to the magnitude of the dipoles and inversely proportional to the third power of the distance between both molecules. The critical distance increases with the fluorescence quantum yield of the donor ϕ_D and the overlap of the spectra:

$$R_0^6 = \frac{9000 \ln 10 \kappa^2 \phi_D}{128\pi^6 n^4 N} \int f_D(\nu)\varepsilon_A(\nu)\frac{d\nu}{\nu^4} \ . \tag{1}$$

SCINTILLATING COCKTAILS

Most of the cocktails for α scintillation contain a solvent (like toluene) and a derivative naphthalene molecule as an α/β discriminator. Following the toluene solvent radiolysis and the recombination of the resulting ions, the longer-lived S_1 excited singlet states and the T_1 triplet states undergo radiative and nonradiative deactivation processes (fluorescence, internal conversion, intersystem crossing, singlet-singlet energy transfer to naphthalene). The T_1 states undergo triplet-triplet annihilation mainly in the spurs and triplet-triplet energy transfer to naphthalene through an exchange mechanism when the toluene triplets have diffused away from the spurs. The naphthalene S_1 singlet states resulting from the toluene to naphthalene singlet-singlet energy transfer undergo a very efficient intersystem crossing to the triplet state. A large concentration of long-lived naphthalene triplet states is thus obtained through two different pathways.

The most important tool to estimate the efficiency of scintillating cocktails is the overlap calculation between the fluorescence spectrum of the donor and the absorption spectrum of the acceptor. The higher the overlap, the better the probability for an energy transfer [Equation (1)]. By calculating overlaps for a large number of donor-acceptor couples (see Table 1), it is possible to find more efficient cocktails. For instance, by replacing toluene with p-xylene, the resolution has been improved by about 12% (Figure 1).[1]

CP673, *Plutonium Futures — The Science*, edited by G. D. Jarvinen

TABLE 1. Overlap Calculations for Some Donor-Acceptor Transfers $\int f_D(\nu)\cdot\varepsilon_A(\nu)\frac{d\nu}{\nu^4}\times10^{-27}$ (in m^6 $mole^{-1}$) (this work).

Acceptor → ↓ Donor	Φ-CH_3	$C_{10}H_8$	CH_3-$C_{10}H_8$	PBBO	PPO	3-HF
Φ-CH_3	0.002 *	2.12	5.32	24.74	23.82	11.39
$C_{10}H_8$		0.005 *		92.50	26.53	37.77
CH_3-$C_{10}H_8$		0.001	0.008*	100.07	24.54	43.50
DIN				96.25	21.20	44.26
PBBO				3.54		
PPO					1.54	
3-HF						0.13

*In cyclohexane, or else in toluene.

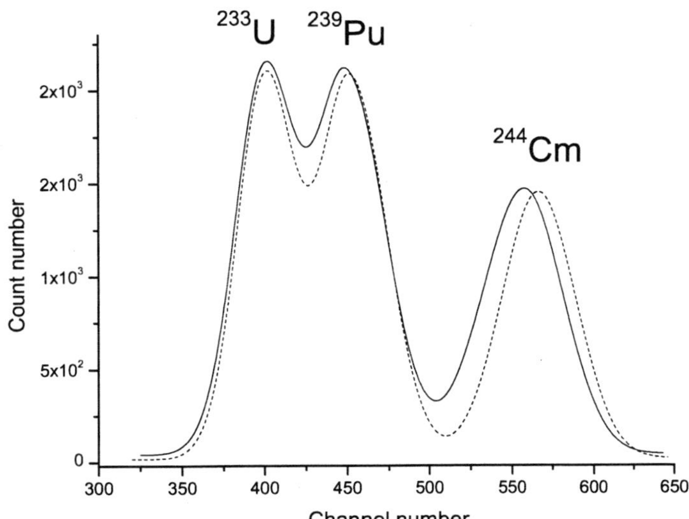

FIGURE 1. Actinides spectrum using Alphaex (solid) and p-xylene-$C_{10}H_8$-PBBO mixture (dash).

It seems difficult to improve to a large extent the resolution in alpha liquid scintillation.[4] This is mainly due to a lack of knowledge of the triplet behavior involved in the delayed component. Nevertheless, the Förster theory describes quite well the energy transfer for singlet states even at the very high concentration encountered in the scintillating cocktails. The knowledge of the fluorescence quantum yield, the fluorescence spectrum of the donor, and the absorption spectrum of the acceptor allow the test of many virtual scintillating cocktails without the need of experiment. The derivative naphthalene molecules seem to be the key for the temporal discrimination as a result of a large singlet conversion after the triplet-triplet annihilation. Nonetheless, no information is available for the other aromatic molecules. Practically, it seems that the technological way is the more appropriate fashion to improve the resolution greatly.

DETECTORS

The hybrid photomultiplier presents a quantum efficiency higher than 40% (≈30% for a common P.M.) although it is near 100% for avalanche diodes. Theoretically, it means that an improvement by a factor 2 of the resolution is possible. The main drawbacks are the small size of the avalanche diode (∅ max = 16 mm) and a wavelength sensitivity in the 500–800 nm range instead of 350–450 nm for the P.M. used in an α/β scintillation. It must also be noticed that the response is more constant along a large range of wavelengths, which is not the case for the common P.M. (Figure 2).

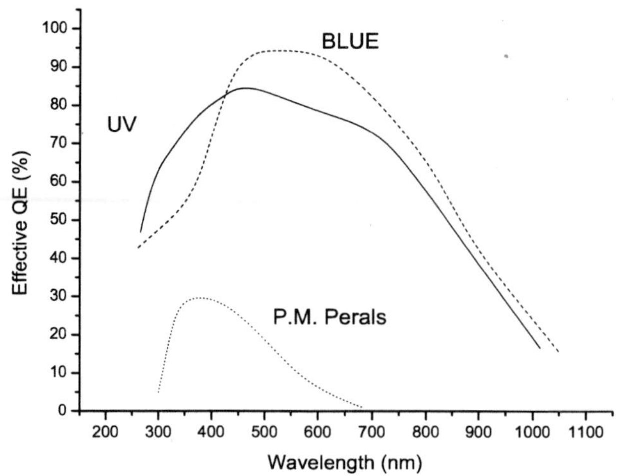

FIGURE 2. Quantum efficiency of the P.M. used in the Perals spectrometer and 2 avalanche diodes; UV and blue sensitive.

The use of blue sensitive avalanche diodes requires scintillators having a very high Stoke shift (like 3-hydroxyflavone, $\lambda_{fluo} \approx 550$ nm, as shown in Figure 3) having an absorption spectrum in the UV range and a fluorescence spectrum in the 500–700 nm range. Indeed, aromatic molecules composing the cocktails have generally fluorescence and absorption spectra in the UV range. In the case of 3-HF, its absorption spectrum overlaps quite well the fluorescence spectrum of naphthalene, suggesting a high efficiency and its possible use jointly with blue-sensitive avalanche diodes. For a UV-sensitive avalanche diode, Alphaex can be used without alteration.

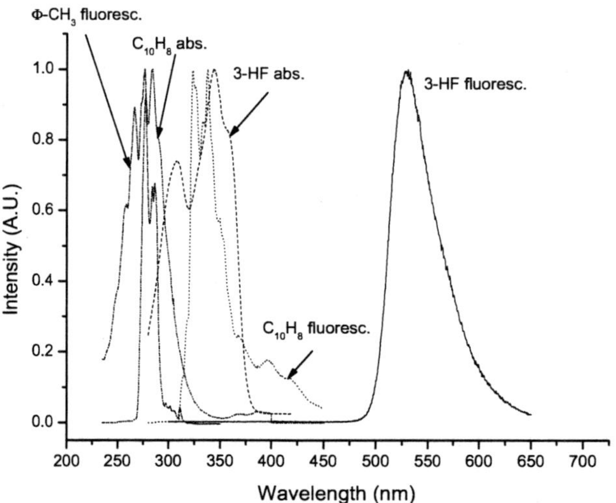

FIGURE 3. Scintillation mixture—toluene-naphthalene-3-HF.

In conclusion, the improvement of the resolution by a factor 2 is possible. From a technological point of view, it is necessary to optimize the geometry of counting (shape of quartz cell, etc.). The solution to all technical problems is under way.

REFERENCES

1. Aubert, C., Thesis, University of Paris XI—Orsay (2002), 280 pp.
2. Cassette, P., and Monnard, E., NIM A 422, 119–123 (1999).
3. Förtser, T., Naturforsch. 4a, 321–327 (1949).
4. Aupiais, J., Aubert, C., and Dacheux, N., Radiochim Acta (to be published).

Interface Development for Coupling Capillary Electrophoresis (CE) and Inductively Coupled Plasma–Mass Spectrometry (ICP-MS): Application to Plutonium Speciation

N. Baglan, A. Delorme, and J. Aupiais

CEA, Département Analyses et Surveillance de l'Environnement, Centre de Bruyères le Chatel, B.P. 12, 91 680 Bruyères le Chatel, France

Over the past few years, speciation studies have gained more and more popularity because its knowledge is necessary for understanding the transfer mechanism observed for various elements. Among them, radionuclide or chemical pollutants become a subject of interest, especially for predicting their behavior in the biosphere and their potential impact on health and population. Indeed, the migration in the environment and the toxicity depend on the chemical species with respect to the thermodynamic conditions encountered in nature such as pH, pE, ionic strength and temperature. Furthermore, the very low concentration of these radionuclides, and their repartitions in several chemical species necessitate the use of a sensitive detector, but the separation of different species requires efficient chromatographic techniques with high separation factors. The most promising combination seems the coupling between capillary electrophoresis and ICP-MS. Indeed, the univocal attribution of a peak to an element is then only possible if the reproducibility in terms of migration time is high enough to attribute any peak without ambiguity to a unique specie. In addition, the detector may exhibit very low detection limits for actinides and a resolution high enough to discriminate between neighboring masses.. However, for each technique, the sample flow rate is different: below 1 $\mu l.min^{-1}$ for EC compared to an introduction flow rate varying from 20 to 1000 $\mu l.min^{-1}$ for ICP-MS depending on the nebulizer. In order to couple both apparatuses, an interface has been built and its performance checked before the first studies on actinides speciation. Then, to improve our understanding for plutonium behavior, a preliminary study was initiated to determine with sufficient accuracy the absolute mobility of all species of interest. We have focused on chemical analogue existing under a single oxidation state to get their absolute mobility extrapolated at infinite dilution. The theory will shortly be presented hereafter, and the first determination of the absolute mobility of Np(V) at low concentration will be presented.

Before building the interface, several nebulizers; Meinhardt (Thermo-Optek, Winsford, Cheschire, UK), MCN100 (CETAC Technologies, USA), PFA 50 and PFA 20 (CPI, Amsterdam, The Netherlands) were used in order to determine which one is the more suited for the coupling. Among them, the PFA 50 presents the best performance in terms of sensitivity and stability at a low flow rate. Then an interface was designated and built using stainless steel (Figure 1). To ensure a sufficient introductory flow rate, a madeup solution was used.[1] After checking performance (separation on lanthanides), the interface was used to investigate actinide migration.

Practically, the determination of the migration time of electroosmotic flow is almost impossible by ICP-MS because there is no neutral marker or inorganic elements comigrating with the solvent. Therefore, internal markers are required. We have used a fast cation Cs^+ ($\mu_0(Cs) = 8.01\ 10^{-4}\ cm^2.V^{-1}.s^{-1}$) and a slow one Li^+ ($\mu_0(Li) = 4.01\ 10^{-4}\ cm^2.V^{-1}.s^{-1}$) in order to match a high range of mobilities, from $\mu^0 \approx 10^{-3}$ to $\approx 10^{-4}\ cm^2.V^{-1}.s^{-1}$. It must be noted that it is not possible to calculate an effective mobility, but only a difference of mobility between the ion i and one of the internal markers j according to the relation,

$$\Delta\mu_{app.,i,j} = \mu_{app.,i} - \mu_{app.,j} = \frac{L^2}{V}\left(\frac{1}{t_i} - \frac{1}{t_j}\right). \quad (1)$$

CP673, *Plutonium Futures — The Science,* edited by G. D. Jarvinen

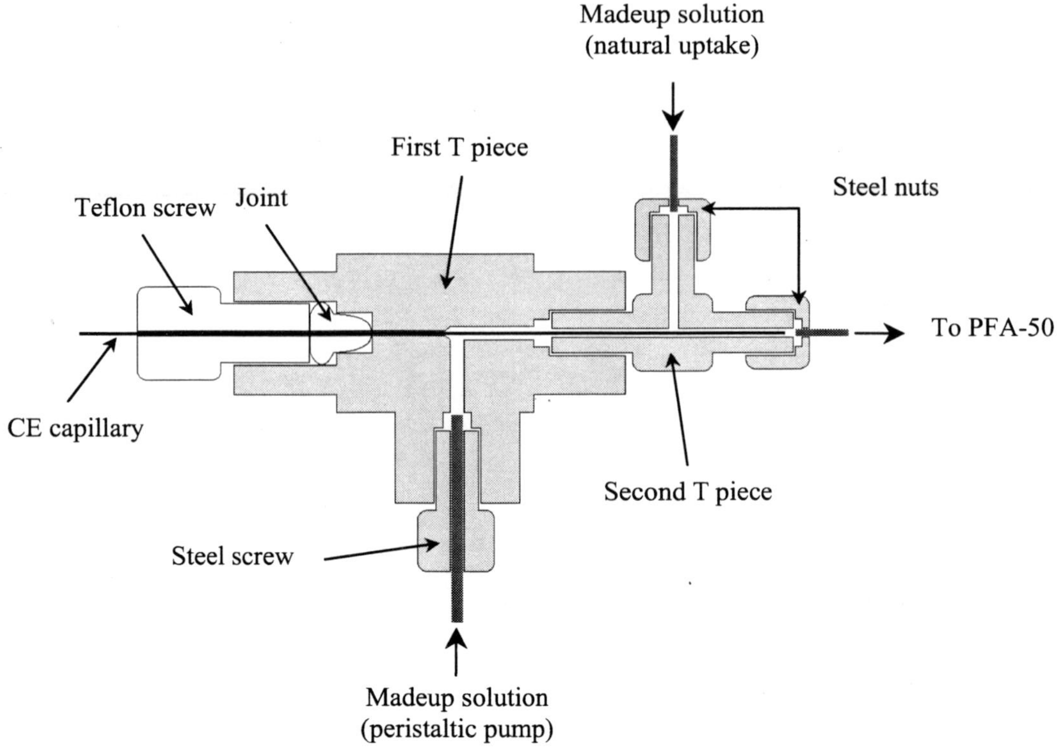

FIGURE 1. Interface capillary electrophoresis / inductively coupled plasma—mass spectrometry.

At a low concentration of ions ($C_i \leq 10^{-3}$ M) and for a totally dissociated 1:1 electrolyte, the ionic conductivity λ_i is assumed to obey the Debye-Hückel-Onsager's equation,

$$\lambda_i = \lambda_i^0 - S_i\sqrt{I}\,. \qquad (2)$$

Because of a simple relation between the limiting equivalent conductivity and the effective mobility, a linear trend should be observed between the effective mobility and the square root of the ionic strength according to the relation,

$$\mu_i = \mu_i^0 - S_i'\sqrt{I}\,, \qquad (3)$$

with $S_i' = S_i \cdot F$, where F is the Faraday.

Using internal standards j like Cs^+ or Li^+, the relative mobility of an ion i is

$$\Delta\mu_{i,j} = \Delta\mu_{i,j}^0 - \mathrm{Const}\cdot\sqrt{I}\,. \qquad (4)$$

It should noticed that Equation (2) is assumed to be valid up to I = 0.1 M according to Reference 2. The determination of the absolute mobility is then determined similarly by Equation (4). Several conditions must be satisfied for obtaining a reliable value: (a) no complexing agent is present in the electrolyte solution, (b) the electrolyte itself has no interaction with the cation, (c) the electrolyte must have a buffer capacity to stabilize the electroosmotic flow, (d) the pH of the electrolyte must be compatible with the thermodynamical stability of the hydrated cation, and (e) the temperature must be controlled. Practically, we have chosen the well-known perchlorate anion as counter-ion, creatinine as the main electrolyte buffer with a pK_a = 4.85. Therefore, near pH = 4–5, Np(V) is far from the hydrolysis region and is present under the hydrated NpO_2^+ species.

Two internal markers are used for the determination of the absolute mobility of Np(V). This approach allows a better determination and allows the verification of the method by the determination of the absolute mobility of one of both markers by using the other as an internal marker. In practice, it is not recommended using an internal marker with mobility close to that of the studied species. Therefore, we have chosen the determination having the lower uncertainty using cesium, thanks to a higher difference between the effective mobility of Np(V) and Cs(I). The experimental data are gathered in Figure 2. The extrapolation at zero ionic strength at 25°C gives the following results (95% confidence level):

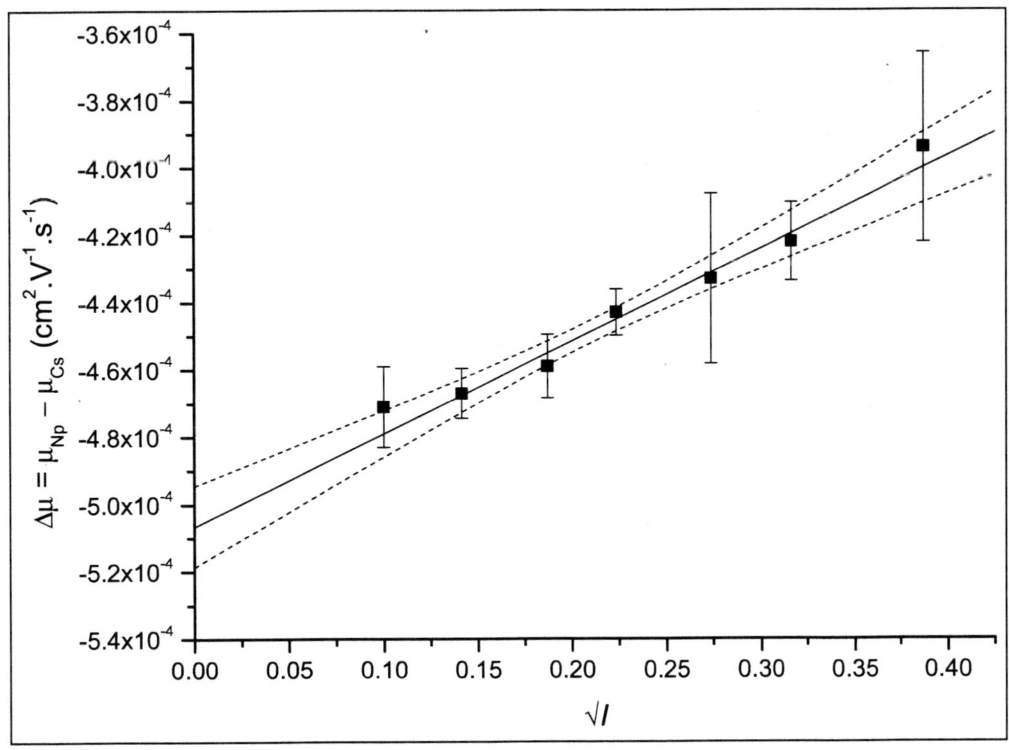

FIGURE 2. Variation of $\Delta\mu_{Np,\ Cs} = \mu_{Np} - \mu_{Cs}$ in function of $\sqrt{I}$, buffer creatinine pH = 5.00 by $HClO_4$, voltages = 5-30 kV, n = 9 determinations per point, intercept = −(5.07 ± 0.12) 10^{-4} $cm^2.V^{-1}s^{-1}$, slope = (2.75 ± 1.33) 10^{-4} $cm^2.V^{-1}s^{-1}.M^{-1/2}$, r = 0.979, 95% confidence level.

- Marker Cs^+: $\mu^0_{NpO_2^+} = (2.94 \pm 0.07)\times 10^{-4}\ cm^2.V^{-1}.s^{-1}$.

Thus, the equivalent ionic conductivity at 25°C is $\Lambda^0_{NpO_2^+}(\times 10^4) = 28.3 \pm 0.7\ m^2.S.mol^{-1}$. The absolute mobility determined herein for Np(V) is one brick that will act as a foundation in a large project related to actinides comportment in environmental samples. Indeed, it will serve to identify pentavalent actinides in a mixture but also to predict their behavior.

REFERENCES

1. Lu et al., Anal. Chem. 67, 2949 (1995).
2. Dudek, D. A., and Fedkiw, P. S., Anal. Chem. 71, 1469 (1999).

New Hermetic Sample Holder for X-Ray Diffraction on Radioactive Materials: Application to the Rietveld Analysis of Plutonium Compounds

Renaud Belin,[1] Philippe Valenza,[1] and Philippe Raison[2]

[1] *Commissariat à l'Energie Atomique*
CEA-Cadarache DEN/DEC/SPUA/LMPC
13108 Saint-Paul-lez-Durance, FRANCE

[2] *European Commission*
Joint Research Center - Institute for Energy
Postbus Nr 2
1755 ZG Petten (N.H.), THE NETHERLANDS

INTRODUCTION

When handling radioactive materials, two types of radiation-induced injury can occur: external irradiation or contamination. In the case of plutonium isotopes being mostly α-emitters, the irradiation is stopped by a simple sheet of paper, but contamination is of major concern. Accordingly, radioactive materials have to be kept in a confined environment. Based on nuclear safety authorities requirements and standard regulations in Europe, at least two containment barriers are necessary to ensure proper protection.

Rietveld analysis on radioactive polycrystalline samples is somewhat uncommon because it calls for the collection of a high-quality data complex, even using an instrument implemented in a glovebox. This latter solution, although handy, is costly and restricting because, most of the time, a glovebox has to be built and dedicated to the instrument. Moreover, all the equipment is contaminated, making servicing and maintenance difficult, if not impossible.

For room-temperature Rietveld analysis, nuclearization of the diffractometer is needless if a specifically designed hermetic sample holder can be used. In that case, the diffractometer is never to be contaminated, a worthwhile solution because maintenance is easier and because the equipment can still be used with nonradioactive samples. Of course, hermetic holders for X-ray diffraction are widely available.[1,2,3] However, to our knowledge, no holder designed for radioactive samples and fitting to Siemens D-5000 and Bruker D8 diffractometers exists.

RESULTS

The object of this presentation is to introduce a new hermetic sample holder designed for use with flat-geometry radioactive samples (Figure 1). This holder can also be used with any kind of air- and moisture-sensitive material.

In the case of radioactive materials, bulk or powder samples are embedded with epoxy resin using a specific device. The epoxy blend, considered as the first containment barrier, is then decontaminated and transferred to an uncontaminated glovebox or compartment. It is introduced into the holder, being the second barrier, and then transported to the diffractometer for measurement.

CP673, *Plutonium Futures — The Science,* edited by G. D. Jarvinen

FIGURE 1. Hermetic sample holder for radioactive samples fitting to Siemens/Bruker X-ray diffractometers.

Thanks to its mechanical conception, the sample holder allows a high accuracy vertical positioning of the sample (±0.005 mm). Thus, correction of the data is needless, lowering experimental uncertainties. As with the standard Bruker/Siemens sample stage, sample positioning of the holder is magnetic, and rotation of the sample is possible, a very useful feature to improve statistics. Besides, no specific tool is needed to lock and unlock the holder because a bayonet pin device is employed. As a result, handling and sample loading are much easier, a valuable benefit when working in a glovebox.

The sample holder is made of two parts. The upper part includes an X-ray window. It is made of Mylar but can be replaced by beryllium if needed. In that particular case and if the sample is an α emitter, a 0.1 mm Mylar or Kapton foil is inserted between both to avoid (α,n) reactions. Of course, the foil thickness is taken into account when dimensioning the holder in order to keep the sample surface precisely in the diffraction plane. The lower part receives the sample, a springbox allowing to keep it in contact with the window. The window tightness is achieved by an O-ring maintained by a screwed flange.

The design offers very accurate and reliable sample positioning on the instrument, allowing the acquisition of high-quality experimental data suitable for Rietveld analysis. So as to illustrate this feature, the Rietveld analysis of plutonium dioxide was carried out. PuO_2 crystallographic structure is well known[4,5] but our ambition at this stage was only to validate the whole process with a representative radioactive material.

The quality of a Rietveld refinement is also very reliant on the data collection strategy:[6] choice of the instrument, wavelength resolution, step width, and step intensity. The fundamental measured quantities are the intensities of the Bragg peaks. The precision of the intensity measurement can be improved by increasing the step time or decreasing the step interval, but this is useful only up to the point where counting variance becomes negligible in relation to other error sources.

Our measurements were performed with a high-resolution Siemens D5000 X-ray diffractometer using a curved quartz monochromator and copper radiation from a conventional tube source. The pattern was obtained by scanning with counting steps of: (a) 30 seconds, from 25° to 58° 2Θ; and (b) 60 seconds, from 58° to 145° 2Θ, at 0.02° step-intervals. The total scan took 86 hours.

The powder diffraction pattern obtained at room temperature was refined using the Rietveld program XND[7] (Figure 2).

The refinement of the experimental data gave very satisfactory results. The method was then broadened to new plutonium compounds such as the pyrochlore oxide $Pu_2Zr_2O_7$ described in another presentation.[8]

We are presently studying americium-based compounds. Handling such materials involves the use of very small amounts of powder resulting from the high radiation level of the isotope employed (^{241}Am). Typically, we are currently working on some ten milligram samples involving specific X-ray specimens' preparation.

This presentation will include a precise description of the sample preparation and of the sample holder. Also, the data collection and Rietveld analysis strategies will be described, and results will be discussed and compared to that of the literature.

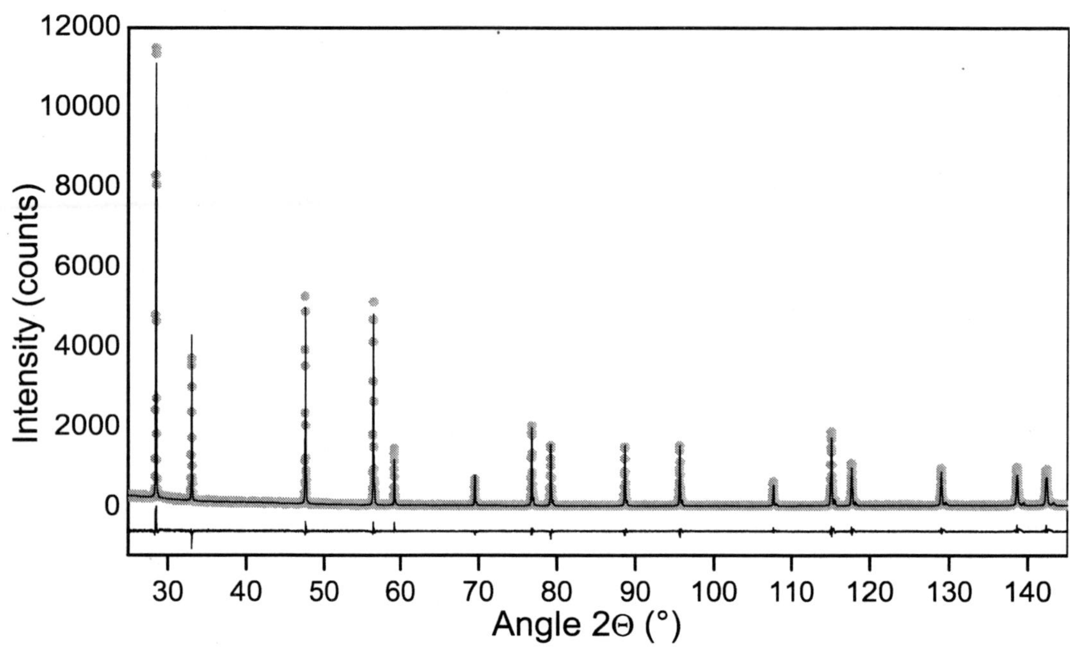

FIGURE 2. View of the best-fit profile of the Rietveld refinement of the PuO_2 phase.

ACKNOWLEDGMENTS

The authors are grateful to Muriel Rouault for her contribution in the X-ray specimens preparation and data collection.

REFERENCES

1. Rink, W. J., Mathias, H. G., and Schlenoff, J. B., J. Appl. Cryst. 27, 666–668 (1994).
2. Sing, M., Peirlberger, H., Boller, H., and Klepp, K., J. Appl. Cryst. 29, 607 (1996).
3. Buhrke, V. E., Jenkins, R., and Smith, D. K., "A Practical Guide for the Preparation of Specimens for X-Ray Fluorescence and X-Ray Diffraction Analysis," John Wiley & Sons, 1997.
4. Zachariasen, W. H., Phys. Rev. 73, 9, 1104–1105 (1948).
5. Zachariasen, W. H., Acta Cryst. 2, 388 (1949).
6. Hill, R. J., "The Reitveld Method," Chapter 5, edited by R. A. Young, International Union of Crystallography, Oxford Science Publications, 1993.
7. Bérar, J. F., and Garnier, P., 2nd APD Conf. NIST Special Publication, Vol. 846, Washington, DC: U.S. Government Printing Office, 1992, p. 212.
8. Belin, R. C., Raison, P. E., and Haire, R. G., "Structural Investigations by X-Ray Diffraction of Plutonium Zirconia-Based Materials Using the Rietveld Method" (this conference).

Exploration of Plasma Source Cavity Ring-Down Spectroscopy for Highly Sensitive Elemental and Isotope Measurements

Yixiang Duan,[1] Chuji Wang,[2] and Christopher B. Winstead[2]

1C-ACS, MS K484, Los Alamos National Laboratory, Los Alamos, NM 87545
[2]Diagnostic Instrumentation & Analysis Laboratory, Mississippi State University, Starkville, MS 39759

We are exploring sensitive techniques for elemental and isotope measurements using cavity ring-down spectroscopy (CRDS) combined with a compact plasma source as an atomic absorption cell. The research work marries the high sensitivity of CRDS with a low power plasma source to develop a new instrument that gives high sensitivity and has the capability of elemental and isotope measurements. CRDS, which uses a single laser pulse ringing down inside the cavity over a thousand times, is several orders of magnitude more sensitive than conventional absorption techniques. An additional benefit is gained from a compact microwave plasma source that possesses the advantages of low power and low plasma gas-flow rate, which benefit for atomic absorption measurement. A laboratory CRDS system consisting of a tunable dye laser is used in this work for testing and building a science base and to demonstrate the feasibility of the new instrument. A laboratory designed and built sampling system for solution sample introduction is used for the new instrument testing. The ring-down signals are monitored using a photomultiplier tube (PMT) and recorded using a digital oscilloscope interfaced to a computer. Several elements were chosen for the system optimization and characterization. Baseline noise of the plasma source has been thoroughly studied in this work. A favorable detection limit is obtained with such a device.

CP673, *Plutonium Futures — The Science,* edited by G. D. Jarvinen

High-Precision Assay of Uranium and Plutonium by Glovebox-Enclosed Automated Titration Systems

N. S. Howard, T. J. Piper,* M. A. Thomas, and N. Wainwright

Atomic Weapons Establishment, Aldermaston, Berkshire RG7 4PR, United Kingdom

INTRODUCTION

There is a requirement for the accurate and precise determination of uranium and plutonium to be carried out on a wide range of metal, alloy, oxide and other actinide containing material. Specifications can demand that the total random error, i.e., precision, associated with the measurement technique must be less than $0.05^{w}/_{0}$ relative. Also, the systematic error, i.e., bias or inaccuracy, of the technique must be insignificant when compared with the random error. There is an additional requirement for the capability to process relatively large numbers of samples that necessitate the use of automated techniques.

To meet these requirements, the Actinides Analysis section at AWE Aldermaston has successfully installed proprietary autotitration systems into glovebox environments. Titration methods have been developed and validated for use on these instruments, which allow the high-precision assay of U and Pu materials.

PLUTONIUM ASSAY

For Pu assays, a Metrohm 730 autotitrator was installed into a glovebox, incorporating a titration station and 16-position multisample carousel (Figure 1). Reagent addition occurs from outside the box from model 685 dosimats and a 736 Titrino, which controls the dosimats and ancillary equipment. A PC running Metrohm's Windows-based TiNet software controls the system. Reagent lines feed into the glovebox through breakthroughs in the service panel. The lines are fitted with solenoid-controlled check valves which, together with anti-diffusion tips on the titration head, prevent egress of contaminated solution from the glovebox.

The titration method employed is a modified Corpel method.[1] In this method, Pu samples are dissolved in hydrochloric acid and Ti(III) chloride is use to reduce all plutonium to Pu(III). Excess Ti(III) is oxidized with nitric acid in the presence of sulphuric acid. Sulphamic acid is then added to remove interfering nitrite ions. Pu(III) is then titrated against Ce(IV) until a potentiometric end point has been slightly exceeded. The excess Ce(IV) is then back-titrated against Fe(II).

A typical run for a batch of samples takes 20 minutes per sample to complete. Duplicate analysis of the laboratory's plutonium control material typically gives results with a standard error of 0.03%.

URANIUM ASSAY

Uranium assays are performed using a Radiometer VIT 90 autotitrator system installed in a glovebox and also a Metrohm Titrino system installed in a fume cupboard (Figure 2). The Radiometer system consists of a model SAC80 10-position sample changer and associated electrode head inside the glovebox. Reagents are delivered by Radiometer ABU93 triburettes situated outside the glovebox. Reagent delivery lines are fed through the glovebox service panel and are protected by solenoid-controlled check valves and antidiffusion tips. The Metrohm system

CP673, *Plutonium Futures — The Science,* edited by G. D. Jarvinen
2003 American Institute of Physics 0-7354-0140-3

comprises a model 730 16 position sample changer and associated electrode head inside the fumecupboard. Reagent delivery is through a Titrino 736 and five model 685 dosimats with antidiffusion tips on delivery lines.

The titration used on both titrators is a Modified Davies and Gray method.[2] A solution of uranium in nitric or nitric/hydrofluoric acid is fumed with sulphuric acid and the resultant U(VI) reduced with Fe(II) in concentrated phosphoric acid. The excess Fe(II) is oxidized to Fe(III) with nitric acid catalyzed by Mo(VI), and sulphamic acid is used to remove nitrite ion and oxides of nitrogen. The U(IV) solution is diluted, resulting in oxidation to U(VI) by Fe(III). The resultant Fe(II) is then titrated with Cr(VI) catalyzed by V(IV).

A typical run for a batch of eight samples takes 20 minutes per sample to complete. Duplicate analysis of the laboratory uranium control material typically gives results with a standard error of 0.05%.

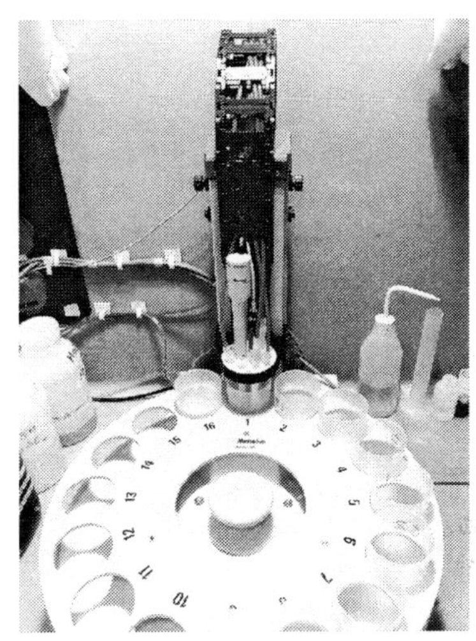

FIGURE 1. Glovebox system.

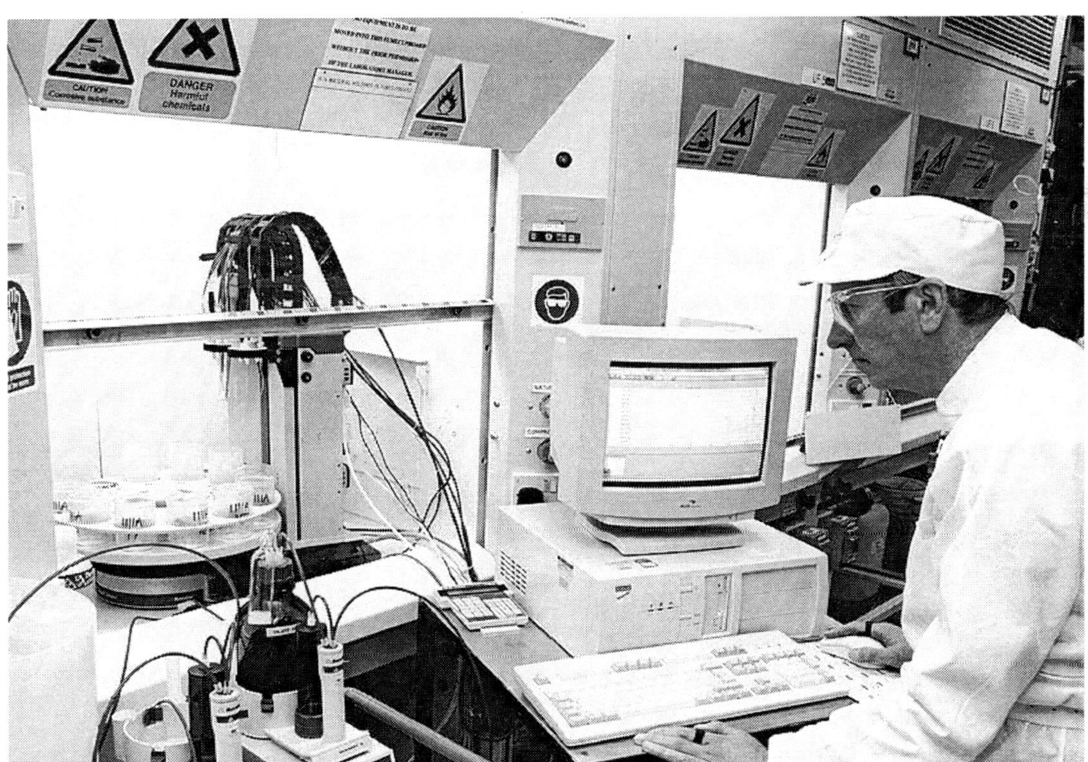

FIGURE 2. Fume-cupboard system.

REFERENCES

1. Wainwright, N., "Glovebox Installation and Evaluation of an Automated Titration System for the High Precision Assay of Plutonium," AWE DTech Report 113/99 (October 1999).
2. Howard, N., "Commissioning of a Metrohm Titrino Based Autotitrator for the High Precision Assay of Uranium," AWE DRAS Report 176/99 (October 2000).

A Technique for Determination of the Age of Weapons-Grade Plutonium

R. P. Keegan and R. J. Gehrke

Mail Stop 2114, Idaho National Engineering and Environmental Laboratory, P.O. Box 1625, Idaho Falls, Idaho 83415-2114, United States of America.

INTRODUCTION

This paper describes a noninvasive technique that has been developed at the INEEL to determine the time since last purification or "age" of weapons-grade plutonium. For this application, the initial isotopic composition by weight has been assumed to be weapons-grade plutonium[1] with about 0.65% ^{241}Pu, 93% ^{239}Pu, and the remainder ^{240}Pu. Details of the mathematical derivations and analysis process will be described.

The method uses the Bateman equations and the ^{241}Pu decay chain, namely $^{241}\text{Pu} \rightarrow {}^{237}\text{U}$ and $^{241}\text{Am} \rightarrow {}^{237}\text{Np} \rightarrow {}^{233}\text{Pa}$. Age can be determined by assuming that the ^{237}Np and ^{233}Pa progeny are in secular equilibrium and that the plutonium is of sufficient age (>one year) so that the ^{241}Pu decay through the ^{237}U branch can be neglected.[2] Initial amounts of all other progeny (i.e., ^{241}Am and ^{237}Np) are allowed to vary. The authors' method is more sophisticated than previous studies[3] because it includes the effects of ^{241}Pu that initiate the decay chain, and fewer assumptions are made about the initial amounts of the progeny. The algorithm uses γ-peaks for ^{241}Pu, ^{241}Am, and ^{233}Pa to yield the age of the sample.

It is expected that the present method could be applied not only to weapons-grade plutonium, but also to any medium that contains plutonium, provided that sufficient statistics for the peaks of the ^{241}Pu and progeny of interest exist in the spectrum.

RESULTS/TESTS

Applying the Bateman equations to the first four members of the ^{241}Pu decay chain, and assuming that the contribution of the ^{237}U branch is negligible, and that secular equilibrium exists between the ^{233}Pa and ^{237}Np, results in the following formula for the initial amount of ^{237}Np at time t denoted by N_3^0:

$$
\begin{aligned}
N_3^0 &= \frac{\lambda_1 \lambda_2 N_1}{(\lambda_2 - \lambda_1)(\lambda_3 - \lambda_1)} e^{\lambda_1 t} \\
&+ \left[\frac{\lambda_2 N_2}{(\lambda_3 - \lambda_2)} - \frac{\lambda_1 \lambda_2 N_1}{(\lambda_2 - \lambda_1)(\lambda_3 - \lambda_2)} \right] e^{\lambda_2 t} \\
&+ \left[\frac{\lambda_1 \lambda_2 N_1}{(\lambda_2 - \lambda_1)(\lambda_3 - \lambda_2)} - \frac{\lambda_2 N_2}{(\lambda_3 - \lambda_2)} - \frac{\lambda_1 \lambda_2 N_1}{(\lambda_2 - \lambda_1)(\lambda_3 - \lambda_1)} + \frac{\lambda_4 N_4}{\lambda_3} \right] e^{\lambda_3 t}.
\end{aligned} \tag{1}
$$

The subscripts 1 through 4 represent ^{241}Pu, ^{241}Am, ^{237}Np, and ^{233}Pa in that order. The symbols λ and N represent the associated decay constant, and number of nuclei respectively.[4]

CP673, *Plutonium Futures — The Science,* edited by G. D. Jarvinen

The method involves measuring the number of nuclei of ^{241}Pu, ^{241}Am, and ^{233}Pa (N_1, N_2, and N_4) in the sample, substituting the values into the above equation and then letting the right-hand side be equal to a function $f(t)$.

The function $f(t)$ is then graphed versus time as shown in Figure 1. The curve has a minimum at the time when the sample was last purified. At the minimum of the function $f(t)$ becomes equal to N_3^0, and the corresponding value of t is the age of the sample. This is only true at the minimum of the function. This function can be differentiated with respect to time and then graphed versus time to locate the minimum more accurately. If ^{237}Np is not initially present in the sample, the curve will touch the time axis at one point, otherwise it will have a systematic offset. Experimentally, it is possible for the curve to dip below the time axis as a result of experimental uncertainties.

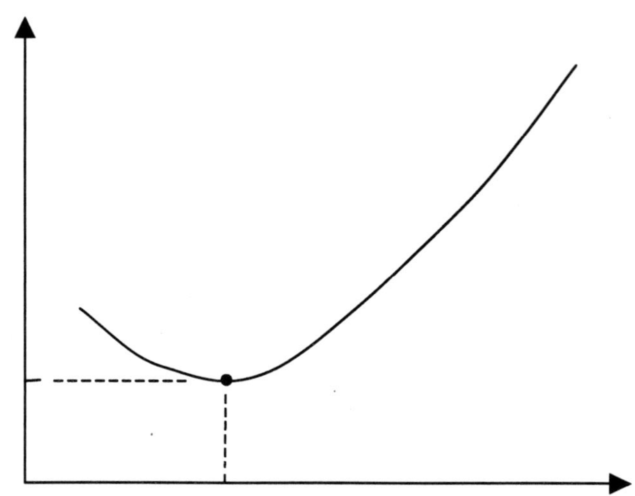

FIGURE 1. A graph of $f(t)$ versus t where $f(t)$ is equal to the right hand side of Eq. (1). The age of the source is given by the minimum of the function where the magnitude of $f(t)$ becomes equal to N_3^0.

Two tests of the algorithm were carried out by first using an almost pure ^{241}Pu source and second a weapons-grade plutonium source. The first test involved a 97% pure ^{241}Pu unshielded source, separated on March 7, 1983, namely 19.1 years before the count. The peaks chosen for this study were those located at 148.5 keV for ^{241}Pu, 312.2 keV for ^{233}Pa, and 322.5 keV for ^{241}Am. The source was counted for 10 days using a well-calibrated HPGe gamma spectrometer to obtain good statistics (about 2500 counts) in the ^{241}Am and ^{233}Pa peaks. The method gives the time since last purification to be 18.97 ± 1.00 years, which places the time of last purification in January 1983 in good agreement with the last purification date of March 1983.

The second test involved using a disk source consisting of weapons-grade plutonium. It was about 25 mm in diameter and 3 mm thick with the plutonium encased between plates of 0.3-mm-thick copper soldered together, and tinned with solder on the outside. It is anticipated that the method will work even though the efficiency curve used was determined for a point source. The analysis procedure was the same as for the first test, except this time an attenuation correction was applied to the spectral data. The measured data are fed into the algorithm, which indicates that the age of the source was 41.1 ± 2.4 years ago, placing the time of purification during July 1961. The source assay date was January 1, 1960, 42.8 years ago, and it is believed that it was last purified during the last quarter of 1959.

DISCUSSION

The results are consistent with our expectations, which implies that our assumptions were correct. Theory and experiment indicate that a point source efficiency calibration could be used for other geometries. The technique would find use where plutonium fissile material is found and its history is unknown, and therefore it may have application in the identification of plutonium material of uncertain or unknown origin. The technique could also be used to date a plutonium mass or possibly residue in a storage container when a complete manifest is not available. It might also have environmental application after further development, where such material might be recovered from a waste burial pit and needs to be characterized. Knowledge of the sample age would enable the calculation of the abundance of each progeny in the sample. This information could be used to verify the accuracy of any records that do exist, and provide historical information if it does not. A more detailed description of this method has been submitted for publication earlier this year.[5]

REFERENCES

1. Reilly, D., Enselin, N., Smith, H., and Kreiner S., "Passive Nondestructive Assay of Nuclear Materials," United States Nuclear Regulatory Commission, NUREG/CR-5550, (1991), p. 264.
2. West, D., and Sherwood, A.C., "Gamma-Rays from an Unirradiated Plutonium-Uranium Oxide Fuel Pin and a Method of Measuring the Age of Plutonium Since Chemical Processing," Technical Report AERE-R 10020, Nuclear Physics Division, A.E.R.E., Harwell (1981).
3. Gehrke, R. J., East, L. V., and Harker Y.D., "Information in Spectra from Sources Containing "Aged" ^{241}Am as from TRU Waste," Waste Management 20, 555–559 (2000).
4. Kaplan, I., 1963. Nuclear Physics, Addison-Wesley Publishing Company, Second Edition (1963), pp. 239–247.
5. Keegan, R. P., and Gehrke, R. J, "A Method to Determine the Time Since Last Purification of Weapons Grade Plutonium," Applied Radiation and Isotopes, submitted for publication (2003).

Rapid Determination of ^{237}Np and Pu Isotopes in Environmental Samples by ICP-MS

C. S. Kim, C. K. Kim, B. W. Rho, and K. J. Lee*

Korea Institute of Nuclear Safety (KINS),
P.O. Box 114, Yusong, Daejeon 305-338, Korea, k254kcs@kins.re.kr
** Department of Nuclear & Quantum Engineering, Korea Advanced Institute of Science and Technology, Korea*

INTRODUCTION

Recently ICP-MS instrumentation is actively applied in the determination of long-lived actinides at sub fg ml^{-1} levels[1] by the advance of a sensitivity and sample introduction system. Contrary to Pu, the application of ICP-MS in the analysis of ^{237}Np is very rare owing to the limitation of tracers and low concentrations in environmental samples. Np and Pu usually were analyzed by independent chemical separation. Accordingly, a long analysis time is needed to get information on the concentration of Np and Pu and the atomic ratio between these actinides in environmental samples. In order to shorten the analysis time and automate the chemical separation, diverse on-line sequential injection (SI) techniques have been recently developed for the analysis of actinides.[2] In our previous studies,[3] an on-line SI system associated with ICP-MS was successfully applied for the determination of Pu isotopes and its atomic ratio in environmental samples. Meanwhile, the solid-phase extraction chromatography resin has been successfully used in the purification of actinide elements. Especially, TEVA-Spec shows a high affinity to the tetravalent Np and Pu in the strong nitric acid, but it is very low in weak hydrochloric acid.[4] Accordingly, common chemical separation schemes and tracers could be used in the analysis of Pu and Np. The aim of the present study is to develop a rapid analytical method for simultaneous determination of Np and Pu using ICP-MS.

RESULTS AND DISCUSSION

Elution of Np and Pu on TEVA-Spec

A typical elution profile obtained from a 0.17-ml TEVA column, depicted in Figure 1, shows that elution patterns of Np and Pu in elution coincide with 0.5 M HCl. Although the elution behaviors of Np and Pu are slightly different, the final chemical recoveries for Np and Pu were very similar, about 98% and 99% in 1 ml of 0.5 M HCl, respectively.

Effect of Ascorbic Acid in the Oxidation of Np and Pu

In this study, ascorbic acid was used as a redox agent to adjust the oxidation state of Np and Pu among various agents. To find the optimum amount of ascorbic acid, the chemical recovery of Np and Pu and the recovery ratio between both actinides were examined and compared with the amount of ascorbic acid. As shown in Figure 2, the chemical recovery of Np and Pu was reached about 90%; the recovery ratio increased from 19 mg to 225 mg.

CP673, *Plutonium Futures — The Science,* edited by G. D. Jarvinen

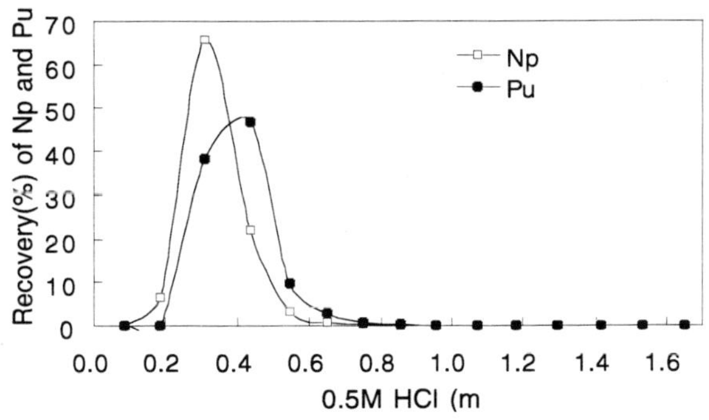

FIGURE 1. Elution profile for Np and Pu in TEVA-Spec resin.

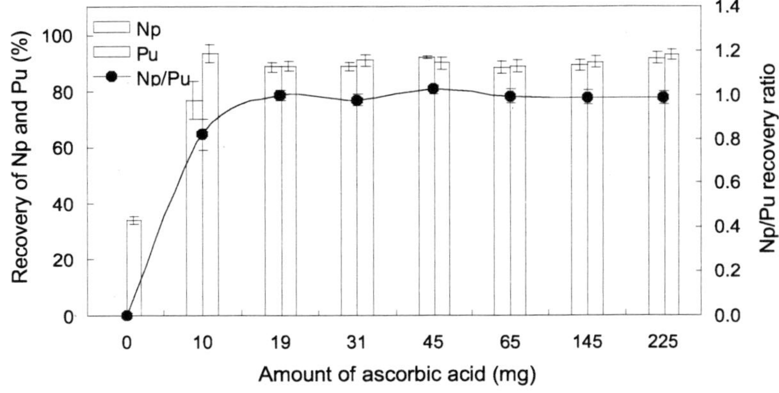

FIGURE 2. Effect of the ascorbic acid on the chemical recovery of Np and Pu in on-line SI.

On-Line SI Separation of Np and Pu

The whole purification process of Np and Pu in the SI system was designed with a total of 9 steps from the rinsing step to the collection of the final elution in the microautosampler (Figure 3). A rinsing solution of 5 M HNO_3, 1 M HNO_3 and 9 M HCl was used for U, Th, and bulk matrix elements. The operation of a two-way valve system was programmed in an independent control system to match each operation of PrepLab. The entire on-line SI procedure was automatically processed by the ICP-MS operation program. In the final step, step 9, the purified Np and Pu on TEVA-Spec were transferred to a 2-ml conical vial in the microautosampler and then was injected into the plasma through MCN-6000. It took approximately 24 minutes to treat one sample with the SI system.

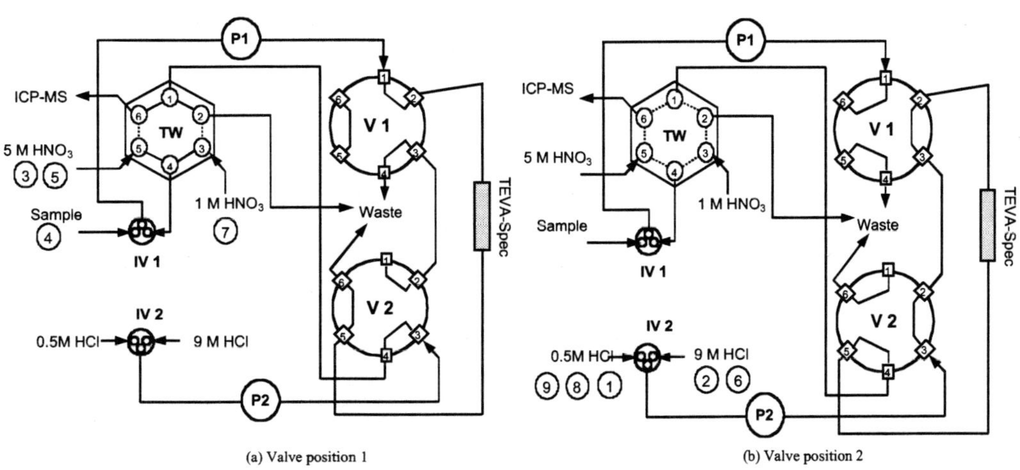

FIGURE 3. Schematic diagram of the on-line SI system in a simultaneous analysis of Np and Pu.

Validity Test and Application to the Environmental Samples

To verify the accuracy and reproducibility of the proposed analytical method, IAEA standard reference material, IAEA-135, was analyzed. As shown in Table 1, the total100 Pu activity fell within the certified interval suggested by IAEA[5] and the $^{240}Pu/^{239}Pu$ atom ratio agreed well with the reported value (0.21 ± 0.006) by MEL laboratory.[6] The result of ^{237}Np was very similar with the reported value (0.85 ± 5.3 %) by Chen et al.[7] The average chemical recovery was above 90% as well as the concentrations of U that ranged from 0.9 to 1.9 ppt, which is a very low negligible level in the interference effect resulting from U. Table 2 shows the results for the Np and Pu concentration and the atomic ratio in soil collected from Jeju island. Both actinides showed similar behavior along depths resulting from the peak concentration of around 2 to 4 cm found below the surface. Meanwhile, $^{240}Pu/^{239}Pu$ atom ratios were uniformly distributed along depths with an average value of 0.17, which was similar to global fallout at level (0.18), but the distribution pattern of the $^{237}Np/^{239}Pu$ atom ratio is different from the former with a large variation with depth.

TABLE 1. Analytical Results of ^{237}Np and Total Pu and Pu Atom Ratio in IAEA-135.

No.	^{238}U Conc. ($pg\ ml^{-1}$)	Chemical Recovery (%)	$^{240}Pu/^{239}Pu$ Atom Ratio [a]	Total Pu ($Bq\ kg^{-1}$)[b]	^{237}Np ($Bq\ kg^{-1}$)[c]
1	1.0	99.7	0.207	211	0.84
2	0.9	89.5	0.202	217	0.86
3	1.6	93.4	0.205	210	0.81
4	1.9	77.1	0.201	215	0.83
5	0.7	81.2	0.215	238	0.83
6	1.4	79.3	0.217	237	0.79
Mean	1.3	86.7	0.212 ± 0.005	206 ± 8.96	0.81 ± 0.045

[a] The reported value by S. H. Lee et al. is 0.207 ± 0.006.
[b] The 95 % confidence interval by IAEA is 205 - 225.8 Bq kg^{-1}.
[c] The reported value by Chen et al. is 0.85 ± 5.3 %.

TABLE 2. Analytical Results of ^{237}Np and Pu Isotopes and Atomic Ratios Between Pu Isotopes and Np in Soil.

Depth (cm)	Chemical Recovery (%)	^{239}Pu ($Bq\ kg^{-1}$)	^{240}Pu ($Bq\ kg^{-1}$)	$^{240}Pu/^{239}Pu$ Atom Ratio	^{237}Np ($mBq\ kg^{-1}$)[b]	$^{237}Np/^{239}Pu$ Atom Ratio
0–2	97	2.05 ± 0.07	1.23 ± 0.01	0.17 ± 0.00	5.01 ± 0.22	0.21 ± 0.01
2–4	78	4.16 ± 0.14	2.45 ± 0.04	0.16 ± 0.00	17.93 ± 0.23	0.37 ± 0.01
4–6	87	3.02 ± 0.29	1.87 ± 0.01	0.17 ± 0.01	16.84 ± 0.95	0.48 ± 0.04
6–8	93	0.72 ± 0.03	0.44 ± 0.00	0.17 ± 0.00	2.59 ± 0.15	0.31 ± 0.02
8–10	95	0.23 ± 0.01	0.15 ± 0.00	0.17 ± 0.00	1.98 ± 0.01	0.73 ± 0.01
10–12	90	0.10 ± 0.01	0.05 ± 0.00	0.16 ± 0.01	0.85 ± 0.01	0.77 ± 0.04
12–14	99	0.04 ± 0.00	0.03 ± 0.00	0.18 ± 0.01	0.40 ± 0.01	0.81 ± 0.03
14–16	87	0.09 ± 0.00	0.06 ± 0.00	0.19 ± 0.01	0.41 ± 0.01	0.37 ± 0.01
16–18	99	0.11 ± 0.00	0.07 ± 0.00	0.18 ± 0.00	0.28 ± 0.01	0.23 ± 0.01
18–20	85	0.09 ± 0.01	0.06 ± 0.00	0.18 ± 0.02	0.22 ± 0.01	0.22 ± 0.01

REFERENCES

1. Becker, J. S., Soman, R. S., Sutton, K. L., Caruso, J. A., and Dietze, H. J., J. Anal. At. Spectrom. 14, 933 (1999).
2. Grate, J. W., Egorov, O. B., and Fiskum, S. K., Analyst 124, 1143 (1999).
3. Kim, C. S., Kim, C. K., Lee, J. I., and Lee, K. J., J. Anal. At. Spectrom. 15, 247 (2000).
4. Bunzl, K., and Kracke, W., J. Radioanal. Nucl. Chem. Letters 186, 401 (1994).
5. IAEA, "Report on the Intercomparison Run IAEA-135," IAEA/AL/063, IAEA, Vienna, 1993.
6. Lee, S. H., Gastaud, J., La Rosa, J. J., Liong, L., Kwong, W., Povince, P. P., Wyse, E., Fifield, L. K., Hausladen, P. A., Di Tada, L. M., and Santos, G. M., J. Radioanal. Nucl. Chem. 248, 757 (2001).
7. Chen, Q., Dahlgaard, H., Nielsen, S. P., and Aarkrog, A., J. Radioanal. Nucl. Chem. 253, 451 (2002).

Preparation of Traceable Working Reference Material Standards for the National TRU Waste Performance Demonstration Program

S. L. Mecklenburg,[†] D. L. Thronas,[†] A. S. Wong,[†] R. S. Marshall,[†] and G. K. Becker[‡]

[†]*Los Alamos National Laboratory, C-AAC, MS G740, Los Alamos, NM 87545*
[‡]*Custom Surrogate Matrices, Inc., Idaho Falls, ID 83405*

Traceable nondestructive assay (NDA) standards containing a variety of radionuclides (including uranium, americium, and plutonium oxides) were fabricated and certified for use in the U.S. Department of Energy's National TRU Waste Program (NTWP). The NTWP requires traceable nuclear material standards of the Working Reference Material (WRM) class for use in the independent assessment of performance of NDA systems that assay nuclear material contained in DOE-generated waste before transport and final disposition at the Waste Isolation Pilot Plant (WIPP) in Carlsbad, New Mexico.[1] The performance assessment, qualification, and approval process for waste assay systems is accomplished in part through successful participation in the Non-Destructive Assay (NDA) Performance Demonstration Program (PDP)[2,3] and is required for DOE and EPA regulatory compliance, as well as repository permitting and performance evaluation. An overview of the PDP program highlighting the role of the certified WRMs fabricated at LANL is presented, as well as a summary of the WRM fabrication process and an overview of the inventory of 187 WRMs fabricated and deployed to DOE NDA measurement facilities to date.

Along with traceable nuclear material standards, the other component of the PDP is the PDP matrix drum set—surrogate waste drums that have been manufactured to represent the predominant waste forms of the DOE waste inventory. The PDP drum set is comprised of empty, combustible, metal, glass, and sludge matrix drums. PDP Test Cycles are conducted in which selected PDP standards are inserted into PDP matrix drums in a specified manner to produce a PDP test sample. The test sample is presented to NDA analysts for measurement in a blind test, and results are reported for scoring. Satisfactory performance according to established criteria contributes to the qualification of the NDA measurement system, permitting it to be used to quantify waste bound for WIPP. The NDA PDP provides independent objective method performance data for the waste NDA techniques employed across the DOE complex.

The process for fabrication of the required nuclear material standards incorporates extensive quality-assurance and control measures and detailed documentation. Criteria for WRM content, composition, and allowable uncertainty are specified in a Statement of Work. WRMs in each production phase span a range of Curie contents specified in program data quality objectives.[2,3] Seven PDP standard fabrication campaigns were conducted over a nine-year period, producing seven phases of certified, traceable nuclear material standards, each with distinct properties (see Table 1). The majority of the standards fabricated contained plutonium or uranium in the form of powdered oxides (PuO_2 or U_3O_8), where the oxide was rigorously characterized by analytical chemistry before being homogeneously blended with diatomaceous earth (an inert matrix material). The blend was then packed into a stainless-steel cylinder. For the Phase II.B standards, macroscopic plutonium oxide granules were embedded in graphite felt disks, and the disks were stacked in layers in the stainless-steel cylinder. All cylinder end caps were welded in place using a glovebox-mounted welding stand and TIG welder, after which the cylinders were decontaminated, helium leak-checked, and welded into a second, outer stainless-steel cylinder using a second welding stand outside the glovebox to ensure that the outer cylinder remained cold and uncontaminated. After final dimensional and radiological inspection, the certified gram loading and material distribution in the completed WRMs was verified by gamma-ray spectroscopy. Holdup measurements on emptied blend bottles and associated filling materials were carried out in order to make the holdup correction necessary to accurately determine the net

CP673, *Plutonium Futures — The Science,* edited by G. D. Jarvinen

weight of nuclear material contained in each WRM. When fabrication, final inspections and verifications were complete, content uncertainties were propagated in detail, *Certificates of Content and Traceability* were prepared for each WRM, and the PDP WRM sets were distributed to DOE NDA characterization facilities according to their respective needs for assay of various waste forms.

TABLE 1. Summary of NDA PDP Production Phases.

Phase	Content	Total Number of WRMs	Number of Sets	Number per Set	Approximate Mass Range
I, IIA	WG Plutonium (^{239}Pu)*	104	8	13	0.02–75 g
IIB	WG Plutonium, Increased Particle Size (^{239}Pu)	32	8	4	0.3–30 g
IIC	Highly Enriched Uranium (^{235}U)	20	4	5	1–75 g
IID	Depleted Uranium (^{238}U)	6	1	6	0.5–1 kg
IIIA	Increased Am/Pu Ratio ($^{241}Am/^{239}Pu$)**	15	5	3	5–250 mg Am, 0.10–10 g Pu
IIIB	Heat Source Plutonium (^{238}Pu)	10	2	5	0.5–350 mg

* WG = Weapons Grade

** Zirconium encapsulation used (all other phases used stainless steel)

Stainless-steel cylinders were used for double encapsulation of the majority of the standards, although one set of standards was encapsulated in zirconium cylinders to take advantage of that material's low neutron attenuation characteristics. Both stainless-steel and zirconium containers were certified for containment of Radioactive Sealed Sources according to the ANSI Standard N43.7-1997.

Required traceability to national standards of critical WRM properties such as nuclear material mass, assay, isotopic composition, and activity was established during analytical characterization of the nuclear material feedstock used in each set of standards. The analytical procedures used to determine the critical quantities were each calibrated with primary Standard Reference Materials (SRMs) or Certified Reference Materials (CRMs). In addition, weighing of feedstock/blend components was performed using balances certified by the LANL metrology group; balance performance was verified before use with certified check weights.

The critical quantities of mass, assay, isotopic composition, and activity must also be bound by uncertainty limits. Careful consideration was given to random and systematic factors contributing to uncertainty. Standard methods for propagation of uncertainty were employed to arrive at final bounds for all quantities given on the WRM Certificates. All processes were designed to meet fabrication criteria, which specified that uncertainty in content and total transuranic alpha activity be $<0.5\%$.

The NDA PDP is essential to the flow of properly characterized and documented DOE-generated waste to WIPP for final disposition. The program has contributed to the overall evaluation of TRU waste NDA system capability and performance for the assay of DOE wastes. Finally, the NDA PDP provides valuable technical feedback to NDA facilities on system capabilities and applications, allowing for the continual refinement of waste NDA methods and instrumentation employed across the DOE complex.

REFERENCES

1. "Waste Acceptance Criteria for the Waste Isolation Pilot Plant," DOE/WIPP-069, Revision 7, Carlsbad, New Mexico, Waste Isolation Pilot Plant, U.S. Department of Energy (1999), with updates.
2. "Performance Demonstration Program Plan for Nondestructive Assay of Drummed Wastes for the TRU Waste Characterization Program," DOE/CBFO-01-1005 Rev. 0.1.
3. "Performance Demonstration Plan Program Management Plan," DOE/CBFO-01-3107 Rev.1, http://www.wipp.carlsbad.nm.us/library/pdp/DOE-CBFO-01-3107Rev1.pdf.

Furthering the Science—ASTM International

Donivan R. Porterfield

Actinide Analytical Chemistry Group
Los Alamos National Laboratory

The American Society for Testing and Materials (ASTM) International is a voluntary consensus standards organization that develops standard methods, practices, and guides in a wide array of areas. Most of the standard development activities occur within a committee structure, each having a defined jurisdiction. Two ASTM International committees develop standards with application to Plutonium materials. These two committees are the Nuclear Fuel Cycle committee and Nuclear Technology and Applications committee.

Each ASTM International committee has a number of subcommittees, which address specific needs within the overall jurisdiction of the committee. The Nuclear Fuel Cycle committee has the following standard developing subcommittees:

C26.01	Editorial and Terminology
C26.02	Fuel and Fertile Material Specifications
C26.03	Neutron Absorber Materials Specifications
C26.05	Methods of Test
C26.07	Waste Materials
C26.08	Quality Assurance, Statistical Applications, and Reference Materials
C26.09	Nuclear Processing
C26.10	Nondestructive Assay
C26.12	Safeguard Applications
C26.13	Repository Waste
C26.14	Remote Systems

The Nuclear Technology and Applications committee has the following standard developing subcommittees:

E10.01	Dosimetry for Radiation Processing
E10.02	Behavior and Use of Nuclear Structural Materials
E10.03	Radiological Protection for Decontamination and Decommissioning of Nuclear Facilities and Components
E10.04	Radiation Protection Methodology
E10.05	Nuclear Radiation Metrology
E10.06	Food Irradiation Processing and Packaging
E10.07	Radiation Dosimetry for Radiation Effects on Materials and Devices
E10.08	Procedures for Neutron Radiation Damage Simulation

An important component of ASTM International and any other consensus standards organization is the process by which each member's technical input is received and used in the development of any new standard or revision to an existing standard. For any ASTM International standard, there are typically three levels of balloting that must be accomplished before a standard is made available to the public. These three levels of balloting correspond to the subcommittee, the committee, and society (the whole of the ASTM International membership). At each level, the desire for consensus must be addressed before a standard advances to the next level. A member at each of these levels may cast a negative vote on a technical aspect of the draft standard. The other members at that level must then give due consideration to the reasoning for the negative and the suggested changes to address it before the draft

CP673, *Plutonium Futures — The Science,* edited by G. D. Jarvinen

standard can proceed to the next level of balloting. Through this process, the final content of any ASTM International standard comes to represent the vast technical expertise of all the members of the subcommittee and committee that developed the standard.

Also of importance to ASTM International is that all standard methods, practices, and guides be current in order to insure that they serve their intended purpose. In order to accomplish this, ASTM requires that all current standards be reviewed by the committee in which they originated at a frequency of every five years. Any needed revisions or the simple reapproval of the current content of each standard is also balloted at the same three levels mentioned above.

The value of standards developed by voluntary consensus standard organizations such as ASTM International is recognized by the U.S. Federal government through the 1996 National Technology Transfer and Advancement Act (P.L. 104-113). This act requires Federal agencies to achieve greater reliance on voluntary standards and lessened dependence on in-house standards. The National Institute of Standards and Technology (NIST) is the Federal agency directed to coordinate and monitor the implementation of this requirement by all Federal agencies through the Interagency Committee on Standards Policy. On an annual basis, NIST issues the Annual Report on Federal Agency Use of Voluntary Consensus Standards as an assessment of agency adherence to this requirement.

Although both the ASTM International Nuclear Fuel Cycle and Nuclear Technology and Applications committees have significant participation by non-U.S. citizens, the developed standards are sometimes mistakenly considered only useful to U.S. organizations. ASTM International is addressing this issue by developing a working relationship with the International Standards Organization (ISO). Through this working relationship, 25 ASTM International standards developed by the Nuclear Technology and Applications committee have gone on to be accepted and distributed as ISO standards. ASTM International also provides the official U.S. representation to the ISO Nuclear Energy Technical Committee (TC 85). Thus, the expertise of the members of the Nuclear Fuel Cycle and Nuclear Technology and Applications committees of ASTM International is directly shaping the content of many other ISO standards.

ASTM International (formerly the American Society for Testing and Materials)
100 Barr Harbor Drive, PO Box C700, West Conshohocken, PA 19428-2959
http://www.astm.org/

Plutonium Process Monitoring (PPM) System for Plutonium-238 in Aqueous Scrap Processing

D. L. Thronas, A. S. Wong, C. E. VanPelt, and J. S. Gower

Nuclear Materials Technology Division
Los Alamos National Laboratory
Los Alamos, NM 87545

INTRODUCTION

The ^{238}Pu Science and Engineering Group at Los Alamos National Laboratory have implemented an on-line plutonium process monitoring system (PPM) for ^{238}Pu Aqueous Scrap Processing. This paper describes the PPM system integrated into an anion-exchange process and its effectiveness in determining the separation of actinide impurities as well as other notable impurities in plutonium feed-stock materials.

AQUEOUS SCRAP PROCESS

The Aqueous Scrap Processing of ^{238}Pu allows for the separation of americium, uranium, and trace impurities from plutonium. One effective method for purification is through nitrate anion exchange resin. The plutonium-bearing materials (impure oxides, ashes, residue, and various miscellaneous compounds) are dissolved in nitric acid. The next step is to run the solution, containing the dissolved plutonium, through an ion-exchange column using Reillex-HPQ resin to separate the americium, uranium, and trace impurities from plutonium. The purified plutonium is then removed from the resin, recovered using an oxalate precipitation where the Pu(III) is collected, and calcined into the oxide.

PLUTONIUM PROCESS MONITORING (PPM) SYSTEM

The PPM system uses a high-purity germanium detector and digital electronics to collect data. The digital data and advanced software provide the capability to obtain real-time information and anion-exchange elution profiles within a few seconds. This setup also provides monitoring from a remote location through the local-area network. Below are the descriptions of the components of the PPM system.

Hardware

The high-purity germanium detector is placed in the geometric center of a 1-in. stainless-steel pipe tube that carries the solution from the ion exchange resin to an effluent storage tank. The detector provides high sensitivity and optimum resolution in the 30 to 250-keV region of the gamma-ray spectrum. The detector is shielded with 1-in. lead and 1-in. borated polyethylene, and inside the shielding is graded with copper plate.[1] An electromechanical cooling unit is used to cool the germanium detector. The advantages of an electromechanical cooling unit are a diminished amount of time required for maintenance and a compact space-saving arrangement. A digital spectrometer (DSPEC^{Plus}) is selected over the traditional NIM bin electronic modules. The spectrometer is a digital-to-signal process system. The connection between the DSPEC^{Plus} and the computer is a single Ethernet cable. The

CP673, *Plutonium Futures — The Science,* edited by G. D. Jarvinen

combination of resolution, throughput, and stability in count rate and temperature are all encompassed in the DSPEC^{Plus}. The main functions are administered and optimized by computer software that provides a high degree of accuracy and operational flexibility required in routine operation and maintenance.

Software

To link the hardware, a Visual Basic software program was written by ORTEC to provide a user-friendly interface for operation. The software integrates the detector and digital spectrometer setup, data acquisition (Gamma Vision), and real-time monitoring. To set up the detection system, the operator uses Gamma Vision software to set the DSPEC^{plus} detector for gain and energy calibration. To construct the elution profiles, the regions of interest (ROI) are defined and stored as ROI files for the plutonium processes. An additional software feature is to set up an alarm level based on the count rate of predefined gamma-ray peaks. The software will then administer both an audible and visible alarm warning, should the count rate exceed the threshold limits. Once the basic operating parameters for the on-line gamma system are defined, the PPM software will acquire the gamma-ray counting data directly from the DSPEC^{plus} and translate the information into a trend plot (Figure 1).[1] This type of real-time trend plot is displayed and updated at predefined time intervals. This gives the operator the ability to change the acquisition time, add comments, and start/pause/resume/stop during the process run. In addition, it allows the operator to print and save the gamma-ray spectra at a moment's notice during the elution process. A built-in ^{133}Ba check source is used for Quality Assurance (QA) checks and is performed before each run. A summarized data sheet with the QA and specific run information is printed along with the real-time trend plot at the end of each process run.

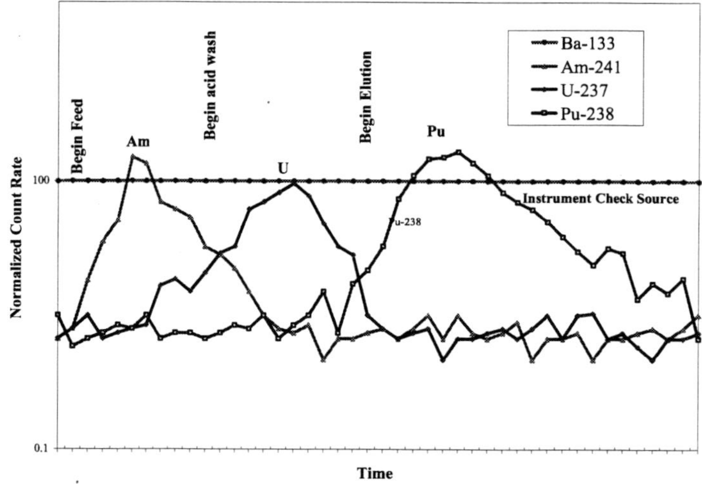

FIGURE 1. Simulated trend plot data.[1]

SUMMARY

The use of the integrated PPM system provides several enhanced features and flexible operations for monitoring of the aqueous plutonium recovery process. The system provides real-time monitoring of elution steps and is a critical element in reduction of plutonium losses in waste streams and aids in the minimization of waste volume.

REFERENCES

1. Wong, A. S., Ricketts, T. E., Pansoy-Hjelvik, M. E., Ramsey, K. B., Hansel, K. M., and Romero, M. K., "Plutonium Process Monitoring (PPM) System," Los Alamos National Laboratory document LA-UR-00-0406 (2000).
2. Wong, A. S., Ramsey, K. B., and Pansoy-Hjelvik, M. E., "On-Line Gamma Monitoring System for Plutonium-238 Aqueous Recovery Process," Los Alamos National Laboratory document LA-UR-99-3646 (1999).

3. Marsh, S. F., and Miller, M. C., "Plutonium Process Control Using an Advanced On-Line Gamma Monitor for Uranium, Plutonium, and Americium," Los Alamos National Laboratory report LA-10921 (1987).
4. Pope, N. G., and Marsh, S. F., "An Improved, Computer-Based On-Line Gamma Monitor for Plutonium Anion Exchange Process Control," Los Alamos National Laboratory report LA-10975 (1987).

A

B

C

D

E

F

G

H

I

J

K

L

M

N

O

P

R

S

T

U

V

W

X

Y

Z